# *Computational Methods in Chemical Engineering*

PRENTICE HALL INTERNATIONAL SERIES
IN THE PHYSICAL AND CHEMICAL ENGINEERING SCIENCES

NEAL R. AMUNDSON, SERIES EDITOR, *University of Houston*



BALZHIZER, SAMUELS, AND ELLIASSEN *Chemical Engineering Thermodynamics*
CROWL AND LOUVAR *Chemical Process Safety*
DENN *Process Fluid Mechanics*
FOGLER *Elements of Chemical Reaction Engineering, 2nd Edition*
HANNA AND SANDALL *Computational Methods in Chemical Engineering*
HIMMELBLAU *Basic Principles and Calculations in Chemical Engineering, 5th Edition*
HINES AND MADDOX *Mass Transfer*
JACKSON *Process Engineering in Biotechnology*
KYLE *Chemical and Process Thermodynamics, 2nd Edition*
NEWMAN *Electrochemical Systems, 2nd Edition*
PAPANASTASIOU *Applied Fluid Mechanics*
PRAUSNITZ, LICHTENTHALER, AND DE AZEVEDO *Molecular Thermodynamics of Fluid-Phase Equilibria, 2nd Edition*
PRENTICE *Electrochemical Engineering Principles*
STEPHANOPOULOS *Chemical Process Control*

# Computational Methods in Chemical Engineering

**OWEN T. HANNA**
**ORVILLE C. SANDALL**

Department of Chemical Engineering
University of California, Santa Barbara

*PRENTICE HALL PTR, Upper Saddle River, New Jersey 07458*

*Library of Congress Cataloging-in-Publication Data*

Hanna, Owen T.

Computational methods in chemical engineering / Owen T. Hanna

p. cm.

Includes bibliographical references and index.

ISBN 0-13-307398-X

1. Chemical engineering—Data processing. I. Sandall, Orville C. II. Title.

TP184.H36 1995 94-45872

511'.8'02466—dc20 CIP

*Editorial/Production Supervision:* Lisa Iarkowski
*Acquisitions Editor:* Bernard Goodwin
*Manufacturing Manager:* Alexis R. Heydt

Prentice-Hall, Inc.
A Simon & Schuster Company
Upper Saddle River, NJ 07458



The publisher offers discounts on this book when ordered in bulk quantities. For more information, contact:

Corporate Sales Department
Prentice Hall PTR
1 Lake Street
Upper Saddle River, NJ 07458
Phone: 1-800-382-3419
FAX: 201-236-7141
email: corpsales@prenhall.com



Printed in the United States of America

10 9 8 7 6 5 4 3 2 1

ISBN 0-13-307398-X

Prentice-Hall International (UK) Limited, London
Prentice-Hall of Australia Pty. Limited, Sydney
Prentice-Hall of Canada, Inc., Toronto
Prentice-Hall Hispanoamericana S.A., Mexico
Prentice-Hall of India Private Limited, New Delhi
Prentice-Hall of Japan, Inc., Tokyo
Simon & Schuster Asia Pte. Ltd., Singapore
Editora Prentice-Hall do Brasil, Ltd., Rio de Janeiro

*To Our Wives,*
*ZETTA HANNA*
*and*
*PAT SANDALL*

# Contents

# Preface

We have written this book for the purpose of presenting a clear, concise exposition of practical computational methods in chemical engineering. It is our belief that successful mathematical modeling, analysis, and computation require a *blend* of analytical and numerical approaches; they are complementary and reinforce each other. An important objective is to provide continuity in going from the introductory applications of mathematics to more advanced levels. The presentation has been designed primarily for the undergraduate level, but even graduate students should be able to profit from various parts. The only prerequisite is the standard course in calculus.

An important part of the present work is the inclusion of a number of powerful, user-friendly microcomputer programs (included on the accompanying diskette) which are demonstrated and used for many of the problems discussed. Detailed information concerning the computer programs follows below in a separate section. The broad coverage of topics generally includes material at both the beginning (Sophomore-Junior) and somewhat more advanced levels. Our selection of methods and algorithms has been guided by criteria of simplicity, generality, and efficiency. The general topics covered include analytical and numerical procedures for solving problems in:

Mathematical modeling
Linear algebraic equations and matrix analysis
Approximation of functions (series, interpolation, least squares)
Nonlinear algebraic equations

Optimization
Evaluation of integrals
Ordinary differential equations (ODE)
Partial differential equations (PDE)

Several special features of our presentation include broad use of convergence acceleration techniques such as Padé approximation for series; Shanks transformation for series, linear, and nonlinear systems of algebraic equations; systematic use of global Richardson extrapolation for integrals and ODE systems to monitor the *overall* error; and discussion of methods for the solution of stiff ODE. Some additional topics, which go beyond the customary undergraduate curriculum, include:

1. The interpolation method for matrix eigenvalues
2. Direct computer error bounds for Taylor expansions
3. Advanced series manipulations: division, reversion, and so forth
4. Spline, rational, general linear, and nonlinear interpolation
5. Linear and nonlinear multidimensional least squares
6. Fitting equations to numerical functions with rational least squares
7. Asymptotic methods for nonlinear algebraic equations
8. Method of continuation in nonlinear algebraic equations
9. The asymptotic expansion of integrals
10. Approximate analytical methods for ODE
11. A new explicit numerical method for moderately stiff ODE

The computer programs discussed in this book are available in IBM format on the enclosed diskette in the languages FORTRAN 77 *and* True BASIC. Included are some 25 user-friendly but powerful programs which cover virtually all of the standard computational problems encountered in chemical engineering. We realize that many instructors will prefer the FORTRAN 77 version, since that venerable language is available and used in many engineering and scientific computing environments. On the other hand, we decided some years ago to adopt True BASIC for our computational courses. True BASIC, which was developed by Kemeny and Kurtz, the originators of the BASIC language, has a number of important attributes that caused us to choose it for instructional purposes. In particular, True BASIC follows the new American National Standards (ANS) guidelines, is inexpensive for students (there is a student edition for about $20!), and contains excellent language structures, local and global variables, and both internal and external subroutines. The editing is simple but powerful, the intermediate code is very fast, and there are built-in matrix calculations and graphics. Also, the same True BASIC programs run on IBM, Apple Macintosh, and some other microcomputers; a version is also available for certain UNIX workstations. The well-known and extensive collection of computer programs from the book *Numerical Recipes* is also available on diskette in True BASIC.

To achieve our objectives in this work, emphasis is placed much more on mathematical ideas and practical applications than on proofs. However, on the theoretical side, an attempt is made to accomplish error analysis or error estimation for all methods discussed, both analytical and numerical. Emphasis is also placed not only on the use of computer programs but also on the appropriate validation and checking of results. A special feature of our presentation is a selection of practical example problems taken from the spectrum of subjects in the chemical engineering curriculum.

This book is intended primarily for courses in computational methods in chemical engineering at the undergraduate level. It should also be useful for courses involving mathematical modeling, numerical methods, applications of mathematics, and computer programming. The book could be used as a computer/problem-solving supplement throughout the entire chemical engineering curriculum. Because of its breadth of coverage, the book may also be of interest to other engineers and scientists. The concise approach, involving both example problems and easy-to-use computer programs, may be valuable to engineers on the job as well as for self-study.

A number of people have contributed indirectly to this book, and we wish to acknowledge their assistance. First we want to express special appreciation to our families for their support. We also thank our UCSB colleagues (particularly former Dean John E. Myers) and our former students (especially Dr. Sami Ashour, Dr. Richard Davis, and Mr. John Rickey). We are also grateful for assistance provided by the UCSB Office of Instructional Development (especially Dr. Stan Nicholson and Mr. Rick Johnson). Thanks are also due the staff of Prentice-Hall for their assistance, especially Betty Sun, Mike Hays, and Bernard Goodwin. We particularly appreciate all the help given to us by our production manager, Lisa Iarkowski.

# Computer Programs

An important feature of this work is that it comes with some 25 powerful computer programs on diskette (IBM format), which are discussed in the text and used to solve various example problems. The programs on the diskette are written in two separate versions, True BASIC and FORTRAN 77. Most of these programs have been used successfully in class and research work for several years (a few are more recent). We expect that they should prove to be extremely valuable in various courses, practical engineering work, and research. The programs are designed to be *very user-friendly.* Details regarding True BASIC and FORTRAN 77 are covered in Appendix A.

We now wish to indicate the nature and scope of these programs. At the end of the appropriate chapter or appendix in the text, the various programs are introduced. A "concise listing" in True BASIC is given for each, which includes essential program information, a brief how-to-run box that gives instructions to the user, and any user subroutines that must be modified for a particular problem by the user. Quantities that must be specified by the user are clearly marked with three asterisks, !***. The user subroutines already contain a problem that can be run directly to get a feeling for how the program operates. After running the original problem, the user can, in general, easily edit any required user subroutines to make them correspond to his/her problem. The programs available and what they can accomplish are now indicated.

## CHAPTER 2 LINEAR EQUATIONS AND MATRIX ANALYSIS

*SIMLE (simple linear equations).* This program solves a system of linear algebraic equations and provides an error estimate.

*TRID (tridiagonal)*. This is a subroutine for rapidly solving the special tridiagonal linear algebraic system. The subroutine is used later as part of the spline interpolation program and in finite difference programs.

*MATXEIG (matrix eigenvalue)*. This program computes the $n$ eigenvalues (real and complex) of a real matrix (symmetric or unsymmetric). It can also determine eigenvectors (for real eigenvalues only).

## CHAPTER 3 TAYLOR EXPANSIONS

*SERIES* This program (located in Appendix D) has options to calculate the product of two series, the quotient of two series, the logarithm or exponential of a series, raising a series to a power, reverting a series, doing an Euler transformation, and computing direct error bounds.

## CHAPTER 4 ACCELERATION TECHNIQUES FOR SERIES

*PADÉ* This program does Padé approximation, given the Taylor coefficients of a series. A direct error bounds option is available, similar to that in program SERIES (Chapter 3)

*SHANKS* This program allows multiple Shanks transformations for either a given sequence of numbers or a given power series. Error estimation is provided.

## CHAPTER 5 INTERPOLATION

*NEWTINT (Newton interpolation)* This program does simple linear or quadratic interpolation for given data and calculates any optimum and the derivative and integral of the interpolation function at any $x$.

*CUBSPL (cubic spline)* This program fits a cubic spline to either given data or a specified analytical function, and calculates $y$, $y'$, and $y''$ of the spline at any $x$, as well as the integral between any limits. It also shows a graph of the result.

*RATINTG (rational interpolation graphics)* This program carries out rational interpolation (a polynomial is a special case) for given data, and calculates $y$ and $y'$ of the interpolation function at any $x$, as well as the integral between any limits. It also shows a graph of the result.

## CHAPTER 6 LEAST SQUARES

*LSTSQGR (least squares graphics)* This program has the option for linear (in parameters) or nonlinear least squares fitting in one independent variable. The general linear model has the form $y = g_0(x) + C_1 g_1(x) + \ldots + C_p g_p(x)$. The nonlinear model is anything else, such as $y = f(x, C_1, C_2, \ldots, C_p)$. For a given set of data, the program calculates the unknown parameters $C_i$, the least-squares error and evaluates $y$ at any

$x$. It also allows evaluation of $y'$ and the integral between any limits. This program can also be used to evaluate, differentiate, or integrate any specified analytical function, as explained in the concise listing. For either data or an analytical function, the program can provide a graph of the result.

*LSTSQMR (least squares multiple regression)* This program does linear (in parameters) or nonlinear multiple-regression least squares fitting for $k$ independent variables. The general linear model has the form $y = g_0(x_1, \ldots, x_k) + C_1 g_1(x_1, \ldots, x_k) + \ldots + C_p g_p(x_1, \ldots, x_k)$. The nonlinear model is anything else, such as $y = f(x_1, \ldots, x_k, C_1, \ldots, C_p)$. The program calculates the unknown parameters $C_i$, the least-squares error and evaluates $y$ at any $x$.

*RATLSGR (rational least squares graphics)* This program determines a (nonlinear) rational least squares approximation for a function of one independent variable. *No starting values of the parameters are required.* Either data or an analytically defined function can be approximated. The program calculates the unknown parameters, the least-squares error, the maximum error, and the maximum relative error. The prediction function can be evaluated, differentiated, or integrated. A graph of the result is available. Beyond its obvious applicability to problems involving noisy data, this program is particularly valuable for determining a rational analytical approximation to a *deterministic* function or data.

## CHAPTER 7 NONLINEAR ALGEBRAIC EQUATIONS

*DELSQ (delta-square)* This program, which is for a system of nonlinear equations, gives the option of using ordinary iteration or ordinary iteration plus the modified "delta square" acceleration.

*NEWTNLE (Newton nonlinear equations)* This program uses the Newton method to solve a system of nonlinear algebraic equations. The partial derivatives must be specified analytically.

*NEWTONVS (Newton variable step)* This program uses either the full Newton or modified Newton method to solve a system of nonlinear equations. The partial derivatives are calculated numerically by means of a variable increment algorithm. *The program may be used to determine partial derivatives only.*

*POLYRTS (polynomial roots)* This program solves for all roots (including complex) of a real polynomial using the Bairstow method.

*METHCON (method of continuation)* This program implements the differential continuation method for a system of nonlinear algebraic equations using the IEXVS integration method. The program calculates the required partial derivatives numerically. This is the most powerful of the nonlinear equations programs included here.

## CHAPTER 8 OPTIMIZATION

*OPTFRPR (optimization Fletcher-Reeves-Polak-Ribiere)* This program uses a conjugate gradient algorithm to find the unconstrained minimum of a function of $n$ variables. The user must furnish the partial derivatives of the objective function.

*OPTLP (optimization linear programming)* This program solves a general linear programming problem (minimizing a linear function subject to linear constraints) using the simplex method of G. Dantzig.

*OPTSRCH (optimization search)* This program solves general discrete constrained minimization problems by a direct search method. It is also useful for the approximate solution of continuous constrained problems.

## CHAPTER 9 EVALUATION OF INTEGRALS

*TRMPQUAD (trapezoidal/midpoint quadrature)* This subroutine does numerical quadrature using either the trapezoidal or midpoint rule plus Richardson extrapolation. It gives accuracy to $O(h^6)$ and includes error information. It is used to evaluate integrals in programs RATINTG, LSTSQGR, and RATLSGR.

*IEX ($q = 4$) (improved Euler/extrapolation)* This version of ODE solver IEX (Chapter 10) accomplishes trapezoidal quadrature plus Richardson extrapolation for analytically defined functions. It gives accuracy to $O(h^6)$ and includes error information.

## CHAPTER 10 ANALYTICAL METHODS FOR ODE

*OC (orthogonal collocation)* This is an orthogonal collocation program for a single linear or nonlinear differential equation. It allows the use of up to six unknown coefficients and shows a comparison of results. There is an option for Shanks acceleration.

## CHAPTER 11 NUMERICAL METHODS FOR IVP IN ODE

*IEX ($q = 3$) (improved Euler/extrapolation)* This is the improved Euler/extrapolation program for a system of ordinary differential equations, which accomplishes up to two single and one double *global* Richardson extrapolations. Global error estimation is provided, and an "error ratio" quantity is calculated for validation of the extrapolation process.

*IEXVS (improved Euler/extrapolation variable-step)* This is a variable step-size extension of program IEX which allows local error control along with global error estimation.

*IIEX* This program is specifically for *stiff* systems of ODE. It offers options for the implicit methods Backward Euler or Implicit Improved Euler. For each method, there is an option for local error control and global Richardson extrapolation, similar to IEXVS.

## CHAPTER 12 NUMERICAL METHODS FOR BVP IN ODE

*IEXVSSH (improved Euler/extrapolation variable-step shooting)* This is the IEXVS program from Chapter 11 altered to provide a "shooting" capability for an ODE boundary value problem having one guessed boundary condition.

*FINDIF (finite difference)* This is a simple "brute force" finite difference program for ODE boundary value problems based on program DELSQ of Chapter 6. It can be applied to either linear or nonlinear problems. It is very inefficient compared to program FDTRID.

*FDTRID* This program uses the tridiagonal algorithm to solve ODE boundary value problems by means of finite differences. It incorporates an option for "delta square" acceleration and is applicable to either linear or nonlinear problems.

## CHAPTER 14 NUMERICAL METHODS FOR PDE

*PDEHYBD* This program uses the stability-enhanced explicit hybrid method (discussed in Chapter 11) to implement a method-of-lines algorithm for the solution of PDE marching problems.

*PDEFD* This is an implicit, tridiagonal PDE method for one-spatial-dimension PDE marching problems.

## Appendix D Advanced Series Manipulations

*SERIES* This program (referred to in Chapter 3) has options to calculate the product of two series, the quotient of two series, the logarithm or exponential of a series, raising a series to a power, reverting a series, doing an Euler transformation, and computing direct error bounds.

## MISCELLANEOUS

For differentiation or integration of a function or data, see programs NEWTINT, CUBSPL, RATINTG, LSTSQGR, RATLSGR, and NEWTONVS.

For automatic computation of *partial derivatives* of functions of several variables, see program NEWTONVS.

For plots of functions, see programs CUBSPL, RATINTG, LSTSQGR, and RATLSGR.

# Guide to Instructors

Portions of the material in this book have been used for the instruction of freshmen through beginning graduate students in various courses for a period of some 20 years. Over this period of time, the emphasis of computation has moved more and more in the direction of numerical solutions implemented on a computer.

Our treatment of the topics included here generally involves rather elementary introductions, followed by a gradual progression to more advanced material. We hope that this approach will make the book useful to students at various stages of their education. The general level of the book is intended for the Junior engineering student who has completed the calculus course sequence. However, as discussed below, a substantial portion of the material should be accessible to Sophomores, while a few topics are somewhat more advanced than Junior-level.

A succinct coverage of the material in this book could be achieved in a one-semester course, while a selective coverage makes sense for a one quarter (one-third-year) course. A quite thorough coverage of all the material, together with substantial applications, could be achieved in a one-year course.

A solutions manual, which gives solutions to the problems, is available to instructors. To assist an instructor who might wish to use this book, we offer the following table, which indicates our opinion of what material may be appropriate for different course levels. For instance, we believe that the Sophomore material in the table will be accessible to most Sophomore students. On the other hand, the material appropriate for Juniors would include that contained in *both* the Sophomore and Junior columns of the table. Finally, the Advanced column refers to material of a level appropriate for Seniors or beginning graduate students.

SUGGESTED MATERIAL FOR DIFFERENT COURSE LEVELS

| Chapter | Sophomore | Junior | Advanced |
|---|---|---|---|
| Math. Model | 1.1–1.5 | 1.6–1.7 | |
| Lin. Eqns./Mtx. | 2.1–2.3 | 2.4–2.5 | |
| Taylor Series | 3.1–3.3 | 3.4–3.5 | |
| Accel. of Series | | | 4.1-4.4 |
| Interpolation | 5.1–5.4 | 5.5–5.7 | |
| Least Squares | 6.1, 6.3 | 6.2, 6.4–6.5 | |
| Nonlinear. Eqns. | 7.1, 7.3 | 7.4, 7.6 | 7.2, 7.5 |
| Optimization | 8.1 | 8.2–8.3 | 8.4 |
| Eval. Integrals | 9.1–9.2 | 9.3–9.4 | 9.5–9.6 |
| Analyt. ODE | 10.1 | 10.2–10.4 | 10.5 |
| Num. ODE IVP | 11.1 | 11.2–11.3 | 11.4–11.5 |
| Num. ODE BVP | 12.1–12.2 | 12.3–12.5 | |
| Analyt. PDE | | 13.1–13.3 | 13.4 |
| Num. PDE | 14.1–14.2, 14.4 | 14.3, 14.5–14.6 | |

| Appendices | Sophomore | Junior | Advanced |
|---|---|---|---|
| Computer Programming | x | | |
| Difference Equations | x | | |
| Review of Infinite Series | x | | |
| Advanced Series Manipulations | | | x |
| Limits and Convergence Acceleration | | x | |
| Euler-Maclaurin Summation Formulas | | | x |
| Numerical Differentiation | x | | |

# 1

# Mathematical Modeling

## 1.1 FUNDAMENTAL CONSIDERATIONS

Mathematical modeling plays a very important role in contemporary chemical engineering. This has come about because of recent theoretical advances in chemical engineering combined with advances in computing machines. It is now possible in many cases to develop and solve sophisticated and accurate mathematical descriptions of many chemical engineering processes. Since the economics usually favors mathematical modeling and simulation of complex chemical and physical phenomena, as compared to carrying out these processes experimentally, the simulation approach seems to be of increasing value. Of course, it is neither possible nor desirable to eliminate careful experimentation. The best approach to most chemical engineering problems involves a combination of mathematical modeling and the use of data from carefully planned and executed experiments.

Most mathematical modeling in chemical engineering involves some combination of mass balances, energy balances, momentum balances, associated rate equations, equilibrium relations, and equations of state. The mass balances can be overall or species balances. Energy balances are important if the problem involves any type of energy transfer. Momentum considerations are of use in flow problems. Rate equations apply when there are chemical reactions. Equilibrium principles may be either physical (for equilibrium between phases) or chemical (for reaction-rate equilibrium). Equations of state are used primarily for problems involving gases.

Although the above-mentioned principles are easy to enunciate, their actual implementation in particular problems varies so much that general equations can only be written in a vague way. Against such a broad background, each special case must be handled in its own way. However, there are a number of important "prototypes" of various relatively common problems that can be used as guides.

Before embarking on a discussion of particular modeling problems, a few comments are in order. One important general consideration involves the concept of "degrees of freedom," which we define as the number of unknowns minus the number of independent equations. The degrees of freedom represent the *number* of variables that can and must be prescribed in order for the model to be algebraically determinate. Which of the variables make up the degrees of freedom depends on the problem being solved. For a given model, a particular variable may sometimes be specified and sometimes unknown.

Another topic related to modeling is that of interpretation of experimental data. Often, a model must somehow be "reconciled" to corresponding inconsistent data, and this usually requires some kind of calculation to provide a "best fit" to the data by minimizing the error between the data and the model prediction. Finally, it should be mentioned that another practical aspect of modeling concerns the development of useful formulas to correlate the results of numerical solutions. The most common types of mathematical problems encountered in modeling include algebraic equations, difference equations, ordinary differential equations, optimization, integrals, and partial differential equations. All of these topics are given consideration in this book.

### 1.1.1 Nondimensionalization

At some point in the development of a mathematical model, it is usually desirable to *nondimensionalize* the equations and boundary conditions. There are several reasons for this. Obviously, this is convenient for checking the dimensional consistency of a calculation, but it is also very important for presenting results in a concise and useful form. Furthermore, through nondimensionalization we can *reduce the number of parameters in the problem* in order to effect an efficient numerical solution. Finally, the dimensionless form of the problem is very useful for seeking limiting or "asymptotic" forms of the solution for some large or small parameter value. For the above reasons, we will usually work in terms of dimensionless variables in example problems.

The material in this chapter addresses the *formulation* of mathematical models. Methods for determining appropriate solutions of these models will be given in subsequent chapters.

## 1.2 ALGEBRAIC EQUATIONS

Algebraic equations, both linear and nonlinear, occur frequently in modeling, particularly for the expression of steady-state mass and energy balances as well as for equations of state and thermodynamic equilibrium.

**Example 1.1 Linear Equations—Plant Mass Balance**

A process to recover acetone from air is shown in Figure 1.1. Acetone from air is absorbed into water in a packed column. Essentially all of the acetone is removed from

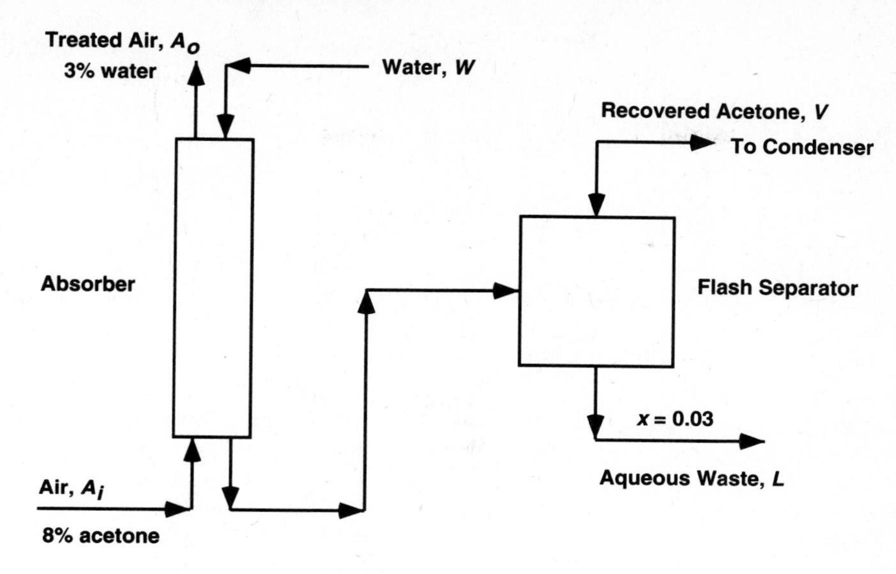


**Figure 1.1.** Plant to Recover Acetone from Air

the air in the absorber and the water leaving the absorber which contains the acetone is treated in a flash separator to recover the acetone. The air entering the absorber contains no water vapor but becomes saturated with water vapor in the absorber. The air leaving the absorber contains 3 mass % water vapor. The flash separator acts as a single equilibrium stage so that the composition of the vapor stream, $V$, and the liquid stream, $L$, leaving the flash separator are related by

$$y = 20.5x \tag{a}$$

where

$y$ = mass fraction of acetone in the vapor stream leaving
$x$ = mass fraction of acetone in the liquid stream leaving

If 600 lb/hr of an air stream containing 8 mass % acetone is to be treated with 500 lb/hr of water in the absorber, and if the waste water stream ($L$) is specified to have an acetone content of 3 mass % acetone, we wish to determine the concentration of acetone in the recovered product stream ($V$) and the flow rates of both product streams leaving the flash separator.

Since there are three components (air, water, and acetone), we can write three independent material balance equations around the complete plant.

Let

$A_0$ = flow rate of treated air, lb/hr
$A_i$ = flow rate of dry air to be treated, lb/hr
$V$ = flow rate of acetone stream from flash separator, lb/hr
$L$ = flow rate of waste water stream from flash separator, lb/hr
$W$ = flow rate of water to absorber, lb/hr

The mass balances are:

air: $$0.92A_i = 0.97A_0 \tag{b}$$

acetone: $$0.08A_i = yV + 0.03L \tag{c}$$

water: $$W = 0.03A_0 + (1-y)V + 0.97L \tag{d}$$

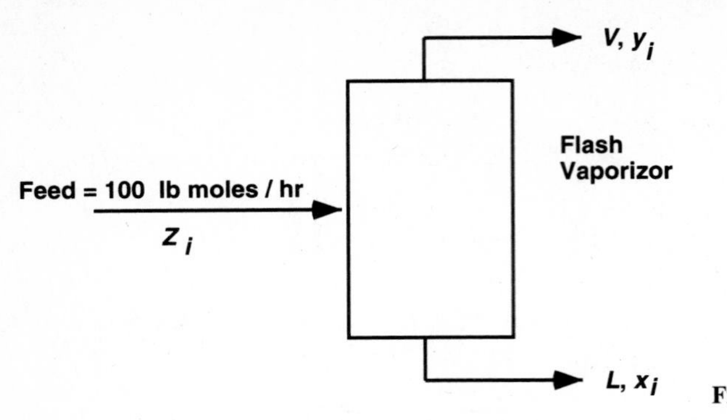


**Figure 1.2.** Flash Vaporization

**TABLE 1.1** FEED COMPOSITION

| Component | Mole Fraction in Feed $z_i$ | $K_i$ |
|---|---|---|
| n-butane | 0.25 | 2.13 |
| n-pentane | 0.45 | 1.10 |
| n-hexane | 0.30 | 0.59 |

Since $x$ is known to be 0.03, the value of $y$ is obtained from the equilibrium relation, Equation (a), as $y = 20.5(0.03) = 0.615$. Noting that $A_i$ is given to be 600 lb/hr and $W$ is given as 500 lb/hr, we finally reduce Equations (b), (c), and (d) to a system of three linear equations in the three unknowns $A_0$, $V$, and $L$. These equations may be solved using the matrix techniques of Chapter 2 (e.g., program SIMLE) to give

$$A_o = 569\,\text{lb/hr}$$
$$V = 54.8\,\text{lb/hr}$$
$$L = 476\,\text{lb/hr}$$

This process recovers $0.615(54.8)100\%/(0.08)(600) = 70.2\%$ of the acetone in the inlet air stream.

**Example 1.2 Non-Linear Algebraic Equations—Flash Vaporization**

A hydrocarbon mixture containing 25 mole % n-butane, 45 mole % n-pentane and 30 mole % n-hexane is to be separated in a simple flash vaporization process, shown in Figure 1.2. The process will operate at 270°F and 10 atm. The equilibrium K-values at these conditions are given in Table 1.1. We wish to calculate the composition and flow rates of the vapor and liquid products if the feed rate to the flash vaporizer is 100 lb moles/hr.

Flash vaporization processes are designed on the basis that they act as an equilibrium stage; that is, the vapor and liquid mole fractions for each component in the product streams are in equilibrium. Thus, the vapor and liquid mole fractions in the product stream are related by the equilibrium $K$-value, obtained from thermodynamics.

$$K_i = y_i/x_i \tag{a}$$

where

$x_i$ is the mole fraction of component $i$ in the liquid

$y_i$ is the mole fraction of component $i$ in the vapor

An overall and three component material balances can be written (only three of these balance equations are independent).

**TABLE 1.2** PRODUCT COMPOSITIONS FROM FLASH VAPORIZER

| Component | $x_i$ | $y_i$ |
|---|---|---|
| n-butane | 0.129 | 0.275 |
| n-pentane | 0.415 | 0.457 |
| n-hexane | 0.456 | 0.268 |

Overall balance:

$$F = V + L \tag{b}$$

Component balances:

$$z_i F = y_i V + x_i L \tag{c}$$

If we substitute into the component mass balance for $y_i$ from the equilibrium relationship and for $L$ from the overall mass balance, and then solve for $x_i$, we obtain

$$x_i = z_i / \left\{1 + \frac{V}{F}(K_i - 1)\right\} \tag{d}$$

A similar equation can be obtained for $y_i$.

$$y_i = K_i z_i / \left\{1 + \frac{V}{F}(K_i - 1)\right\} \tag{e}$$

This problem could be set up by taking either $\sum_i y_i = 1$ or $\sum_i x_i = 1$ to get a nonlinear equation to solve for $V/F$. The approach used here is that recommended by Rachford and Rice [1952], which is to to take $\sum_i y_i - \sum_i x_i = 0$ to give

$$f\left(\frac{V}{F}\right) = \sum_i \frac{z_i(K_i - 1)}{(K_i - 1)(V/F) + 1} = 0 \tag{f}$$

This equation is particularly convenient for computer solution since $f(V/F)$ monotonically decreases with $V/F$ with no minimum or maximum in the range of physically possible values of $V/F, 0 \leq V/F \leq 1$. Since we have three components in this problem, the equation for $V/F$ is a nonlinear third-degree equation that we need to solve for $V/F$. Once $V/F$ is obtained, $V$ and $L$ can be calculated from the overall mass balance.

The equation for $V/F$, Equation (f), may be solved using any of the systematic iterative procedures discussed in Chapter 7 to give

$$\frac{V}{F} = 0.834 \tag{g}$$

Thus, we get $V = 83.4$ and $L = 16.6$ lb moles/hr, and the liquid and vapor compositions, obtained by back substitution, are given in Table 1.2.

## 1.3 DIFFERENCE EQUATIONS

Difference equations, which are sometimes also called recurrence relations, can be considered to be discrete forms of differential equations. That is, the argument or independent variable in a difference or recurrence relation is generally an integer. In modeling and computation, difference equations are used principally to model discrete stagewise processes (such as separation processes) and to conduct error-propagation analyses for various numerical algorithms.

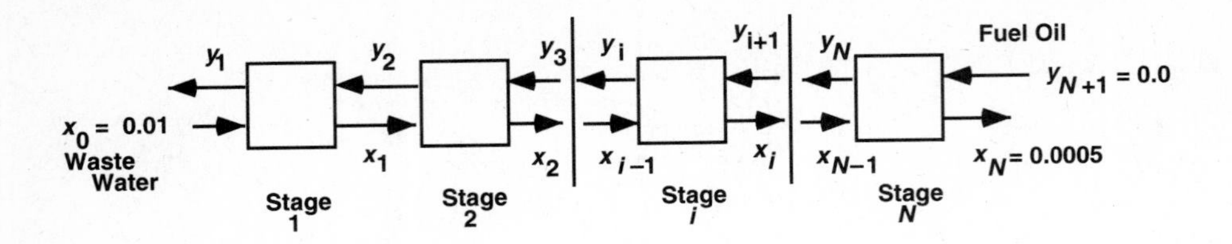


**Figure 1.3.** Counter-Current Liquid Extraction

**Example 1.3 Difference Equations—Liquid-Liquid Extraction of Hazardous Wastes from Waste Water**

The waste water from a chemical processing plant contains a mixture of organic acids with a total concentration of 1.0% by weight. This acid concentration is too large to be acceptable for discharge. The acids have a relatively high solubility in a fuel oil that is used for process heating, and it is planned to use this fuel oil to extract the mixed acids out of the waste water stream. It is desired to reduce the acid concentration in the waste water to 0.05 wt. % by a counter-current liquid extraction process, as shown schematically in Figure 1.3. There is no chlorine, nitrogen, or sulfur in the acids, so when they are burned with the fuel oil, they will not contribute to any additional air pollution.

We wish to determine the number of equilibrium stages necessary for this separation process. The equilibrium distribution of the organic acids between the fuel oil and the water may be expressed as

$$y_i = Kx_i \tag{a}$$

where

$y_i$ is the weight fraction of acids in the fuel oil

$x_i$ is the weight fraction of acids in the water

$K$ is the equilibrium distribution constant

the subscript $i$ refers to the the stage number

For this system, the equilibrium distribution constant was found to be

$$K = 2.2 \tag{b}$$

The flow rate of water to be treated is 150 gallons per hour, and the flow rate of fuel oil to be used in this extraction process is 95 gallons per hour. The specific gravity of the fuel oil is 0.90, and the fuel oil enters the process with no dissolved organic acids. This problem is similar to one described in *Safety, Health, and Loss Prevention in Chemical Processes* [1990].

There are different approaches that could be taken to solve this problem. The method developed here combines a component balance with the equilibrium relationship to form a difference equation. This difference equation can then be solved for $N$, the number of equilibrium stages required for the specified separation.

Since the fuel oil is essentially insoluble in water, and water has a very low solubility in the fuel oil, we can assume that the waste water stream (the raffinate phase), $R$ (lb/hr), and the fuel oil stream (the extract phase), $E$ (lb/hr), remain constant as they flow through the extraction battery.

A mass balance around the complete extraction process for the acids gives an equation for $y_i$, the concentration of acids in the exiting fuel oil stream.

$$y_{N+1}E + x_0R = y_iE + x_NR \tag{c}$$

$R$ and $E$ may be obtained from the problem statement as

$$R = \left[150\frac{\text{gal}}{\text{hr}}\right]\left[8.337\frac{\text{lb}}{\text{gal}}\right] = 1251\frac{\text{lb}}{\text{hr}} \tag{d}$$

$$E = \left[95\frac{\text{gal}}{\text{hr}}\right][0.90]\left[8.337\frac{\text{lb}}{\text{gal}}\right] = 713\frac{\text{lb}}{\text{hr}} \tag{e}$$

Since the concentrations $y_{N+1}, x_0$, and $x_N$ are specified, the component mass balance may be solved to give $y_1 = 0.0167$.

A mass balance for the acids around the $i^{\text{th}}$ stage gives

$$x_{i-1}R + y_{i+1}E = x_iR + y_iE \tag{f}$$

The $y_i$'s can be eliminated from this equation by using the equilibrium relationship, to give, after some rearrangement,

$$x_{i-1} - (1+S)x_i + Sx_{i+1} = 0 \tag{g}$$

where

$$S = \frac{KE}{R} \tag{h}$$

The boundary conditions are $x_0 = 0.01$ and $x_N = 0.0005$, as shown in Figure 1.3. Equation (g) is in the form of a difference equation which can be solved using the methods discussed in Appendix B to give an equation for the number of stages as

$$\frac{x_0 - x_N}{x_0 - x_{N+1}} = \frac{S^{N+1} - S}{S^{N+1} - 1} \tag{i}$$

This equation may be solved iteratively for $N$ using the values $S = 1.254, x_0 = 0.01, x_N = 0.0005, x_{N+1} = y_{N+1}/K = 0$ to give

$$N = 7.0 \tag{j}$$

## 1.4 ORDINARY DIFFERENTIAL EQUATIONS: INITIAL VALUE PROBLEMS

Probably the majority of mathematical modeling problems in chemical engineering involve ordinary (and/or partial) differential equations. A distinction is made between initial value problems (where all of the boundary conditions are expressed at the *same* point) as opposed to boundary value problems where the boundary conditions are specified at more than one point. In modeling, initial value problems are often connected to the "perfect mixing" idea, so we now stop to discuss and illustrate this idea.

### 1.4.1 Perfect Mixing Processes

One important and rather general approach in mathematical modeling can be referred to as the "perfect mixing" assumption. This involves the assumption that there are no variations with respect to one or more particular coordinates within the interior of the system being considered. Conditions at the system boundaries in these coordinates, however, are required to behave as if there really were interior gradients, as far as overall balances are concerned. These "perfect mixing" relationships are often equivalent to "averaging" or integrating over a more detailed relationship. In many engineering systems, such as in stirred vessels and similar equipment, the perfect mixing concept is found to hold experimentally to a good degree of accuracy.

In other situations, the "perfect mixing" approximations are often used more for mathematical expediency than for their supposed reality. However, these models often give surprisingly good results in practice. This can be explained from the viewpoint that the "perfect mixing" models represent approximate integrated forms of more exact models.

### 1.4.2 The Mass Balance

For any given system, we can write a general overall mass balance in the form

$$I - O = \frac{dm}{dt} \tag{1-1}$$

Here, $I$ is the total rate of input of mass at some arbitrary instant of time, $O$ is the corresponding output rate, and $m$ is the total system mass at the same time (see Figure 1.4). Here $I, O$, and $m$ may have several different contributions, and all of these quantities may depend on time. For any particular component species of the system, we can write a component mass balance as

$$Ix_i - Ox_o + G = \frac{dmx}{dt} \tag{1-2}$$

Here, $x_i$ and $x_o$ are the inlet and outlet mass fractions of the particular component, $x$ is the mass fraction of the component in the system, and $G$ is the generation rate of the component for the case where it participates in one or more chemical reactions. Also, $x_o, x_i$, and $x$ may be functions of time. Note here that Equation (1-2) already implies a certain "perfect-mixing" assumption; that is, we tacitly wrote that there is a *single* composition $x$ which is characteristic of the system at a particular instant. In reality, it would be expected that at any instant, the composition would vary from one location to another within the system, so that one could speak of a single system composition only in some average sense. It seems quite evident, however, that the use of a single average system composition would vastly reduce the effort required to solve the model equations. If the composition was considered to vary explicitly with position within the system, the component mass balance would become a partial differential equation.

To complete the perfect-mixing assumption for the component mass balance, we suppose that the exit composition, $x_o$, is equal to the system composition $x$. Then Equation (1-2) becomes

$$Ix_i - Ox + G = \frac{dmx}{dt} = m\frac{dx}{dt} + x\frac{dm}{dt} \tag{1-3}$$

It is sometimes desirable to combine Equations (1-1) and (1-3) to eliminate $dm/dt$; this causes the $Ox$ term in Equation (1-3) to cancel. Equation (1-3) then becomes

$$m\frac{dx}{dt} = I(x_i - x) + G \tag{1-4}$$

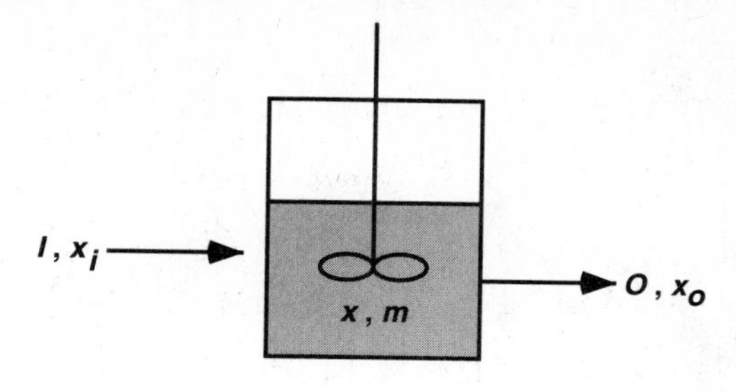


**Figure 1.4.** Unsteady Mass Balance

The generation (or depletion) rate expression for $G$ depends on $x$ and must normally be specified from reaction kinetics. If $G$ varies linearly with $x$, then Equation (1-4) is a first-order linear differential equation, since $m, I$, and $x$ vary only with time.

**Example 1.4 Concentration Buffering Tank**

A simple but instructive illustration of the preceding ideas can be made for a so-called concentration buffering tank similar to that shown in Figure 1.4. Such a device could be used to damp upstream oscillations in flow velocity and/or composition prior to entry into a stability-sensitive reactor. This would be accomplished by placing a tank containing mass $m$ of liquid in the line. The fundamental question is to relate the outlet state of the system to the inlet and system parameters.

To keep matters simple, we consider that the flow rates are fixed and equal, having no fluctuations, and that the inlet stream $I$ has a composition variation $x_i = x_{io}(1 + b \sin t)$, where $x_{io}$ is a constant and $b$ is the amplitude parameter. It is desired to express the outlet composition from the tank as simply as possible in terms of other problem parameters. In this case the overall and component balances, in terms of Equations (1-1)–(1-4), become

$$I - O = \frac{dm}{dt}, \quad m = \text{constant}, \text{ since } I = O \tag{a}$$

and

$$\frac{dx}{dt} + \frac{I}{m}x = \frac{I}{m}x_i = \frac{I}{m}x_{io}(1 + b \sin t) \tag{b}$$

It is convenient to work in terms of the ratio $I/m = a$. Equation (b) is a first-order linear equation which can be integrated analytically in the standard way by using the integrating factor approach (discussed in Chapter 10) in the form

$$xe^{at} - x_{i0} = \int_{z=0}^{t} a(x_{i0} + bx_{i0} \sin z)e^{az}\, dz \tag{c}$$

This is useful to do, since it allows development of simple relations between maximum outlet perturbations and the system parameters. It is convenient to choose the time scale so that at $t = 0$, $x = x_{i0}$. Integrating Equation (c) yields the solution for $x$ as

$$x = x_{i0}\left[1 + \frac{ba}{1 + a^2}(e^{-at} + a \sin t - \cos t)\right] \tag{d}$$

To keep $x$ within specified limits of $x_{i0}$ for *all* time, it is convenient to generate a simple upper bound to the second term in Equation (d). This can be accomplished by using the inequalities $e^{-at} \leq 1$, $\sin(t) \leq 1$, and $-\cos(t) \leq 1$.

For example, if $b$ is equal to 0.1, and if it is required that the deviation of $x$ from $x_{i0}$ be no more than 0.01, then we must specify the inequality $0.1a(2+a)/(1+a^2) \leq 0.01$. This results in the requirement that $a = I/m$ must be less than 0.0487. In more complicated situations involving this model, for instance if $I$, $m$, or $x_i$ varied in a

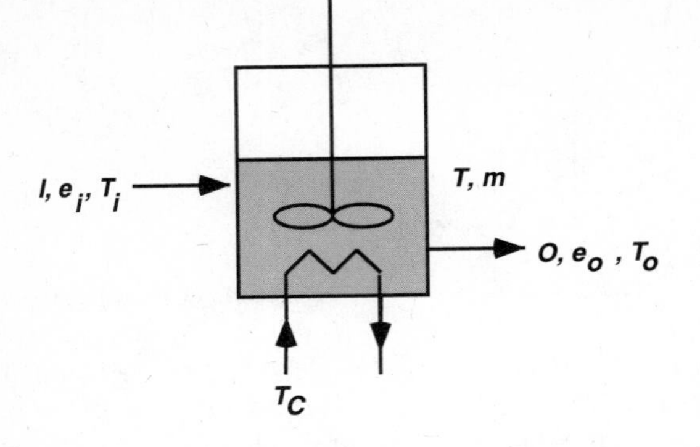


**Figure 1.5.** General Energy Balance

complicated way with time, a numerical solution of Equation (a) would probably be necessary or desirable.

### 1.4.3 The Energy Balance

We now consider, in an analogous way, the problem of making an overall energy balance on a stirred vessel system. Figure 1.5 shows the system considered, which includes one input and one output stream, along with a heating (or cooling) coil within the tank. There may also be external heat losses from the tank to the surroundings. We also assume that kinetic and potential energy effects, as well as pressure work, may be neglected in the overall energy balance. This is normally a very good approximation in problems involving thermal mixing and/or heat transfer in incompressible flows. On the other hand, if the flow from the vessel is by gravity, the kinetic and potential energy changes nearly balance each other; this is accounted for in the mechanical energy balance. This is the basis, for example, of the so-called Torricelli law in physics for the flow from an orifice (i.e., $u = (2gh)^{1/2}$).

In words, we may write the overall energy balance as (Rate of Energy Input) – (Rate of Energy Output) = (Rate of Accumulation of Energy within System). It is supposed that the flow here is not essentially by gravity, so we need not worry about the mechanical energy balance. The overall mass balance, as before, is just $I - O = dm/dt$. The overall energy balance becomes

$$Ie_i + U_c A_c (T_c - T) - U_o A_o (T - T_o) - Oe = \frac{d(me)}{dt} \tag{1-5}$$

Here the perfect-mixing assumption is invoked in order to say that (a) the internal energy (per unit mass) of the system has a uniform value at any instant, and (b) the internal energy of the exit stream is the same as that of the vessel contents at any instant. In the energy balance, Equation (1-5), $e_i$ is the internal energy of the input stream, while the $c$ subscript refers to quantities associated with the heating (or cooling) coil, such as the overall heat transfer coefficient, $U$, between coil and vessel, and the heat transfer area, $A$. The subscript 0 refers to heat transfer between the vessel and the external environment. If we eliminate $dm/dt$ using the mass balance equation, we get

$$I(e_i - e) + U_c A_c (T_c - T) - U_o A_o (T - T_o) = m\frac{de}{dt} \tag{1-6}$$

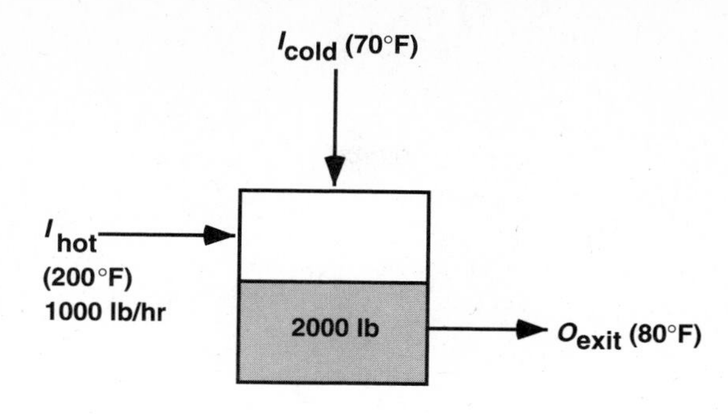


**Figure 1.6.** Thermal Mixing Tank at Steady State Conditions

Also, $(e_i - e) = C_p(T_i - T)$ and $de/dt = C_p dT/dt$ for incompressible liquids. Then Equation (1-6) can finally be written as

$$\frac{dT}{dt} = \frac{I}{m}(T_i - T) + \left(\frac{U_c A_c}{mC_p}\right)(T_c - T) - \left(\frac{U_o A_o}{mC_p}\right)(T - T_o) \tag{1-7}$$

In this form, Equation (1-7) will be first-order and linear provided that all of the coefficients vary only with time. On the other hand, if $C_p$ or $U$ are taken to vary with $T$, Equation (1-7) is nonlinear. In this latter instance, approximation or (especially) *numerical* methods of solution would generally be required for the solution.

**Example 1.5 Thermal Mixing**

One thousand pounds/hr. of a hot water stream at 200°F are mixed with a large amount of 70°F cold water in a "thermal mixing tank" as shown in Figure 1.6. Under normal, steady-state operating conditions, the discharge from the mixer is at 80°F. At these steady-state conditions, the inflow and outflow rates are constant, and the amount of material in the tank is constant at 2000 pounds. If a pump failure suddenly reduces the input rate of cold water by a factor of two, how long will it take the discharge temperature to reach 85°F? The pump failure affects only the cold-water stream; the hot-water and discharge flow rates remain fixed. The heat capacity of water can be taken to be constant at 1 Btu/lb°F.

To determine the initial steady-state operating conditions of the mixer, it is necessary to write overall mass and energy balances. The overall mass balance, in words, can be expressed as "the rate of input of the hot stream, $I_{hot}$, plus the rate of input of the cold stream, $I_{cold}$, minus the rate of output of the exit stream, $O_{exit}$ equals the rate of accumulation within the system (which is zero because of steady state)." In equation form, this becomes

$$I_{hot} + I_{cold} - O_{exit} = 0 \tag{a}$$

The overall energy balance at the initial steady-state can be written as

$$I_{hot}e_{hot} + I_{cold}e_{cold} - O_{exit}e = 0 \tag{b}$$

In this equation $e_{hot}$, $e_{cold}$, and $e$ represent the internal energy per unit mass of the hot, cold, and exit streams, respectively. It is convenient to eliminate $O_{exit}$ from the energy balance by using the mass balance equation. We also use the thermodynamic relationships $e_{hot} - e = C_p(T_{hot} - T)$ and $e_{cold} - e = C_p(T_{cold} - T)$ to express the internal energies in terms of temperatures. Since the fluid is incompressible, the heat capacities at constant volume and pressure are the same. Using the preceding relationships and substituting into the energy balance equation yields

$$\frac{I_{hot}}{I_{cold}} = \frac{T - T_{cold}}{T_{hot} - T} = \frac{80 - 70}{200 - 80} = \frac{1}{12} \tag{c}$$

Since $I_{hot}$ is given to be 1000 lb/hr, this produces $I_{cold} = 12,000$ lb/hr and $O_{exit} = 13,000$ lb/hr. Thus, the energy balance allows the mass balance to be completed.

After the pump failure, the change in the cold-water flow rate disrupts the steady-state condition of the system, and this must be analyzed by considering new time-dependent mass and energy balances which use the former steady-state values as initial conditions. The mass balance now contains the rate of accumulation term $dm/dt$ as follows:

$$I_{hot} - \frac{I_{cold}}{2} - O_{exit} = \frac{dm}{dt} = 1000 + \frac{12{,}000}{2} - 13{,}000 \tag{d}$$

Thus, $dm/dt = -6000$ lb/hr, and by integration $m = 2000 - 6000t$, where we have used the initial condition $m = 2000$ at $t = 0$ (the time of the disturbance). The overall energy balance after the pump failure now contains the rate-of-accumulation of energy within the system, which is $d(me)/dt = m\,de/dt + e\,dm/dt$. This new unsteady overall energy balance takes the form

$$I_{hot}e_{hot} + \frac{I_{cold}}{2}e_{cold} - O_{exit}e = \frac{dme}{dt} = m\frac{de}{dt} + e\frac{dm}{dt} \tag{e}$$

When we eliminate $dm/dt$ in this equation using the mass balance, and incorporate the thermodynamic relations to express internal energies in terms of temperatures, we get

$$I_{hot}(T_{hot} - T) + \frac{I_{cold}}{2}(T_{cold} - T) = m\frac{dT}{dt} \tag{f}$$

When we substitute the known quantities into the above equation, we obtain

$$1000(200 - T) + 6000(70 - T) = (2000 - 6000t)\frac{dT}{dt} \tag{g}$$

This last equation can be solved most easily by separating variables according to

$$\frac{dT}{620 - 7T} = \frac{dt}{2 - 6t} \tag{h}$$

When this is integrated (and exponentiated), we finally get

$$1 - 3t = \left(\frac{620 - 7T_{final}}{620 - 7T_{initial}}\right)^{\frac{6}{7}} \tag{i}$$

For $T_{final} = 85°F$, $T_{initial} = 80°F$, we find that the time required to reach a final temperature of 85°F is $t = [1 - (25/60)^{6/7}]/3 = 0.176\,\text{hr} = 10.5\,\text{min}$.

We note from our final equation what at first appears to be peculiar behavior. The left-hand side goes to zero for $t = 1/3$ hr and then becomes negative for any larger time. This apparently "strange" behavior occurs because the unsteady state balances only apply up to the time when the tank goes dry, that is, when $m$ goes to zero. Beyond this time, $dm/dt$ must be zero, since material cannot leave the tank faster than it comes in. The exit temperature at the time that the mass in the tank goes to zero is just $T_{final} = 620/7 = 88.57°F$, which first occurs at $t = 1/3$ hr. This ultimate exit temperature can be obtained from either (i) the *new* steady-state balances based on the new flow rates or (ii) from the $T$ versus $t$ relation when mass runs out in the tank. The *time* required for this to happen can only be obtained from (ii). After mass has run out in the tank, the outlet temperature remains at the "mixed" value 88.57°F.

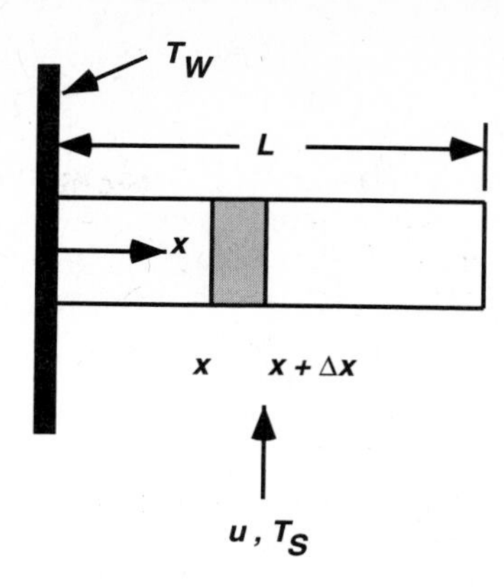


**Figure 1.7.** Heat Transfer from an Extended Surface

## 1.5 ORDINARY DIFFERENTIAL EQUATIONS: BOUNDARY VALUE PROBLEMS

**Diffusion Processes.** Diffusion processes (mass, energy, or momentum) are often the source of boundary value problems, where the boundary conditions are specified at different locations. In contrast to our discussion of modeling for so-called perfect mixing processes, in many instances it is necessary to take into account local gradients of temperature, concentration, or velocity within a system. In such situations there is always a possibility of a diffusion of mass, energy, or momentum, so we refer to such problems as "diffusion processes." It is very common for convective effects to accompany diffusion and, indeed, to often dominate diffusion when both are present. One important and general aspect of diffusion processes is that they generally raise the order of the model, and hence, require additional boundary conditions. This, in turn, frequently changes the character of the model from an initial value problem to a boundary value problem. This distinction will be seen to be particularly important as regards numerical solutions. Models of complex systems can often be composed of various "perfect-mixing" and "diffusion" components. The modeling of diffusion processes will now be illustrated by some examples.

**Example 1.6 Heat Transfer From an Extended Surface**

Consider the problem of steady-state heat transfer from a cylindrical rod which protrudes from a wall maintained at temperature $T_w$ into a moving fluid stream having a uniform temperature $T_s$. The situation is depicted in Figure 1.7. We expect that the cooler fluid will cause the rod to cool, so that at $x = L$ the rod will be nearer in temperature to $T_s$ than to $T_w$. The problem is to calculate the total rate of energy loss from the rod.

In the first place, note that if the "perfect-mixing" model is adopted, we would say that the rod has the uniform temperature $T_w$, and the heat loss would be given by the simple expression $q = hA_T(T_w - T_s)$, where $A_T$ is the lateral area. Since we expect that $T \leq T_w$, this should represent an upper bound to the heat loss. To set up a differential equation describing temperature variation in the rod, we consider some arbitrary element of width $\Delta x$ which lies entirely within the *interior* of the rod and write an appropriate energy balance. The boundaries of the system must be treated separately. We start by assuming that there is no internal *radial* temperature variation within the rod. At first, this assumption appears unreasonable, since there will apparently be lateral convective heat losses from the surface of the rod to the surrounding fluid. These losses, however, will be included in the overall energy balance so that there will be no inconsistency. By assuming the absence of internal radial temperature gradients, we are saying that we wish to model the axial temperature variation by an ordinary differential equation. If *internal* radial variations had also been allowed, the model would be a partial differential equation.

To proceed, we consider the element of width $\Delta x$ shown in Figure 1.7 and make an energy balance. Note that we assume that the element is being cooled and that the conductive fluxes are in the positive $x$ direction (from left to right) and that there is a convective loss from the lateral surface of the element to the surrounding stream. In words, the steady-state energy balance is simply Rate In − Rate Out = 0, since there is no generation. We recall from heat conduction that the energy flux by conduction can be written as $q_{\text{cond}} = \pm k dT/dx$, where $q_{\text{cond}}$ is the rate of energy transfer by conduction per unit area, and $k$ is the thermal conductivity. The question of the correct sign for the conduction expression is handled as follows. We wish for the conductive flux in the $x$ direction to be positive when heat is being transferred by conduction in the positive $x$ direction. Thus, the quantity $q_{\text{cond}}$ should be positive when $T$ is decreasing with $x$, that is, when $dT/dx$ is negative. Thus, we need the minus sign in our expression $q_{\text{cond}} = -k dT/dx$. In a similar way, from heat transfer we know that we can express the convective flux from the lateral surface as $q_{\text{conv}} = h(T - T_s)$, where $h$ is the heat transfer coefficient. According to this equation, if $q_{\text{conv}}$ is to be taken as a loss or output, the loss will be positive when heat is actually transferred from system to surroundings (i.e., when $T > T_s$).

Our energy balance now has a conduction input at $x$, a conduction output at $x + \Delta x$, and a convection loss from the lateral surface. We can write this in equation form as

$$\left(-kA_c \frac{\partial T}{\partial x}\right)_x - \left(-kA_c \frac{\partial T}{\partial x}\right)_{x+\Delta x} - hA_T(T - T_s) = 0 \left(= \frac{\partial(\rho V \cdot e)}{\partial t}\right) \tag{a}$$

Here, $A_c$ and $A_T$ are the respective cross-sectional and lateral areas of our element. $A_T$ can be written as $P\Delta x$, where $P$ is the element perimeter at $x$. As is customary in diffusion problems, our equation contains two terms which may be viewed as $[u(x + \Delta x) - u(x)]$; by the mean value theorem for derivatives (or Taylor's theorem), we write $u(x + \Delta x) - u(x) = [du(\xi)/dx]\Delta x$. In our equation, the role of $u(x)$ is played by the quantity $(kA_c dT/dx)$. Using this relationship and passing to the limit as $\Delta x \to 0$, we get

$$\frac{d}{dx}\left(kA_c \frac{dT}{dx}\right) = hP(T - T_s) \tag{b}$$

Equation (b) can now be further simplified. But first, it is important to note that in the form presented, Equation (b) is really applicable to a quite general problem. That is, if our rod geometry was a function of $x$, so that $A_c$ and $P$ varied with $x$, Equation (b) would still be appropriate. It also applies if the conductivity $k$ varies with distance $x$ or with temperature, as well as if $h$ varies with distance or temperature. In such complicated cases, a *numerical* rather than analytical solution may be necessary.

For the case of the cylindrical rod considered here, $A_c = \pi D^2/4$ and $P = \pi D$. We further assume, for simplicity, that $k$ and $h$ are constant. Then Equation (b) becomes

$$\frac{d^2 T}{dx^2} = \frac{4h}{kD}(T - T_s) \tag{c}$$

We next consider the question of boundary conditions. Since Equation (c) is second-order, we need two conditions. One condition is obtained very easily, by noting that the rod temperature is the same as the wall temperature at $x = 0$; thus, $T(0) = T_w$. The second condition is not so obvious. If the rod length $L$ is sufficiently long, we would expect that nearly all the heat conducted into the rod from the wall would be lost out the sides; in this event, $\partial T/\partial x = 0$ at $x = L$. However, we don't know how large $L$ would have to be before this condition would be reasonable. Instead, we consider an element of the rod having thickness $\Delta x$ whose right-hand side coincides with $x = L$. Then we make an energy balance on this *end* element. The rate of conduction into this element approaches the rate of convection from the end as $\Delta x \to 0$. Thus, at $x = L$ we must have $-kdT/dx = h(T - T_s)$. Note that the balance on the end element is different from the balances on an arbitrary *interior* element. Alternatively, we could derive the end condition by saying that the conductive flux into the boundary plane must equal the convective flux from this plane into the fluid stream. By looking back over the development given, it should be noted that the previous equation and boundary conditions would also hold if the fluid stream were heating the rod instead of cooling it.

It is normally useful to nondimensionalize the equations and boundary conditions in a mathematical model ([Aris, 1993] discusses a set of rules for nondimensionalization of model equations). This allows a simple dimensional check and, more important, reduces the number of parameters involved. We therefore introduce the new variables $\theta = (T - T_s)/(T_w - T_s)$, $\Delta T = T_w - T_s$ and $z = x/L$. Then $dT/dx = (\Delta T/L)d\theta/dz$ and $d^2T/dx^2 = (\Delta T/L^2)d^2\theta/dz^2$. Then Equation (c), together with the boundary conditions, becomes

$$\frac{d^2\theta}{dz^2} = \frac{4hD}{k}\left(\frac{L}{D}\right)^2\theta \tag{d}$$

$$\theta(0) = 1, \quad \left[\frac{d\theta}{dz} = -\frac{hD}{k}\frac{L}{D}\theta\right]_{z=1} \tag{e}$$

In this form, we see that the equation for $\theta$ is second-order, linear, constant-coefficient, and homogeneous. Equation (d) may be integrated to give

$$\theta = C_1\, exp(Mz) + C_2\, exp(-Mz) \tag{f}$$

where

$$M = 2\left(\frac{hD}{k}\right)^{\frac{1}{2}}\frac{L}{D}$$

The solution parameters are $hD/k$, which is similar to a Nusselt number, and $L/D$, a dimensionless length parameter. The constants, $C_1$ and $C_2$, are determined from the boundary conditions.

To illustrate the use of the above solution, we consider the problem of a fin of length $L = 4$ cm and diameter $D = 1$ cm whose base temperature $T_w = 150$°C. The fluid stream temperature $T_s = 50$°C. The value of $hD/k$ corresponding to the heat transfer coefficient $h$ and fin thermal conductivity $k$ is 0.2. It is desired to calculate the temperature at the tip of the fin ($z = 1$). For this problem, the value of $M = 2(0.2)^{1/2}/4 = 0.2236$. Substitution of Equation (f) for $\theta$ into the boundary conditions of Equation (e), and solving two linear algebraic equations for $C_1$ and $C_2$ finally results in $\theta(1) = 0.0465$, so that the temperature at the tip of the fin is $T(1) = 50 + 100\theta(1) = 54.65$°C.

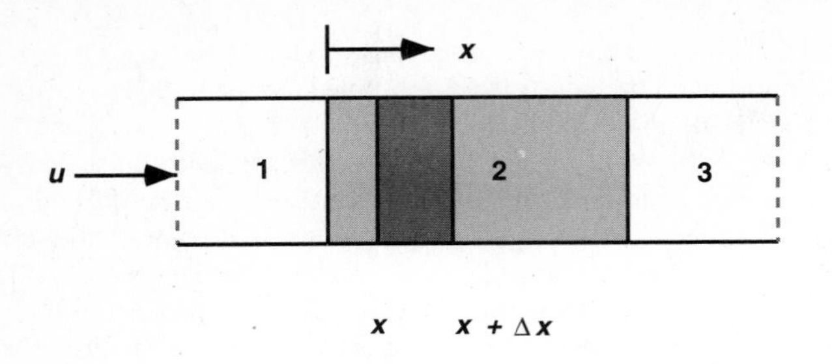


**Figure 1.8.** Flow Reactor

**Example 1.7 Chemical Flow Reactor with Axial Dispersion**

We consider the steady-state operation of a packed chemical flow reactor, where the reactor packing (in section 2 of Figure 1.8) contains a catalyst which promotes reaction. The occurrence of chemical reaction necessitates a consideration of the flow sections ahead of and following the packed catalyst section, because of the possibility of axial diffusional transport into these sections.

A simple model for this problem that has been used much in the past is the so-called pseudo-homogeneous model. Using this approach, the species mass balance in the reaction section is written as if no packing is present. The effects of the packing are incorporated into the "dispersion" (or diffusion) parameter $D$. The $D$ quantity generally is not known *a priori,* but rather must be determined from some combination of theory and experiment.

We consider the simplest case of only one reaction species, constant density, a constant flow velocity $u$, and isothermal operation. Referring to Figure 1.8, we divide the transport region into three sections; 1 is ahead of the packed reaction section, 2 is the reaction section, and 3 follows the reaction section. Sections 1 and 3 may also contain some kind of packing, but it is assumed that there is no catalyst (and hence no reaction) in these sections. We now focus attention on an arbitrary interior element of section 2 (Figure 1.8) and make a species mass balance. In words, we have that "the net rate of species [input − output + generation] = rate of accumulation. At location $x$ (where we *assume* that $dC/dx$ is negative) we have

$$\text{input rate by convection} = (uAC)_x \tag{a}$$

$$\text{input rate by diffusion} = (-DAdC/dx)_x \tag{b}$$

where $A$ is the duct cross-sectional area, $C$ is the concentration of the species of interest, and the negative sign is chosen in the diffusion relationship to correspond to the assumption that $dC/dx$ is negative. As in our earlier discussion of extended surface heat transfer, the sign choice based on the *assumption* that $dC/dx$ is negative will also be valid when $dC/dx$ is positive or zero. The reaction rate per unit volume (assumed here to be a depletion) is $r$. We then finally have

$$(uAC)_x - (uAC)_{x+\Delta x} + \left(-DA\frac{\partial C}{\partial x}\right)_x - \left(-DA\frac{\partial C}{\partial x}\right)_{x+\Delta x} - rA\Delta x = \frac{\partial(CA\Delta x)}{\partial t} \tag{c}$$

As in the case of extended surface heat transfer, we include the rate-of-accumulation term and work in terms of partial derivatives in order to illustrate the approach for the case of an *unsteady-state* model. If we now use the mean value theorem for derivatives to express the convection and diffusion terms, and if we cancel the $A$ and $\Delta x$ values and pass to the limit of $\Delta x \to 0$, we obtain the following differential equation *for the reaction section* (where we now assume a steady state, so that $\partial C/\partial\theta = 0$). Here we have also taken $u$, $A$, and $D$ to be constant.

$$D\frac{d^2C}{dx^2} - u\frac{dC}{dx} - r = 0 \quad \text{(section 2)} \tag{d}$$

In a similar way, we can make a species balance for both the upstream and downstream sections of the system; since these sections contain no reaction, their equations are merely the same as the previous one, except that there is no reaction term. These equations are

$$D\frac{d^2C}{dx^2} - u\frac{dC}{dx} = 0 \quad \text{(sections 1 and 3)} \tag{3}$$

The point of special interest regarding the above model for this system concerns appropriate boundary conditions. If we denote quantities for the upstream section with an overbar, those for the reaction section with no bar, and those for the downstream section with an underbar, we get the following "matching conditions" between the sections:

$$\overline{C}(0-) = C(0+) = C(0);\ \overline{D}\frac{d\overline{C}}{dx}\Big)_{0-} = D\frac{dC}{dx}\Big)_{0+} \quad \text{(between sections 1 and 2)} \tag{f}$$

$$C(L-) = \underline{C}(L+) = C(L);\ D\frac{dC}{dx}\Big)_{L-} = \underline{D}\frac{d\underline{C}}{dx}\Big)_{L+} \quad \text{(between sections 2 and 3)} \tag{g}$$

where, for example, $\overline{C}(0-)$ refers to the value of $\overline{C}$ at zero approached from the left.

The above boundary conditions correspond to the requirements that (i) the concentration must be continuous at the interface between sections, and (ii) the rate of mass transfer out of one section must equal that into the next section. For convective transport, this is satisfied by continuity of concentrations; for diffusive transport, the diffusional fluxes must be equal.

To satisfy the above boundary conditions, it is convenient to perform one integration of the equations for the upstream and downstream sections in order to ultimately derive an equation involving the reaction section only. These integrations take the form

$$\overline{D}\frac{d\overline{C}}{dx}\Big)_{0-} - \overline{D}\frac{d\overline{C}}{dx}\Big)_{-\infty} - u[\overline{C}(0-) - \overline{C}(-\infty)] = 0 \quad \text{(upstream)} \tag{h}$$

$$\underline{D}\frac{d\underline{C}}{dx}\Big)_{\infty} - \underline{D}\frac{d\underline{C}}{dx}\Big)_{L+} - u[\underline{C}(\infty) - \underline{C}(L+)] = 0 \quad \text{(downstream)} \tag{i}$$

At $x = -\infty$ in the upstream section, $\overline{C} = C_o$, the inlet concentration, and $\overline{D}(d\overline{C}/dx)_{-\infty} = 0$. By combining the integrated form of the upstream equation with the boundary conditions between sections 1 and 2, we get

$$D\frac{dC}{dx}\Big)_0 = u[C(0) - C_o] \tag{j}$$

This condition then represents the initial or beginning condition for the reaction section. At $x = \infty$ in the downstream section, the concentration, $\underline{C}(\infty)$, is not known, although $d\underline{C}/dx)_\infty = 0$. Instead, we consider a geometrical analysis of the following equation, which is the result of combining the integrated form of the downstream equation with the boundary conditions between sections 2 and 3.

$$D\frac{dC}{dx}\Big)_{L-} = u[C(L-) - \underline{C}(\infty)] \tag{k}$$

Since the equation $D d^2 \underline{C}/dx^2 - u d\underline{C}/dx = 0$ for the downstream section cannot have a local maximum or minimum (since if $d\underline{C}/dx = 0$ at some point, all higher derivatives must be zero also), we see that the previous equation leads to a contradiction unless $(dC/dx)$ at $L-$ equals zero. This is evident from considering the individual cases of $(dC/dx)$ being positive or negative, both of which would require that either a maximum or minimum occur in the downstream section. Thus, $dC/dx)_{L-} = 0$ is the appropriate boundary condition for the end of the reaction zone, and it shows that $\underline{C}(\infty) = C(L-)$, so that no further change of concentration occurs in the downstream section.

If we use the dimensionless variables $\phi = C/C_o$, $z = x/L$, and $\varepsilon = D/Lu$, we get the following dimensionless differential equation and boundary conditions representing the model.

$$\varepsilon \frac{d^2\phi}{dz^2} - \frac{d\phi}{dz} - \left(\frac{L}{uC_o}\right) r = 0 \tag{l}$$

$$\varepsilon \left.\frac{d\phi}{dz}\right)_0 = \phi(0) - 1, \quad \left.\frac{d\phi}{dz}\right)_1 = 0 \tag{m}$$

In the above equation, $r$ is a function of concentration. For instance, in the case of first-order reaction $r = k_1 C$; for second-order reaction $r = k_2 C^2$; and so on.

As an example, suppose we consider a problem involving a second-order reaction, with $r = k_2 C^2$ so that $(L/uC_0)k \equiv B$. The values of $\varepsilon$ and $B$ are taken to be 0.1 and 7.5, respectively. It is desired to determine the outlet concentration $C(1)$. This nonlinear boundary value problem does not have a known analytical solution. A numerical solution (using the shooting method discussed in Chapter 12) gives $C(1) = 0.150$.

## 1.6 INTEGRALS

It occasionally happens that a modeling problem requires the formulation of an integral. We consider next an example to illustrate this type of problem.

**Example 1.8 Estimation of Effect of Radial Velocity Profile on Conversion in a Flow Reactor**

For steady-state transport in a flow reactor having no diffusional effects, the following species balance equation applies if the flow velocity $u$ is constant and the species is being depleted due to reaction.

$$(uAC)_x - (uAC)_{x+\Delta x} - r\Delta V = 0 \tag{a}$$

Here, $A$ is the reactor cross-section (assumed constant), $\Delta V = A\Delta x$, $C$ is species concentration, and $r$ is the species rate of depletion per unit volume. For purposes of illustration, we suppose that the reaction is first-order so that $r = kC$. Then, by using the mean value theorem to write $(uAC)_{x+\Delta x} - (uAC)_x = uA(dC/dx)_\xi \Delta x$ and passing to the limit of $\Delta x \to 0$, we get

$$u\frac{dC}{dx} = -kC \tag{b}$$

This simple differential equation has the solution $C = C_o \exp(-kx/u)$, where $C_o$ is the concentration entering the reactor. If our reactor corresponds to the situation where the velocity *does* vary over the cross-section, then we can see from the preceding discussion that, since $C$ depends on $u$, at a given $x$ station the concentration $C$ will also vary with transverse position. Thus, we observe that even if we assume that transverse diffusion is *not* present in our system, the transverse variation of velocity will induce a transverse

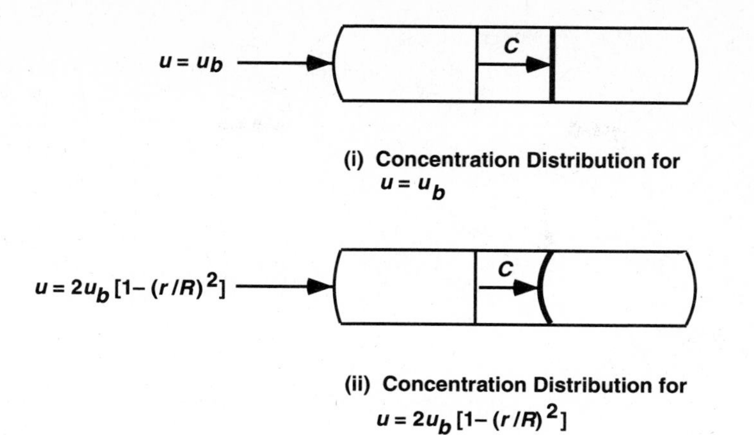


**Figure 1.9.** Effect of Velocity Profile on Concentration Distribution in Tubular Reactor

variation of concentration, as indicated in Figure 1.9 We wish to ascertain approximately the effect of this induced transverse composition variation on the average conversion in the reactor. Of course, to be precise we would need to account for *transverse diffusion,* which would tend to diminish concentration variations, but this would require a much more complex partial differential equation model. Instead, we will be content with the simpler calculation of the mean concentration at any particular downstream location.

To conveniently achieve our objective, it is necessary to define the mean velocity and concentration. The bulk velocity $u_b$ is defined to be that velocity which, when multiplied by the cross-sectional area $A_o$, is equal to the actual volumetric flow rate, $\int u\,dA$, where the integration is over the entire flow cross-section. The bulk concentration $C_b$ is defined as that concentration which, when multiplied by the bulk velocity $u_b$ and the cross-sectional area $A_o$, produces the correct mass balance at a given location. Thus, $C_b u_b A_o = \int CudA$ over the cross-section. If $C$ is constant across the cross-section, then $C_b = C =$ constant.

Now let us consider, for concreteness, the specific case of a cylindrical tube reactor having a fully developed laminar flow. For this situation $A_o = \pi R^2$, $dA = 2\pi r dr$, and $u = 2u_b[1-(r/R)^2]$. The expression for the bulk concentration becomes

$$C_b = \frac{2}{R^2}\int_{r=0}^{R} r\frac{u}{u_b}Cdr = \frac{2}{R^2}C_0\int_{r=0}^{R} r\frac{u}{u_b}e^{-\frac{kx}{u}}\,dr \tag{c}$$

This expression shows that if $u = u_b =$ constant, then $C_b = C = C_o\exp(-kx/u)$. It is convenient to introduce the dimensionless variables $z = r/R$, $\phi = u/u_b$, and $a = kx/u_b$. It is also useful, then, to change the integration variable from $r$ to $\phi$ by noting from the velocity distribution that $\phi = 2[1-z^2]$, and therefore, $zdz = -d\phi/4$. This finally yields

$$C_b = 2\int_{z=0}^{1} z\phi C_0 e^{-\frac{a}{\phi}}\,dz = \frac{C_0}{2}\int_{\phi=0}^{2} \phi e^{-\frac{a}{\phi}}\,d\phi \tag{d}$$

The first part of this final equation again shows that if $\phi = 1 =$ constant, $C_b/C_o = \exp(-a)$. The second part of the equation, which is the most convenient form for either analytical or numerical approximation, shows that if $a \to 0$, $C_b \to C_o$. This corresponds to the limiting case of no reaction. For other values of $a$, $C_b/C_o$ is a function only of the parameter $a = kx/u_b$.

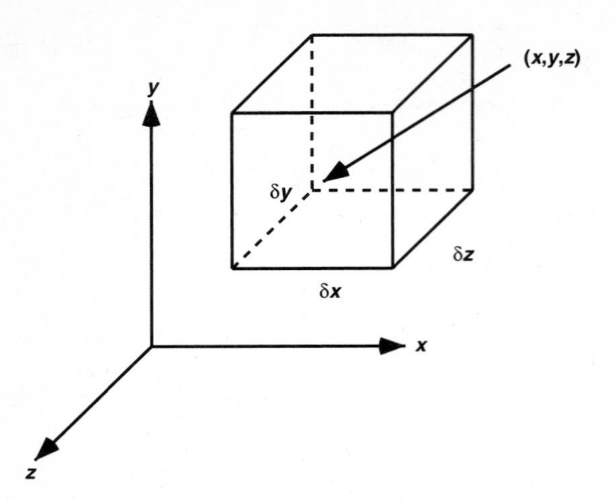


**Figure 1.10.** Volume Element for Mass Balance

## 1.7 PARTIAL DIFFERENTIAL EQUATIONS

Partial differential equations arise frequently in modeling problems, generally where a system must be described by more than one independent variable (for example, time and one space dimension, or two space dimensions). Such equations can be formulated by procedures similar to, but more complicated than, those used in deriving ordinary differential equations. Since solution methods are considerably more lengthy for PDEs than for ODEs, in the interests of simplicity, "perfect-mixing" assumptions are sometimes made in order to convert a system model from a PDE to an ODE. In complex situations involving a number of system components, the overall model may consist of a combination of PDEs, ODEs, and algebraic equations.

**Example 1.9 The Differential Component Mass Balance for Convective Mass Transfer**

We wish to derive the general component mass balance for a single component, $A$, in a multicomponent flowing fluid. Consider the small volume element shown for Cartesian coordinates in Figure 1.10.

The mass of component $A$ in the volume element is

$$\rho_A \delta x\, \delta y\, \delta z \tag{a}$$

where $\rho_A$ is the mass density of component $A$ in the mixture. The rate of accumulation of mass of $A$ in the volume element is

$$\frac{\partial \rho_A}{\partial t} \delta x\, \delta y\, \delta z \tag{b}$$

Consider the net rate of input of $A$ into our volume element across faces perpendicular to the $x$-axis:

by convection,

$$u_x \rho_A \delta y\, \delta z - \{u_x \rho_A \delta y\, \delta z + \frac{\partial u_x \rho_A \delta y\, \delta z}{\partial x} \delta x\} \tag{c}$$

by diffusion,

$$J_{Ax}\,\delta y\,\delta z - \{J_{Ax}\,\delta y\,\delta z + \frac{\partial J_{Ax}\,\delta y\,\delta z}{\partial x}\,\delta x\} \tag{d}$$

Thus, the net rate of input across faces perpendicular to the $x$-axis is

$$-\left\{\frac{\partial u_x \rho_A}{\partial x} + \frac{\partial J_{Ax}}{\partial x}\right\}\delta x\,\delta y\,\delta z \tag{e}$$

where $u_x$ is the $x$-component of the mass average velocity, and $J_{Ax}$ is the $x$-component of the diffusion flux with respect to the mass average velocity.

In a similar way, the net rate of input of mass of $A$ across faces perpendicular to the $y$-axis is

$$-\left\{\frac{\partial u_y \rho_A}{\partial y} + \frac{\partial J_{Ay}}{\partial y}\right\}\delta x\,\delta y\,\delta z \tag{f}$$

and the net rate of input of $A$ across faces perpendicular the $z$-axis is

$$-\left\{\frac{\partial u_z \rho_A}{\partial z} + \frac{\partial J_{Az}}{\partial z}\right\}\delta x\,\delta y\,\delta z \tag{g}$$

The rate of generation of $A$ due to chemical reaction in the volume element is

$$r_A\,\delta x\,\delta y\,\delta z \tag{h}$$

where $r_A$ is the rate of generation of mass of $A$ per unit volume due to chemical reaction. In words, the mass balance of $A$ is

$$\text{net rate of input} + \text{rate of generation} = \text{rate of accumulation} \tag{i}$$

Substitution of the derived terms into Equation (i) and division by $\delta x\,\delta y\,\delta z$ gives

$$\frac{\partial \rho_A}{\partial t} + \frac{\partial u_x \rho_A}{\partial x} + \frac{\partial u_y \rho_A}{\partial y} + \frac{\partial u_z \rho_A}{\partial z} + \frac{\partial J_{Ax}}{\partial x} + \frac{\partial J_{Ay}}{\partial y} + \frac{\partial J_{Az}}{\partial z} - r_A = 0 \tag{j}$$

For binary diffusion, the diffusion flux with respect to the mass average velocity, $u$, may be expressed as

$$J_{Ax} = -D_{AB}\,\rho\frac{\partial x_A}{\partial x} \tag{k}$$

where

$D_{AB}$ is the diffusion coefficient.

$\rho$ is the mass density of the fluid.

$x_A$ is the mass fraction of $A$.

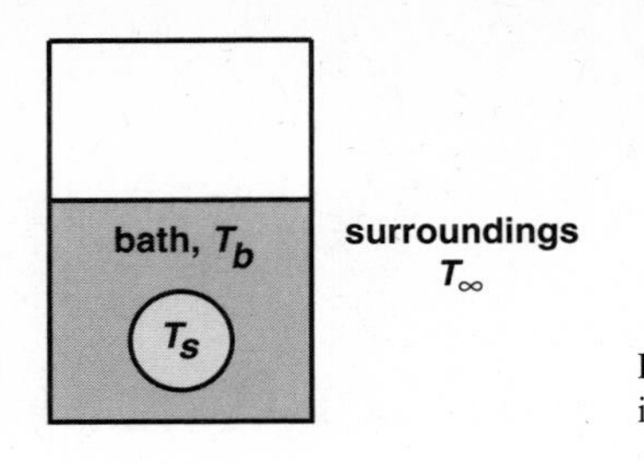


**Figure 1.11.** Quenching of a Sphere in a Bath

**Example 1.10 A PDE/ODE Model-Quenching of a Sphere in a Finite Bath**

It is desired to develop a model for the time-dependent behavior of the process of quenching a sphere (which is initially at a high temperature) in a finite bath of coolant which can exchange heat with the surroundings. The model, which governs the quenching process, should be based on the assumptions of (i) angular symmetry within the sphere and (ii) uniform spatial temperature of the bath at any instant of time. The quantity of special interest is the center temperature of the sphere, which is to be determined as a function of time over the whole process until the sphere, bath, and surroundings have thermally equilibrated. This process is illustrated in Figure 1.11.

Here we will analyze the problem in two different ways. The simplest, or lumped, parameter model is based on assuming that at any instant of time the temperature of the sphere is uniform in the radial direction. This corresponds to a "perfect-mixing" assumption. The simple lumped model leads to a system of two coupled ordinary differential equations for the sphere and bath temperatures. On the other hand, the distributed parameter model accounts for the radial variation of the sphere temperature with time and is described by a partial differential equation for the sphere temperature coupled to an ordinary differential equation for the bath temperature.

### *Lumped Model*

For the lumped model, an energy balance on the sphere yields

$$\text{input} - \text{output} = \text{accumulation}$$

$$0 - h_i A_T(T_s - T_b) = \rho V C_{p_s} \frac{dT_s}{dt} \tag{a}$$

Here, $h_i$ is the inside convective heat transfer coefficient (between sphere and bath); it must be determined independently, for example, from a handbook or experiments. The quantity $A_T$ is the sphere surface area, $T_b$ is the bath temperature, and $\rho$, $V$, $C_p$, and $T_s$ are the sphere density, volume, heat capacity, and temperature, respectively. For a sphere, we have $A_T = 4\pi R^2$ and $V = (4/3)\pi R^3$. The energy balance for the bath can be written as

$$\text{input} - \text{output} = \text{accumulation}$$

$$h_i A_T(T_s - T_b) - h_o A_T(T_b - T_\infty) = m_b C_{P_b} \frac{dT_b}{dt} \tag{b}$$

Here, the subscript $b$ refers to the bath, $h_o$ is the outside convective heat transfer coefficient between bath and surroundings, and $T_\infty$ is the surroundings temperature. The quantities $h_i$ and $h_o$ *may* vary with their respective $\Delta Ts$, making the system nonlinear. We see that the lumped model is described by two coupled ODEs which may be nonlinear.

### *Distributed Model*

For the distributed model, in which radial temperature variations in the sphere are allowed, we proceed as follows. To make an energy balance on a thin spherical shell of material having thickness $\Delta r$, surface area $A_r = 4\pi r^2$, and volume $\Delta V = 4\pi r^2 \Delta r$, we write (rate of input by conduction at $r$) − (rate of output by conduction at $r + \Delta r$) = rate of accumulation within volume element. The conductive flux (energy rate per unit area normal to transfer direction) is expressed as $-k(\partial T/\partial r)$ since if the actual flux is in the positive $r$ direction, $T$ is decreasing with $r$ and $\partial T/\partial r$ is negative. In equation form, we have

$$\left(-kA_r\frac{\partial T_s}{\partial r}\right)_r - \left(-kA_r\frac{\partial T_s}{\partial r}\right)_{r+\Delta r} = \rho\Delta V C_p\frac{\partial T_s}{\partial r} \tag{c}$$

Using the mean value theorem for derivatives, the left-hand side of this equation can be written as $\partial(kA_r\partial T/\partial r)/\partial r$. Then we finally get

$$\frac{l}{r^2}\frac{\partial}{\partial r}\left(kr^2\frac{\partial T_s}{\partial r}\right) = \rho C_p\frac{\partial T_s}{\partial t} \tag{d}$$

This is the one-dimensional, radially symmetric, unsteady-state partial differential equation for heat conduction in a sphere. It requires specification of two radial boundary conditions and one initial condition for its solution. The boundary conditions are (i) $(\partial T/\partial r)_{r=0} = 0$ for all $t$, and (ii) $(-kA_R\partial T_s/\partial r)_{r=R} = h_i A_R[T_s(R) - T_b]$.

The first of these two conditions can be viewed as a boundedness condition, since if $\partial T_s/\partial r$ did not go to zero for $r = 0$, the left-hand side of the previous equation would be infinite. The second boundary condition above expresses the fact that at the outer boundary of the sphere, the energy transferred from the sphere by conduction is equal to the energy transferred to the bath by convection. The initial temperature of the whole sphere is equal to its temperature at the beginning of the quench process.

The energy balance for the bath in the case of the distributed model is the same as for the lumped model, except the temperature $T_s$, which previously depended only on time, must now be specified as $T_s(R)$, the temperature at the outer surface of the sphere. This quantity depends implicitly only on time. The present distributed model, which may be nonlinear due to temperature dependencies of $k$, $h_i$, and $h_o$, is often solved most conveniently by the method of lines, discussed in "Partial Differential Equations."

## *1.8 DIMENSIONAL ANALYSIS*

It is almost always advantageous to transform the partial differential equations that describe a problem of interest into a dimensionless form before attempting solution. By casting the equations into a dimensionless form, we obtain a minimum parametric representation of the problem. This is particularly useful when carrying out numerical solutions, because the number of solutions needed to cover a given range in the dimensional parameters will also be minimized if we solve the equations in dimensionless form. To accomplish the nondimensionalization, we use characteristic dimensional constants that appear naturally in the equation or in the boundary conditions. Even when we are not going to solve the differential equation describing our physical model, useful information can often be obtained from a dimensional analysis of the equations.

**Example 1.11 Dimensional Analysis of the Diffusion Equation**

To illustrate these ideas, we consider the differential mass balance derived in Example 1.9. For constant density and diffusion coefficient, this equation may be written as

$$\frac{D\rho_A}{Dt} = D_{AB}\left(\frac{\partial^2 \rho_A}{\partial x^2} + \frac{\partial^2 \rho_A}{\partial y^2} + \frac{\partial^2 \rho_A}{\partial z^2}\right) \tag{a}$$

where the substantial derivative $D/Dt$ is defined as

$$\frac{D}{Dt} = \frac{\partial}{\partial t} + u_x \frac{\partial}{\partial x} + u_y \frac{\partial}{\partial y} + u_z \frac{\partial}{\partial z} \tag{b}$$

Equation (a) describes the binary mass transfer problem. Since the velocity appears in this equation, we need to include the continuity equation and the equation of motion in our model. For constant density, the continuity equation is

$$\frac{\partial u_x}{\partial x} + \frac{\partial u_y}{\partial y} + \frac{\partial u_z}{\partial z} = 0 \tag{c}$$

For constant density and viscosity, the equation of motion in the $x$-direction is

$$\rho \frac{Du_x}{Dt} = \mu\left(\frac{\partial^2 u_x}{\partial x^2} + \frac{\partial^2 u_x}{\partial y^2} + \frac{\partial^2 u_x}{\partial z^2}\right) - \frac{\partial p}{\partial x} + \rho g_x \tag{d}$$

where $\mu$ is viscosity, $\rho$ is density, $p$ is pressure, and $g_x$ is the body force per unit volume in the $x$ direction (usually due to gravity). For the general case, we also have equations similar to Equation (d) for $u_y$ and $u_z$.

To make the equations dimensionless, we assume that there are some characteristic parameters of the system such as $U$ (velocity), $p_0$ (pressure), $D$ (length), $g$ (acceleration due to gravity), $\rho_{A0}$ and $\rho_{A1}$ (mass fractions), and define our dimensionless variables in the following manner:

$$t^* = \frac{Ut}{D}, \quad x^* = \frac{x}{D}, \quad y^* = \frac{y}{D}, \quad z^* = \frac{z}{D}$$

$$u_x^* = \frac{u_x}{U}, \quad p^* = \frac{p - p_0}{\rho U^2}, \quad \rho_A^* = \frac{\rho_A - \rho_{A0}}{\rho_{A1} - \rho_{A0}}, \quad g_x^* = \frac{g_x}{g} \tag{e}$$

In terms of these dimensionless variables, our equations become

continuity:
$$\frac{\partial u_x^*}{\partial x^*} + \frac{\partial u_y^*}{\partial y^*} + \frac{\partial u_z^*}{\partial z^*} = 0 \tag{f}$$

conservation of A:
$$\frac{D\rho_A^*}{Dt^*} = \frac{l}{\text{Re Sc}}\left(\frac{\partial^2 \rho_A^*}{\partial x^{*2}} + \frac{\partial^2 \rho_A^*}{\partial y^{*2}} + \frac{\partial^2 \rho_A^*}{\partial z^{*2}}\right) \tag{g}$$

motion:
$$\frac{Du_x^*}{Dt^*} = \frac{1}{\text{Re}}\left\{\frac{\partial^2 u_x^*}{\partial x^{*2}} + \frac{\partial^2 u_x^*}{\partial y^{*2}} + \frac{\partial^2 u_x^*}{\partial z^{*2}}\right\} - \frac{\partial p^*}{\partial x^*} + \frac{g_x^*}{\text{Fr}} \tag{h}$$

where

$$\text{Re} = \frac{DU\rho}{\mu}, \quad \text{Sc} = \frac{\mu}{\rho D_{AB}}, \quad \text{Fr} = \frac{U^2}{gD} \tag{i}$$

By inspection of Equations (f) to (h), we see that a solution of these equations will depend on the parameters Re (Reynolds number), Sc (Schmidt number), Fr (Froude number), and any additional parameters that arise through the particular boundary conditions.

We are normally interested in the mass flux at a particular boundary, $J_{A0}$. For the purposes of this discussion, we assume that $J_{A0}$ is given by

$$J_{A0} = -D_{AB} \left.\frac{\partial \rho_A}{\partial z}\right)_0 \tag{j}$$

If we define a mass transfer coefficient as

$$k_\rho = J_{A0}/(\rho_{A1} - \rho_{A0}) \tag{k}$$

then, in terms of our dimensionless variables, we have

$$\text{Sh} = \frac{k_\rho D}{D_{AB}} = -\left.\frac{\partial \rho_A^*}{dz^*}\right)_0 \tag{l}$$

where Sh is the Sherwood number.

From Equation (l), we can see that the Sh will be a function of Re, Sc, Fr, and any other parameters that arise from the boundary conditions.

$$\text{Sh} = f(\text{Re}, \text{Sc}, \text{Fr}, \text{boundary conditions}) \tag{m}$$

This information does not give us the form of the functional relationship, but is very useful for designing experiments to determine this relationship.

## 1.9 SCALING

In chemical engineering modeling we almost never solve the complete partial differential equation system that describes the physical problem. This is the case even when we are seeking a numerical solution. The reason for this is that a complete solution may be unnecessarily complicated when some terms in the equation are small compared to others. The question that needs to be addressed is, under what conditions can the original differential equation be simplified? The procedure that we undertake to formally answer this question is generally referred to as scaling. To implement the method, we need to have some idea about which terms we expect to be small. The scaling procedure then tells us the conditions under which these terms can be neglected. Very briefly, the method consists of casting the equations in dimensionless form in such a way that the dimensionless derivatives are approximately of the same order of magnitude. The dimensionless parameters in the equation will then indicate under what conditions some terms can be neglected. The method will be introduced and explained here through use of an example. Further examples of the application of scaling may be found in [Krantz, 1970].

**Example 1.12 Mass Transfer between a Wall and a Falling Liquid Film**

Consider the problem illustrated in Figure 1.12. A liquid film falls in steady flow down a vertical wall with a fully developed velocity profile given by

$$u = u_m \left\{ \frac{2y}{H} - \left(\frac{y}{H}\right)^2 \right\} \tag{a}$$

where $u$ is the maximum velocity at $y = H$.

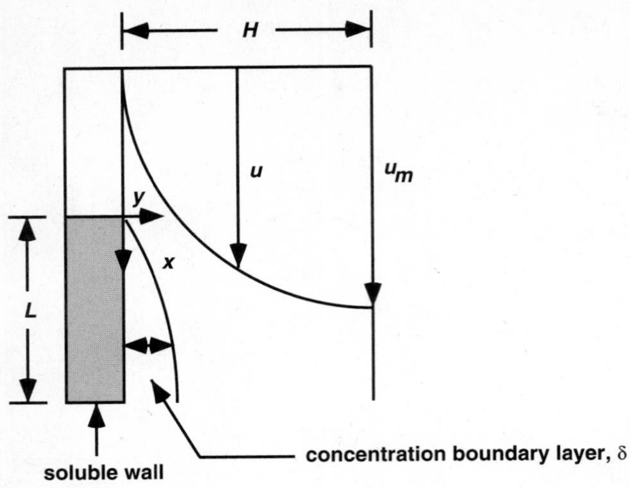


**Figure 1.12.** Solid Dissolution into a Falling Liquid Film

The differential equation that describes the mass transfer process is

$$u_m\left\{\frac{2y}{H}-\left(\frac{y}{H}\right)^2\right\}\frac{\partial \rho_A}{\partial x}=D_{AB}\left\{\frac{\partial^2 \rho_A}{\partial x^2}+\frac{\partial^2 \rho_A}{\partial y^2}\right\} \tag{b}$$

with the following boundary conditions:

1. $y=0,\quad \rho_A=\rho_{As}$ (a constant)
2. $y=H,\quad \dfrac{\partial \rho_A}{\partial y}=0$ (no mass transfer across gas-liquid interface) (c)
3. $x=0,\quad \rho_A=\rho_{Ai}$ (uniform solute concentration at $x=0$)

Since diffusion coefficients in the liquid phase are small, our intuition would tell us that there will be a relatively thin concentration boundary layer of thickness $\delta(x)$ over which the concentration change will occur. Within this boundary layer, it might be permissible to simplify the velocity expression by using a one-term Taylor series approximation near the wall,

$$u \simeq 2u_m\left(\frac{y}{H}\right) \tag{d}$$

Also, if we are sufficiently far from $x=0$, we might expect that axial diffusion would be very small compared to transverse diffusion. Thus, the simplified equation describing this mass transfer process would be

$$2\left(\frac{u_m y}{H}\right)\frac{\partial \rho_A}{\partial x}=D_{AB}\frac{\partial^2 \rho_A}{\partial y^2} \tag{e}$$

We wish to find the conditions under which these simplifications can be justified.

We begin by defining dimensionless variables as

$$\rho_A^*=\frac{\rho_A-\rho_{Ai}}{\rho_{As}-\rho_{Ai}}\qquad y^*=\frac{y}{\delta(x)}\qquad x^*=\frac{x}{L} \tag{f}$$

$\delta(x)$ is the concentration boundary layer thickness, and $L$ is the downstream $x$-value where the simplified differential equation is valid. $\rho_A^*$, $y^*$ and $x^*$ have been defined so that they vary between 0 and 1 within the boundary layer. In terms of these dimensionless variables our original differential equation, Equation (a), becomes, after some rearrangement,

$$\left\{\frac{2\delta}{H}y^* - \left(\frac{\delta}{H}\right)^2 y^{*2}\right\}\frac{u_m\delta^2}{D_{AB}L}\frac{\partial\rho_A^*}{\partial x^*} = \frac{\partial^2\rho_A^*}{\partial y^{*2}} + \left(\frac{\delta}{L}\right)^2\frac{\partial^2\rho_A^*}{\partial x^{*2}} \tag{g}$$

In order that the convective term (the L.H.S.) be the same order of magnitude as the dominant diffusion term $(\partial^2\rho_A^*/\partial y^{*2})$, we must have

$$\frac{u_m\delta^2}{D_{AB}L}\left(\frac{2\delta}{H}\right) \simeq 1 \tag{h}$$

Equation (h) implies that the unknown boundary layer thickness has the approximate relationship

$$\delta = \left(\frac{D_{AB}LH}{2u_m}\right)^{1/3} = \left(\frac{LH^2}{2Pe}\right)^{1/3} \tag{i}$$

where the Peclet number is defined as

$$Pe = \frac{u_m H}{D_{AB}} \tag{j}$$

In order for the axial diffusion term $\left(\dfrac{\delta^2}{L^2}\dfrac{\partial^2\rho_A^*}{\partial x^{*2}}\right)$ to be negligible, we see from Equation (g) that we must have

$$\left(\frac{\delta}{L}\right)^2 \ll 1 \tag{k}$$

Upon substituting our relationship for $\delta$, Equation (i), into Equation (k), we get

$$\left(\frac{H}{L}\right)^{4/3}\left(\frac{1}{Pe^{2/3}}\right) \ll 1, \qquad \text{for axial diffusion to be negligible}$$

or

$$\left(\frac{H}{L}\right)^2\left(\frac{1}{Pe}\right) \ll 1 \tag{l}$$

In order for the velocity profile to be well approximated by the linear Taylor series expansion, we see that we must have

$$\frac{\delta}{H} \ll 1 \tag{m}$$

Substituting for $\delta$ in Equation (m) gives

$$\frac{1}{Pe^{1/3}}\left(\frac{L}{H}\right)^{1/3} \ll 1 \tag{n}$$

To summarize these arguments we can state that if the inequality expressed by Equation (l) is valid, then axial diffusion may be neglected; and if the inequality of Equation (n) is true, then the velocity profile may be approximated by Equation (d). Under these conditions, the original differential equation reduces to

$$\frac{2u_m}{H}y\frac{\partial\rho_A}{\partial x} = D_{AB}\frac{\partial^2\rho_A}{\partial y^2} \tag{o}$$

This form of the equation is much easier to solve than the original form, Equation (b). This simplified equation is conveniently solved using the method of combination of variables discussed in Chapter 13.

## REFERENCES


AMUNDSON, N. R. (1966), *Mathematical Methods in Chemical Engineering: Matrices and Their Application*, Englewood Cliffs, NJ: Prentice-Hall.

ARIS, R. (1993), *Chem. Eng. Sci.*, **48**, p. 2507.

BENNETT, C. O. and MYERS, J. E. (1982), *Momentum, Heat and Mass Transfer*, 3rd ed., New York: McGraw-Hill.

BEVERDIGE, G. S. G. and SCHECHTER, R. S. (1970), *Optimization: Theory and Practice*, New York: McGraw-Hill.

BIRD, R. B., STEWART, W. E., and LIGHTFOOT, E. N. (1960), *Transport Phenomena*, New York: Wiley.

BUNDAY, B. D. (1984), *Basic Linear Programming*, Baltimore, MD: Arnold.

DANBY, J. M. (1985), *Computing Applications to Differential Equations*, Reston, VA: Reston Publishing.

DENN, M. M. (1986), *Process Modeling*, New York: Longman.

EDGAR, T. F. and HIMMELBLAU, D. M. (1988), *Optimization of Chemical Processes,* New York: McGraw-Hill.

FRANKS, R. G. E. (1972), *Modeling and Simulation in Chemical Engineering*, New York: Wiley.

FRIEDLY, J. C. (1972), *Dynamic Behavior of Processes*, Englewood Cliffs, NJ: Prentice-Hall.

HERSHEY, D. (1973), *Transport Analysis*, New York: Plenum.

HIMMELBLAU, D. M. and BISCHOFF, K. B. (1968), *Process Analysis and Simulation: Deterministic Systems*, New York: Wiley.

JELEN, F. C. (1970), *Cost and Optimization Engineering*, New York: McGraw-Hill.

JENSON, V. G. and JEFFREYS, G. V. (1977), *Mathematical Methods in Chemical Engineering*, 2nd ed., New York: Academic Press.

KRANTZ, W. B. (1970), *Chem. Eng. Educ.,* **4**, p. 145.

LUYBEN, W. L. (1973), *Process Modeling, Simulation, and Control for Chemical Engineers*, New York: McGraw-Hill.

MICKLEY, H. S., SHERWOOD, T. K., and REED, C E. (1957), *Applied Mathematics in Chemical Engineering*, New York: McGraw-Hill.

RAMIREZ, W. F. (1989), *Computational Methods for Process Simulation*, Stoneham, MA: Butterworth.

RATKOWSKY, D. A. (1990), *Handbook of Nonlinear Regression Models*, New York: Marcel Dekker.

RIGGS, J. B. (1988), *An Introduction to Numerical Methods for Chemical Engineers*, Lubbock, TX: Texas Tech. Univ. Press.

RACHFORD, H. H., Jr., and RICE, J. D. (1952), *J. Petrol. Technol.* **4**, No. 10; Sec. 1, p. 19, Sec 2, p. 3.

*Safety, Health and Loss Prevention in Chemical Processes*, (1990), New York: American Institute of Chemical Engineers.

SILEBI, C. A., and SCHIESSER, W. E. (1992), *Dynamic Modeling of Transport Process Systems*, New York: Academic Press.

WALAS, S. M. (1991), *Modeling with Differential Equations in Chemical Engineering*, Boston, MA: Butterworth.

# 2

# Linear Algebraic Equations and Matrix Analysis

Matrix analysis is primarily concerned with the properties of systems of linear algebraic equations. Linear equations are equations in which the unknown variables appear only raised to the first power. Most students are familiar with the procedure for solving up to several linear equations by means of eliminating the unknowns one by one, or by using "determinants." However, in general, it is very important to know whether solutions exist or not for *much larger systems* of equations. In addition, we need to know the properties of the solutions as well as the means of calculating them. The development of matrix analysis allows an entire system of linear equations to be represented by a single matrix equation. Using matrix algebra, this system can be operated on in a direct manner to produce various desired results. The systematic approach afforded by the replacement of any number of linear equations by a single matrix equation allows us to obtain results that would be extremely difficult to achieve in other ways.

This organization and unification of the theory of linear equations represents great mathematical progress, since it brings into focus the common features and essential results that characterize any linear system, regardless of size. Moreover, it turns out that the appropriate modifications and extensions of matrix analysis permit unification of results for virtually all kinds of linear equations, such as differential equations, difference equations, integral equations, and so on. In addition, much of the analysis of nonlinear systems depends on comparison with appropriate linear equations. For more detail concerning matrix analysis, the reader is referred to Amundson [1966], Johnson and Riess [1982], and Kahaner *et al.* [1989].

## 2.1 INTRODUCTION TO LINEAR SYSTEMS

We are interested in considering a system of linear algebraic equations such as the following:

$$
\begin{aligned}
a_{11}x_1 + a_{12}x_2 + \cdots + a_{1n}x_n &= b_1 \\
a_{21}x_1 + a_{22}x_2 + \cdots + a_{2n}x_n &= b_2 \\
&\vdots \\
a_{n1}x_1 + a_{n2}x_2 + \cdots + a_{nn}x_n &= b_n
\end{aligned}
\tag{2-1}
$$

or, in more compact notation,

$$\sum_{j=1}^{n} a_{ij}x_j = b_i \qquad (i = 1, 2, \ldots, n) \tag{2-2}$$

Here, we have used single subscripts for the $x$ and $b$ quantities and a double subscript notation for the $a$ quantities, in order to avoid running out of variable names in large problems. The $a_{ij}$ subscript notation follows the rule that the first subscript refers to the row and the second to the column in which the element is found. We define a matrix in general to be simply a rectangular array of numbers having the property that two matrices are equal only when all of their elements are equal. We denote a matrix with a capital letter in italics, such as $A$. Thus, the matrix $A$, having three rows and four columns, is written as

$$A = \begin{pmatrix} a_{11} & a_{12} & a_{13} & a_{14} \\ a_{21} & a_{22} & a_{23} & a_{24} \\ a_{31} & a_{32} & a_{33} & a_{34} \end{pmatrix}$$

A matrix having only one column is called a column vector and is denoted by a boldface small letter, such as $\mathbf{x}$. A column vector is a special case of a matrix and has only one subscript, which denotes its row. The *sum* of two matrices, $C = A + B$, is the matrix whose elements are given by $c_{ij} = a_{ij} + b_{ij}$. Scalar multiplication of any matrix is defined to give a new matrix, all of whose elements are multiplied by the scalar.

If $A$ is a matrix having $i$ rows and $j$ columns, and $B$ is a matrix of $j$ rows and $k$ columns, the *product* $C = AB$ is defined to be the matrix of $i$ rows and $k$ columns having as its $i, k$ element

$$c_{ik} = \sum_{j=1}^{n} a_{ij}b_{jk} \tag{2-3}$$

For the product to be defined, the number of columns in matrix $A$ must be the same as the number of rows in matrix $B$. For example, if we multiply the two matrices $A$ and $B$ to get matrix $C$, as below, we have

$$AB = \begin{pmatrix} 1 & 2 \\ 3 & 4 \end{pmatrix} \begin{pmatrix} 5 & 6 \\ 7 & 8 \end{pmatrix} = C = \begin{pmatrix} 19 & 22 \\ 43 & 50 \end{pmatrix} \tag{2-4}$$

where the matrix product $C = AB$ is calculated as $c_{11} = (1)(5) + (2)(7) = 19$; $c_{12} = (1)(6) + (2)(8) = 22$; $c_{21} = (3)(5) + (4)(7) = 43$, and $c_{22} = (3)(6) + (4)(8) = 50$.

Using the definition of matrix multiplication in Equation (2-2), we can write the system of equations (2-1) in matrix form as

$$A\mathbf{x} = \mathbf{b} \tag{2-5}$$

where

$$A = \begin{pmatrix} a_{11} & a_{12} & \cdots & a_{1n} \\ \vdots & & & \\ a_{n1} & a_{n2} & \cdots & a_{nn} \end{pmatrix}; \quad \mathbf{x} = \begin{pmatrix} x_1 \\ \vdots \\ x_n \end{pmatrix} \quad \mathbf{b} = \begin{pmatrix} b_1 \\ \vdots \\ b_n \end{pmatrix} \tag{2-6}$$

The fundamental problem in linear algebra is concerned with the existence and uniqueness of solutions of systems of linear equations. It turns out that it is possible to characterize the behavior of *square* linear systems ($n$ equations, $n$ unknowns) in terms of a single number, called the *determinant*, which depends only on the elements of the coefficient matrix $A$. The solvability of a linear system of equations, $A\mathbf{x} = \mathbf{b}$, depends on the value of this determinant. If the determinant is zero, the system does not have a unique solution; otherwise, it does. Closely connected with the determinant is the concept of the *inverse matrix*, $A^{-1}$, which is used in matrix analysis as the analog of division with real numbers. The inverse concept allows for the theoretically elegant solution of the system $A\mathbf{x} = \mathbf{b}$ in the form $\mathbf{x} = A^{-1}\mathbf{b}$.

The inverse matrix $A^{-1}$ is defined according to the equation $AA^{-1} = I$, where $I$ is the *identity matrix* of order $n$. The identity matrix has unity for all diagonal elements, and zero for all others. When any matrix is multiplied by the identity matrix, the product is just the original matrix. Multiplying a matrix by the identity matrix is quite analogous to multiplying a real number by unity. To clarify this important concept, we show the details of writing out the fundamental inverse relationship $AA^{-1} = I$ for the case $n = 2$.

$$AA^{-1} = \begin{pmatrix} a_{11} & a_{12} \\ a_{21} & a_{22} \end{pmatrix} \begin{pmatrix} a_{11}^{-1} & a_{12}^{-1} \\ a_{21}^{-1} & a_{22}^{-1} \end{pmatrix} = I = \begin{pmatrix} 1 & 0 \\ 0 & 1 \end{pmatrix} = \sum_{j=1}^{2} a_{ij} a_{jk}^{-1} = \delta_{ik} \tag{2-7}$$

In the above equation for the matrix inverse, we have used the superscript $-1$ on the elements of the inverse. Also, the convenient symbol $\delta_{ik}$ is known as the Kronecker delta; it is equal to unity for $i = k$ and zero otherwise. Equation (2-7) shows not only the definition of the inverse matrix, but also how to calculate it. For the case of two dimensions ($n = 2$) shown, the inverse has four elements (for $n$ dimensions it has $n^2$ elements). By partitioning the inverse matrix into two columns, we see that we can solve for each column of the inverse separately by solving a system of two linear equations. Each of these two systems has the same coefficient matrix $A$. Apart from the unknown elements of the matrix inverse, the systems vary only in their right-hand-side vectors. We now write out the specific system of equations for the *first column* of the inverse matrix.

$$\begin{aligned} a_{11}a_{11}^{-1} + a_{12}a_{21}^{-1} &= 1 \\ a_{21}a_{11}^{-1} + a_{22}a_{21}^{-1} &= 0 \end{aligned} \tag{2-8}$$

By solving for the two unknowns $a_{11}^{-1}$ and $a_{21}^{-1}$ in Equation (2-8), we obtain the elements of the first column of the $n = 2$ inverse matrix. The equations for the second column of the inverse matrix differ from Equations (2-8) only in that (i) the subscripts 11 and 21 on the inverse elements are replaced by 12 and 22, respectively, and (ii) the right-hand side values are now 0 in the first equation and 1 in the second equation. This calculation suggests that the existence of a unique inverse matrix $A^{-1}$ is closely related to the solvability of the linear system.

A summary of these definitions and operations is discussed next.

## 2.2 FUNDAMENTAL DEFINITIONS AND OPERATIONS OF MATRIX ANALYSIS

### DEFINITIONS

The *determinant* of the square matrix $A$, denoted by the symbol $|A|$, is usually calculated by reduction of the square matrix $A$ to upper or lower triangular form. The value of the determinant is just the product of the diagonal elements of the reduced matrix. The determinant is primarily a theoretical tool: if its value is zero, the system has no unique solution; otherwise, the system does have a unique solution. Although determinants can be used for the numerical solution of a system of equations or for computation of the inverse matrix, it is normally *not practical* to do so. However, various efficient procedures for solving linear equations can easily produce the value of the determinant with no extra work.

The determinant is not needed for the solution of a system of equations or for computation of the inverse matrix.

The *transpose* of a matrix, $A^t$, is defined as the matrix formed by interchanging rows and columns in the original matrix ($a_{ji}^t = a_{ij}$).

A *vector* $\mathbf{x}$ (boldface) is a one-dimensional matrix. A column vector is denoted by $\mathbf{x}$, a row vector by $\mathbf{x}^t$.

The *identity matrix*, $I$, is defined as a square matrix of any dimension and has diagonal values ($i = j$) of unity and off-diagonal values ($i \neq j$) of zero. The product of matrix $A$ with $I$ yields $A$; $AI = A$

The *zero matrix* has all elements equal to zero.

The *inverse matrix* to $A$, denoted by $A^{-1}$, is defined for a square matrix $A$ by the product $A^{-1}A = I = AA^{-1}$. Thus, if $A^{-1}$ exists (which requires that $|A| \neq 0$), we can obtain a unique solution to the system $A\mathbf{x} = \mathbf{b}$ by multiplying on the left by $A^{-1}$ to get $\mathbf{x} = A^{-1}\mathbf{b}$.

The particular linear homogeneous system $A\mathbf{x} = \lambda\mathbf{x}$ is of special importance in matrix analysis. This equation clearly possesses the so-called trivial solution ($\mathbf{x} = \mathbf{0}$, where all solution components $x_i = 0$). A result of fundamental importance is that the equation $A\mathbf{x} = \lambda\mathbf{x}$ generally possesses *nontrivial* solutions (where at least one solution component $x_i \neq 0$) only for certain particular values of the parameter $\lambda$, called *eigenvalues*.

An *eigenvector* is the nontrivial solution vector $\mathbf{x}$ corresponding to an eigenvalue $\lambda$ for the system $A\mathbf{x} = \lambda\mathbf{x}$. It is determined only up to a constant multiplicative factor.

**OPERATIONS AND RESULTS [SEE AMUNDSON, 1966]**

$A + B = B + A$.
If $C = kB$, then $c_{ij} = kb_{ij}$.
$A(B + C) = AB + AC$.
$A(BC) = (AB)C = ABC$.
$AB \neq BA$ in general.
$AB = 0$ does *not* necessarily imply that $A$ or $B = 0$.
$|AB| = |A|\,|B|$ if both $|A|$ and $|B|$ exist.
$(a^t)_{ji} = (a)_{ij}$.
$(AB)^t = B^t A^t$.
$A$ has a unique inverse if and only if $|A| = 0$.
$AA^{-1} = A^{-1}A = I$. If $A^{-1}$ exists, it is unique.
$(AB)^{-1} = B^{-1}A^{-1}$ if all three inverses exist.

$A\mathbf{x} = 0$ has a nontrivial solution for $\mathbf{x}$ if and only if the determinant $|A|$ of the square matrix $A$ is equal to zero.

$A\mathbf{x} = \mathbf{b} \neq 0$ has a unique solution for $\mathbf{x}$ if and only if $|A| \neq 0$.

$A\mathbf{x} = \lambda\mathbf{x}$ has nontrivial solutions for $\mathbf{x}$ only for special values of $\lambda$ (called eigenvalues) corresponding to the determinantal equation $|A - \lambda I| = 0$.

Matrix calculus: $dA/dt = (da_{ij}/dt)$; $\int A dt = (\int a_{ij}\,\mathrm{dt})$
That is, to differentiate or integrate a matrix, we perform these operations componentwise.

## 2.3 SOLUTIONS OF LINEAR ALGEBRAIC SYSTEMS

As we have mentioned above, the theory of matrix analysis offers an elegant theoretical solution of the linear system $A\mathbf{x} = \mathbf{b}$ in the form $\mathbf{x} = A^{-1}\mathbf{b}$. Thus, in principle, we can solve linear systems by computing the inverse matrix $A^{-1}$. However, as discussed in virtually all textbooks on the subject, direct computation of the inverse matrix is generally not the most efficient way to solve linear equations, since computation of the inverse requires more arithmetical operations than the direct algebraic solution of the equations.

A very important exception to the rule of avoiding inversion corresponds to the situation where a good built-in matrix inversion routine is part of the computer language being used. This is the case for the True BASIC language we employ for our computer programs. For problems that are not too large (say < 50 equations), the matrix inversion "inefficiency" is not a difficulty on a microcomputer. Moreover, calculation of the inverse (the determinant is also computed) in these cases also provides important information to assist in monitoring the accuracy of the solution, as we shall discuss later. Since, in our discussions, we never have occasion to solve problems involving a dense matrix larger than 20 × 20 or so, we utilize the matrix inversion capability of True BASIC. Although we do not use it in our True BASIC computer programs, we discuss below the concepts of Gauss elimination, which we believe should be familiar to all students.

### 2.3.1 Gauss Elimination

A useful method, both theoretical and practical, for studying a system of linear equations is the Gauss elimination method. The method involves eliminating each variable, one at a time, from all equations below it in the system array. This is accomplished by simply solving for some particular variable in the first equation, and using it to eliminate that variable in all succeeding equations. After this variable has been eliminated from all succeeding equations, we just move down to the second equation and repeat the process to eliminate a new particular variable from all equations below the second, and so on. That is, suppose we have the $n$th order system shown below.

$$\begin{aligned} a_{11}x_1 + a_{12}x_2 + \cdots + a_{1n}x_n &= b_1 \\ a_{21}x_1 + a_{22}x_2 + \cdots + a_{2n}x_n &= b_2 \\ &\vdots \\ a_{n1}x_1 + a_{n2}x_2 + \cdots + a_{nn}x_n &= b_n \end{aligned} \tag{2-9}$$

Now assuming $a_{11} \neq 0$, select $x_1$ in the first equation and solve for it in terms of the other variables. Using this solution of the first equation for $x_1$, we can eliminate $x_1$ from all of the *succeeding* equations in which it appears. This then yields a new $(n-1)$ by $(n-1)$ system below the first equation, which includes equations 2 through $n$ and variables 2 through $n$. The original equations then take the new form

$$\begin{aligned} x_1 + \frac{a_{12}x_2}{a_{11}} + \frac{a_{13}x_3}{a_{11}} + \cdots &= \frac{b_1}{a_{11}} \\ \left(a_{22} - \frac{a_{21}a_{12}}{a_{11}}\right)x_2 + \left(a_{23} - \frac{a_{21}a_{13}}{a_{11}}\right)x_3 + \cdots &= b_2 - \frac{a_{21}b_1}{a_{11}} \\ \left(a_{32} - \frac{a_{31}a_{12}}{a_{11}}\right)x_2 + \left(a_{33} - \frac{a_{31}a_{13}}{a_{11}}\right)x_3 + \cdots &= b_3 - \frac{a_{31}b_1}{a_{11}} \\ &\vdots \\ \left(a_{n2} - \frac{a_{n1}a_{12}}{a_{11}}\right)x_2 + \left(a_{n3} - \frac{a_{n1}a_{13}}{a_{11}}\right)x_3 + \cdots &= b_n - \frac{a_{n1}b_1}{a_{11}} \end{aligned} \tag{2-10}$$

or

$$\begin{aligned} x_1 + a'_{12}x_2 + a'_{13}x_3 + \cdots &= b'_1 \\ a'_{22}x_2 + a'_{13}x_3 + \cdots &= b'_2 \\ &\vdots \\ a'_{n2}x_2 + a'_{n3}x_3 + \cdots &= b'_n \end{aligned} \tag{2-11}$$

where we are using primes to denote the alteration of the $a_{ij}$ and $b_i$ values due to this process. To continue this process, if $a'_{22} \neq 0$, we could start with $x_2$ in the second equation and eliminate $x_2$ from all equations below the second equation. This would cause Equations (2-11) to take the form

$$
\begin{aligned}
x_1 + a'_{12}x_2 + a'_{13}x_3 + \cdots &= b'_1 \\
x_2 + \frac{a'_{23}x_3}{a'_{22}} + \cdots &= \frac{b'_2}{a'_{22}} \\
\left(a'_{33} - \frac{a'_{32}a'_{23}}{a'_{22}}\right)x_3 + \cdots &= \left(b'_3 - \frac{a'_{32}b'_2}{a'_{22}}\right) \\
&\vdots \\
\left(a'_{n3} - \frac{a'_{n2}a'_{23}}{a'_{22}}\right)x_3 + \cdots &= \left(b'_n - \frac{a'_{n2}b'_2}{a'_{22}}\right)
\end{aligned}
\tag{2-12}
$$

Equations (2-12) show the results of two steps of the Gauss elimination process. Below the second equation, we have a modified $(n-2)$ by $(n-2)$ system remaining, containing variables 3 through $n$. Thus, if the coefficient of $x_3$ in the third equation is not zero, we could solve for $x_3$ in terms of variables 4 through $n$ and eliminate $x_3$ from all of the remaining equations, and so on.

How long can this be continued? The most common situation corresponds to the case where the $n$th degree system has a unique solution for the $n$ unknowns. To illustrate this common and favorable case, we show below the *final result* of the Gauss elimination procedure as applied to a system of three equations.

$$
\begin{aligned}
x_1 + a'_{12}x_2 + a'_{13}x_3 &= b'_1 \\
x_2 + a''_{23}x_3 &= b''_2 \\
x_3 &= b'''_3
\end{aligned}
\tag{2-13}
$$

The primes in Equation (2-13) denote the fact that the original coefficients $a_{ij}$ and $b_i$ have been altered by the Gauss reduction. We see that the elimination process has resulted in an "upper triangular" structure for the system. When the system is in such a form, it is apparent that we can solve it easily by "back-substitution." That is, we start with the last equation whose solution is $x_3 = b'''_3$. Then we move up one equation and, knowing $x_3$, solve directly for $x_2$. Finally, knowing $x_3$ and $x_2$, we move up to the first equation and solve for $x_1$. This is the typical result of the Gauss elimination process for a square $(n \times n)$ system when the solution exists and is unique.

However, there are also other, possibly less desirable, outcomes of the Gauss procedure.

(i) For instance, it may happen that we run out of equations before we have eliminated all of the variables. In this case, we can solve uniquely for a certain number of the variables in terms of arbitrary values of the others. Then we have an infinity of solutions, by virtue of reduction to the identity $0 = 0$ in all of the remaining equations.

(iia) Second, we may reach a point where we suddenly (all at once, in one step) find that all remaining variables have been eliminated and we have $0 = b_i$ in all of the remaining equations, where $b_i$ is the right-hand side when this point is reached. If it happens that all of these remaining $b_i$ are equal to zero, then everything is consistent and we can simply move back up to the last equation (before all the variables dropped out) to find a $p$-fold infinity of solutions.

(iib) On the other hand, if any of the remaining $b_i$ values are *not* zero, the equations are inconsistent, and there is no solution.

If we run out of variables before equations, we will be left with several equations of the form $x_j = b_k$ $(k = 1, \ldots)$. If these $b_k$ values are all the same, we have case (iia); otherwise, we have case (iib). This covers the various possibilities.

Thus, the only case where we will get a *unique* solution of the system is where we either actually have a square system to begin with, or the system has more equations than variables but is "redundant," so that it is effectively square. Another way to look at this is that a square nonhomogeneous system has a unique solution if and only if the corresponding homogeneous system has only the trivial solution $x_i = 0$.

It is important to realize that the *computer implementation* of the Gauss elimination procedure must be accomplished with an appreciation of possible *round-off errors* in the computations. To diminish roundoff errors, the Gauss procedure is commonly implemented using so-called "partial pivoting," whereby we exchange equations (rows) during the calculation so that the first equation in the new system is the one for which the coefficient of the first variable has the largest magnitude. This procedure is is discussed in detail by Kahaner *et al.* [1989].

**Example 2.1 Gauss Elimination**

We consider the following system of four equations.

$$\begin{aligned} x_1 + 2x_2 - x_3 - 2x_4 &= -1 \\ 2x_1 + x_2 + x_3 - x_4 &= 4 \\ x_1 - x_2 + 2x_3 + x_4 &= 5 \\ x_1 + 3x_2 - 2x_3 - 3x_4 &= -3 \end{aligned} \tag{a}$$

We first solve the first equation in (a) for $x_1$ and then eliminate $x_1$ from all succeeding equations. This gives, for the first stage,

$$\begin{aligned} x_1 + 2x_2 - x_3 - 2x_4 &= -1 \\ -3x_2 + 3x_3 + 3x_4 &= 6 \\ -3x_2 + 3x_3 + 3x_4 &= 6 \\ x_2 - x_3 - x_4 &= -2 \end{aligned} \tag{b}$$

Next, we solve for $x_2$ in the second equation in (b) and use this to eliminate $x_2$ from the succeeding equations. This gives, for the second stage,

$$\begin{aligned} x_1 + 2x_2 - x_3 - 2x_4 &= -1 \\ x_2 - x_3 - x_4 &= -2 \\ 0x_3 + 0x_4 &= 0 \\ 0x_3 + 0x_4 &= 0 \end{aligned} \tag{c}$$

Since the last two equations are identically zero, the system (c) reduces to

$$\begin{aligned} x_1 + 2x_2 &= x_3 + 2x_4 - 1 \\ x_2 &= x_3 + x_4 - 2 \end{aligned} \tag{d}$$

Thus, if we take $x_3$ and $x_4$ to be *any* particular constant values, the unique solution of Equations (d) is directly determined by back-substitution. If we attempt to solve this problem using the computer program SIMLE (simple linear equations), we get the error message "can't invert singular matrix." This is because of the redundancy of the equations as seen in the Gauss reduction process, Equations (d). Program SIMLE is discussed at the end of the chapter.

To illustrate the most common situation, suppose we change the following elements of the original coefficient matrix in Equations (a): Let $a_{31} = 10$, $a_{32} = -5$, $a_{41} = -1$, $a_{42} = 5$. Then the new system becomes

$$\begin{aligned} x_1 + 2x_2 - x_3 - 2x_4 &= -1 \\ 2x_1 + x_2 + x_3 - x_4 &= 4 \\ 10x_1 - 5x_2 + 2x_3 + x_4 &= 5 \\ -1x_1 + 5x_2 - 2x_3 - 3x_4 &= -3 \end{aligned} \tag{e}$$

The Gauss reduction process applied to the system in (e) yields

$$\begin{aligned} x_1 + 2x_2 - x_3 - 2x_4 &= -1 \\ x_2 - x_3 - x_4 &= -2 \\ x_3 + \frac{4}{13}x_4 &= \frac{35}{13} \\ x_4 &= -1 \end{aligned} \tag{f}$$

Back-substitution readily shows that the unique solution to Equations (f) is $x_1 = 0$, $x_2 = 0$, $x_3 = 3$, $x_4 = -1$. If we apply computer program SIMLE to Equations (b), we get the solution $x_1 = 0$, $x_2 = -6.10e - 16$ ($\approx 0$), $x_3 = 3$, $x_4 = -1$, which is virtually the same as the hand calculation. The miniscule error in the computer value for $x_2$ is due to roundoff. The value the program gives for the determinant of the coefficient matrix of Equations (e) is 30.

### 2.3.12 Over and Under Determined Systems: Least Squares

We have briefly discussed the (unusual) cases where we have a "non-square" system; that is, where we have fewer equations than unknowns or vice versa. When we have more equations than unknowns, for a *unique* solution to exist there must be some redundancy in the system.

Another approach to the over-determined problem, which we shall consider in a later chapter in connection with data analysis, can be used for the problem where there are more equations than unknowns and no exact solution is possible (e.g., no redundancy). This is the so-called least-squares solution of the over-determined problem. For example, suppose we have the following system of three equations in two variables:

$$\begin{aligned} a_{11}x_1 + a_{12}x_2 &= b_1 \\ a_{21}x_1 + a_{22}x_2 &= b_2 \\ a_{31}x_1 + a_{32}x_2 &= b_3 \end{aligned} \tag{2-14}$$

Here, we suppose that there is no redundancy in Equations (2-14), and hence no unique solution. We wish to find the best values for the $x_j$ in the *least-squares sense.*

To do this, we define the *residual* of equation $i$ as $r_i = a_{i1}x_1 + a_{i2}x_2 - b_i$. The residual is the amount by which the equation fails to be satisfied. We then minimize the sum of the squares of the residuals of all equations with respect to $x_1$, $x_2$. Formally, we have

$$\min_{x_1, x_2} \sum_{i=1}^{3} (b_i - a_{i1}x_1 - a_{i2}x_2)^2 \tag{2-15}$$

To carry out the minimization, we simply differentiate Equation (2-15) with respect to $x_1$ and $x_2$ and set the results equal to zero. This generates two linear algebraic equations for the optimal values of $x_1$ and $x_2$. To generalize this idea, we consider the following non-square linear system of $n$ equations involving $p$ unknowns, where $p < n$.

$$\sum_{j=1}^{p} a_{ij}x_j = b_i + r_i \quad \text{or} \quad A\mathbf{x} = \mathbf{b} + \mathbf{r} \tag{2-16}$$

Since the equations generally cannot be satisfied exactly, the $r_i$ and $\mathbf{r}$ in Equations (2-16) represent residuals. The least-squares criterion requires that "Error" $= \sum r_i^2$ be minimized with respect to the $x_j$. The required necessary condition is

$$\frac{\partial E}{\partial x_k} = 2\sum_{i=1}^{n} r_i \frac{\partial r_i}{\partial x_k} = 0 \qquad (k = 1, 2, \ldots, p) \tag{2-17}$$

where $E$ represents "error".

Since we have

$$r_i = \sum_{j=1}^{p} a_{ij}x_j - b_i; \qquad \frac{\partial r_i}{\partial x_k} = a_{ik} \tag{2-18}$$

It follows that we must have

$$\sum_{i=1}^{n} a_{ik}r_i = 0 \qquad (k = 1, 2, \ldots, p) \quad \text{or} \quad A^t\mathbf{r} = \mathbf{0} \tag{2-19}$$

In matrix notation, the condition $A^t\mathbf{r} = \mathbf{0}$ is written $A^t(A\mathbf{x} - \mathbf{b}) = \mathbf{0}$ or $A^tA\mathbf{x} = A^t\mathbf{b}$. Finally, the formal solution for the general least-squares solution vector $\mathbf{x}$ is expressed as

$$\mathbf{x} = (A^tA)^{-1}A^t\mathbf{b} \tag{2-20}$$

Equation (2-20) is used in the computer programs used to solve least-squares problems, which are discussed in the chapter on least squares. A very simple yet significant example of least-squares over-determined equations is as follows. Consider the following three inconsistent equations in one unknown: $x_1 = 1$, $x_1 = 1.5$, $x_1 = 2$. This might be the result of attempting to fit the model, $x_1 = K$ = a constant, to three inconsistent data values, such as three measurements of a viscosity. In this case, we hardly need the formalism of Equation (2-17); instead, the direct minimization of $\sum r_i^2 = (x_1 - 1)^2 + (x_1 - 1.5)^2 + (x_1 - 2)^2$ shows that $x_1 = (1 + 1.5 + 2)/3$, or the *arithmetic mean* of the inconsistent estimated values (a general result even when there are more estimated values).

### 2.3.3 Iterative Solutions

For linear systems that have a large number of zeros in the coefficient matrix (a "sparse" system), it is often useful to use an indirect *iterative* method of solution. The idea is to *partially solve* each equation for a different variable so as to form a new system in which each unknown appears exactly once on the left-hand side. Any equation may contain any number of unknowns on the right-hand side. Under appropriate conditions (to be discussed later), a repetition or *iteration* of this partial solution process eventually converges to the true solution. To illustrate the approach, consider the system of two equations

$$\begin{aligned} a_{11}x_1 + a_{12}x_2 &= b_1 \\ a_{21}x_1 + a_{22}x_2 &= b_2 \end{aligned} \tag{2-21}$$

An intuitively natural way to set up an iteration is to solve the first equation of (2-21) for $x_1$ and the second equation for $x_2$ as follows:

$$\begin{aligned} x_1^{[n+1]} &= (-a_{12}x_2^{[n]} + b_1)/a_{11} \\ x_2^{[n+1]} &= (-a_{21}x_1^{[n+1]} + b_2)/a_{21} \end{aligned} \tag{2-22}$$

Here the superscripts refer to the iteration number. Notice that in Equations (2-22), the quantity $x_1^{[n+1]}$ is used as soon as it becomes available; this is called updating the variables, or the Gauss-Seidel procedure. This generally works better than the no-update or Jacobi iteration. To solve Equations (2-22), we first take $n = 0$ and select values of $x_i^{[0]}$ as the initial estimates to be used in the right-hand side of the equations. Starting with these initial estimates of any variables required (only $x_2^{[0]}$ in this example), we calculate $x_1^{[1]}$ from the first equation. We then use this value on the right-hand side of the second equation to compute $x_2^{[1]}$. This completes the first cycle of the iterative process. In the next cycle, we take $n = 1$ and use the values $x_1^{[1]}$, $x_2^{[1]}$ as necessary in the right-hand sides of Equations (2-22) to give $x_1^{[2]}$, $x_2^{[2]}$, and so on until (hopefully) convergence is achieved.

Sufficient conditions for convergence are considered later when we discuss vector and matrix norms. We may note, however, that one of the main applications of iterative procedures is in the finite difference solution of ordinary or partial differential equations. In these applications, we often have *diagonal dominance* in the coefficient matrix; this corresponds to the condition where the magnitude of each diagonal element is larger than the magnitude of the sum of the nondiagonal elements in the same row. If we then solve each equation for its diagonal variable (e.g., variable $i$ in equation $(i)$), we obtain a convergent iteration. It is important to keep in mind that the iterative procedure may *diverge* if a proper selection of the variables is not made.

**Example 2.2 Solution of Linear Equations by Iteration**

Consider the following linear system (which happens to be based on a simple discretization of the Laplace partial differential equation for heat conduction; the $x_i$ values represent temperature):

$$\begin{aligned} x_1^{[n+1]} &= (x_2^{[n]} + x_3^{[n]} + 50)/4 \\ x_2^{[n+1]} &= (x_1^{[n+1]} + x_4^{[n]} + 100)/4 \\ x_3^{[n+1]} &= (x_1^{[n+1]} + x_4^{[n]} + 150)/4 \\ x_4^{[n+1]} &= (x_2^{[n+1]} + x_3^{[n+1]} + 200)/4 \end{aligned} \tag{a}$$

We show the behavior of the iterative procedure in this problem by using the Gauss-Seidel method, using each new value for $x_i$ within a given cycle as soon as it is available. As starting values, we take all $x_i^{[0]} = 100$. We now show explicitly the first step of the iterative calculations.

$$\begin{aligned} x_1^{[1]} &= (100 + 100 + 50)/4 = 62.5 \\ x_2^{[1]} &= (62.5 + 100 + 100)/4 = 65.62 \\ x_3^{[1]} &= (62.5 + 100 + 150)/4 = 78.125 \\ x_4^{[1]} &= (65.62 + 78.125 + 200)/4 = 85.156 \end{aligned} \tag{b}$$

The calculated values of $x_i^{[1]}$ in Equations (b) are given to five significant figures. To carry out the entire calculation, we utilize computer program DELSQ, since it has convenient built-in bookkeeping for iterative solutions of either linear or nonlinear algebraic systems. This program is discussed in the chapter on nonlinear algebraic equations. For this problem, the Gauss-Seidel method converges in 10 iterations to the solution $x_1 = 43.75$, $x_2 = 56.25$, $x_3 = 68.75$, and $x_4 = 81.25$, correct to four figures.

Convergence of the general iterative procedure described here depends on the so-called diagonal dominance of the arrangement of the equations. That is, if the sum of the magnitudes of the coefficients of the $x_i$ on the right-hand side of each equation is less than unity, the process normally converges. Further discussion of conditions of convergence follows the introduction of matrix norms. To accelerate convergence of the iterative procedure, a technique known as "successive over-relaxation" (SOR) has been used, as discussed by Scraton [1987]. In our experience, the use of delta-square acceleration (a form of Shanks transformation), as discussed by Fadeev and Fadeeva [1963], is usually better than SOR. The delta-square calculation option is available automatically in program DELSQ.

### 2.3.4 Tridiagonal Problems

A very important special case of a "sparse" system of linear equations (where the coefficient matrix has many zeros) is the so-called *tridiagonal* problem. This kind of problem is important for two reasons. First, the tridiagonal problem occurs naturally in many applications, such as in numerical solutions of ordinary and partial differential equations, difference equations, stagewise separation processes, and spline approximation. Second, by taking advantage of the special mathematical structure, it is possible to solve these problems much more easily than using the standard Gauss elimination, which would necessarily have to process a large number of zero elements in the coefficient matrix.

The typical tridiagonal system, using a standard notation, is shown below.

$$\begin{aligned}
b_1x_1 + c_1x_2 \qquad\qquad\qquad\qquad &= d_1 \\
a_2x_1 + b_2x_2 + c_2x_3 \qquad\qquad &= d_2 \\
a_3x_2 + b_3x_3 + c_3x_4 \qquad &= d_3 \\
\vdots \qquad\qquad \vdots \qquad\qquad & \\
a_{n-1}x_{n-2} + b_{n-1}x_{n-1} + c_{n-1}x_n &= d_{n-1} \\
a_nx_{n-1} + b_nx_n &= d_n
\end{aligned} \tag{2-23}$$

As seen from Equations (2-23), the pattern is that the first equation has one term to the right of the diagonal ($b_i$ is the diagonal coefficient of equation ($i$)), the last equation has one term to the left of the diagonal, and all equations in between have the "tridiagonal" structure; namely three terms, including the diagonal and one on each side of it. The typical "in between" equation (Equation (k)) is $a_kx_{k-1} + b_kx_k + c_kx_{k+1} = d_k$ (for $k = 2, 3\ldots, n-1$).

Note that after we eliminate $x_1$ from the second equation (only) in (2-23) and consider the remaining system, the new first equation has the same form as the original first equation. Thus, for one elimination cycle, after dividing by the coefficient of $x_k$, we are always working with the recurrent system

$$\begin{aligned}
x_k + h_kx_{k+1} \qquad\qquad &= p_k \\
a_{k+1}x_k + b_{k+1}x_{k+1} + c_{k+1}x_{k+2} &= d_{k+1}
\end{aligned} \tag{2-24}$$

The computational procedure is to start at the top of the current system (Equations (2-24)) and eliminate $x_k$ from the second equation in order to find $h_{k+1}$ and $p_{k+1}$ in the *new* first equation, which will be of the form

$$x_{k+1} + h_{k+1}x_{k+2} = p_{k+1} \tag{2-25}$$

To determine the new $h_{k+1}$ and $p_{k+1}$ in terms of the new coefficients $a_{k+1}$, $b_{k+1}$, $c_{k+1}$, and $d_{k+1}$, as well as the old $h_k$ and $p_k$, we multiply the first equation of (2-24) by $(-a_{k+1})$ and add the result to the second equation. When we then divide by the coefficient of $x_{k+1}$, we get

$$x_{k+1} + \frac{c_{k+1}}{(-a_{k+1}h_k + b_{k+1})}x_{k+2} = \frac{(-a_{k+1}p_k + d_{k+1})}{(-a_{k+1}h_k + b_{k+1})} \tag{2-26}$$

Comparing Equations (2-26) and (2-27), we see that the desired recurrence relations for $h_{k+1}$ and $p_{k+1}$ are

$$\begin{aligned}
h_{k+1} &= \frac{c_{k+1}}{b_{k+1} - a_{k+1}h_k} \qquad h_0 = 0 \; (k = 1, 2, \ldots, n-1) \\
p_{k+1} &= \frac{d_{k+1} - a_{k+1}p_k}{b_{k+1} - a_{k+1}h_k} \qquad p_0 = 0 \; (k = 1, 2, \ldots, n-1)
\end{aligned} \tag{2-27}$$

After all $h_k$ and $p_k$ have been computed (up to $h_{n-1}$, $p_n$ ($h_n = 0$)), the system has been reduced to the form

$$x_k + h_kx_{k+1} = p_k \qquad (k = 1, 2, \ldots, n) \tag{2-28}$$

where $h_n = 0$ so that $x_n = p_n$. Then we work backwards to compute $x_{n-1}$, $x_{n-2}$, and so on from

$$x_k = p_k - h_k x_{k+1} \qquad (k = n-1, n-2, \dots, 1) \tag{2-29}$$

This completes our discussion of the tridiagonal algorithm. We now consider an example of its use.

**Example 2.3 Solution of a Tridiagonal Problem**

As a simple illustration of the tridiagonal algorithm, we consider the equations generated by using the finite difference method to solve the boundary value problem $d^2y/dx^2 + y = 0$ subject to the boundary conditions $y(0) = 0.5$ and $y(1) = 0.69$. Following the finite difference procedure discussed in the chapter on ODE boundary value problems, the difference equations to be solved if 10 subintervals are employed are

$$y_1 = 0.5$$
$$y_{i-1} - 1.99y_i + y_{i+1} = 0 \qquad (i = 2, 3, \dots, 10) \tag{a}$$
$$y_{11} = 0.69$$

A computer program that can be used easily to solve the above equations (or any set of tridiagonal equations) is TRID, which is discussed in detail at the end of the chapter. To use it for Equations (a), we merely need to identify and program the coefficients $a_i$, $b_i$, $c_i$, and $d_i$ for Equations (2-23). These are

$$b(1) = 1 \qquad c(1) = 0 \qquad d(1) = 0.5$$

$$\begin{aligned} a(i) &= 1 \\ b(i) &= -1.99 \qquad (i = 2, 3, \cdots, 10) \\ c(i) &= 1 \\ d(i) &= 0 \end{aligned} \tag{b}$$

$$a(11) = 0 \qquad b(11) = 1 \qquad d(11) = 0.69$$

For all tridiagonal problems, $a_1 = c_n = 0$ (here, $n = 11$). The solution to this problem is given in Table 2.1. The computed value of $y_6 = 0.678$ at $x = 0.5$ may be compared to the corresponding exact solution to the ODE boundary value problem at $x = 0.5$, which is 0.679. The reason for the slight difference is *not* the tridiagonal algebraic solution, but rather the relatively coarse discretization corresponding to the use of 10 spatial increments.

The example above illustrates the ease with which the user can solve a tridiagonal problem on the computer. It is interesting to solve a slight modification of the above tridiagonal system corresponding to the use of a different number of subintervals. Using a Macintosh SE 30 personal computer, the solution for the above 11 equations, as well as for 101 equations, is almost instantaneous. The solution for 1001 equations takes a couple of seconds, while that for 10,001 equations takes closer to 15 seconds. If we were using the *full matrix* with Gauss elimination, computation times for even 101 equations would take considerably longer (by our experiment,

**TABLE 2.1** SOLUTION OF A TRIDIAGONAL SYSTEM

| |
|---|
| $y_1 = 0.500$ |
| $y_2 = 0.547$ |
| $y_3 = 0.589$ |
| $y_4 = 0.625$ |
| $y_5 = 0.655$ |
| $y_6 = 0.678$ |
| $y_7 = 0.694$ |
| $y_8 = 0.704$ |
| $y_9 = 0.706$ |
| $y_{10} = 0.702$ |
| $y_{11} = 0.690$ |

nearly five minutes). This is because for a system of $n$ equations, the approximate computational effort for Gauss elimination is proportional to $n^3$, while that for the tridiagonal algorithm is proportional to $n$. Thus, for large systems there is a *very distinct* advantage to using the tridiagonal method wherever possible.

### 2.3.5 Sensitivity Analysis

It is sometimes desired in applications of linear equations to determine the rate of change of the solution vector $\mathbf{x}$ with respect to some coefficient element $a_{ij}$ or right-hand-side element $b_k$. Derivatives such as these (for example, $\partial x_2/\partial a_{41}$ or $\partial x_1/\partial b_3$) are called sensitivity coefficients. The question concerns how to calculate these quantities, both formally and numerically. First, let us consider the case of sensitivity with respect to a right-hand-side element. We write our original system of equations as

$$\sum_{j=1}^{n} a_{ij}x_j = b_i \quad \text{or} \quad A\mathbf{x} = \mathbf{b} \tag{2-30}$$

Thus, if we change *any* $a_{ij}$ or $b_i$ value, *all* of the $x_i$ values will change. If we want the rate of change of any particular $x_i$, we must necessarily calculate the rate of change of all the $x_i$. The solution of Equations (2-30) is written as

$$x_i = \sum_{j=1}^{n} a_{ij}^{-1}b_j \quad \text{or} \quad \mathbf{x} = A^{-1}\mathbf{b} \tag{2-31}$$

Using the component form of Equations (2-31), we can differentiate $x_i$ with respect to $b_k$ to get $\partial x_i/\partial b_k = a_{ik}^{-1}$. Thus, for any $(i,k)$, we have that $\partial x_i/\partial b_k$ is equal to the $(i,k)$ element of the inverse $A^{-1}$. This might be written in the suggestive matrix notation $\partial\mathbf{x}/\partial\mathbf{b} = A^{-1}$. Since $A^{-1}$ is computed in program SIMLE, we could simply print out those values and use those that are pertinent. Alternatively, as seen by differentiating Equations (2-30) partially with respect to $b_k$, we could solve the system

$$\begin{aligned}\sum_{j=1}^{n} a_{ij} \frac{\partial x_j}{\partial b_k} &= 0, \quad \text{for } i \neq k \\ \sum_{j=1}^{n} a_{kj} \frac{\partial x_j}{\partial b_k} &= 1\end{aligned} \tag{2-32}$$

Equations (2-32) are just the original system with the right-hand sides replaced by zero, except in row $k$, where the element is unity. Naturally, consistent with Equation (2-31), this calculation just produces the $k$th column of the inverse $A^{-1}$. Note that since $x_i$ is *linear* in $b_k$, all higher derivatives are identically zero.

We now consider the calculation of the sensitivity coefficient $\partial x_i / \partial a_{km}$. In this case, the $x_i$ do *not* vary linearly with the $a_{km}$, so that the higher derivatives are not generally zero. The most convenient procedure is to differentiate Equations (2-30) with respect to $a_{km}$ to get the following:

$$\begin{aligned}\sum_{j=1}^{n} a_{ij} \frac{\partial x_j}{\partial a_{km}} &= 0 \qquad \text{(all } i \text{ except } i = k) \\ \sum_{j \neq m}^{n} a_{kj} \frac{\partial x_j}{\partial a_{km}} + \left( a_{km} \frac{\partial x_m}{\partial a_{km}} + x_m \right) &= 0\end{aligned} \tag{2-33}$$

Equations (2-33) can be written in the convenient matrix form

$$A \frac{\partial \mathbf{x}}{\partial a_{km}} = \begin{pmatrix} 0 \\ \vdots \\ -x_m \text{ (row k)} \\ \vdots \\ 0 \end{pmatrix} \tag{2-34}$$

Equations (2-34) are the same as the original system with the right-hand sides replaced by zero, except in row $k$, where $b_k = -x_m$. Now $x_m$ must be found first through the solution of the original system $A\mathbf{x} = \mathbf{b}$. If higher derivatives are desired (for example, for a Taylor or Padé approximation), they can be found in an analogous way.

## 2.4 ERROR ANALYSIS AND NORMS

In this section, we discuss the question of the *accuracy* of a numerical solution of a system of linear equations, along with other related matters. It turns out that the concept of *matrix norms* is very important for these problems, so we give a very brief introduction to this topic along with some applications.

Since the Gauss elimination algorithm (and its variants) involves only a finite combination of arithmetic operations, it does, *in principle*, yield exact results. However, the computer implementation invariably introduces rounding errors, which fortunately are often held in check by using partial pivoting. Thus, not only do we need to keep errors "small," but, perhaps more important, we need to know roughly what the errors are in a given problem. An indication of poor accuracy will be valuable as a warning to either use the results with caution, or somehow improve the calculation to increase the accuracy.

For a given algorithm/computer program, how much error is introduced into the numerical solution depends on the problem itself. Some problems are naturally well-conditioned and lead to excellent results, while others are "ill-conditioned" and difficult to solve accurately. As a simple illustration of this idea, take the case of the graphical representation of the solution (intersection) of two linear equations (lines). If the two lines are nearly perpendicular, a slight change in one line does not change the solution appreciably. On the other hand, if the two lines are nearly parallel, a slight change in one line may *drastically* change the solution. This latter case is a simple example of an ill-conditioned linear system.

### 2.4.1 Iterative Improvement and Accuracy Estimation

It is possible to devise a simple but effective scheme for both estimating the accuracy and improving the solution of a system of linear equations. The procedure is as follows. For the system $A\mathbf{x} = \mathbf{b}$, we calculate from our computer program the necessarily *approximate* numerical solution $\mathbf{x}_c$ ($= A^{-1}\mathbf{b}$ in program SIMLE). We then compute the so-called residual vector $\mathbf{r}$, which is defined by $A\mathbf{x}_c = \mathbf{b} + \mathbf{r}$. By subtracting these two matrix equations, we get

$$A(\mathbf{x}_c - \mathbf{x}) \equiv A\mathbf{e} = \mathbf{r} \quad or \quad (\mathbf{x}_c - \mathbf{x}) \equiv \mathbf{e} \cong A^{-1}\mathbf{r} \tag{2-35}$$

The first equation in (2-35) is general, and suggests that if we have retained high accuracy for the residual vector $\mathbf{r}$, we may solve this equation (using the same Gauss decomposition of matrix $A$ or the same inverse $A^{-1}$) to calculate a necessarily approximate value for the error vector $\mathbf{e} = \mathbf{x}_c - \mathbf{x}$. By a simple rearrangement, we write the improved solution in component form as $x_i = x_{ci} - e_i$. Thus, for any solution component $x_i$, an improved approximation is just the original calculated value minus the calculated "error" value. The new error in this improved approximation should be smaller in magnitude than the current estimated error; thus, we would expect that $|e_i|$ would be a conservative estimate of the error in the new approximation. We would also expect that the relative error in the *true* (unknown) $x_i$ value would generally be less than $|e_i|/|x_{ci} - e_i|$. The above process could be repeated, but this is usually not necessary. The first step is *crucial*, however, since it results in an indication of the accuracy of the solution. This calculation procedure is followed in program SIMLE.

**Example 2.4 Iterative Improvement and Error Estimation**

As mentioned above, computer program SIMLE uses one step of the above iterative procedure to both improve accuracy and estimate errors. Here we will test this program on a small but interesting three-dimensional *ill-conditioned* system. The problem considered is due to Johnson and Riess [1982], and is as follows:

$$\begin{aligned} 6x_1 + 6x_2 + 3.00001x_3 &= 30.00002 \\ 10x_1 + 8x_2 + 4.00003x_3 &= 42.00006 \\ 6x_1 + 4x_2 + 2.00002x_3 &= 22.00004 \end{aligned} \tag{a}$$

**TABLE 2.2** SOLUTION OF EQUATIONS (a) USING PROGRAM SIMLE

| Estimated Solution | Estimated Relative Error |
|---|---|
| $x_1 = 1$ | $2 \times 10^{-14}$ |
| $x_2 = 3$ | $5 \times 10^{-10}$ |
| $x_3 = 2$ | $1 \times 10^{-9}$ |
| Determinant = 0.00004 | |

The exact solution of Equations (a) is $x_1 = 1$, $x_2 = 3$, $x_3 = 2$. This system is very sensitive to small changes in the numerical values; for instance, if the right-hand side is replaced by $b_1 = 30$, $b_2 = 42$, $b_3 = 22$, corresponding to chopping the small $b_i$ decimal contributions in Equations (a), the exact solution changes by a *large* amount to $x_1 = 1$, $x_2 = 4$, $x_3 = 0$. We now solve Equations (a) using program SIMLE. The results are shown in Table 2.2.

We see from the table that program SIMLE does produce the exact solution (to six significant figures, which is all that were printed). This is because True BASIC does arithmetic using double precision. We also see that the determinant of the coefficient matrix is quite small, so we expect that the system is ill-conditioned. But even more useful is the information on *estimated relative errors*. The values computed suggest the estimates of $x_2$ and $x_3$ to be some four or five orders of magnitude less accurate than that of $x_1$. This certainly makes sense, since in the perturbed solution quoted above, $x_1$ didn't change at all, while $x_2$ and $x_3$ changed considerably.

To give an idea of computation times, program SIMLE running on a Macintosh SE/30 personal computer (68030 processor) solves 50 equations in about 40 seconds. For scaling purposes, the computing time increases as the cube of the system size. Thus, solving 100 equations takes about $40 \times 2^3 = 320$ seconds.

### 2.4.2 Matrix Norms

In working with matrix equations, it is very useful to introduce the concept of matrix norms. Matrix norms in relation to matrix equations play the same role as absolute values in ordinary scalar algebraic equations. Any norm of a matrix is just a *number* that in some sense reflects a kind of size of the matrix. By using matrix norms, we can develop a formalism of *inequalities* for matrix equations; this is of paramount importance in dealing with approximations and error estimates. Our discussion of matrix norms will be very brief; for further details at an elementary level, the book of Johnson and Riess [1982] is highly recommended.

The concept of matrix norms can actually be implemented in various ways. Here we will focus on only one form, the so-called maximum row sum norm. Vector norms can naturally be considered to be a subset of matrix norms, and we adopt that viewpoint here. The desired properties of matrix norms are as follows:

$$\begin{aligned}
&\|A\| \geq 0 \text{ with } \|A\| = 0 \text{ only if } A \text{ is the zero matrix} \\
&\|kA\| = |k| \cdot \|A\| \qquad \text{for any scalar } k \\
&\|A + B\| \leq \|A\| + \|B\| \qquad \text{triangle inequality for addition} \\
&\|AB\| \leq \|A\| \cdot \|B\| \qquad \text{triangle inequality for multiplication}
\end{aligned} \tag{2-36}$$

One matrix norm that satisfies all the conditions of Equations (2-45) is the *maximum row sum norm*, which is defined for a general matrix as

$$\|A\| = \underset{1 \leq i \leq n}{\text{Max}} \sum_{j=1}^{n} |a_{ij}| \tag{2-37}$$

In words, the matrix norm in Equation (2-37) is just the *largest* of the $n$ row sums generated by adding the absolute values of each element in a row. Since an ordinary column vector $\mathbf{b}$ of dimension $n$ is a special case of a matrix having $n$ rows but only one column, the definition (2-37) reduces for a vector to

$$\|\mathbf{b}\| = \underset{1 \leq j \leq n}{\text{Max}} |b_j| \tag{2-38}$$

The vector norm in Equation (2-38) is referred to as the *maximum norm*, for obvious reasons. It must be emphasized that there are various other alternative definitions for vector-matrix norms which satisfy all the conditions of Equations (2-36). The common ones are discussed by Johnson and Riess [1982], but they will not be needed in our discussion.

It is very easy to verify that the norm we have defined in Equation (2-37) satisfies the first two conditions quoted in Equations (2-36). For the third condition, the triangle inequality for addition, we have that

$$\|A + B\| \equiv \underset{i}{\text{Max}} \sum_{j=1}^{n} |a_{ij} + b_{ij}| \leq \underset{i}{\text{Max}} \left( \sum_{j=1}^{n} |a_{ij}| + \sum_{j=1}^{n} |b_{ij}| \right) \leq \|A\| + \|B\| \tag{2-39}$$

To show the validity of the fourth condition, the triangle inequality for multiplication, we proceed as follows.

$$\begin{aligned}
\|AB\| &\equiv \underset{i}{\text{Max}} \sum_{k=1}^{n} \left( \left| \sum_{j=1}^{n} a_{ij} b_{jk} \right| \right) \leq \underset{i}{\text{Max}} \sum_{j=1}^{n} \sum_{k=1}^{n} |a_{ij}| |b_{jk}| \\
\|AB\| &\leq \underset{i}{\text{Max}} \sum_{j=1}^{n} |a_{ij}| \left( \sum_{k=1}^{n} |b_{jk}| \right) \leq \underset{j}{\text{Max}} \sum_{k=1}^{n} |b_{jk}| \cdot \underset{i}{\text{Max}} \sum_{j=1}^{n} |a_{ij}| = \|A\| \cdot \|B\|
\end{aligned} \tag{2-40}$$

The manipulations carried out in Equations (2-40) include using the ordinary triangle inequality for sums, reversing the order of the sums, taking $|a_{ij}|$ out of the double sum, and finally maximizing the $|b_{jk}|$ row sum.

**2.4.2.1 Relationships and Applications Involving Matrix Norms.** As a simple example of a matrix norm relationship, we can cite the expression $I = A^{-1}A$, from which we obtain $1 \leq \|A^{-1}\| \cdot \|A\|$. That is, the product of the norms of $A$ and $A^{-1}$ is greater than or equal to one. This is based on the triangle inequality for products and the fact that the norm of the identity matrix is always equal to unity, since this matrix has only one nonzero element (which is unity) on the diagonal. The product $\|A^{-1}\| \cdot \|A\|$ is called the "condition number" of a square matrix; it is useful in characterizing the ill-conditioning of the system $A\mathbf{x} = \mathbf{b}$. If the condition number is relatively large, even when the error residual is small, the solution may not be accurate.

For iterative solutions of linear systems that we have considered earlier, the matrix norm is important for theoretical error analysis. Suppose we are iterating according to the scheme

$$\mathbf{x}^{[n+1]} = B\mathbf{x}^{[n]} + \mathbf{c} \tag{2-41}$$

where $B$ is a matrix. Since the exact solution satisfies $\mathbf{x} = B\mathbf{x} + \mathbf{c}$, by subtraction we find that if the error vector at the $n$th stage is defined to be $\mathbf{e}^{[n]} \equiv (\mathbf{x}^{[n]} - \mathbf{x})$, the error satisfies the equation $\mathbf{e}^{[n+1]} = B\mathbf{e}^{[n]}$. By taking the norm of this error equation, we find that $\|\mathbf{e}^{[n+1]}\| \leq \|B\| \cdot \|\mathbf{e}^{[n]}\|$. Thus, if $\|B\| < 1$, the error norms necessarily diminish as the iterations proceed, so that the process converges. This gives a sufficient condition for the convergence of the iterative solution procedure.

There are many other applications of matrix norms in linear algebra computations. Two more that we mention concern errors arising in the numerical calculation of the inverse matrix, as well as development of bounds for matrix eigenvalues (which we shall discuss shortly).

## 2.5 THE ALGEBRAIC EIGENVALUE PROBLEM

The algebraic eigenvalue problem occurs in numerous practical applications, so we now give an elementary discussion of this problem. The mathematical problem may be stated as follows: "find the eigenvalues ($\lambda$) and the eigenvectors ($\mathbf{x}$) corresponding to a *nontrivial* solution of the homogeneous system"

$$A\mathbf{x} = \lambda\mathbf{x} \tag{2-42}$$

The quantity $\lambda$ is a scalar parameter. Of course, we can easily see that Equation (2-42) always admits the *trivial* solution corresponding to $\mathbf{x} = \mathbf{0}$ (all solution components = 0). The problem is to find the particular *special values of* $\lambda$ (called eigenvalues) for which a nontrivial solution (called an eigenvector), having at least one nonzero component, is possible. If these $\lambda$ values were known, the eigenvector solution components could be determined by solving Equation (2-42). From the homogeneous nature of the system, it is apparent that if we have any nontrivial solution of Equation (2-42), any nonzero constant $k$ times this solution would *also* be a solution. Thus, we can determine eigenvectors only up to a multiplicative constant.

We first note that we can determine upper and lower bounds for the magnitudes of the eigenvalues by using matrix norms. Since the basic eigenvalue equation is $A\mathbf{x} = \lambda\mathbf{x}$, taking norms and dividing by $\|\mathbf{x}\|$ gives the upper bound $|\lambda| \leq \|A\|$. By multiplying the original eigenvalue equation by $A^{-1}$, we find that $A^{-1}\mathbf{x} = (1/\lambda)\mathbf{x}$, so that $A^{-1}$ has eigenvalues $(1/\lambda)$. Taking norms gives the *lower* bound $1/\|A^{-1}\| \leq |\lambda|$ This latter result is of practical significance if $A^{-1}$ is being calculated (as in program SIMLE). Another general result which can be useful for checking purposes is that *the sum of the eigenvalues is equal to the trace (sum of diagonal components) of the matrix.*

In general, it can be shown that eigenvalue problems for a matrix of order $n$ lead to an $n$th degree polynomial for the eigenvalues, called the *characteristic equation*. Thus, there are normally $n$ distinct roots (eigenvalues) of the characteristic equation and $n$ distinct corresponding eigenvectors (exceptions occur when there are repeated roots). Sophisticated methods for solving the general $n$th degree eigenvalue problem, such as the QR algorithm, are quite complicated. We refer the reader to Johnson and Riess [1982] and Press *et al.* [1986] for more details on this method. We will instead consider a conceptually simple and general procedure, known as the interpolation method, which is also computationally useful. This method has the advantages of being applicable to both symmetric and unsymmetric matrices, and the ability to produce all of the eigenvalues.

We will first, for illustrative purposes, consider the analytical solution of the second-degree eigenvalue problem. The general second-degree eigenvalue problem has the form

$$\begin{aligned} a_{11}x_1 + a_{12}x_2 &= \lambda x_1 \\ a_{21}x_1 + a_{22}x_2 &= \lambda x_2 \end{aligned} \tag{2-43}$$

This homogeneous system will presumably have no *nontrivial* (nonzero) solution unless one of the equations is redundant. We eliminate $x_2$ from the first equation and substitute into the second equation to get

$$0 = x_1[a_{21}a_{12} - (a_{22} - \lambda)(a_{11} - \lambda)] \tag{2-44}$$

If the quantity in brackets in Equation (2-44) (which the reader may identify as the determinant of system (2-43)) is *not* equal to zero, then the only solution to Equations (2-43) is $x_1 = 0$, $x_2 = 0$, which is the trivial solution. On the other hand, *if the determinant is equal to zero,* then $x_1$ may have *any* value, and the corresponding $x_2$ value follows from the solution of one of the equations in (2-43). From Equation (2-44), we see that the condition for the determinant to vanish corresponds to $\lambda$ being a root of the quadratic equation. Thus, the determinant will generally vanish for two distinct values of $\lambda$ (unless $\lambda$ is a repeated root).

### 2.5.1 The Interpolation Method

The situation for higher-order eigenvalue problems is qualitatively similar to that for the second-degree one; the difference is that it is much more difficult to find the eigenvalues in higher-dimensional problems. The fundamental eigenvalue problem $A\mathbf{x} = \lambda\mathbf{x}$ can also be expressed by the requirement that

$$D(\lambda) = |A - \lambda I| = 0 = a_0\lambda^n + a_1\lambda^{n-1} + \cdots + a_{n-1}\lambda + a_n \tag{2-45}$$

Here $D(\lambda)$ is just an abbreviation for the determinant of the matrix $(A - \lambda I)$. It can be shown that the vanishing of the determinant in Equation (2-45) leads, in general, to the characteristic equation for the eigenvalues, which is the $n$th-degree polynomial shown in Equation (2-45). The idea of the interpolation method [Fadeev and Fadeeva, 1963] is as follows. By selecting $(n + 1)$ arbitrary values of $\lambda$ and calculating the corresponding $D(\lambda)$, we get a system of $(n + 1)$ linear equations in the $(n + 1)$ unknowns $(a_0, a_1, \ldots, a_n)$. When this system is solved, we then know the characteristic polynomial. Following determination of the characteristic equation, we then use the polynomial solver program POLYRTS to solve for the $n$ eigenvalues. This program finds all $n$ roots $\lambda_i$ of the polynomial, both real and complex. Program POLYRTS is discussed in the chapter on nonlinear algebraic equations.

After the $\lambda_i$ are known, the eigenvectors $\mathbf{x}_i$ can be computed by solving Equations (2-42), if they are required. For a given $\lambda_i$, this is accomplished by arbitrarily setting one of the eigenvector components, $x_k$, equal to unity, dropping one of the equations (any one, to form a square $(n-1)$ order system), and solving for the remaining $x_i$. This is permissible because the eigenvectors can be determined only up to any multiplicative constant. The only way this process can fail is if it happens that the component $x_k$ actually has a value of zero. In this case, we choose a different component to be unity and repeat the process. Eventually this must work, since by the nature of the eigenvalue problem, at least one solution component cannot be zero (otherwise, we would not have a nontrivial solution). A computer program that carries out these calculations, called MATXEIG (matrix eigenvalues), is discussed at the end of the chapter.

**Example 2.5 Calculation of Eigenvalues and Eigenvectors**

We consider here the three-by-three matrix

$$A = \begin{pmatrix} -1.8 & 2.1 & 1.7 \\ 1.2 & -3.5 & 0.3 \\ 0.6 & 1.4 & -2.0 \end{pmatrix} \tag{a}$$

To determine the eigenvalues of this matrix using program MATXEIG, we merely need to specify the above elements $a_{ij}$ in subroutine Matrix, along with $n$ (size of matrix) = 3. The values of lambda and d_lambda used were −2 and 1. The eigenvalues were calculated as $\lambda_1 = 0$, $\lambda_2 = -3.03153$ and $\lambda_3 = -4.26847$ (note as a check that the sum of the three eigenvalues is equal to the trace of the matrix $-1.8 - 3.5 - 2 = -7.3$). By running the program again, this time choosing the eigenvector option (with known $\lambda_i$), we obtain the corresponding eigenvectors as

$$\mathbf{x}_1 = \begin{pmatrix} 1 \\ 0.392097 \\ 0.574468 \end{pmatrix}, \mathbf{x}_2 = \begin{pmatrix} 1 \\ 1.17116 \\ -2.17116 \end{pmatrix}, \mathbf{x}_3 = \begin{pmatrix} 1 \\ -1.92114 \\ 0.921135 \end{pmatrix} \tag{b}$$

In this calculation we selected the option $k = 1$ (taking first eigenvector components to be unity for each $\lambda_i$).

### 2.5.2 Constant Coefficient Differential Equations

One important but simple application of matrix eigenvalues and eigenvectors is to the analytical solution of a system of linear, homogeneous, constant coefficient, ordinary differential equations. We may write such a system as

$$\frac{d\mathbf{y}}{dt} = A\mathbf{y} \tag{2-46}$$

If a solution is attempted in the natural form $\mathbf{y} = e^{\lambda t}\mathbf{x}$, substitution into Equation (2-46) gives

$$\lambda e^{\lambda t}\mathbf{x} = Ae^{\lambda t}\mathbf{x} \quad \text{or} \quad A\mathbf{x} = \lambda\mathbf{x} \tag{2-47}$$

Equation (2-47) is just the fundamental eigenvector/eigenvalue equation. If there are $n$ distinct eigenvalues $\lambda_i$, we get for *each* eigenvalue a solution of the assumed form $\mathbf{y} = e^{\lambda t}\mathbf{x}$. Because of the linearity of the differential equations, it is easy to show that a complete solution to the above problem can be obtained by adding together each of the $n$ fundamental solutions, with each one being multiplied by an arbitrary constant $k_i$. Thus, for this case, after the eigenvalues and eigenvectors have been computed, the complete solution can be written as $\mathbf{y} = \sum k_i \exp(\lambda_i t)\mathbf{x}_i$. The $n$ arbitrary constants $k_i$ are determined from the $n$ initial values of the $y_i(0)$ by solving a set of $n$ algebraic equations.

When the eigenvalues are distinct and all real, the solution components are real exponentials. When the eigenvalues are distinct, but some real and some complex-conjugate, the components are either exponentials or exponentials multiplied by sines and cosines. In this case the eigenvectors will also be complex, so that complex algebra must be used to obtain the final form of the solution. If some eigenvalues are repeated, the solution contains an extra power factor for each repeated eigenvalue [Amundson, 1966; Noble, 1969; Franklin, 1968].

### 2.5.3 Qualitative Behavior of Solutions: Stability

Another important application of matrix eigenvalues is to the *qualitative* behavior of the solution of constant-coefficient differential equations. In many problems (especially in process control), it is not necessary to know the full explicit solutions of the differential equations, just their qualitative behavior for large values of the independent variable (usually time). In such a situation, it is only necessary to determine the eigenvalues of the matrix $A$ (and not the eigenvectors). For constant coefficient problems of the type $d\mathbf{y}/dt = A\mathbf{y}$, there are three possibilities, which trace back to the *form* of the general solution for *distinct eigenvalues*, $\mathbf{y} = \sum k_i \exp(\lambda_i t)\mathbf{x}_i$. First, if the real parts of *all* the eigenvalues $\lambda_i$ are negative, then the solutions decay toward zero for large $t$, and the system is called *stable*. Second, if *any* eigenvalue has a positive real part, the solution eventually grows without bound for large $t$, and the system is said to be *unstable*. Finally, if no eigenvalues have a positive real part, but one or more have a zero real part, the solution ultimately remains bounded for large $t$.

The foregoing results correspond, as mentioned above, to the case of *distinct eigenvalues*. These results also hold for the situation where there are one or more repeated eigenvalues, with the following (unusual) exception for the case where we encounter a *repeated complex eigenvalue having a zero real part*. This gives rise to a solution component which is a polynomial in $t$ multiplied by an oscillatory factor; thus, this solution component ultimately becomes unbounded for large $t$ and hence yields *unstable* behavior.

**Example 2.6 Solution of a System of Kinetic Differential Equations**

We consider the following system of three constant coefficient differential equations of the form $dC/dt = AC$, where $C$ is a three-dimensional column vector $(dC_1/dt, dC_2/dt, dC_3/dt)^t$ and $A$ is the matrix of Example 2.5. The system is typical of a problem arising in chemical kinetics, where it is desired to model the concentrations $C_i$ as a function of time. This system of differential equations can be written as

$$\begin{aligned}\frac{dC_1}{dt} &= -1.8C_1 + 2.1C_2 + 1.7C_3 \\ \frac{dC_2}{dt} &= 1.2C_1 - 3.5C_2 + 0.3C_3 \\ \frac{dC_3}{dt} &= 0.6C_1 + 1.4C_2 - 2C_3\end{aligned} \tag{a}$$

Since the matrix $A$ in this problem is identical to that of Example 2.5, we already know its eigenvalues and eigenvectors from that calculation. Since all the eigenvalues are real (typical of most reaction kinetics problems), the solution can be expressed as $\mathbf{y} = \sum k_i \exp(\lambda_i t)\mathbf{x}_i$, where the eigenvector components $x_{ij}$ are also real (see Example 2.5). If we write this expression in terms of components, we have

$$\begin{pmatrix} C_1 \\ C_2 \\ C_3 \end{pmatrix} = k_1 e^{\lambda_1 t} \begin{pmatrix} x_{11} \\ x_{12} \\ x_{13} \end{pmatrix} + k_2 e^{\lambda_2 t} \begin{pmatrix} x_{21} \\ x_{22} \\ x_{23} \end{pmatrix} + k_3 e^{\lambda_3 t} \begin{pmatrix} x_{31} \\ x_{32} \\ x_{33} \end{pmatrix} \tag{b}$$

If we now substitute into Equation (b) the eigenvalues and eigenvectors computed in Example 2.5, we get the following explicit solutions for $C_1(t)$, $C_2(t)$, and $C_3(t)$.

$$\begin{pmatrix} C_1(t) \\ C_2(t) \\ C_3(t) \end{pmatrix} = k_1 e^{0t} \begin{pmatrix} 1 \\ 0.392097 \\ 0.574468 \end{pmatrix} + k_2 e^{-3.03153t} \begin{pmatrix} 1 \\ 1.17116 \\ -2.17116 \end{pmatrix} + k_3 e^{-4.26847t} \begin{pmatrix} 1 \\ -1.92114 \\ 0.921135 \end{pmatrix} \tag{c}$$

The constants $k_1$, $k_2$, and $k_3$ must be determined from known values of the $C_i(t)$ at some particular time (usually initial conditions at time zero).

The great advantage of the matrix solution for the system given above is that it eliminates the tedious necessity of algebraically reducing the system to a single higher-order equation for purposes of solution. Two extensions of the above problem that are easily handled by the matrix approach include the differential equation systems $\mathbf{y}' = A\mathbf{y} + \mathbf{b}$ and $D\mathbf{y}' = B\mathbf{y} + \mathbf{c}$, where $A$, $B$, and $D$ are constant matrices, and $\mathbf{b}$ and $\mathbf{c}$ are constant vectors. In the first of these two extensions, a particular solution to the full equation is obtained by setting $\mathbf{y}$ equal to the unknown constant vector $\mathbf{d}$. In this case $\mathbf{y}' = A\mathbf{y} + \mathbf{b}$ reduces to $\mathbf{0} = A\mathbf{d} + \mathbf{b}$, and we see that $\mathbf{d} = -A^{-1}\mathbf{b}$. This particular solution can simply be added to the vector complementary solution derived earlier in order to obtain the complete solution. For the problem $D\mathbf{y}' = B\mathbf{y} + \mathbf{c}$, we may note that multiplication of both sides by the inverse matrix $D^{-1}$ produces $\mathbf{y}' = D^{-1}B\mathbf{y} + D^{-1}\mathbf{c}$, which is of the same form as the first extension. The foregoing discussion shows that the matrix technique is a general and powerful tool for the solution of a system of constant coefficient ordinary differential equations.

## 2.6 MISCELLANEOUS TOPICS

Since our discussion of matrix analysis is only an introduction to the subject, there are a number of important topics we do not cover that should be mentioned. Discussion of these topics can be found in the references at the end of the chapter; for numerical purposes, see especially Press *et al.* [1986]. The first topic in this category is the LU decomposition, which is an alternative to Gauss elimination for the solution of systems of linear equations. The advantage of the LU decomposition is that it easily allows the solution of many different right-hand-side vectors **b** in the equation $A\mathbf{x} = \mathbf{b}$, while only requiring one decomposition of matrix $A$. This is quite useful for the inversion process, as well as for other applications. The LU decomposition is discussed, among others, by Press *et al.* [1986] and Johnson and Riess [1982].

A "sparse" matrix is one that contains a relatively large number of zero elements. We have encountered the most famous sparse matrix, the tridiagonal, in our earlier discussion. However, even when there is a less regular pattern of zeros in the matrix, it is often possible to deal with manipulation and storage in a particularly efficient way. For a discussion of sparse matrix techniques, see Press *et al.* [1986].

The singular value decomposition (SVD) is a very valuable technique for dealing with a variety of matrix problems, particularly those of nearly singular systems and ill-conditioning. Such considerations arise often in the case of larger least-squares problems. The SVD is discussed by Press *et al.* [1986] and Golub and Van Loan [1983].

Finally, we must note that there exist a number of powerful software packages for accomplishing various matrix calculations. These are discussed, for example, by Press *et al.* [1986] and Kahaner *et al.* [1989].

## 2.7 COMPUTER PROGRAMS

In this section we discuss the computer programs applicable to various problems in the chapter. For each program we give a brief discussion and a concise listing, which includes a program description, how to run it, and user-specified subroutines, including a sample problem. Detailed program listings can be obtained from the diskette.

### Linear Equation System Solver SIMLE

Instructions for running the program are given in the concise listing. This program uses the True BASIC built-in matrix inversion routine to estimate the solution vector. One step of an iterative improvement is conducted, using the original inverse, to produce an improved solution and (*especially!*) an estimate of the relative error in the solution components. The problem considered in the concise listing is that of Example 2.4.

### PROGRAM SIMLE

```
!This is program SIMLE.tru, a True BASIC program for solving the
!linear equation system Ax = b by O.T.Hanna. A is the coefficient
!matrix, b is the right-hand side vector, n is the number of
!equations and x is the solution vector. The program does one
!iterative improvement and uses this to estimate accuracy.
```

```
!***************************************************************
!*To run, specify n, A and b in Subroutine Equations. Program  *
!*prints A,b, solution vector x, estimated relative errors and *
!*the determinant                                              *
!***************************************************************

Sub Eqns(n,A(,),b())    !user specified subroutine
!***denotes user specified quantities

Let n = 3               !***no. of eqns.
mat redim A(n,n),b(n)   !redimension A,b arrays

!***Put in matrix A(i,j) by rows
Let a(1,1) = 6
Let a(1,2) = 6
Let a(1,3) = 3.00001
Let a(2,1) = 10
Let a(2,2) = 8
Let a(2,3) = 4.00003
Let a(3,1) = 6
Let a(3,2) = 4
Let a(3,3) = 2.00002

!***Put in right-hand-side vector b(i)
Let b(1) = 30.00002
Let b(2) = 42.00006
Let b(3) = 22.00004

End Sub
```

### *Tridiagonal Linear System Solver TRID*

This program, in contrast to most others that we present, is organized as a main calling program for the tridiagonal subroutine. Instructions for running the program are given in the concise listing. The problem considered is that of Example 2.3.

### PROGRAM TRID

```
!This is program TRID.tru for solving a tridiagonal linear system
!of algebraic equations by O.T.Hanna.
!The following calling program utilizes Subroutine Trid which is
!given at the end of the main program.
!Program solves a(i)*x(i-1) + b(i)*x(i) + c(i)*x(i+1) = d(i)
!***************************************************************
!*To run,specify n,and coefficients a(i),b(i),c(i),d(i)        *
!*in program below; program prints solution                    *
!***************************************************************

!*** denotes user specified quantity
```

```
DIM a(1),b(1),c(1),d(1),x(1)
LET n = 11                                          !***
MAT redim a(n),b(n),c(n),d(n),x(n)

!***Define coefficients a(i),b(i),c(i),d(i) !***

LET b(1) = 1                                        !***, Etc.
LET c(1) = 0
LET d(1) = 0.5

FOR i = 2 to n-1                                    !***, Etc.
   LET a(i) = 1
   LET b(i) = -1.99
   LET c(i) = 1
   LET d(i) = 0
NEXT i

LET a(n) = 0                                        !***, Etc.
LET b(n) = 1
LET d(n) = 0.69

CALL Trid(n,a(),b(),c(),d(),x())

!Now print results
```

### Matrix Eigenvalue/Eigenvector Program MATXEIG

Instructions for running the program are given in the concise listing. To employ this program, the user need only specify the square matrix $A$ and its size $n$ in subroutine Matrix. Program MATXEIG uses the interpolation method to calculate the eigenvalue characteristic equation of the $(n \times n)$ real matrix $A$ by evaluating $Det(\lambda)$ for $(n+1)$ user-selected values of $\lambda$. Here $Det$ is the determinant of the matrix $(A-\lambda I)$. The matrix can be either symmetric or unsymmetric. The subroutine Polyrts (polynomial roots) is used to solve the characteristic polynomial for all roots, both real and complex-conjugate.

For the case of any calculated real eigenvalue $\lambda_i$, the program can be run again, using the eigenvector option, to find eigenvector $i$. By selecting some eigenvector component $(k)$ to be unity, the remaining $(n-1 \times n-1)$ equations (subproblem) are solved for the remaining eigenvector components. *This portion of the program only works provided that the subproblem has a unique solution.* The reason for the option to choose $k$ is that it may happen that, unbeknownst to the user, some eigenvector component may be *zero* and therefore not permitted to be set to unity. In such a (favorable) case, the user merely repeats the calculation, choosing a different value of $k$. If no solution is obtained for *any* possible $k$ value, the subproblem does not have a *unique* solution, and hence the program does not work.

## PROGRAM MATXEIG

```
!This is program MATXEIG.tru by O.T.Hanna
!Program computes the n eigenvalues (real and complex) of a real
!matrix (symmetric or unsymmetric), as well as the eigenvectors
!for real eigenvalues only. The program computes the
!coefficients of the characteristic polynomial for given a
!matrix A  corresponding to the equation |A - lambda*I| = 0.
!This polynomial is given by
!0 = p(0)v^n + p(1)v^(n-1) ... + p(n-1)v + p(n).
!The calculation of the polynomial is carried out by the
!interpolation method, which evaluates the above determinant for
!various user-specified lambda values, and then computes the
!coefficients of the characteristic polynomial. When the
!eigenvalues [lambda(i)] are known, the program can be re-run and
!the eigenvector option selected in order to compute real
!eigenvectors. This works only if the sub-problem corresponding
!to x(k) = 1 has a unique solution.
!*************************************************************
!*To run, specify in Subroutine Matrix at end of prgm the      *
!*input matrix A, n = size of matrix, initial value of         *
!*lambda [lambda(1)],and the increment d_lambda to be used.    *
!*Program prints A, and asks for eigenvector option. This      *
!*should not be used until any real eigenvalues have been      *
!*calculated. Program provides option to over-ride             *
!*recursively specified lambda values. Polynomial solver       *
!*tries for real roots first, with prompt for new starting     *
!*value(if desired). When user does not elect new starting     *
!*value, program seeks pairs of complex conjugate roots.       *
!*After eigenvalues are determined, program can be run         *
!*again to determine eigenvectors which correspond to real     *
!*eigenvalues.                                                 *
!*************************************************************
SUB Matrix(n,A(,),lambda(),d_lambda)
   !***denotes user specifed quantities

   LET n = 4              !***size of square A matrix
   LET lambda(1) = -2     !***initial value of lambda;
   LET d_lambda = 1       !***others computed recursively

   MAT redim A(n,n)       !redimension A array

   !***Put in matrix A(i,j) by rows
    LET A(1,1) = 1
    LET A(1,2) = 1
    LET A(1,3) = 1
    LET A(1,4) = 1
    LET A(2,1) = 1
    LET A(2,2) = 2
    LET A(2,3) = 2
```

```
    LET A(2,4) = 2
    LET A(3,1) = 1
    LET A(3,2) = 2
    LET A(3,3) = 3
    LET A(3,4) = 3
    LET A(4,1) = 1
    LET A(4,2) = 2
    LET A(4,3) = 3
    LET A(4,4) = 4
END SUB
```

## PROBLEMS

**2.1.** Find the matrix product $AB$ if

$$A = \begin{pmatrix} 5 & 6 \\ 1 & 6 \end{pmatrix} \text{ and } B = \begin{pmatrix} 1 & 4 \\ 3 & 2 \end{pmatrix}$$

**2.2.** Show for the column vector $\mathbf{x} = (x_1, x_2, x_3)'$ that $\mathbf{x}'\mathbf{x} = x_1^2 + x_2^2 + x_3^2$.

**2.3.** Find $A'A$ if $A$ is given by

$$A = \begin{pmatrix} 2 & 7 & 5 \\ 3 & 1 & 6 \\ 4 & 2 & 5 \end{pmatrix}$$

Check to see that the result is symmetric, as it should be.

**2.4.** Calculate by hand the inverse of the matrix

$$A = \begin{pmatrix} 2 & 8 \\ 6 & 4 \end{pmatrix}$$

Check this by multiplying the calculated $A^{-1}$ by $A$ to verify that $A^{-1}A = I$. Also verify by using program SIMLE to calculate $A^{-1}$.

**2.5.** Use program SIMLE to find the inverse of the matrix

$$A = \begin{pmatrix} 1.2 & 6.5 & 2.7 \\ 2.1 & 7.8 & 10.2 \\ 9.3 & 6.1 & 2.6 \end{pmatrix}$$

Then invert the calculated inverse to see how close we are to the original $A$.

**2.6.** Consider the set of linear equations

$$\begin{pmatrix} 10 & 4 & 7 \\ 2 & 6 & 8 \\ 1 & 4 & 3 \end{pmatrix} \cdot \begin{pmatrix} x_1 \\ x_2 \\ x_3 \end{pmatrix} = \begin{pmatrix} b_1 \\ b_2 \\ b_3 \end{pmatrix}$$

where the $b_i$ are arbitrary right-hand-side values. Use the Gauss reduction process (by hand) to see whether or not these equations have a unique solution. Also carry out a calculation using program SIMLE to show the same thing.

**2.7.** It is desired to fit the equation $y = c_1 + c_2e^{-2x}$ through the pair of points $(x = 0, y = 1)$ and $(x = 1, y = 0.5)$. Determine the appropriate coefficients $c_1$ and $c_2$ and verify the calculation.

**2.8.** The equation $Sh = aRe^b Sc^c$ is to be fitted to the three points

| pt. | Sh | Re | Sc |
|---|---|---|---|
| 1 | 43.7 | 10800 | 0.6 |
| 2 | 21.5 | 5290 | 0.6 |
| 3 | 24.2 | 3120 | 1.8 |

By taking natural logarithms in the above modeling equation, find the required values of $a$, $b$, and $c$. Verify that the equation does fit the points.

**2.9.** Consider the three *inconsistent* (due to measurement errors) mass balance equations

$$m_1 + m_2 = 2.2$$
$$m_2 = 0.31(2.2)$$
$$m_1 = 0.67(2.2)$$

(i) Formulate a least-squares solution for $m_1$ and $m_2$.
(ii) Solve for the appropriate numerical values of $m_1$ and $m_2$.

**2.10.** It is desired to solve the following system of equations using the Gauss-Seidel iterative method.

$$x_1 = (1/4)[0 + 100 + x_2 + x_3]$$
$$x_2 = (1/4)[x_1 + 100 + 0 + x_4]$$
$$x_3 = (1/4)[0 + x_1 + x_4 + 100]$$
$$x_4 = (1/4)[x_3 + x_2 + 0 + 100]$$

Use three different sets of starting values and verify the solution.

**2.11.** The following system of tridiagonal linear equations represents a simple model of an $n$ element equilibrium-stage process (e.g., liquid extraction or gas absorption).

$$-1.8x_1 + 0.8x_2 = -0.08$$
$$x_{i-1} - 1.8x_i + 0.8x_{i+1} = 0 \qquad (i = 2, \cdots, n-1)$$
$$x_{n-1} - 1.8x_n = -0.6$$

Take $n = 30$ and solve for $x_1$ and $x_{30}$.

**2.12.** Consider the system of equations from problem 2-6 and take the right-hand-side vector to be $(2, 9, 6)^t$. Use both matrix methods and direct numerical differentiation (Appendix G) to compute
(i) $\partial x_2/\partial b_3$
(ii) $\partial x_1/\partial a_{23}$

**2.13.** The following system of linear equations has the exact solution $x_1 = 3$, $x_2 = 1$.

$$x_1 - 2x_2 = 5$$
$$x_1 - 2x_2 = 1$$

Suppose an approximate numerical solution has been calculated as $x_1 = 2.9$, $x_2 = 1.5$. Carry out by hand one step of the iterative improvement procedure for correcting these approximate values. This procedure is used in computer program SIMLE.

**2.14.** In the text, the iterative procedure $\mathbf{x}^{[n+1]} = B\mathbf{x}^{[n]} + \mathbf{c}$ is shown to converge if $\|B\| < 1$. Calculate $\|B\|$ for problem 2-10 to see if this condition is met.

**2.15.** Calculate the condition number $\|A\|\|A^{-1}\|$ for the matrix in problem 2-4.

**2.16.** Determine the eigenvalues of the matrices

$$\text{(i)} \quad A = \begin{pmatrix} 4 & 10 \\ -6 & 20 \end{pmatrix} \qquad \text{(ii)} \quad A = \begin{pmatrix} -4 & 2 & 1 \\ 2 & -3 & 1 \\ 2 & 1 & -2 \end{pmatrix}$$

**2.17.** Determine the eigenvectors of the matrices in problem 16(i) and problem 16(ii).

**2.18.** Find the general solution of the differential equation system $d\mathbf{C}/dt = A\mathbf{C}$, where the matrix $A$ is that given in problem 16(ii).

## REFERENCES

AMUNDSON, N. R. (1966), *Mathematical Methods in Chemical Engineering: Matrices and Their Application*, Englewood Cliffs, NJ: Prentice-Hall.

BELLMAN, R. (1960), *Introduction to Matrix Analysis*, New York: McGraw-Hill.

FADEEV, D. K. and FADEEVA, V. N. (1963), *Computational Methods of Linear Algebra*, San Francisco: W. H. Freeman.

FRANKLIN, J. N. (1968), *Matrix Theory*, Engelwood Cliffs, NJ: Prentice-Hall.

GOLUB, G. H. and VAN LOAN, C. F. (1983), *Matrix Computations*, Baltimore, MD: The Johns Hopkins University Press.

JOHNSON, L. W. and RIESS, R. D. (1982), *Numerical Analysis,* 2nd ed., Reading, MA: Addison-Wesley.

KAHANER, D., MOLER, C., and NASH, S. (1989), *Numerical Methods and Software*, Englewood Cliffs, NJ: Prentice-Hall.

NOBLE, B. (1969), *Applied Linear Algebra*, Englewood Cliffs, NJ: Prentice-Hall.

PRESS, W. H., FLANNERY, B. P., TEUKOLSKY, S. A., and VETTERLING, W. T. (1986), *Numerical Recipes*, New York: Cambridge University Press.

SCRATON, R. E. (1987), *Further Numerical Methods in Basic*, Baltimore, MD: Edward Arnold.

SCHEID, F. (1988), *Numerical Analysis*, 2nd ed., New York: McGraw-Hill.

TROPPER, A. M. (1962), *Matrix Theory*, Reading, MA: Addison-Wesley.

# 3

# Taylor Expansions

Taylor's formula is probably the most important single result in the application of mathematics to engineering problems. The Taylor formula provides a means for the approximation of some complicated function in terms of a simpler polynomial. It can be applied either directly or indirectly in a variety of different ways. It is this combination of simplicity plus versatility that makes Taylor's formula so important in practical application.

## 3.1 DERIVATION AND ERROR ANALYSIS

By way of review, if we wish to approximate $f(x)$ near $x = a$ we write $f(x) = c_0 + c_1(x-a) + c_2(x-a)^2 \ldots$, and so on. If we require that $f(x)$ and its derivatives up to some order agree exactly at the particular point $a$, we find that $f(a) = c_0$, $f'(a) = c_1$, $f''(a) = 2!c_2$, $f'''(a) = 3!c_3$, and, in general, $f^{(k)}(a) = k!\,c_k$. Substitution of these values for $c_k$ back into the assumed polynomial representation yields

$$f(x) = f(a) + f'(a)(x-a) + f''(a)\frac{(x-a)^2}{2!} + f'''(a)\frac{(x-a)^3}{3!} \cdots \tag{3-1}$$

The foregoing procedure shows how to compute the polynomial coefficients and suggests that the Taylor approximation should be valid when $x$ is sufficiently close to $a$ (assuming, of course, that the derivatives occurring in the approximation exist). A Taylor expansion having an expansion point of zero is known as a *Maclaurin* expansion.

### 3.1.1 Derivation

This procedure does *not* tell us, however, about the goodness of the approximation or the effect of adding more terms. In order to answer such questions, it is necessary to proceed more carefully in the development of the Taylor approximation. A convenient way to do this is to write, from the fundamental theorem of integral calculus,

$$f(x) = f(a) + \int_a^x f'(t)\,dt \tag{3-2}$$

This formula is exact and can be interpreted as $f(x) = f(a) + R$, where $R$ is the remainder in a one-term Taylor approximation. To continue, we integrate the integral remainder by parts to get

$$f(x) = f(a) + (t + c)f'(t)|_a^x - \int_a^x (t + c)f''(t)\,dt \tag{3-3}$$

Here $c$ is an arbitrary constant of integration. We wish to choose the constant, if possible, so that the integrated part of the above expression coincides with the second term of the Taylor approximation, which is $f'(a)(x - a)$. It can be observed that the choice $c = -x$ satisfies this requirement; the preceding equation then becomes

$$f(x) = f(a) + f'(a)(x - a) + \int_a^x (x - t)f''(t)\,dt \tag{3-4}$$

A comparison of this equation with a two-term Taylor approximation shows that the integral represents the remainder, or error, of the two-term Taylor approximation. At this point, it seems natural to continue the integration-by-parts procedure in hopes of developing an expression for the remainder in a Taylor approximation of any order. If this is done again, setting the constant of integration equal to zero, we find that

$$f(x) = f(a) + f'(a)(x - a) + f''(a)\frac{(x - a)^2}{2!} + \int_a^x \frac{(x - t)^2 f'''(t)\,dt}{2!} \tag{3-5}$$

This result strongly suggests a pattern. To establish that there is indeed a general formula of the type suggested, we proceed by mathematical induction. Thus, assume that for any integer $k$,

$$f(x) = \sum_{i=0}^{k} \frac{f^{(i)}(a)(x - a)^i}{i!} + \int_{t=a}^x \frac{(x - t)^k f^{(k+1)}(t)\,dt}{k!} \tag{3-6}$$

We have already seen directly that this formula is true for $k = 0$, 1, and 2. If we now integrate the integral in this expression by parts and take the constant of integration equal to zero, we see that we recover the *assumed* expression with $k$ replaced by $(k + 1)$. Thus, if the assumed expression is true for any integer $k$, it must also be true for $(k + 1)$. But since it is true for $k = 2$, it must be true for $k = 3$; and if for $k = 3$, then also for $k = 4$, and so on. Thus, mathematical induction shows that the previous expression is true for any integer $k$, provided, of course, that all of the quantities appearing in the equation exist. This requirement is not merely mathematically convenient, but has important implications in the practical application of Taylor's formula.

### 3.1.2 Alternate Forms of the Remainder

The remainder, or error, in Taylor's formula has been shown to be equal to

$$R_k = \int_{t=a}^{x} \frac{(x-t)^k f^{(k+1)}(t)\,dt}{k!} \tag{3-7}$$

While this result is true, it is normally not very useful in its present form. Rather, we use the mean value theorem for integrals, which says that if $p(t)$ is of uniform sign in the interval of integration, then

$$\int_a^x f(t)p(t)\,dt = f(\xi)\int_a^x p(t)\,dt$$

where $\xi$ is between $a$ and $x$. Since $(x-t)$ has a uniform sign on the interval of integration from $t=a$ to $t=x$, we may take $(x-t)$ as the function $p(t)$ in the mean value theorem. Then we can write

$$R_k = \int_{t=a}^{x} \frac{(x-t)^k f^{(k+1)}(t)\,dt}{k!} = \frac{f^{(k+1)}(\xi)(x-a)^{k+1}}{(k+1)!} \tag{3-8}$$

In this form the remainder is seen to be just the next term in the approximation, except that the $(k+1)$st derivative is evaluated at the unknown intermediate point, $\xi$, rather than at the expansion point, $a$. Thus, to obtain a "mathematically guaranteed" bound on the remainder $R_k$, we need to bound the $(k+1)$st derivative of $f(t)$ on the interval $(a, x)$. Many times this is exceedingly difficult to do. We may therefore be satisfied with *estimating* this error to be of the order of the next term, which will be closer to the truth the nearer $x$ is to the expansion point, $a$.

### 3.1.3 Finite Versus Infinite Interpretation

Another important point concerns the interpretation of Taylor's formula as either a finite or infinite expansion. Either of these interpretations may be appropriate in various circumstances. Thus, as we have seen, we can always consider Taylor's formula to be a finite expansion, terminating with a remainder. The goodness of this finite approximation depends on the size of the relative error (i.e., the magnitude of the ratio of the error divided by the approximation). On the other hand, by considering the limit, as $k \to \infty$, of the remainder $R_k$, Taylor's formula may represent the function $f(x)$ as a convergent power series, if $R_k \to 0$ as $k \to \infty$. What often happens is that for a certain range of $x$ values, $R_k \to 0$ for $k \to \infty$ and the infinite expansion converges, while for other $x$ values the infinite interpretation does not apply.

### 3.1.4 Implementation of Taylor's Theorem

We now consider some examples of Taylor approximation. First, consider the exponential function $f(x) = e^x$. All derivatives are also equal to $e^x$ so that we may write a three-term Taylor approximation about the point $a = 0$ $(e^0 = 1)$ as

$$e^x = 1 + x + \frac{x^2}{2!} + e^{\xi}\frac{x^3}{3!} (0 < \xi < x) \tag{3-9}$$

When $x$ is small, $\xi$ is necessarily small, and thus, $e^{\xi}$ must be near unity. In this case the remainder is approximately $x^3/3!$, and for small $x$ this is clearly small compared to the approximation. For large $x$ values, all we know is that $e^{\xi} < e^x$ (since the exponential is an increasing function), so the relative error will be substantial when $x$ is too large. Note that since $e^{\xi}$ is always greater than 1, the error of the above expansion is actually *larger* than the first neglected term. By examining $\lim_{k\to\infty} R_k$, it can be shown that this limit is zero for any $x$, so that for $f(x) = e^x$ the infinite Taylor expansion is valid for any $x$. However, note that when $x$ is much larger than unity, many terms of the expansion will be required to give good accuracy.

As a second example, consider the function $g(x) = (1 + x)^{1/2}$ with $a = 0$. By differentiation we find that

$$g'(0) = 1/2, g''(0) = -1/4, \text{ and } g'''(x) = (3/8)(1 + x)^{-5/2} \tag{3-10}$$

Substitution in Taylor's formula gives, for positive $x$,

$$g(x) = (1 + x)^{1/2} = 1 + \frac{x}{2} - \frac{x^2}{8} + R \tag{3-11}$$

where $R = g'''(\xi)x^3/3! = x^3/[(16)(1 + \xi)^{5/2}]$

In this case we see that, for positive $x$, the remainder $R$ is less in magnitude than the first neglected term $x^3/16$. Once again, the Taylor approximation will have good accuracy when $x$ is small. For $x = 1$, the above approximation gives $2^{1/2} = 1.375$ versus the true value, 1.414. Although it is not evident just from the appearance of this approximation, consideration of $\lim_{k\to\infty} R_k$ shows that the *infinite* Taylor expansion converges only for $x$ less than or equal to 1. Thus, if we were to use, for example, $x = 1.1$, the Taylor approximation would yield terms that first decrease in magnitude but ultimately become arbitrarily large. Nevertheless, the *finite* Taylor expansion for this case could produce useful results.

If we wished to compute $3^{1/2}$ by using $x = 2$ in the previous expansion, the results would not be very useful, since the terms begin to increase in magnitude very quickly. However, note that we could still use Taylor's formula in the following modified way. We write $3^{1/2} = (4 - 1)^{1/2} = 4^{1/2}(1 - 1/4)^{1/2} = 2(1 - 1/4)^{1/2}$. Thus, $3^{1/2} = 2(1 - x)^{1/2}$ for $x = 1/4$. The Taylor expansion about $a = 0$ for this new function converges rapidly for $x = 1/4$.

Finally, in connection with square roots and similar fractional powers, it is interesting to note that the function $x^{1/2}$ does *not* have a Taylor expansion about the point $a = 0$. This is because the derivatives of $x^{1/2}$ at $x = 0$ do not exist.

Figure 3.1 compares the one-term ($x$), two-term ($x - x^3/3!$), and three-term ($x - x^3/3! + x^5/5!$) Taylor series approximations for the function $y = \sin x$. It is seen that the one-term approximation is valid to about $x = 0.5$, the two-term approximation to about $x = 1.5$, and the three-term approximation to about $x = 2.2$. These observations would change according to the accuracy desired in the approximation. In this example, each particular successive approximation (1 term, 2 terms, 3 terms) happens to be uniformly on one side of the true value. This is not always the case in Taylor approximation.

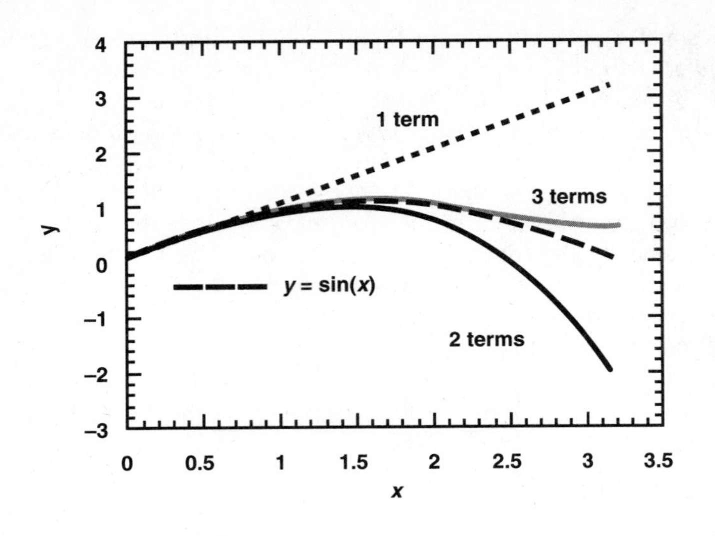


**Figure 3.1.** Sin($x$) Compared to its 1-Term, 2-Term, and 3-Term Taylor Series Approximations

## 3.2 FUNCTIONS OF SEVERAL VARIABLES

We now consider the extension of Taylor's formula to functions of two variables. This is done by application of the one-dimensional result to a suitably defined function. Let

$$F(t) = f(x_0 + ht, y_0 + kt) = f[x(t), y(t)] \tag{3-12}$$

We can now find the Taylor expansion relating $f(x_0 + h, y_0 + k)$ to $f$ and its partial derivatives at $(x_0, y_0)$ by writing the Taylor expansion for $F(t)$ and evaluating it at $t = 1$. Thus,

$$F(t) = F(0) + F'(0)t + \frac{F''(0)t^2}{2!}\ldots \tag{3-13}$$

The derivatives of $F(t) = f[x(t), y(t)]$ must be computed by the chain rule. We therefore have

$$F'(t) = \frac{\partial f}{\partial x}\frac{dx}{dt} + \frac{\partial f}{\partial y}\frac{dy}{dt} = h\frac{\partial f}{\partial x} + k\frac{\partial f}{\partial y} \tag{3-14}$$

and

$$F''(t) = h\left(\frac{\partial^2 f}{\partial x^2}h + \frac{\partial^2 f}{\partial x \partial y}k\right) + k\left(\frac{\partial^2 f}{\partial y \partial x}h + \frac{\partial^2 f}{\partial y^2}k\right) \tag{3-15}$$

and so on. If we assume that the second mixed partial derivatives $f_{xy}$ and $f_{yx}$ are continuous, and hence equal, then substitution into Taylor's formula and setting $t = 1$ gives

$$\begin{aligned} f(x_0 + h, y_0 + k) = f(x_0, y_0) &+ \left(h\frac{\partial f}{\partial x} + k\frac{\partial f}{\partial y}\right)_{x_0, y_0} \\ &+ \frac{1}{2!}\left(h^2\frac{\partial^2 f}{\partial x^2} + 2hk\frac{\partial^2 f}{\partial x \partial y} + k^2\frac{\partial^2 f}{\partial y^2}\right)_{x_0, y_0} + \ldots \end{aligned} \tag{3-16}$$

In a similar way, Taylor's formula can be extended to functions of more variables with analogous results. For example, if we retain only linear terms of the approximation for functions of three variables, we get

$$f(x_0+h, y_0+k, z_0+l) = f(x_0, y_0, z_0) + \left(h\frac{\partial f}{\partial x} + k\frac{\partial f}{\partial y} + l\frac{\partial f}{\partial z}\right)_{x_0, y_0, z_0} + \ldots \tag{3-17}$$

For functions of several variables, normally only the linear, or at the most quadratic, terms in the approximation are retained. It should be noted that, for the purposes of sensitivity analysis discussed below, we can easily derive the multidimensional *finite* form of the linear Taylor expansion as follows. We can write Equation (3-13) in the finite form $F(t) = F(0) + F'(\xi)t$ with $0 \le \xi \le t$. Using Equations (3-12) and (3-14) with $t = 1$, this becomes (illustrated here for the two-dimensional case)

$$f(x_0+h, y_0+k) = f(x_0, y_0) + \left(h\frac{\partial f}{\partial x} + k\frac{\partial f}{\partial y}\right)_{x_0+\xi h, y_0+\xi k} \tag{3-18}$$

This form is *exact* (there is no error term), and the partial derivatives $\partial f/\partial x$ and $\partial f/\partial y$ are each evaluated at the (unknown) intermediate point $(x_0 + \xi h, y_0 + \xi k)$, where $0 \le \xi \le 1$ since $t = 1$. Extension of this result for the case of many variables or higher-order terms follows directly.

## 3.3 APPLICATIONS

Applications of Taylor's theorem are extremely widespread in computational problems. Here we shall be content to illustrate some examples of the use of the Taylor expansion in sensitivity analysis, approximation of integrals, and checking computer results for the solution of ordinary differential equations. Many other applications will be demonstrated later.

### 3.3.1 Sensitivity or Error Analysis

In many engineering problems, it is important to perform what is often called an "error" or "sensitivity" analysis. This situation arises whenever it is necessary to assess what overall change may be caused in some dependent quantity $y(x_1, x_2, \ldots)$ by incremental changes, $\Delta x_i$, in the independent variables $x_i$. An important application of this type occurs in the analysis of experimental errors. Analysis of experimental errors can be approached in two different but related ways. The simplest procedure corresponds to a deterministic, or "worst-case," approach, which will be discussed below. An alternative, more sophisticated method is based on a model involving statistical considerations [Ross, 1987]. This method (which will not be considered here) requires the use of random variables; its principal advantage is that it produces more realistic results because the theory allows for the possibility of some error cancellation.

Suppose we are given a relationship $y = f(x_1, x_2, \ldots)$. The development of a formula to determine the maximum value of $|\Delta y|$ corresponding to given maximum increments $|\Delta x_i|$ about some nominal point proceeds as follows. We wish to ascertain, at the given point ($y = Y, x_1 = X_1, x_2 = X_2, \dots$), how much $y$ will deviate from $Y$ when the $x_i$ deviate from $X_i$ by a certain amount. This is given *approximately*, when the $\Delta x_i$ increments are "small," by the formula

$$\Delta y = \left(\frac{\partial f}{\partial x_1}\right)_{X_1,X_2\dots} \Delta x_1 + \left(\frac{\partial f}{\partial x_2}\right)_{X_1,X_2\dots} \Delta x_2 \dots \tag{3-19}$$

This formula is known as the fundamental increment formula or linear Taylor approximation. Note that in Equation (3-19), the partial derivatives are all evaluated at the base point ($X_1, X_2, \dots$), so that Equation (3-19) is only approximate since the error terms have been neglected. We now wish to apply Equation (3-19) to a worst-case situation in which the $\Delta x_i$ are merely known to lie in some given range on either side of the nominal values $X_i$, $[(X_i - \Delta x_i) < x_i < (X_i + \Delta x_i)]$. To do this we apply the triangle inequality, which states that $|a + b + c| \le |a| + |b| + |c|$. We then obtain from Equation (3-19) the *approximate* inequality

$$|\Delta y| \le \sum_{i=1}^{n} \left|\frac{\partial f}{\partial x_i}\right|_{X_1,X_2\dots} \cdot |\Delta x_i| \tag{3-20}$$

Since Equation (3-20) does not allow for any possible cancellation in the sum $\sum \partial f/\partial x_i \cdot \Delta x_i$, it is generally thought to be a conservative or worst-case result. *However, it is important to realize that this is not a truly rigorous result.* This is because Equation (3-20) is strictly valid only when the $\Delta x_i$ are sufficiently small in magnitude. In many engineering applications this is *not* the case. To be sure of a *guaranteed* result, it is necessary to alter the foregoing procedure as follows. As in Equation (3-19), we write the *exact* form of the linear Taylor expansion as

$$\Delta y = \sum_{i=1}^{n} \left(\frac{\partial f}{\partial x_i}\right)_{X_i+\xi\Delta x_i} \cdot \Delta x_i \tag{3-21}$$

Here the partial derivatives are evaluated at the unknown intermediate $x_i$ values corresponding to the $\xi$ value, where $\xi$ is located somewhere between zero and unity. Therefore, to get a strict inequality from Equation (3-21), when we apply the triangle inequality we must replace each $|\partial f/\partial x_i|$ by the *maximum magnitude* it can have when $\xi$ varies between zero and one. In a given problem, this rigorous result will approach that of the approximate Equation (3-20) as the range of the $\Delta x_i$ are restricted to be smaller and smaller. That is, the rigorous and approximate results will be very close provided the "error tolerances" are sufficiently small. But how "small" is small enough depends on how strongly the partial derivatives vary at the point under consideration. This question can only be answered by exploring this variation in a particular problem.

**Example 3.1 Use of Taylor's Theorem in Sensitivity Analysis**

In fluid mechanics there are a number of problems where the friction factor in a circular pipe, $f$, is related to the Reynolds number, Re, by the relation $f = K\text{Re}^{-a}$, where $K$ and $a$ are positive constants ($a \approx 0.25$) [Bennett and Myers, 1982]. The Reynolds number is a composite variable which is given by $\text{Re} = Du\rho/\mu$, where $D$ is the diameter of the pipe, $u$ is the velocity, $\rho$ is the fluid density, and $\mu$ is the fluid viscosity. We consider, in a purely deterministic way, how uncertainties or variations in $u$ will affect

$f$. These variations could be attributed to experimental limitations or differences in design conditions.

Suppose that we consider a two-dimensional situation, that is, where $f$ may be affected by errors or variations in the two variables $u$ (velocity) and $\mu$ (viscosity). In this calculation, the density and diameter are assumed to be known exactly. Thus, for our purposes $f = f(u, \mu)$. To do the standard approximate sensitivity analysis, we wish to implement Equation (3-20). To calculate the required partial derivatives, it is convenient to take logarithms of the equation for $f$ above to get

$$\ln f = \ln K - a[\ln D + \ln u + \ln \rho - \ln \mu] \tag{a}$$

Then the partial derivatives are calculated as

$$\frac{1}{f}\frac{\partial f}{\partial u} = \frac{-a}{u}, \quad \frac{1}{f}\frac{\partial f}{\partial \mu} = \frac{a}{\mu} \tag{b}$$

It is common for the estimated maximum errors or deviations in the variables to be reported on a *relative* basis with respect to the nominal values. For instance, we might have the specifications $u = u_0 \pm \epsilon_u u_0$ and $\mu = \mu_0 \pm \epsilon_\mu \mu_0$ where the nominal values $u_0, \mu_0$ and the maximum relative errors $\epsilon_u, \epsilon_\mu$ are both known. In this case we would have $|\Delta u| \leq \epsilon_u u_0$ and $|\Delta\mu| \leq \epsilon_\mu \mu_0$. If we now substitute these relations along with the partial derivatives of Equations (b) into Equation (3-20) we obtain

$$|\Delta f| \leq \frac{af_0}{u_0}|\Delta u| + \frac{af_0}{\mu_0}|\Delta \mu| = af_0[\epsilon_u + \epsilon_\mu] \tag{c}$$

This can be expressed in the concise and convenient dimensionless form

$$\left|\frac{\Delta f}{f_0}\right| \leq a\epsilon_u + a\epsilon_\mu \tag{d}$$

For example using $a = 0.25$, if $\epsilon_u = 0.07$ and $\epsilon_\mu = 0.11$, then the approximate maximum relative error in $|\Delta f/f_0|$ is $(0.25)(0.07) + (0.25)(0.11) = 0.045$. In this problem, it happens that error propagation is diminished because of the factor 0.25, which is less than unity. In other problems, the situation may be reversed.

### 3.3.2 Evaluation of Integrals

Taylor's theorem is important for developing series approximations for integrals. As an example, consider the integral $I = \int_0^x e^{-t^2}\,dt$ which is closely related to the famous "Error function." This integral *cannot* be expressed exactly as a finite combination of elementary functions. Notice that a "brute force" Taylor expansion of $\exp(-t^2)$ (by computing derivatives) is decidedly lengthy and tedious. However, by taking $t^2 = z$ and using the well-known expansion $\exp(-z) = 1 - z + z^2/2!\ldots$, we find that $\exp(-t^2) = 1 - t^2 + (t^2)^2/2!\ldots$ where the error is less in magnitude than the first neglected term and of the same sign. Substituting the expansion for the integrand produces the result

$$I = x - \frac{x^3}{3} + \frac{x^5}{5 \cdot 2!} - \frac{x^7}{7 \cdot 3!} \cdots \tag{3-22}$$

For $x = 1$ the above equation gives $I = 1 - 1/3 + 1/10 - 1/42 + 1/216\ldots$, which is a very useful result. For $x = 2$ we get $I = 2 - 8/3 + 32/10 - 64/21 + 64/27\ldots$. This is marginally useful, since not too many more terms are needed before the magnitudes of the terms decrease enough to be insignificant. On the other hand, for $x = 10$ we obtain $I = 10 - 1000/3 + 10,000 - 10^7/42 + 10^9/216....$ This series does (eventually) converge, but it is not *useful*, even on a fast computer, because so many terms of large magnitude are required that there may be disastrous roundoff error. For large $x$, a different formulation in terms of an asymptotic expansion is more suitable; this approach will be discussed later.

### 3.3.3 Solution of Ordinary Differential Equations

Another important application of Taylor's formula is in developing approximate solutions to ordinary differential equations. This general topic will be taken up in detail later, but here we illustrate how the procedure can be used to assist in checking a computer solution. Consider the first-order, nonlinear differential equation

$$\frac{dy}{dx} = \frac{2 + 4xy}{3 + y} \tag{3-23}$$

We take the initial condition as $y(1/2) = 5$. This equation is not one of the "standard" types, so we expect that either an approximate analytical solution or a numerical solution (or both) may be desirable. A three-term Taylor approximation of the solution in the vicinity of $x = 1/2$ is given by

$$y(x) = y(1/2) + y'(1/2)(x - 1/2) + \frac{y''(1/2)(x - 1/2)^2}{2!} \tag{3-24}$$

The quantity $y'(1/2)$ is calculated by evaluating the given differential equation at the known values $x = 1/2, y = 5$. This gives $y'(1/2) = [2 + 4(1/2)(5)]/8 = 3/2$. The second derivative, $y''(1/2)$, is calculated by differentiating $y'(x)$ with respect to $x$, taking care to note that $y$ is a function of $x$, and then evaluating the result at $x = 1/2, y = 5, y' = 3/2$. The differentiation gives

$$y'' = \frac{(3 + y)[0 + 4(xy' + 1 \cdot y)] - (2 + 4xy)y'}{(3 + y)^2} \tag{3-25}$$

Evaluation of $y''(x)$ at $x = 1/2, y = 5, y' = 3/2$ yields

$$y''(1/2) = \frac{(3 + 5)[4(1/2) \cdot (3/2) + 5] - (2 + 4 \cdot (1/2) \cdot 5)(3/2)}{(3 + 5)^2} = 83/32 \tag{3-26}$$

Thus, for the three-term Taylor approximation to $y(x)$, we finally get

$$y(x) = 5 + \frac{3}{2}(x - 0.5) + \frac{83}{32 \cdot 2!}(x - 0.5)^2 \tag{3-27}$$

This formula will yield a good approximation to $y(x)$ when $x$ is sufficiently close to $1/2$. As usual, the error of approximation could be *estimated* by the magnitude of the last term retained. Thus, we could use this formula to partially check a numerical solution evaluated very close to $x = 1/2$. If more terms of the Taylor approximation are desired, they may be obtained by calculating the higher derivatives sequentially from the differential equation. That is, for example, to get $y'''$ we would differentiate $y''$ with respect to $x$, and so forth.

## 3.4 DIFFICULT TAYLOR EXPANSIONS

The utility of Taylor expansions is vastly enhanced if we can fairly easily (i) determine a reasonable number of terms for complicated functions and (ii) establish tight error bounds for these approximations. Fortunately, a combination of analysis and computing power (program SERIES) allows this to be accomplished in many problems. Several examples of "difficult" Taylor expansions will be given below. Error estimation and bounding approaches are given later. Details of the "advanced series calculations" together with program SERIES are given in Appendix D. Here, we illustrate some sample computations.

**Example 3.2** $y = \tan x = \sin x/\cos x = d_0 + d_1 x + d_2 x^2$

This function is perhaps the most common one that does not possess an easy Taylor expansion. Since we know the expansions of $\sin x$ and $\cos x$, the series division algorithm (see Appendix D) is probably the simplest way to compute terms of this series. It is convenient to work in terms of the variable $x^2$ by letting $x^2 = z$ and noting that we can write

$$\tan x = x\left\{\frac{\sin x}{x}\right\}/\cos x = x\left\{\frac{1 - z/3! + z^2/5!\ldots}{1 - z/2! + z^2/4!\ldots}\right\} \tag{a}$$

Thus, after cross-multiplication, we can use the division algorithm in terms of $z$ as

$$(c_0 + c_1 z + c_2 z^2 \ldots)\left(1 - \frac{z}{2!} + \frac{z^2}{4!}\ldots\right) = 1 - \frac{z}{3!} + \frac{z^2}{5!}\ldots \tag{b}$$

By equating powers of $z$ on both sides of this equation, the result for the first three nonzero terms, expressed back in terms of $x$, is $\tan x = x + x^3/3 + (2/15)x^5 + O(x^7)$. We could easily obtain more terms of this expansion by using program SERIES, although this would give decimal approximations to the coefficients rather than integer values.

**Example 3.3** $y = \ln(5 + 2\cos x) = b_0 + b_1 x + b_2 x + b_3 x + \cdots$

In a composite function problem such as this, it is often useful to differentiate the function to see if the series for some derivative of the function can be obtained easily. In this problem, the derivative $y'$ is given by

$$y' = \frac{-2\sin x}{5 + 2\cos x} = b_1 + 2b_2 x + 3b_3 x^2 \ldots \tag{a}$$

Thus, we see that $y'$ is expressed as the quotient of the two simple series $-2\sin x = -2x + 2x^3/3!\ldots$ and $(5 + 2\cos x) = 5 + 2(1 - x^2/2! + x^4/4!\ldots) = 7 - x^2 + x^4/12\ldots$. We therefore use the division algorithm with these series and equate with the coefficients of $y'$. In this way, we obtain

$$\ln(5 + 2\cos x) = \ln 7 - \frac{x^2}{7} + \frac{x^4}{(21)(28)} + O(x^6) \tag{b}$$

The notation $O(x^6)$ here means "proportional to $x^6$ when $x$ is small." Further terms could be obtained easily with the use of program SERIES. We could also directly use the program option for the logarithm of a series.

**Example 3.4** $y = \exp(\sin x) = b_0 + b_1x + b_2x^2 + b_3x^3 + \cdots$

Here, we first take logarithms to get $\ln y = \sin x$ and then differentiate to get $y'/y = \cos x$. Then we have

$$y' = b_1 + 2b_2x + 3b_3x^2 \ldots = y\cos x = (b_0 + b_1x \ldots)(1 - \frac{x^2}{2!}\ldots) \tag{a}$$

This approach gives recurrence relations for the $b_k$; through terms of $x^4$, we get

$$y = e^{\sin x} = 1 + x + \frac{x^2}{2} - \frac{x^4}{8} + O(x^5) \tag{b}$$

Once again, we could obtain further terms of this series using program SERIES. Also, we could directly use the program option for the exponential of a series.

Error bounds for the previous three expansions will be developed below by the direct method. Taylor series for many difficult compound functions can be derived through a combination of the algorithms for multiplication, division, raising to a power, and differentiation. Some combination of the options in program SERIES can be used in many of these problems.

## 3.5 ADVANCED ERROR ANALYSIS AND ESTIMATION

In problems involving Taylor expansions, the ideal is to develop an approximation with associated error bounds. In many cases, bounding the approximation error is either impossible or impractical, so only an error estimate (not a guaranteed bound) can be obtained.

### 3.5.1 Sign Change of the Remainder

One simple but important error bound is applicable whenever it is known that the remainder changes sign when some new quantity $a_{n+1}$ is added to the current approximation. In this case, if $S = S_n + R_n = S_n + a_{n+1} + R_{n+1}$ and *if we know that* $\mathrm{R_n}$ *and* $\mathrm{R_{n+1}}$ *have opposite signs*, then both $|R_n|$ and $|R_{n+1}|$ are less than $|a_{n+1}|$, which follows from the fact that $a_{n+1} = R_n - R_{n+1}$ and $R_n$ and $R_{n+1}$ have opposite signs. This principle is independent of whether $S_n$, the sequence of partial sums, eventually converges to $S$. Such a situation occurs, for example, in many alternating series, both convergent and asymptotic.

**Example 3.5 The Geometric Expansion ($x$ Positive)**

By using the identity $1/(1+x) = (1+x-x)/(1+x) = 1 - x/(1+x)$ *recursively*, we obtain, for example, $1/(1+x) = 1 - x + x^2/(1+x) = 1 - x + x^2 - x^3/(1+x)$ and so on. When $x$ is positive, since the remainder changes sign when the new term $x^2$ is added to the approximation $1-x$, we have that the magnitude of the error in either of the two approximations $(1-x)$ or $(1-x+x^2)$ is less than $x^2$. For the first of these approximations, the error is less than the first neglected term, while for the second, it is less than the last term retained. Note that this is true for *any* positive $x$, although the infinite geometric series converges only for $|x| < 1$.

Unfortunately, the sign change of error principle is just that; it does *not* refer to a sign change in a term of a series. Thus, if at a certain point in a series a term changes sign, that in itself says nothing about the error. However, if we have an alternating series *with terms that decrease monotonically in magnitude*, it is easily shown that the sign change of error principle does apply.

### 3.5.2 Use of Inequalities to Bound the Remainder in a Series

As discussed in Appendix C, the two most common tests for establishing the convergence of power series are the ratio test and the integral test. In many problems, it is also possible to bound the error in a series computation using these techniques.

To use the geometric series/ratio test approach, we bound the tail of a power series by construction of a geometric series that is larger than the tail and can be summed exactly. To do this, we need to know an upper bound for the ratio $|a_{n+1}/a_n|$ for all terms in the tail, $\sum_{i=N}^{\infty} a_i x^i$. In general, in convergent processes where the limiting ratio is less than one, this limit will usually be approached from either below or above the final value. We illustrate both common behaviors with examples. The approach is *not* limited to power series.

**Example 3.6 A Series Ratio Coming In from Below Limit**

Consider the series $\sum_{i=1}^{\infty} x^n/n^2$. Calling $a_n = x^n/n^2$, we examine the ratio $a_{n+1}/a_n = x \cdot n^2/(n+1)^2$. Since $x$ is fixed, this ratio clearly approaches the limiting value of $x$ from below as $n \to \infty$, and thus the series converges for $x < 1$. To bound the error incurred by summing up to some point N, we bound the tail as follows:

$$\sum_{i=N+1}^{\infty} \frac{x^n}{n^2} = \left(\frac{x^{N+1}}{(N+1)^2}\right)\left[1 + x\left(\frac{N+1}{N+2}\right) + x^2\left(\frac{N+1}{N+3}\right)^2 \cdots\right] \tag{a}$$

The right-hand side of this expression is less than the geometric series $[x^{N+1}/(N+1)^2](1 + x + x^2 \ldots) = x^{N+1}/[(N+1)^2(1-x)]$. In this case, where the ratio comes in from below, we select the largest ratio, and this corresponds to $n \to \infty$. If we wish six-place accuracy, for example, the error bound shows that the number of terms required depends entirely on the value of $x$. If $x$ is small the convergence is rapid, but if $x$ is near unity it is slow. Note also that since the terms are positive, the sign of the remainder is positive, and this information can be used to improve the known accuracy.

**Example 3.7 A Series Ratio Coming In from Above Limit**

We consider the series $\sum_{i=1}^{\infty} iR^i$. Here, the ratio test yields $a_{i+1}/a_i = (i+1)R/i$, and this quantity approaches the limit $R$ from above as $i \to \infty$. Thus, the series converges for $R < 1$, but the quantity $R$ cannot be used to bound the error incurred by taking a finite number of terms. Instead, we must proceed as follows. A short calculation demonstrates that the largest value of $a_{i+1}/a_i$ in the tail is $R(N+2)/(N+1)$. Therefore, to bound the tail we write

$$\sum_{i=N+1}^{\infty} iR^i = (N+1)R^{N+1}[1 + R(N+2)/(N+1)\ldots] < \\ (N+1)R^{N+1}/\{1 - [R(N+2)/(N+1)]\} < 1 \tag{a}$$

Of course, this bound has a meaning only if $R(N+2)/(N+1) < 1$. Thus, before we can use this procedure at all we must take enough terms to ensure that this condition is met. Since $R < 1$ this is always possible, but for $R$ near 1, $N$ must be quite large, and some kind of acceleration would be extremely useful. On the other hand, for small values of $R$ the series converges very rapidly.

The integral test is used to bound the tail of a series as follows. As shown in Appendix C, the tail of a series $T = \sum_{N+1}^{\infty} a_n$, where the $a_n$ are positive and monotonically decreasing, is bounded by $T \leq \int_N^{\infty} f(x)dx$ where $f(n) = a_n$. In bounding the tail this way, we can always proceed by taking absolute values of $a_n$ (if some $a_n$ happen to be negative), and we can bound an even larger tail if this is simpler. The technique is now illustrated with an example.

**Example 3.8** $T = \sum_{N+1}^{\infty} 1/(n^3 - n^2 + 1)$

In this problem, we could take $f(x) = 1/(x^3 - x^2 + 1)$, but the integral of this function is *not* obvious. Instead, note that

$$T < \sum_{N+1}^{\infty} \frac{1}{(n^3 - n^2)} = \sum_{N+1}^{\infty} \frac{1}{n^3(1 - 1/n)} < \frac{1}{[1 - 1/(N+1)]} \sum_{N+1}^{\infty} \frac{1}{n^3} \tag{a}$$

With this simplified dominating series, it is easy to calculate that

$$T < \frac{1}{[1 - 1/(N+1)]} \int_N^{\infty} \frac{dx}{x^3} = \frac{1}{[1 - 1/(N+1)]} \cdot \frac{1}{2N^2} \tag{b}$$

By summing enough terms directly so that $N$ is large, the bound on the tail can be made small. To actually estimate the tail much more accurately for a given $N$, we could employ the Euler-Maclaurin formula, which is discussed in Appendix F.

### 3.5.3 Experimental Determination of Error Bounds for Taylor Series

In most problems, it is desirable to have a bound for the error in a finite Taylor expansion, but often this calculation is nearly impossible to carry out analytically. However, a straightforward computer algorithm can often accomplish this task in a practical way. The procedure is as follows. Suppose the Taylor approximation for $f(x)$, say $A_n(x)$, is known through the $x^n$ term. Restricting ourselves to some range, $x_{\min} < x < x_{\max}$, for example, we know that the error is equal to $f^{(n+1)}(\xi)x^{n+1}/(n + 1)!$, where $\xi$ is unknown and depends on $x$. We compute the quantity $|f^{(n+1)}(\xi(x))|/(n + 1)!$ *indirectly*, not by differentiation but by computing the known difference $Z \equiv |f(x) - A_n(x)|/|x|^{n+1}$ at various values of $x$ in the desired range. In this way, the *experimental* value of $\max|f^{(n+1)}(\xi)|/(n + 1)! \equiv M$ is determined. When $M$ is known, the error bound in the specified $x$ range is $M \cdot x^{n+1}$. Naturally, this procedure can only be used when numerical values of the function $f(x)$ are known in the desired range. This idea is the basis of the calculation carried out by the error bounds option of program SERIES. One particularly nice feature of this approach is that it produces *best* bounds; that is, bounds that actually hold as an equality for some $x$ in the desired range.

It should be noted that when $x$ is too close to zero, roundoff error will contaminate the calculation of $|f(x) - A_n(x)|$. In this case, if it appears that the desired value of $M$ corresponds to small $x$, then the $M$ value will usually correspond to

$x = 0$. In cases where more accuracy is needed in the calculation for small $x$, we could generally employ an extrapolation based on $Z$ values at larger $x$. Such a calculation is carried out in program SERIES.

**Example 3.9 $\exp(\sin x) = b_0 + b_1x + b_2x^2 + R$ for $0 \le x \le 1$**

Here, we use the three-term Taylor approximation generated earlier ($b_0 = 1, b_1 = 1, b_2 = 1/2$) to directly evaluate $|Z(x)|$ for 30 equally spaced values of $x$ between $x_{max} = 1$ and $x_{min} = 1/30$, using program SERIES. The calculation shows that in the range $0 \le x \le 1$, the maximum of $|Z(x)|$ occurs at $x = 1$ and is equal to 0.1802. The bound for the error in this range can then be written as $|R| \le 0.19\,x^3$. This corresponds to simplifying the $M$ value by making it slightly larger.

**Example 3.10 $\ln(1 + x) = x - x^2/2 + R$ for $0 \le x \le 1$**

In this simple problem, we already know from direct calculation that the best (actually achieved) bound is $|R| \le x^3/3$, corresponding to $\xi = 0$. It is therefore interesting to try program SERIES in this case. With 20 intervals, we get $\max|Z(x)| = 0.321$ for $0.05 \le x \le 1$. This is just barely above $y'''(0)/3! = 1/3 \approx 0.333$, the true value. If we were to *extrapolate* with numerical data to $x = 0$, we would expect to get very close to 0.333. An easy way to refine a calculation such as this one is to repeat it within a smaller range close to zero, say by taking $x_{min} = 0, x_{max} = 0.1$.

## 3.6 COMPUTER PROGRAM

We give here some information regarding the computer program SERIES, which is discussed in Appendix D on advanced series manipulations. A concise listing of the program is also given there. This program, which is quite powerful, has a number of options. For each option the user must specify, in Subroutine Coefficients, the number of terms to be used and values of the coefficients of the input series. In using this program, note that both the input series and the output series are calculated based on the assumption that all series involved contain terms up to $x^n$.

This program also contains an *option* (in Subroutine Function) for the direct calculation of Taylor error bounds. To utilize this option the user must indicate the range of interest by specifying xmin, xmax, and the number of intervals to be used in the calculation. The function whose series is being calculated must be specified either by using explicit computer library functions or by giving numerical function values at the required locations (see Subroutine Function for examples).

## PROBLEMS

**3.1.** Derive the general form for the *Maclaurin* expansion (expansion point $a = 0$) for the following functions:

(i) $\sin x$

(ii) $\cos x$

(iii) $(1 + x)^p$

(iv) $\ln(1+x)$
(v) $\tan^{-1} x$

**3.2.** Give a bound for the approximation error for the Maclaurin expansions of problem 3-1.

**3.3.** Give a simple, direct derivation of the Maclaurin expansion for a function of two variables by using the one-dimensional result in two steps. (Hint: $y(x_1+h_1, x_2+h_2) = y(x_1, x_2+h_2)\ldots$ and $y(x_1, x_2+h_2) = y(x_1, x_2)\ldots$).

**3.4.** It is desired to determine the uncertainty in the Nusselt number for developed transport in a tube due to uncertainties in the fluid velocity, the tube diameter, and the physical properties of the fluid. Suppose that the formula for the Nusselt number is given by $\text{Nu} = C\text{Re}^a\text{Pr}^b$ where $C, a$, and $b$ are known constants, $\text{Re} = Du\rho/\mu$ is the Reynolds number and $\text{Pr} = \mu C_p/k$ is the Prandtl number. Estimate the relative error in the Nusselt number corresponding to the following possible percentage uncertainties.

| | |
|---|---|
| $\epsilon_u$ | 5% |
| $\epsilon_\mu$ | 10% |
| $\epsilon_k$ | 3% |
| $\epsilon_\rho, \epsilon_{Cp}$ | 1% |
| $\epsilon_D$ | 0.1% |

**3.5.** Determine a bound for the relative error in the Nusselt number of problem 3-4 if only errors in velocity ($\epsilon_u = 5\%$) and viscosity ($\epsilon_\mu = 10\%$) are taken into account.

**3.6.** For turbulent flow in a circular pipe, the friction factor ($f$) versus Reynolds number (Re) relationship is sometimes expressed as $f = 0.046\ \text{Re}^{-0.2}$, where $\text{Re} = (Lu\rho/\mu)$. At a Reynolds number of 50,000, it is desired to calculate the relative uncertainty in the friction factor $(f - f_0)/f_0$ corresponding to relative uncertainties in $L$ and $\mu$ of $(L - L_0)/L_0 = 0.05$ and $(\mu - \mu_0)/\mu_0 = 0.15$.

**3.7.** The nonlinear differential equation $dy/dx = x^2 + y^2$ with $y(0) = 1$ is to be solved numerically (it can't be solved analytically). To check the computer program, it has been decided to compute the numerical solution at $x = 0.01$ and compare this with a three-term Taylor approximation. The numerical solution was found to be $y(0.01) = 1.01010$ to six figures. How does this compare with the Taylor approximation? Does the numerical solution seem to be correct?

**3.8.** Generate the first five terms of the Taylor expansion of the integral $\int_0^z dT/(1+T^4)$. This problem arises in calculations involving cooling by radiation heat transfer.

**3.9.** The normal probability integral, which is used in statistics, is given by the expression

$$I(x) = \frac{1}{\sqrt{2\pi}} \int_0^x e^{-t^2/2} dt$$

(a) Use Taylor's theorem to develop a series approximation for this function.
(b) Use the series approximation to evaluate $I(1)$ and $I(2)$ to three significant figures.

**3.10.** The following differential equation arises in the modeling of a batch reactor with a heterogeneous catalytic reaction.

$$\frac{dC}{d\theta} = \frac{-k_1 C}{1 + k_2 C}$$

where $k_1 = 2, k_2 = 0.1$, and $C(0) = 1$. Show that the first four terms of the Taylor expansion of this differential equation are given by

$$C(\theta) = 1 - 1.8182\theta + 3.0053\frac{\theta^2}{2!} - 3.9739\frac{\theta^3}{3!}$$

**3.11.** Use program SERIES to determine the first five terms of the Maclaurin expansion of the function $e^{\cos x}$. Partially verify this result by calculating the first three terms by hand. Use program SERIES to determine a direct error bound for the five term series on the interval $0 \le x \le \pi/2$.

**3.12.** Use program SERIES to calculate three nonzero terms of the integral $\int_0^z \left(\frac{\tan x}{1+x}\right)^{1/2} dx$. Check the first term of this result by hand.

**3.13.** Produce an upper bound to the tail of the series $\sum_{n=N+1}^{\infty} 1/(n^2 - 3)$.

**3.14.** Given the series $F = 1 - x^2/2 + x^4/4 - x^6/6\ldots$, determine a simple closed-form expression for $F$ using series manipulations. (Hint: Differentiate and try to find a simple differential equation.)

**3.15.** The *reversion* of a Maclaurin series $y = a_1x + a_2x^2\ldots$ is the series expressed for $x$ in powers of $y$ as $x = d_1y + d_2y^2\ldots$. This procedure is important for obtaining asymptotic solutions of nonlinear algebraic equations and asymptotic representations of integrals. Program SERIES contains a reversion option to carry out this calculation. Consider the series for $y = \tan^{-1} x = x - x^3/3 + x^5/5\ldots$, and use the reversion option to determine the first several terms of the series for $x = \tan y$. Compare the results with the series for $\tan x$ given in the text.

## REFERENCES


BENNETT, C. O. and MYERS, J. E. (1982), *Momentum, Heat and Mass Transfer*, 3rd ed., McGraw-Hill, New York

BROMWICH, T. J. I'A. (1964), *Infinite Series*, 2nd ed., Macmillan, New York

PRESS, W. H., FLANNERY, B. P., TEUKOLSKY, S. A., and VETTERLING, W. T. (1986), *Numerical Recipes*, Cambridge Univ. Press, New York

SCHEID, F. (1988), *Numerical Analysis*, 2nd ed., McGraw-Hill, New York

SPIEGEL, M. (1963), *Advanced Calculus*, Schaum, New York

SQUIRE, W. (1970). *Integration for Engineers and Scientists*, Elsevier, New York

WHEELON, A. D. (1968), *Tables of Summable Series and Integrals Involving Bessel Functions*, Holden-Day, San Francisco

# 4

# Acceleration Techniques for Series

The Taylor series representation becomes a much more powerful tool when we can conveniently apply it to complicated functions by means of computer assistance. It is even more powerful when certain "acceleration" methods can easily be applied with a given number of terms to yield improved results. This is especially true when it is very difficult to obtain very many terms of the Taylor expansion. This may occur, for example, in the case of a regular perturbation expansion of a differential equation or in the series solution of a nonlinear differential equation. The acceleration methods to be discussed here can all be applied when only a limited number of terms of the Taylor series is known. They often (but not always!) dramatically improve convergence; sometimes, the radius of convergence of the Taylor series is increased, and the acceleration methods may even cause a divergent (asymptotic) series to become convergent. Sometimes it is useful to combine the transformations. For instance, in the case of an alternating series, an original power series might be Euler-transformed, then converted to a Padé approximation, and then accelerated with a Shanks transformation.

In some problems, although admittedly not as often as we would like, we know *all* the terms of a slowly convergent series. In such cases we can, of course, use acceleration methods, but it is often better to employ summation techniques such as the Euler-Maclaurin summation formula or the Laplace transform method of Wheelon [1968].

**TABLE 4.1** EULER TRANSFORMATION COEFFICIENTS
THE FIRST FIVE COEFFICIENTS OF EULER'S TRANSFORMATION
$a_0 + a_1 t + a_2 t^2 + a_3 t^3 + a_4 t^4 + \ldots = d_0 + d_1 u + d_2 u^2 + d_3 u^3 + d_4 u^4 + \ldots$
$u = t/(p+t)$, $a_0, a_1, a_2, a_3, a_4$ ARE KNOWN

| |
|---|
| $d_0 = a_0$ |
| $d_1 = a_1 p$ |
| $d_2 = a_1 p + a_2 p^2$ |
| $d_3 = a_1 p + 2a_2 p^2 + a_3 p^3$ |
| $d_4 = a_1 p + 3a_2 p^2 + 3a_3 p^3 + a_4 p^4$ |

## 4.1 THE EULER TRANSFORMATION

The Euler transformation is primarily suited to alternating power series or series that involve at least some sign changes, particularly those whose original form is only slowly convergent (or perhaps even divergent). Theoretically, if the original series converges, then the Euler transformed series does also, but not *necessarily* more rapidly. In practice, one usually applies the transformation and either observes whether the convergence seems to be improved or else makes a direct error calculation, as in program SERIES. It is sometimes convenient to apply the transformation to the remainder or tail of the series after some terms have been summed directly. Also, the transformation can be *repeated* if desired, and this is sometimes useful.

The Euler transformation can be developed through differencing procedures or by means of an algebraic change of variable. Here, we will concentrate on the algebraic approach. Given the power series $S = a_0 + a_1 x + a_2 x^2 + \cdots + a_n x^n$, we introduce the change of variable $y \equiv x/(p+x)$. Here, $p$ is an arbitrary positive parameter, which in the classical Euler transformation is taken to be unity. Since we can easily determine the inverse transformation to be $x = py/(1-y)$, we can substitute for $x$ in favor of $y$ in the series for $S$ to get

$$S(x) = a_0 + a_1 p \frac{y}{1-y} + a_2 p^2 \left(\frac{y}{1-y}\right)^2 + \cdots + a_n p^n \left(\frac{y}{1-y}\right)^n \tag{4-1}$$

In this equation we can expand each $1/(1-y)^k$ term in a Taylor series and collect terms to obtain the Euler-transformed sum as

$$s = d_0 + d_1 y + d_2 y^2 + \cdots + d_n y^n \tag{4-2}$$

Here, the factor $y = x/(p+x)$ now varies between zero and unity for all positive values of $x$. The values of the first few $d_k$ Euler coefficients found in this brute force manner are given in Table 4.1.

A considerably neater Euler algorithm can be developed by following the approach of Meksyn [1961]. The derivation given is for the most common case of $p = 1$. Let $S(x) = \sum_{n=0}^{\infty} a_n x^{n+1}$. With $y = x/(1+x)$ so that $x = y/(1-y)$, we have

$$S(x) = \sum_{n=0}^{\infty} a_n x^{n+1} = \sum_{n=0}^{\infty} a_n y^{n+1}(1-y)^{-(n+1)} \tag{4-3}$$

By Taylor's theorem, for small $y$,

$$(1-y)^{-(m+1)} = \sum_{n=0}^{\infty} \binom{m+n}{n} y^n \tag{4-4}$$

where $\binom{n}{i} \equiv n!/[(n-i)! \cdot i!]$ is the usual binomial coefficient. Then

$$S(x) = \sum_{n=0}^{\infty} a_n \sum_{k=0}^{\infty} \binom{n+k}{k} y^{k+n+1} = \sum_{j=0}^{\infty} y^{j+1} \sum_{n=0}^{j} \binom{j}{j-n} a_n \tag{4-5}$$

Equation (4-5) comes from using the new summation index $n + k = j$ in the sum on the left, followed by a careful change in the order of summation. Some further straightforward manipulation finally yields the concise result

$$S(x) = \sum_{j=0}^{\infty} b_j y^{j+1}, \qquad b_j = \sum_{i=0}^{j} \binom{j}{i} a_i \tag{4-6}$$

It should be noted that the above result, when applied to the usual series $a_0 + a_1 x + \cdots + a_n x^n$, starts with the *second* term, since the series on which the formula is based, Equation (4-3), has an extra factor of $x$. The formula in Equation (4-6) is the basis of the Euler transformation option in computer program SERIES. The algorithm given above was modified slightly to permit arbitrary positive values of $p$.

**Example 4.1 Develop an Euler Transformed Series for $y = \ln(1 + x) = x - x^2/2 + x^3/3 \ldots$**

For this simple but important function, we can easily develop two different Euler series that are interesting and useful, corresponding to $p = 1$ and $p = 2$. This could be done by using the Euler transformation option of program SERIES, but for this function it is instructive to see how the Euler algebraic change of variable can be applied directly to the simple integral representation $\ln(1 + x) = \int_0^x dt/(1 + t)$. First, we arbitrarily take $p = 1$ (the standard value of $p$) and introduce the Euler change of variable $t/(1 + t) = u$. This produces

$$\ln(1+x) = \int_{u=0}^{W} \frac{du}{1-u} = W + \frac{W^2}{2} + \frac{W^3}{3} + \cdots + \frac{W^n}{n} + R \tag{a}$$

The integrated result in Equation (a) is obtained by expanding $1/(1-u)$ in a geometric series. Here $W = x/(1 + x)$ and $R$ is positive and less than $(1 + x) \cdot W^{n+1}/(n + 1)$, since $1/(1-u) < 1/(1 - W)$ in the integration range. This new series converges for *all* positive $x$, in contrast to the usual Taylor series for $\ln(1 + x)$, which converges only for $x \leq 1$ in the case of positive $x$.

It is also very interesting to see that if we use the Euler transformation with $p = 2$, we obtain a series which converges much faster than that of Equation (a). Once again, for the function $\ln(1 + x)$, the Euler series is most easily derived directly from the integral representation using the change of variable $t/(2 + t) = v$. Then we get

$$\ln(1+x) = 2\int_{v=0}^{Y} \frac{dv}{1-v^2} = 2\left\{Y + \frac{Y^3}{3} + \frac{Y^5}{5} + \cdots + \frac{Y^{2n+1}}{2n+1} + R\right\} \tag{b}$$

The integrated result in Equation (b) is obtained by expanding $1/(1-v^2)$ in a geometric series. In Equation (b), $Y = x/(2 + x)$, $R$ is positive and $R < (2 + x) \cdot Y^{2n+3}/(2n + 3)$, because $1/(1 - v^2)$ is less than $1/(1 - Y^2)$ in the range of integration. This series also converges for *all* positive $x$. The two different integral representations of $\ln(1 + x)$

**TABLE 4.2** COMPARISON OF TAYLOR APPROXIMATION AND EULER TRANSFORMATION FOR $\ln(1+x) \approx x - x^2/2 + x^3/3 - x^4/4$

| $x$ | Exact | Taylor | Euler ($p = 1$) | Euler ($p = 2$) |
|---|---|---|---|---|
| 0.5 | 0.4055 | 0.4010 | 0.4043 | 0.4053 |
| 1.0 | 0.6931 | 0.5833 | 0.6823 | 0.6914 |
| 2.0 | 1.099 | −1.333 | 1.037 | 1.083 |

**TABLE 4.3** THE EULER TRANSFORMATION APPLIED TO $\tan^{-1} x = y$ RESULTS FROM PROGRAM SERIES

$n = 9$

| | |
|---|---|
| $a_0 = 0$ | $a_5 = 1/5$ |
| $a_1 = 1$ | $a_6 = 0$ |
| $a_2 = 0$ | $a_7 = -1/7$ |
| $a_3 = -1/3$ | $a_8 = 0$ |
| $a_4 = 0$ | $a_9 = 1/9$ |

$x = 1$

| $p$ | $y_{\text{Exact}}$ | $y_{\text{Taylor}}$ | $y_{\text{Euler}}$ | Euler last term |
|---|---|---|---|---|
| 1 | 0.7854 | 0.8349 | 0.7820 | $3.5 \times 10^{-3}$ |
| 2 | " | " | 0.7894 | $-4.1 \times 10^{-3}$ |

$x = 2$

| $p$ | $y_{\text{Exact}}$ | $y_{\text{Taylor}}$ | $y_{\text{Euler}}$ | Euler last term |
|---|---|---|---|---|
| 1 | 1.107 | 44.34 | 1.066 | $4.6 \times 10^{-2}$ |
| 2 | " | " | 1.237 | $-1.6 \times 10^{-1}$ |

for $p = 1$ and $p = 2$ show clearly in this example that $p = 2$ is superior to $p = 1$. In most problems, an integral representation of the function to be expanded is not available. In this case, a "best" value of $p$ in the Euler transformation might be obtained experimentally, if desired. A comparison of the four-term Taylor expansion and the four-term Euler transformation results for $\ln(1 + x)$ is shown in Table 4.2. The improvement due to the Euler transformation is dramatic; this is fairly typical of the behavior of the transformation when it is applied to slowly convergent alternating series.

**Example 4.2 Develop an Euler Transformed Series for $y = \tan^{-1}(x) \approx x - x^3/3 + x^5/5 - x^7/7 + x^9/9$**

For this familiar and important function, we have a simple integral representation $\tan^{-1}(x) = \int_0^x dt/(1 + t^2)$. But in this case, the Euler change of variable in the integral does not yield an especially convenient result. We therefore use computer program SERIES to determine the Euler transformation results corresponding to the Taylor series expansion of $1/(1 + t^2)$ through terms of $t^8$. Since a "good" value of $p$ is not obvious, we do the calculations for both $p = 1$ and $p = 2$ and compare the results in Table 4.3. As seen in the table, for this problem, both at $x = 1$ and $x = 2$, $p = 1$ seems somewhat better than $p = 2$. Apparently, the Euler approximations here are quite superior to those of the ordinary Taylor expansion, although *exactly the same information is used to generate all of the approximations.* The Taylor expansion (if considered as an infinite series) *diverges* for $x > 1$, whereas the Euler approximation does not. Also, the Euler error estimates (last terms) are seen to be conservative and also reflective of the goodness of the approximation.

## 4.2 PADÉ APPROXIMATION

Another very useful acceleration technique for power series is known as Padé approximation. This method is a *rational* approximation (ratio of polynomials) of a function, which, when expanded in a Taylor series, agrees with a certain number of terms of the original Taylor series of the function. To illustrate the Padé procedure, consider the case where we know only the first three terms of the Taylor expansion $a_0 + a_1x + a_2x^2$. Then we write, for example,

$$a_0 + a_1x + a_2x^2 \cong \frac{p_0 + p_1x}{1 + q_1x} \tag{4-7}$$

This is one particular Padé approximation (1,1) which can be developed in the present situation, corresponding to a first-degree polynomial in both the numerator and denominator. In general, the number of coefficients in the given Taylor approximation should correspond to the number of unknown Padé coefficients that must be determined. To proceed, we *could* expand the right-hand side of Equation (4-7), by the geometric series, through terms involving $x^2$, as $p_0 + x(-p_0q_1 + p_1) + x^2(p_0q_1^2 - p_1q_1)$. Then we could equate coefficients on both sides of the resulting equation to obtain nonlinear algebraic equations for $p_0$, $p_1$, $q_1$. In this particular situation these equations are easy to solve, but, in general, it is computationally much easier to find the coefficients by *multiplying* the polynomial of the denominator in Equation (4-7) by the given Taylor expansion and equating powers in the resulting equation *through terms involving* $x^2$. In the present situation, this procedure gives

$$a_0 + x(a_0q_1 + a_1) + x^2(a_1q_1 + a_2) + 0(x^3) = p_0 + p_1x \tag{4-8}$$

This might at first seem different from the previous approach, since now there is a term proportional to $x^3$ on the left and only terms through $x$ on the right in Equation (4-8). But if we now equate coefficients of $x^0$, $x$, $x^2$ in Equation (4-8), we get the following relationships:

$$x^2: \quad a_1q_1 + a_2 = 0 \tag{4-9}$$

$$x\ : \quad a_0q_1 + a_1 = p_1 \tag{4-10}$$

$$x^0: \quad a_0 = p_0 \tag{4-11}$$

This yields a set of uncoupled linear equations for $q_1$, $p_1$, $p_0$ provided we solve them in the given order. The solution is simply $q_1 = -a_2/a_1$, $p_1 = a_0(-a_2/a_1) + a_1$, $p_0 = a_0$. With these coefficients known, substitution back into Equation (4-7) allows the (1,1) Padé approximation to be calculated. It can be verified directly (in this simple case) that the coefficients determined in this way are exactly the same as those which would result from solving the set of nonlinear algebraic equations referred to above.

**Example 4.3 Develop a Padé Approximation for**
**$\exp(-x) \approx 1 - x + x^2/2 \approx (p_0 + p_1 x)/(1 + q_1 x)$**

For this problem, cross-multiplication as illustrated above gives $q_1 = 1/2$, $p_1 = -1/2$, $p_0 = 1$. This in turn yields the (1,1) Padé approximation for $\exp(-x)$, namely $[1-(x/2)]/[1+(x/2)]$. For example, with $x = 1$, the Taylor expansion gives $\exp(-1) \approx 1 - 1 + 1/2 = 0.5$, whereas the Padé approximation gives $\exp(-1) \approx (1 - 1/2)/(1 + 1/2) = 1/3 = 0.333$. The true value of $\exp(-1) = 0.368$, which is much closer to the Padé value.

The preceding simple example was used to illustrate the calculation method for Padé approximations, along with typical behavior. In general, we will be dealing with a Taylor expansion involving $(n + 1)$ terms, such as $a_0 + a_1 x \ldots + a_n x^n$. Then we can normally determine a $(k, m)$ Padé approximation of the form

$$S(x) = a_0 + a_1 x + \cdots + a_n x^n \approx \frac{p_0 + p_1 x + \cdots + p_k x^k}{1 + q_1 x + \cdots + q_m x^m} \tag{4-12}$$

which will agree in its Taylor expansion with all $(n + 1)$ terms of the given Taylor approximation. The $(k, m)$ Padé approximation in Equation (4-12) corresponds to a degree $k$ polynomial in the numerator and a degree $m$ polynomial in the denominator. Since the number of unknowns $(p_i, q_i)$ is necessarily equal to the number of terms of the known Taylor approximation, we must have $(n + 1) = (k + 1) + m$, or $n = k + m$. With this definition, the Taylor approximation is just a special case of Padé, namely $(n, 0)$. By far the simplest way to determine the Padé coefficients $(p_i, q_i)$ is to generate linear equations for them by cross-multiplication and equating coefficients of terms through $x^n$. In this way, it is not difficult to show that

$$\sum_{i=0}^{r} a_i q_{r-i} = p_r \tag{4-13}$$

In Equation (4-13), $q_0 = 1; r = 0, 1, \ldots, n$ with the restrictions that $q_j = 0$ for $j > m$ and $p_j = 0$ for $j > k$. Then, to get $(p_i, q_i)$, we solve the last $m$ equations first (for $q_1, \ldots, q_m$), and subsequently solve the first $(k + 1)$ equations for $p_0, p_1, \ldots, p_k$. This algorithm is the basis of program PADE.

It has been found by experience that Padé approximation is usually no worse than ordinary Taylor expansion, and is often *much better*. It is known empirically that it is best to keep the degrees of the numerator and denominator in a Padé approximation equal or nearly so. Thus, if $n$ (the degree of the known Taylor approximation) is even, we normally take $k = m = n/2$. If $n$ is odd, we usually take $m = k + 1 = (n + 1)/2$. Typical types of Padé approximations are shown in Table 4.4.

**Example 4.4 Develop a (2,2) Padé Approximation for**
**$y = \ln(1 + x) = x - x^2/2 + x^3/3 - x^4/4$**

For this problem the (2,2) Padé approximation, obtained as described above (or from program Pade), is

$$\ln(1 + x) = \frac{x + x^2/2}{1 + x + x^2/6} \tag{a}$$

In Table 4.5 and Figure 4.1, we see a comparison between this Padé approximation $P(2, 2)$ and the corresponding Taylor approximation T. Padé is seen to be much better

**TABLE 4.4** EXAMPLES OF PADÉ APPROXIMATION

| Taylor | Padé | |
|---|---|---|
| $a_0 + a_1 x$ | $p_0 + p_1 x$ (Taylor) | (1, 0) |
| | $p_0/(1 + q_1)$ | (0, 1)* |
| $a_0 + a_1 x + a_2 x^2$ | $p_0 + p_1 x + p_2 x^2$ (Taylor) | (2, 0) |
| | $(p_0 + p_1 x)/(1 + q_1 x)$ | (1, 1)* |
| | $p_0/(1 + q_1 x + q_2 x^2)$ | (0, 2) |
| $a_0 + a_1 x + a_2 x^2 + a_3 x^3$ | $p_0 + p_1 x + p_2 x^2 + p_3 x^3$ (Taylor) | (3, 0) |
| | $(p_0 + p_1 x + p_2 x^2)/(1 + q_1 x)$ | (2, 1)* |
| | $(p_0 + p_1 x)/(1 + q_1 x + q_2 x^2)$ | (1, 2)* |
| | $p_0/(1 + q_1 x + q_2 x^2 + q_3 x^3)$ | (0, 3) |

$p_i, q_i$ in the various approximations are generally different.

*The best approximations in practice usually correspond to cases where numerator and denominator differ in degree by no more than one.

**TABLE 4.5** COMPARISON OF TAYLOR (T) AND PADÉ (P) APPROXIMATION FOR $\ln(1 + x)$; DEGREE 4

$$T = x - x^2/2 + x^3/3 - x^4/4$$

$$P(2,2) = \frac{x + x^2/2}{1 + x + x^2/6}$$

| $x$ | Exact | $T$ | Error in $T$ | $P(2,2)$ | Error in $P(2,2)$ |
|---|---|---|---|---|---|
| 0.2 | 0.1823 | 0.1823 | 0 | 0.1823 | 0 |
| 0.5 | 0.4055 | 0.4010 | $4.5 \times 10^{-3}$ | 0.4054 | $1 \times 10^{-4}$ |
| 1.0 | 0.6931 | 0.5833 | $1.1 \times 10^{-1}$ | 0.6923 | $8 \times 10^{-4}$ |
| 2.0 | 1.099 | −1.333 | bad | 1.091 | $8 \times 10^{-3}$ |
| 9.0 | 2.303 | −1429 | bad | 2.106 | $2 \times 10^{-1}$ |

than Taylor in this problem. It is especially interesting to note that the Padé values are rather good far beyond the point where the Taylor approximation is convergent (i.e., $x \leq 1$).

**Example 4.5 Develop a Padé Approximation for $y = \tan x \approx x + x^3/3 + (2/15)x^5$**

This positive term series would not be suitable for the Euler transformation. However, since the tangent function has a zero denominator for $x = \pi/2 \approx 1.57$, we would expect that Padé approximation might do rather well, since its structure allows this kind of behavior. We illustrate here a procedure that is sometimes useful in modifying the original function prior to doing the Padé approximation. We know that $\tan x = \sin x / \cos x$, so that if we consider the function $(\tan x)/x$ we see that, since $(\sin x)/x$ and $\cos x$ both have series expansions in terms of $x^2$, we can write the Padé approximation in the form

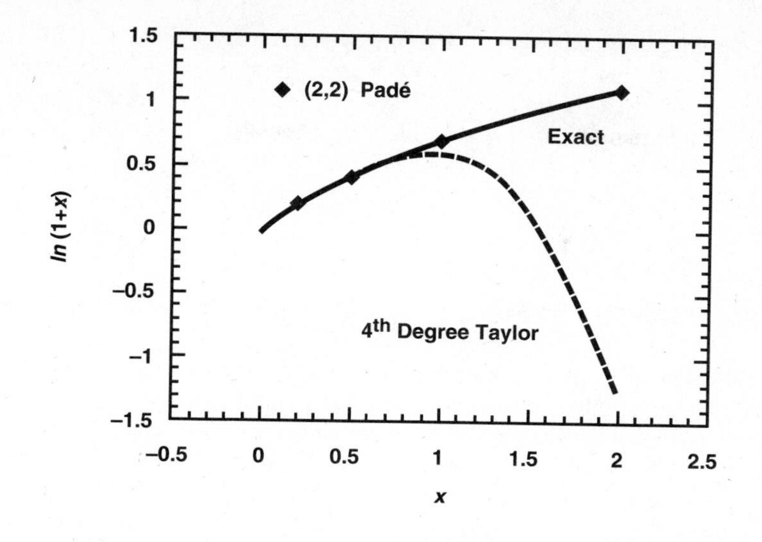


**Figure 4.1.** Comparison of Taylor and Padé Approximations for ln (1 + $x$)

**TABLE 4.6** VALUES OF TAN ($x$) FROM PROGRAM PADE

$$\tan(x) \approx x + x^3/3 + 2x^5/15$$

$$\text{or } \frac{\tan(x)}{x} = 1 + x^2/3 + 2x^4/15 = 1 + z/3 + 2z^2/15, \ z = x^2$$

$$\frac{\tan(x)}{x} = \frac{p_0 + p_1 z}{1 + q_1 z} \qquad p_0 = 1, \ p_1 = -6.66667 \times 10^{-2}, \ q_1 = -0.4$$

| $x$ | $z$ | Taylor | % Error | Padé (1, 1) | % Error | Exact |
|---|---|---|---|---|---|---|
| 0.1 | 0.01 | 0.1003 | 0 | 0.1003 | 0 | 0.1003 |
| 0.5 | 0.25 | 0.5458 | 0.09 | 0.5463 | 0 | 0.5463 |
| 1.0 | 1.00 | 1.467 | 6 | 1.556 | 0.06 | 1.557 |
| 1.2 | 1.44 | 2.108 | 18 | 2.559 | 0.5 | 2.572 |

$$\frac{\tan x}{x} \cong 1 + \frac{z}{3} + \left(\frac{2}{15}\right) z^2 \cong \frac{1 - (1/15)x^2}{1 - (2/5)x^2} \tag{a}$$

where $z = x^2$. The above equation is just the (1,1) Padé result for the $z$ series, with $z$ then replaced by $x^2$. The calculation was done by using program PADE. The results are shown in Table 4.6, where it is seen that the Padé approximation is quite superior to that of Taylor.

Padé approximation can be very useful for series regardless of whether or not there are sign changes (in contrast to the Euler transformation). There is no easy way (even in principle) to bound the error. Generally, errors are estimated either by direct comparison with the function being approximated, or else through the brute-force procedure of comparing consecutive Padé approximations (e.g., (2,2) and (2,3)) and hoping the error in the best approximation (2,3) is less in magnitude than the difference between the two. Program PADE contains a direct error calculation option which is virtually identical to that for Taylor series in program SERIES. It should be mentioned that one principal drawback to a Padé approximation would appear to be that it cannot easily be differentiated or (especially) integrated, in contrast to the original Taylor approximation. While this is true, it is useful to observe that we can easily develop Padé approximations *directly* for the derivative and integral of the original Taylor expansion. A useful discussion of Padé approximation can be found in Bender and Orszag [1978].

## 4.3 SHANKS TRANSFORMATION

This very general sequence-acceleration transformation (which is discussed in Appendix E) can be applied to series through application to the sequence of partial sums. Here, we consider the simplest and most important form of the result, which is based on a "geometric series" model of the error in the convergence of the sequence. Thus, we assume that $S = S_n + C \cdot \lambda^n$, where $S$ is the unknown sum, $S_n$ is the $n$th partial sum of the series, and $C$ and $\lambda$ are constants (hopefully!). As shown in Appendix E, the three unknowns $S$, $C$, and $\lambda$ are determined by using three different known values of $S_n$ (e.g., $S_{n-1}$, $S_n$, and $S_{n+1}$) to give

$$S = S_{n+1} + \frac{\lambda}{1-\lambda}(S_{n+1} - S_n) \quad (\lambda \neq 1) \tag{4-14}$$

where $\lambda = (S_{n+1} - S_n)/(S_n - S_{n-1})$. It is very convenient that this result is independent of a particular value of $n$. The Shanks transformation is mainly useful where convergence is slow but nearly geometric; it can be useful for both positive and sign-changing series. Sometimes, it is useful to *repeat* the transformation.

**Example 4.6 Develop a Shanks Transformation for $y = \ln(1 + x) \approx x - x^2/2 + x^3/3 - x^4/4$**

In this problem, we have five terms of the Taylor expansion (the first term is zero) so that we can apply one Shanks transformation three times. This is done as follows. We first use $S_1$, $S_2$, and $S_3$, obtained from the sum of one, two, and three terms, respectively, to get the first Shanks improvement. Then we use $S_2$, $S_3$, and $S_4$ obtained from the sum of the first two, first three, and first four terms to get the second Shanks improvement. Finally, we get the third Shanks improvement using $S_3$, $S_4$, and $S_5$. This process produces three elements in the sequence of once-transformed partial sums, which is exactly the amount of information needed to apply a second Shanks transformation one time. For $x = 1$, Table 4.7 shows the results of these calculations. We see that a single Shanks transformation produces results which are *dramatically* more accurate than the Taylor approximation. A second transformation produces a considerable refinement. Program SHANKS is set up to allow as many repeated "Shanks transformations" as the user desires, up to the point where the data is exhausted. Although we do not show it here, fewer than ten terms of the Taylor expansion, when used with three Shanks transformations, produce the value of ln 2 with six-figure accuracy. We may observe that, for this problem, Shanks and Padé (Table 4.5) produce nearly the same improvement over Taylor.

The error involved in using Shanks transformation can be estimated by (i) direct calculation, (ii) brute-force comparison of elements of the transformed sequence, or

**TABLE 4.7** COMPARISON OF TAYLOR APPROXIMATION AND SHANKS TRANSFORMATION FOR $\ln(1+x) \approx x - x^2/2 + x^3/3 - x^4/4$

| $x$ | Exact | Taylor | One Shanks | Two Shanks |
|---|---|---|---|---|
| 1 | 0.6931 | 0.5833 | 0.6667<br>0.7<br>0.6905 | 0.6926 |
| 2 | 1.099 | −1.333 | 1.0<br>1.143<br>1.067 | 1.093 |

(iii) by taking the calculated value of $C\lambda^n = [\lambda/(1-\lambda)](S_{n+1} - S_n)$ as a conservative estimate of the error in using the Shanks value, Equation (4-14), as an approximation to the unknown $S$. As with the Euler transformation and Padé approximation, Shanks transformation often causes a divergent series to converge. Useful discussion of Shanks transformation can be found in [Bender and Orszag, 1978].

## 4.4 SUMMATION TECHNIQUES FOR SERIES

Summation techniques refer to methods to calculate a sum based on a knowledge of *all* (or the general term) of a series, in contrast to the previously discussed acceleration techniques which can be used when only a finite number of terms of the Taylor series is known. Also, the summation techniques are not necessarily limited in application to Taylor series expansions, although this will be our primary focus.

### 4.4.1 Integration

There are very important connections between the summation of a positive discrete function and the integral of the continuous function that coincides with the discrete function at integer values. This is discussed in calculus in relation to the integral test for the convergence of series, which is generally more powerful than the ratio test. In Appendix C it is shown that for positive $f(x)$, if $a_n = f(n)$ and if

$$a_{n+1} \le f(x) \le a_n \text{ for all } n > N$$

then

$$\sum_{N+1}^{\infty} a_n < \int_N^{\infty} f(x)\,dx$$

This result is used to bound the tail of a series numerically or to see whether the tail $\to 0$ as $N \to \infty$, as is required for convergence. The foregoing relationship is based on the interpretation of lower-bounding rectangular approximation of the integral. For numerical purposes, a much more useful result is based on interpretation of the integral in terms of both upper- and lower-bounding rectangles. This result uses the two-sided inequality above to show that, for the same assumptions,

$$\sum_{i=N+1}^{\infty} a_i \le \int_N^{\infty} f(x)\,dx \le \sum_{i=N}^{\infty} a_i \tag{4-15}$$

This inequality can be rearranged to give

$$\int_N^\infty f(x)\,dx \le \sum_{i=N}^{\infty} a_i \le \int_N^\infty f(x)\,dx + a_N \tag{4-16}$$

Use of these upper and lower bounds together sharpens the estimate of the tail $\sum_{N+1}^{\infty} a_i$.

**Example 4.7 Determine the Sum for the Series $S = \sum_{i=1}^{\infty} \frac{1}{i^2}$**

If we sum the first nine terms directly, the tail is $\sum_{i=N}^{\infty} 1/i^2$, where $N = 10$. Since $1/x^2$ is uniformly decreasing, we have from Equation (4-16) that $\sum_{i=N}^{\infty} 1/i^2 \le \int_N^\infty dx/x^2 + 1/N^2$, and since $\int_N^\infty dx/x^2 = 1/N$, we see that the tail of the series $\to 0$ as $N \to \infty$, so the series converges. Using the double inequality given in Equation (4-16), we see that $\int_{10}^\infty dx/x^2 \le \sum_{i=10}^{\infty} 1/i^2 \le \int_{10}^\infty dx/x^2 + 1/10^2$. This translates into $0.1 \le \sum_{i=10}^{\infty} 1/i^2 \le 0.11$. By subtracting $(1/2)(0.11 - 0.1) = 0.005$ from all members of this inequality, we can write $\sum_{i=10}^{\infty} 1/i^2 = (0.1 + 0.11)/2 \pm 0.005$. This yields a far more accurate value of the series error than the simple upper bound 0.11. Improved knowledge of the error in turn allows an improved estimate of the sum.

### 4.4.2 The Euler-Maclaurin Summation Formula

The Euler-Maclaurin summation formula is probably the most powerful method for accelerating convergence of a series through integration. This famous classical result, which is derived in Appendix F, is also very useful for the numerical evaluation of integrals. For purposes of *summation*, we express the Euler-Maclaurin formula as

$$\sum_{i=m}^{N} f(i) = \int_{x=m}^{N} f(x)\,dx + \frac{f(m)+f(N)}{2} + \frac{f'(N)-f'(m)}{12} - \frac{f'''(N)-f'''(m)}{720} + \cdots + R \tag{4-17}$$

This is the trapezoidal form of the Euler-Maclaurin formula, which can be continued to include more terms. However, if more terms are included and the result is interpreted as an infinite series, the series often does not converge. For this reason, and also because in practice few terms are usually necessary, we use the result as presented in Equation (4-17). The remainder R can be shown to be less in magnitude than the last term retained, provided that the next-higher derivative has a uniform sign on the interval of integration. That is, in Equation (4-17), if $f^{(iv)}(x)$ has a uniform sign, then the error is less than $|f'''(N) - f'''(m)|/720$; if $f''(x)$ has a uniform sign and if we stop just after the term $(f'(n) - f'(m))/12$, then the error is less than this in magnitude.

**Example 4.8 Determine the Sum for the Series $S = \sum_{i=10}^{\infty} 1/i^2$**

This is the same series as used in the previous example, where simple rectangular integration was employed. Again, we sum the first nine terms directly and consider the tail $\sum_{i=10}^{\infty} 1/i^2$. For this problem, if we wish to bound the error of approximation in Equation (4-17), it is simple to calculate the required derivatives, and both $f''$ and $f^{(iv)}$ are of uniform sign for $10 \le x < \infty$. In particular, $f' = -2/x^3$ and $f''' = -24/x^5$. Substitution in Equation (4-17) then gives $\sum_{i=10}^{\infty} 1/i^2 = 0.1 + 0.005 + 1/6000 - 1/3{,}000{,}000 + R$, where $|R| < 1/3,000,000$. For this example, the Euler-Maclaurin formula is seen to be very powerful.

The key element in the application of the Euler-Maclaurin formula to sums is evaluation of the integral $\int_N^\infty f(x)\,dx$. In many problems this cannot be evaluated exactly, but nevertheless useful large $N$ approximations can often be found, as illustrated in the next example.

**Example 4.9 Determine a Sum for the Series $S = \sum_{i=1}^{\infty} x^i/i^2$**

The previous example was for the case where $x = 1$. In the present case, convergence is rapid if $x$ is less than 1/2 or so, but as $x$ increases toward unity, acceleration would be very desirable. Since the function to be integrated is $f(t) = x^t/t^2 = \exp(t \ln x)/t^2$, we see that $f''$ is positive for all positive $t$. Thus, it is convenient to use Equation (4-17), including the term $[f'(N)-f'(m)]/12$. The integral to be evaluated is $\int_N^\infty \exp(-zt)/t^2\,dt$ where $z = -\ln x$. Here, $z$ is positive since $x \le 1$. This integral cannot be evaluated exactly in closed form, but it is of a general type that can be approximated very accurately for "large" $N$ (say $N \ge 10$) using asymptotic methods. As in this problem, through the use of asymptotic approximation of "difficult" integrals, the Euler-Maclaurin formula can be used to accelerate the convergence of many series, provided *all* their terms are known.

**Example 4.10 Approximation of $N!$**

This problem illustrates Euler-Maclaurin summation of a *growing* function in the development of the famous Stirling formula for the factorial function. We calculate $\ln N!$ by noting that $\ln N! = \ln(2 \cdot 3 \cdots N) = \ln 2 + \ln 3 + \ldots + \ln N$. Thus, we take $f(x) = \ln x$ and perform the summation $\sum_{i=2}^{N} \ln i = \ln N!$ For this summation, substitution into Equation (4-17) yields

$$\sum_{i=2}^{N} \ln i = \int_2^N \ln x\,dx + \frac{\ln 2 + \ln N}{2} + \frac{(1/N - 1/2)}{12} - \frac{(1/N^3 - 1/2^3)}{360} + R \qquad \text{(a)}$$

Since $f^{(iv)}(x) = -6/x^4$, which has a uniform sign for $2 \le x \le N$, we have $|R| < 1/(360 \cdot 8) = 1/2880 < 0.0004$ for any $N \ge 3$. Rearrangement of Equation (a) gives

$$\ln N! = \left(N + \frac{1}{2}\right)\ln N - N - \frac{3\ln 2}{2} + 2 - \frac{1}{24} + \frac{1}{2880} + \frac{1}{12N} - \frac{1}{360N^3} + R \qquad \text{(b)}$$

By exponentiation, Equation (b) shows that $N! \approx C \cdot N^{1/2} \cdot N^N \cdot e^{-N}$ for $N \ge 3$, where $C$ is known to lie between sharp limits. By means of a more sophisticated analysis, it can be shown that $C \to (2\pi)^{1/2}$ for $N \to \infty$ and $1 < C/(2\pi)^{1/2} < 1.1$ for $N \ge 1$. This result is of extreme importance in problems involving limits, series, and integrals; it is also indispensable in the study of probability and statistics.

As will be discussed later, the gamma function, $\Gamma(1 + x) = \int_0^\infty t^x e^{-t}\,dt$, coincides with $N!$ for integral values of $x$; hence, the gamma function is really a generalized factorial function. The Stirling approximation for $N!$ derived above also represents the gamma function for large values of $x$ so that $\Gamma(1 + x) \approx (2\pi x)^{1/2} x^x e^{-x}$ for $x \to \infty$.

### 4.4.3 The Laplace Transform Method of Wheelon

Wheelon [1968] has popularized a useful procedure for summing series which is based on using integral representations, especially the Laplace transform. Here we will briefly indicate a simple application that illustrates the idea. For those unfamiliar with the Laplace transform, Spiegel [1965] can be recommended. Consider the tail of the infinite series $R = \sum_{n=k}^{\infty} f(n)x^n$. To proceed, we assume that $f(s)$ is the Laplace transform of some known function $F(t)$; that is, $f(s) = \int_0^{\infty} e^{-st} F(t)\, dt$. Laplace transform pairs $(f(s), F(t))$ are extensively tabulated, (e.g., in Spiegel [1965]). Then we introduce the integral into the summation, change the order of summation and integration, and sum the geometric series inside the integral to obtain the original sum $R$ expressed as an integral. In term of equations, we have

$$R = \sum_{n=k}^{\infty} x^n \int_{t=0}^{\infty} e^{-nt} F(t)\, dt = \int_{t=0}^{\infty} F(t) \sum_{n=k}^{\infty} (xe^{-t})^n\, dt = x^k \int_{t=0}^{\infty} \frac{e^{-kt} F(t)\, dt}{1 - xe^{-t}} \tag{4-18}$$

To be sure that the interchange of summation and integration is valid (as well as to check the other manipulations), it is always useful to expand the final integral to see that the original series is recovered. Equation (4-18) gives the sum $R$ expressed as an integral. This is very useful for several reasons: there are many more tables of integrals than sums, either numerical or approximate analytical evaluation of integrals is generally easier than for sums, and asymptotic evaluation of the "tail" integral in Equation (4-18) is straightforward.

We note that Equation (4-18) gives an integral for the whole series if $k = 0$. This is very convenient if the integral can be evaluated in closed form. However, even if we cannot evaluate the integral exactly, we can always sum the first several terms of the sum directly, so that Equation (4-18) then represents the tail with $k$ "large" (e.g., 10 or so). Then the integral can be evaluated accurately in general by the Watson lemma large $k$ asymptotic approximation.

**Example 4.11 Evaluate Sum = $\sum_{i=1}^{\infty} x^i/i^2$**

This example was considered earlier by means of the Euler-Maclaurin formula. It is ideally suited for the Wheelon method, since it is well known that $1/s^2 = \int_0^{\infty} te^{-st}\, dt$. Thus, by using Equation (4-18) we find that

$$\text{Sum} = x \int_{t=0}^{\infty} \frac{e^{-t}\, tdt}{1 - xe^{-t}} \tag{a}$$

By expanding $1/(1 - xe^{-t})$ in Equation (a) using the geometric series, and integrating using the known result $\int_0^{\infty} te^{-pt}\, dt = 1/p^2$, we verify that the integral is equivalent to the original sum. It is very interesting to note that the integral representation of the sum in Equation (a) appears to be considerably different from the integral which appears in the Euler-Maclaurin formula for this sum, which is $\int_1^{\infty} e^{(\ln x)t}/t^2\, dt$. The Euler-Maclaurin integral, which is just the integral of the function being summed, only approximates the sum rather crudely. It is when the *correction terms* are developed that the accuracy increases. The integral in Equation (a) cannot be evaluated in closed form, so in this problem it is useful to use Equation (4-18) with a "large" value of $k$, say $k = 10$. Then we would sum the first nine terms directly and have Sum $= \sum_{i=1}^{9} x^i/i^2 + T$ where the tail $T$ is given by

$$T = x^{10} \int_{t=0}^{\infty} \frac{e^{-10t}\,tdt}{1 - xe^{-t}} \tag{b}$$

It can be shown by asymptotic methods that a simple yet fairly accurate approximation to $T$ is given by $T \approx 0.01x^{10}/(1 - xe^{-1/5})$, so long as $x$ is not too close to unity.

To compare the Euler-Maclaurin and Wheelon methods, the following comments are in order. When the Wheelon method is applicable, it is often simpler and neater, but it is restricted to problems where we know the inverse Laplace transform of the function being summed. On the other hand, the Euler-Maclaurin formula, while not as neat, is much more generally applicable.

### 4.4.4 Asymptotic and Miscellaneous Techniques

There are various other techniques that are sometimes useful for summing infinite series. Here, we briefly indicate two of them and then merely mention several others. The first technique essentially uses Taylor's theorem *under the summation sign* to express a more complicated series in terms of series like $\sum_{n=1}^{\infty} n^{-k}$ which are tabulated, plus an error term that can easily be bounded or approximated. The following example illustrates the approach.

**Example 4.12 Determine the Sum of the Series $S = \sum_{n=1}^{\infty} 1/(2 + n^4)^{1/2}$**

For large $n$, this series behaves nearly like $\sum_{n=1}^{\infty} 1/n^2$. To formalize this simplification we factor out $n^4$ and write

$$S = \sum_{n=K}^{\infty} \frac{1}{(2 + n^4)^{1/2}} = \sum_{n=K}^{\infty} \frac{1}{n^2\left(1 + \frac{2}{n^4}\right)^{1/2}} \tag{a}$$

For $n^4 > 2$ (or $n \geq 2)^{1/4}$, by Taylor's theorem we have that $(1+2/n^4)^{-1/2} = 1-(2/n^4)/2+R$, where $|R| \leq (2/n^4)^2 \cdot (3/8)$. Using this result, we can re-express the original series as

$$S = \sum_{n=K}^{\infty} \frac{1}{(2 + n^4)^{1/2}} = \sum_{n=K}^{\infty} \frac{1}{n^2} - \sum_{n=K}^{\infty} \frac{1}{n^6} + R; \quad |R| < \frac{3}{2} \sum_{n=K}^{\infty} \frac{1}{n^{10}} \tag{b}$$

Since the series $\sum_1^{\infty} 1/n^2$ and $\sum_1^{\infty} 1/n^6$ are known and given in Table 4.8, we could take, say, $K = 5$, compute the necessary first few terms of the above series directly, and bound the error sum $\sum_{n=5}^{\infty} 1/n^{10}$ by $\int_4^{\infty} dx/x^{10} = 1/(9 \cdot 4^9) < 5 \times 10^{-7}$. This technique is useful because it makes use of series that have already been computed.

The second technique we wish to mention here is the approach of asymptotic approximation with respect to a parameter. Thus, if the summation is a function of a parameter, we would like to obtain a simple analytical equation expressing this relationship. This can often be accomplished by a simplification based on the parameter being small or large. Frequently, conversion to an integral by either the Euler-Maclaurin or the Wheelon Laplace transform formula is useful.

**TABLE 4.8** VALUES OF THE SUMS $\sum_{k=1}^{\infty} k^{-n}$ AND $\sum_{k=1}^{\infty} (-1)^{k-1} k^{-n}$

| $n$ | $\sum_{k=1}^{\infty} k^{-n}$ | | $\sum_{k=1}^{\infty} (-1)^{k-1} k^{-n}$ | |
|---|---|---|---|---|
| 1 | $\infty$ | | 0.69314 | 71806 |
| 2 | 1.64493 | 40668 | 0.82246 | 70334 |
| 3 | 1.20205 | 69032 | 0.90154 | 26774 |
| 4 | 1.08232 | 32337 | 0.94703 | 28295 |
| 5 | 1.03692 | 77551 | 0.97211 | 99704 |
| 6 | 1.01734 | 30620 | 0.98555 | 10913 |

(from Abramowitz and Stegun [1964])

**Example 4.13 Determine the Sum for the Series $S(A) = \sum_{n=1}^{\infty} 1/(A + n^3)$ for $A \to 0$, $A \to \infty$**

This sum obviously depends on the parameter $A$ so that, more than determining the sum for a particular value of $A$, we wish to develop correlation formulas which would be valid for either small $A$ or large $A$. For small $A$, we can expand $1/(A + n^3)$ as $1/[n^3(1 + A/n^3)] = (1/n^3)[1 - (A/n^3) + (A/n^3)^2 \ldots]$, where the error is less than the first neglected term. Then the sum is

$$S(A) = \sum_{1}^{\infty} \frac{1}{n^3} - A \sum_{1}^{\infty} \frac{1}{n^6} + A^2 \sum_{1}^{\infty} \frac{1}{n^9} - \cdots \tag{a}$$

Evaluation of the simple sums in Equation (a) produces the Maclaurin expansion of $S(A)$ for small $A$. Note that this equation could also be derived directly by calculating derivatives of $S(A)$ with respect to $A$ and substituting in Taylor's formula. The standard sums are evaluated from tables, as in Abramowitz and Stegun [1964], numerical evaluation of Wheelon type integrals, or the Euler-Maclaurin formula. The advantage of developing a formula for $S$ as a function of $A$ is that we then effectively have the value of the sum for *any* $A$ within the range of validity of the approximate formula. As we have discussed earlier, now that we have the Maclaurin expansion for small $A$ from Equation (a), we can attempt to increase the region of validity by using Euler transformation, Padé approximation, or Shanks transformation. In this particular case, the error of the Maclaurin expansion is easily bounded (less than the first neglected term); in other cases, one could use the direct method for the error by calculating the sum directly for several convenient values of $A$.

The above approach, including the acceleration of Padé approximation (for example) should suffice to give results for a considerable range of $A$ values. However, we would generally not expect the above results to be useful *when A is very large*. Rather, for large $A$, we use integration and the Euler-Maclaurin formula, Equation (4-17), with $f(x) = 1/(A + x^3)$. This produces

$$\sum_{n=K}^{\infty} \frac{1}{A + n^3} = \int_{x=K}^{\infty} \frac{dx}{A + x^3} + \frac{1}{2(A + K^3)} - \frac{f'(K)}{12} + \frac{f'''(K)}{720} \cdots \tag{b}$$

The derivative terms in Equation (b) will generally behave as $O(1/A^2)$ (proportional to $1/A^2$). Thus, the main issue is the dependence of $I = \int_K^\infty dx/(A + x^3)$ on $A$. Since $A$ is in the denominator, apparently $I$ gets smaller as $A$ gets larger for a fixed $K$ value. We can obtain the precise dependence of $I$ on $A$ by rescaling. That is, let $x^3 = Au^3$ in the integral of Equation (b) so that

$$I = \left(\frac{1}{A^{2/3}}\right) \int_{u=K/A^{1/3}}^{\infty} \frac{du}{1 + u^3} \tag{c}$$

For $K$ fixed and $A \to \infty$, in the first approximation we get $I \to (1/A^{2/3}) \cdot \int_0^\infty du/(1+u^3)$. Put more simply, for "large" $A$, $I \approx \text{const}/A^{2/3}$ where the infinite integral is just a constant that can be determined numerically or otherwise. This large $A$ result for the integral $I$ can be easily improved by noting that $\int_z^\infty = \int_0^\infty - \int_0^z$. We can write

$$\int_0^z \frac{du}{1 + u^3} = \int_0^z du(1 - u^3 + u^6 \cdots) = z - \frac{z^4}{4} + \frac{z^7}{7} \cdots \quad \left(z \equiv \frac{K}{A^{1/3}}\right) \tag{d}$$

This series converges for $z \leq 1$, which will always be true if $A$ is sufficiently large. However, because of the positive geometric expansion, the error is always less than the first neglected term. Also, we can accelerate convergence of the series in Equation (d) by means of a Euler, Padé, or Shanks transformation. If $K$ is large enough, the error in the Euler-Maclaurin formula, Equation (b), should be less than the $f'''(K)/720$ term, since $f^{(iv)}(x)$ will have a uniform sign for sufficiently large $x$.

Another important point concerns the case where $K$ is large with respect to $A^{1/3}$, or, to be more precise, the case where $K/A^{1/3}$ is large rather than small. This could happen, for example, if $A = 100$ but $K = 10$ or more, when high accuracy is desired. In this situation, we can still use the Euler-Maclaurin formula, Equation (b), with the integral expressed as

$$I = \int_{x=K}^{\infty} \frac{dx}{A + x^3} = \int_{x=K}^{\infty} \frac{dx}{x^3\left(1 + \frac{A}{x^3}\right)} = \int_{x=K}^{\infty} x^{-3}\left(1 - \frac{A}{x^3} + \frac{A^2}{x^6} \cdots\right) dx \tag{e}$$

With the new integration variable $y = K/x$, this becomes

$$I = \frac{1}{K^2} \int_{y=0}^{1} y\left(1 - \frac{y^3}{z^3} + \frac{y^6}{z^6} \cdots\right) dy = \frac{1}{K^2}\left[\frac{1}{2} - \frac{1}{5z^3} + \frac{1}{8z^6} \cdots\right] \tag{f}$$

In Equation (f), $z = K/A^{1/3}$, as before. The series here converges for $z \geq 1$, and again, acceleration can be attempted by using Euler, Padé, or Shanks. Thus, for this example problem, the Euler-Maclaurin formula is useful when $(K/A^{1/3})$ is either large or small.

Several other approaches for summing series which are sometimes used should be mentioned. These include (i) conversion to a differential equation by differentiation or integration with respect to a parameter, (ii) extrapolation of partial sums to $n \to \infty$ $(1/n \to 0)$ by using interpolation methods, (iii) summation by parts using difference equation methods, (iv) the Poisson sum formula, and (v) use of the calculus of residues from complex function theory.

## 4.5 COMPUTER PROGRAMS

In this section we discuss the computer programs applicable to various problems in the chapter. For each program, we give a brief discussion and a concise listing, which includes (i) a program description, (ii) how to run, and (iii) user-specified subroutines, including a sample problem. Detailed program listings can be obtained from the diskette.

### Padé Approximation Program PADE

Instructions for running the program are given in the concise listing. The program prompts for $n$ (the largest power in the Taylor expansion) and $k$ (the largest power of the numerator in the Padé approximation. We must have $(k+m)=n$. To get the ordinary Taylor approximation, set $k$ equal to the value of $n$. The program can do any Padé approximation $(k,m)$ as long as $(k+m)$ is equal to the specified $n$. The program calculates the Padé parameters $(p_i, q_i)$ and evaluates the Padé approximation at any desired $x$. The Taylor coefficients $a_i$ can be specified by either assignment statements or through a recurrence relation. The program has the option to do direct error bounds (as in program SERIES) as illustrated in the concise listing. To calculate direct error bounds, the user must specify the known function in Sub Function. If direct error bounds are not desired, Sub Function is ignored.

#### Program PADE.

```
!This is program PADE.tru by O.T.Hanna
!Program produces Pade approximation of the Taylor expansion a(0) +
!a(1)*x + a(2)*x^2 + ... + a(n)*x^n
!The Pade approx. is [p(0)+p(1)*x+..p(k)*x^k]/[1+q(1)*x+..q(m)*x^m]
!Thus n = k + m
!Program has option to do direct Pade error bounds
!User specified lines denoted by ***
!*****************************************************************
!*To run, generate a(i) coefficients in Sub Init via recurrence *
!*or assignment statements. Program prompts for n = largest     *
!*power of Taylor expansion and k = largest power of Pade       *
!*numerator (then m = n - k). To calculate direct Pade error    *
!*give xmin,xmax,no_intervals,function f(x) in Sub Function;    *
!*Result is used as Abs(Error)<=zmax*Abs(x)^(n+1)               *
!*****************************************************************

Sub Function(n,x,f,f1(),no_intervals,xmin,xmax)  !***user defined
  !This Sub gives original f(x) either analytically or numerically
  !Maximum error is determined in range xmin <= x <= xmax using a
  !number of intervals equal to no_intervals.

  Let no_intervals = 10!      !***
  Let xmin = 0                !***
  Let xmax = .1               !***

  Mat redim f1(no_intervals + 1)

  Let f = Exp(-x)             !***analytically defined function
```

```
   !The following is for function defined by data
   !define b(1)=f(xmax),b(2)=f(xmax-d),...,b(no_intervals+1)=f(xmin)

   !The example below is for xmin=0,xmax=2,no_intervals=5 !***
!  Let f1(1)=Exp(-1)
!  Let f1(2)=Exp(-.8)  !d=(xmax-xmin)/no_intervals = (1-0)/5 = .2
!  Let f1(3)=Exp(-.6)
!  Let f1(4)=Exp(-.4)
!  Let f1(5)=Exp(-.2)
!  Let f1(6)=Exp(-0)
End Sub

Sub Init(n,a())
   !Here we generate a(0) to a(n) by assignment or recurrence ***

   Let a(0) = 1         !assignment
   Let a(1) = -1
   Let a(2) = 1/2
   ! For i = 0 to n     !recurrence
   !    Let a(i) = 0
   ! Next i

End sub
```

### Sequence Acceleration Program SHANKS

Instructions for running the program are given in the concise listing. The program can accomplish Shanks transformations for either specified sequence elements, or for a power series whose coefficients are specified. Program SHANKS allows multiple Shanks transformations until the data are exhausted. The user is prompted for (i) the number of sequence elements (we need a minimum of three for one Shanks, five for two Shanks, seven for three Shanks, etc.), (ii) the method of specification of the sequences (direct or power series coefficients), and (iii) an $x$ value in the case of power series.

#### Program SHANKS.

```
!This is program SHANKS.tru by O.T.Hanna
!Program produces Shanks transformations of the sequence s(1),s(2)...s(n).
!For one Shanks we need 3 s(i), for two Shanks 5 s(i), etc.
!User specified statements are designated by ***
!****************************************************************
!* To run, generate s(i) elements in Sub Init either directly,  *
!* or by a power series (recurrence or assignment)              *
!* Program prompts for n = number of elements of s(i); method   *
!* of specification; and in the case of power series, x value   *
!* Program allows multiple Shanks Trans. until data is exhausted*
!****************************************************************
Sub Init(n,s())
```

```
!Here we generate partial sums s(i) by calculation(series),or by
!assignment statements. The marks *** designate where the
!user must modify program for generation method(meth) being used.

Input prompt"get s(i) by series(1),assignment of s(i)(2) ? ":meth
   Select case meth
      case 1 !generation of s(i) via power series,
             !assignment or recurrence for a(i)***
          Dim a(0 to 1)
          Mat Redim a(0 to n)
          Input prompt"x = ? ":x

          !example prob. below is ln(1 + x) = x - x^2/2 + x^3/3...

          !recurrence option for a(i)***
          !example here is n coeff. of ln (1 + x)
       LET a(0) = 0 !set a(0) here,not in loop
       LET a(1) = 1 !recurrence is useful for variable number of terms
       FOR i = 1 to n-1
          LET a(i+1) = -a(i)*i/(i + 1)
       NEXT i

       !assignment option for a(i)***
       !example here is first 5 coeff. of ln (1 + x)
!      LET a(0) = 0
!      LET a(1) = 1
!      LET a(2) = -1/2
!      LET a(3) = 1/3
!      LET a(4) = -1/4

       LET s(1) = a(0) !initialize partial sum for power series
       FOR i = 1 to n-1
          LET s(i+1) = s(i) + a(i)*x^i
       NEXT i

      case 2 !generation of s(i) via assignment !***
          LET s(1) = 0
          LET s(2) = 0
          LET s(3) = 0
          !Etc.
       END SELECT

END SUB
```

## PROBLEMS

Note that there are a number of both worked examples and homework problems located throughout the book that pertain to the use of acceleration techniques in various applications.

**4.1.** For the function $\ln(1 + x)$, determine numerically the optimal value of $p$ in Euler's transformation, using the SERIES computer program value of "last term" as the error criterion. Use terms of the Taylor series up to $x^5$ and carry out the calculations for both $x = 1$ and $x = 2$ to see how much the optimum changes with $x$. How does the optimal value compare to the simple estimate $p = -a_1/a_2$?

**4.2.** The well-known and important "normal probability integral" can be expressed in the form

$$I(x) = \frac{1}{\sqrt{2\pi}} \int_0^x e^{-t^2/2}\,dt$$

By expanding the exponential and integrating term by term, show that the Maclaurin expansion for this function is expressed as

$$\sqrt{2\pi} \cdot I(x) = x - \frac{x^3}{2 \cdot 3 \cdot 1!} + \frac{x^5}{2^2 \cdot 5 \cdot 2!} - \frac{x^7}{2^3 \cdot 7 \cdot 3!} \cdots = x \sum_{i=0}^{\infty} \frac{(-1)^i x^{2i}}{2^i(2i + 1)i!}$$

Take $x^2 = w$ and find numerically the optimal value of $p$ in Euler's transformation for the series, as in problem 4.1. Use ten terms of the series and $x = 1$. Compare the optimal $p$ value with a simple estimate.

**4.3.** The well-known "exponential integral" can be expressed as

$$J(x) = \int_0^{\infty} \frac{e^{-xt}}{1 + t}\,dt$$

By expanding $1/(1 + t)$ in a Maclaurin series and integrating term by term (see the chapter on integrals), we generate the asymptotic (and divergent!) series

$$J(x) = \frac{0!}{x} - \frac{1!}{x^2} + \frac{2!}{x^3} - \frac{3!}{x^4} + \frac{4!}{x^5} - \frac{5!}{x^6} \cdots$$

This "infinite series" diverges for *all* $x$, yet it is very useful for computation when $x$ is large, since it has the property that the error in its use is less in magnitude than the first neglected term. Take $1/x = w$ and find numerically the optimal value of $p$ in Euler's transformation for the series, as in problem 4.1. Use the six terms of the series shown above and $x = 1$. Compare the optimal $p$ value with a simple estimate.

**4.4.** Carry out a contest between Euler's transformation ($p = 2$), Padé approximation, and multiple Shanks transformations to see which method can best approximate $\ln(1 + x)$ by using its Maclaurin series. Consider $x = 1$ and $x = 9$. How does Euler *plus* Padé compare? What about Euler *plus* Shanks?

**4.5.** Consider the Padé approximation $(p_0 + p_1 x)/(1 + q_1 x) = a_0 + a_1 x + a_2 x^2$. Verify that the solutions of the *nonlinear* algebraic equations for $p_0$, $p_1$, and $p_2$, corresponding to expanding the above expression directly into a Maclaurin series, are the same as those for the *linear* system obtained by cross-multiplying terms.

**4.6.** Consider the integral

$$K(x) = \int_0^x \frac{e^{-t}}{1 + t}\,dt$$

which is closely related to the exponential integral. It is desired to develop and compare Taylor series and various Padé $(k, m)$ approximations to this function at $x = 1, 2, 10$. Use a Taylor approximation of degree 10 and use various combinations of $k, m$ (with $k + m = 10$) to see which Padé approximation is best. The correct values are $K(1) = 0.4634$, $K(2) = 0.5609$, and $K(10) = 0.5963$.

**4.7.** Reconsider the exponential integral function of problem 4.3. Use the six terms of the asymptotic series shown and calculate both Padé and Shanks improvements for $x = 1$,

2, and 10. The numerically computed true values are $J(1) = 0.5963$, $J(2) = 0.3613$, and $J(10) = 0.09156$.

**4.8.** Consider the first three terms of the Maclaurin expansion of a function, $a_0 + a_1 x + a_2 x^2$. For this three-term series, show that the following "improvements" are *identical*: (i) the Euler transformation with $p = (-a_1/a_2)$, (ii) the (1,1) Padé approximation, and (iii) Shanks improvement.

**4.9.** Consider the nonlinear differential equation $dy/dx = x^2 + y^2$, subject to the initial condition $y(0) = 1$. Develop a series solution through terms of $x^4$, and estimate the value of $x$ where $y \rightarrow \infty$. Do this using both Padé approximation [P(2,2)] and Shanks transformation.

**4.10.** Sum the series $\sum 1/i^3$ from $i = 1$ to $\infty$ using the integration method.

**4.11.** Sum the series $\sum 1/i^3$ from $i = 1$ to $\infty$ using the Euler-Maclaurin summation formula.

**4.12.** Develop an integral for the sum of the series $\sum x^i/i^3$ from $i = 1$ to $\infty$ using the Laplace transform method of Wheelon.

**4.13.** Derive approximate expressions for the series $\sum 1/(A + i)^{3/2}$ from $i = 1$ to $\infty$ in the limiting cases $A \rightarrow 0$ and $A \rightarrow \infty$.

## REFERENCES


ABRAMOWITZ, M. and STEGUN, I. (Eds.) (1964), *Handbook of Mathematical Functions*, National Bureau of Standards, Washington, DC.

BENDER, C. and ORSZAG, S. (1978), *Advanced Mathematical Methods for Scientists and Engineers*, New York: McGraw-Hill.

BROMWICH, T. J. I'A. (1964), *Infinite Series,* 2nd ed., Macmillan, New York

MEKSYN, D. (1961), *New Methods in Laminar Boundary Layer Theory*, Pergamon Press, New York

PRESS, W. H., FLANNERY, B. P., TEUKOLSKY, S. A., and VETTERLING, W. T. (1986), *Numerical Recipes*, Cambridge Univ. Press, New York

SPIEGEL, M. (1965), *Laplace Transforms*, Schaum, New York

SQUIRE, W. (1970), *Integration for Engineers and Scientists*, Elsevier, New York

WHEELON, A. D. (1968), *Tables of Summable Series and Integrals Involving Bessel Functions*, Holden-Day, San Francisco

# 5

# Interpolation

Interpolation is an approximation procedure wherein an assumed function is forced to agree at certain designated points with some specified data. There are many engineering applications of this technique, which include estimation of function values, derivatives, and integrals from tabular data, as well as finding formulas to fit numerically defined functions. The interpolation "mathematical model" corresponds to the kind of function chosen to fit the data. Interpolation can be carried out with many types of approximating functions, but *polynomial* interpolation has been the simplest type in practice. This is because polynomial approximation is known to be broadly applicable to many functions, it is conceptually simple and relatively easy to apply, and polynomials can be easily differentiated or integrated. However, now that personal computers are so widely available, spline and rational approximation are becoming increasingly popular. Our discussion will include polynomial interpolation and error analysis, spline interpolation, rational interpolation, general one-dimensional interpolation, and multidimensional interpolation. In addition to a discussion and illustration of these approximation types, computer programs to accomplish these tasks will also be introduced.

## 5.1 LINEAR INTERPOLATION

The mechanics of simple linear interpolation are well known to virtually all students of science. The error associated with linear interpolation is, in contrast, not so well known. However, the error is of fundamental importance for assessing the reliability of interpolation results. It is therefore important in applications of interpolation to

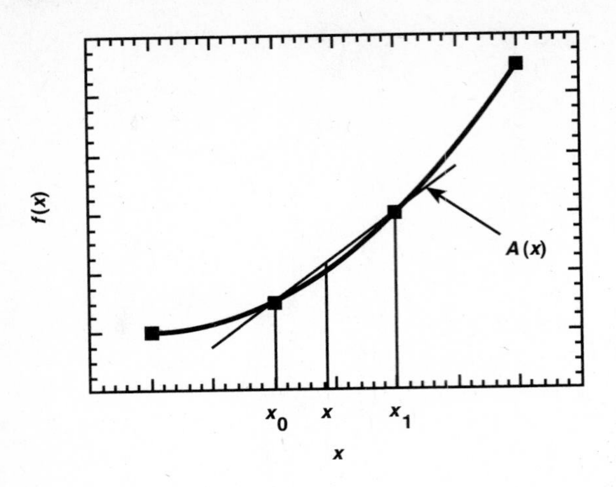


**Figure 5.1.** Linear Interpolation

know something of the nature of the error and how it can be estimated. We now give a brief derivation of a theoretical expression for the error of linear interpolation, using the same reasoning that is needed to extend the result to the general polynomial case. Following that, we indicate a practical procedure for implementing general polynomial interpolation and error estimation.

If we suppose that the two interpolation points are $x_0$ and $x_1$, then the linear interpolation approximation to a function $f(x)$ at any point $x$ is called $A(x)$. $A(x)$ is to be a linear function of $x$ which agrees exactly with $f(x)$ at the interpolation points $x_0$, $x_1$. Figure 5.1 illustrates this idea.

It is easy to verify directly that the linear interpolation approximation $A(x)$ must be of the form

$$A(x) = f(x_0) + \frac{f(x_1) - f(x_0)}{x_1 - x_0}(x - x_0) \tag{5-1}$$

Equation (5-1) shows that $A(x)$ is a linear function of $x$ which agrees exactly with $f(x)$ at $x_0$ and $x_1$. The error, or remainder, of the linear interpolation approximation is called $R$ and is a function of $x$. Thus, $R(x)$ is defined by

$$f(x) = f(x_0) + \frac{f(x_1) - f(x_0)}{x_1 - x_0}(x - x_0) + R(x) \tag{5-2}$$

By definition, $R(x)$ vanishes at the two points $x_0$ and $x_1$. It is therefore convenient to write $R(x)$ as $R(x) = K(x) \cdot (x - x_0)(x - x_1)$. Now, in a rather non-obvious way, a new function $\phi(z)$ is defined as

$$\phi(z) \equiv f(z) - A(z) - K(x)(z - x_0)(z - x_1) \tag{5-3}$$

The function $\phi(z)$ is very convenient for our purposes; note that it is deliberately formed by subtracting $A(x) + K(x)(x - x_0)(x - x_1)$ from $f(x)$ with all of the $x$ values, except that appearing in $K(x)$, replaced by $z$. From the definition of $\phi(z)$, $\phi(x)$ is equal to zero. Also, $\phi(z) = 0$ at the interpolation points $x_0$ and $x_1$. Thus, $\phi(z)$ is zero at the three points $z = x$, $x_0$, and $x_1$. We now appeal to Rolle's theorem, which states that between two zeros of a continuously differentiable function $\phi(z)$, $\phi'(z)$

**TABLE 5.1** VALUES OF THE BESSEL FUNCTION $J_0(x)$

| $x$ | $J_0(x)$ |
|---|---|
| 6.2 | 0.2017 |
| 6.3 | 0.2238 |
| 6.4 | 0.2433 |

must have at least one zero. We see, therefore, that since $\phi(z)$ has three zeros, $\phi'(z)$ must have at least two zeros somewhere in the range separating $x_0$, $x$, and $x_1$. But since $\phi'(z)$ has at least two zeros in this range, again by Rolle's theorem it follows that $\phi''(z)$ must have at least one zero in this same range. If we call the point at which $\phi''(z)$ is zero $z = \xi$, then we have that $\phi''(\xi) = 0 = f''(\xi) - A''(\xi) - 2K$. Note that we have used the fact that $d^2/dz^2[(z - x_0)(z - x_1)] = 2$; also note that since $A(z)$ is a linear function of $z$, $A''(z) = 0$. Therefore, we finally get $K(x) = f''(\xi)/2!$ and

$$R(x) = \frac{f''(\xi)(x - x_0)(x - x_1)}{2!} \tag{5-4}$$

The quantity $\xi$ is unknown but depends on $x$. Note that as $x_1 \to x_0$—that is, as the linear interpolation points come closer together—both the approximation and remainder approach those appropriate for the linear Taylor formula with expansion point $x_0$. By taking the maximum of the product $|(x - x_0)(x - x_1)|$, an upper bound for the error of linear interpolation between the points $x_0$ and $x_1$ can be shown directly to be

$$R \leq \max_{x_0 \leq x \leq x_1} |f''(x)| \frac{(x_l - x_0)^2}{8} \tag{5-5}$$

It is usually difficult to bound this second derivative; however, it is quite easy to *estimate* the remainder; for example, by replacing the $f''$ term by a three-point numerical differentiation approximate value, calculated from the three known points closest to the point in question. For *equally spaced* interpolation points, the preceding estimate gives $\max |R(x)| \approx |y(x_2) - 2y(x_1) + y(x_0)|/8$. We could also estimate $f''(x)$ from spline interpolation. As will be shown later, use of the Newton polynomial interpolation procedure also produces an error estimate.

**Example 5.1 Linear Interpolation**

We now show how the foregoing linear interpolation calculations can be used in practice. Consider the values shown in Table 5.1 for the Bessel function $y = J_0(x)$, which occurs in many engineering problems involving transport in cylindrical geometry.

It is desired to estimate by linear interpolation the value of $J_0(6.33)$, together with an assessment of the accuracy of this estimate. To accomplish linear interpolation, we choose the tabulated values at $x_0 = 6.3$ and $x_1 = 6.4$, which span the desired $x$ value of 6.33. Linear interpolation then gives $y(6.33) \approx 0.2238 + [(0.2433 - 0.2238)/0.1](6.33 - 6.3) = 0.2297$. To estimate accuracy we can use the maximum error estimate $y''(x) \cdot (x_1 - x_0)^2/8$. This yields

$$\text{Max Error} \cong \frac{y''(x_1 - x_0)^2}{8} = \frac{1}{8}\left|\frac{0.2433 - 2(0.2238) + 0.2017}{(0.1)^2}\right|(0.1)^2 = 0.00033 \tag{5-6}$$

where $y''(x)$ is calculated by numerical differentiation. This calculation suggests that we can be confident that $J_0(6.33) = 0.2297 \pm 0.0003$. The true value, for purposes of comparison, is $J_0(6.33) = 0.2299$, which is seen to fall within the estimated limits.

## 5.2 GENERAL POLYNOMIAL INTERPOLATION

In the general case of $n$th-degree polynomial interpolation, it is easy to see that, in principle, we need only require that the assumed polynomial agree with the given continuous function at $(n + 1)$ particular points. This yields a system of linear algebraic equations for the unknown polynomial coefficients. There are alternative methods, such as the Lagrange formula or the Newton algorithm, which accomplish the same thing. It is important to keep in mind that no matter how a polynomial interpolation calculation is organized, in theory the approximation is always the same (there could be some difference due to roundoff error).

A derivation similar to that given for linear interpolation shows that the remainder for $n$th-degree polynomial interpolation can be written as

$$R = f^{(n+1)}(\xi)[(x - x_0)(x - x_1)\ldots(x - x_n)]/(n + 1)! \tag{5-7}$$

Here, $\xi$ is an unknown point somewhere in the range including $(x, x_0, x_1, \ldots, x_n)$; the value of $\xi$ again depends on $x$. This expression for error cannot normally be implemented, even approximately, for $n$ larger than 2 or 3. However, in the theory of approximation, higher-degree interpolation and, especially, consideration of the error play an important role. The polynomial error formula given above reminds us of the error for a Taylor expansion, and illustrates, as we would expect, that polynomial interpolation approximation becomes more like Taylor approximation as the interpolation points all approach $x_0$.

### 5.2.1 The Newton Algorithm

As mentioned earlier, there are several equivalent ways to actually implement polynomial interpolation. For our purposes, the most convenient of these is the so-called Newton algorithm. It is unusual in dealing with data to carry out numerical polynomial interpolation beyond the second or third degree. This is because higher-degree polynomial approximation often tends to have an undesirable oscillatory behavior between the data values. Therefore, our discussion of Newton's algorithm will be restricted to first- or second-degree interpolation. It should be clear how to extend the development to higher approximations. As far as our computer programs are concerned, program NEWTINT implements the first- or second-degree approximation discussed here. This program also calculates the maximum or minimum (if one exists), the integral, and the derivatives $y'$ and $y''$ of the linear or quadratic approximation. Higher-degree polynomial approximations can be derived, should they be desired, from program RATINTG, which is discussed under rational interpolation.

To develop the Newton algorithm, we proceed as follows. We define the particular second-degree polynomial approximation as follows:

$$y = C_0 + C_1(x - x_0) + C_2(x - x_0)(x - x_1) \tag{5-8}$$

We wish Equation (5-8) to be satisfied exactly for the data pairs $(x_0, y_0)$, $(x_1, y_1)$, and $(x_2, y_2)$. To accomplish this, we first observe that for $x = x_0$, $y = y_0$ we must have $C_0 = y_0$. Then we subtract $C_0$ from $y$ in Equation (5-8) and divide by $(x - x_0)$ to get

$$\frac{y - C_0}{x - x_0} = C_1 + C_2(x - x_1) \tag{5-9}$$

To get $C_1$ we substitute $x = x_1$, $y = y_1$ into Equation (5-9) to obtain $C_1 = (y_1 - C_0)/(x_1 - x_0)$. Finally, we subtract $C_1$ from the left-hand side of Equation (5-9) and divide by $(x - x_1)$ to yield

$$\frac{\dfrac{y - C_0}{x - x_0} - C_1}{x - x_1} = C_2 \tag{5-10}$$

Now we substitute $x = x_2$, $y = y_2$ into Equation (5-10) to determine the value of $C_2$. By following the pattern of the calculations for the second-degree approximation, it should be clear that we can simply add on more terms in the approximation (e.g., the next term for third-degree approximation in Equation (5-8) would be $C_3(x - x_0)(x - x_1)(x - x_2)$) and proceed accordingly. It should be noted that the first-degree approximation, obtained by putting $C_2 = 0$ and dropping the data point $(x_2, y_2)$, is exactly the same as that written previously in Equation (5-1).

Before considering an example, we mention the following advantages of the Newton algorithm for polynomial interpolation:

- No algebraic equations need be solved.
- The process yields a formula that can be used to calculate $y$ at any $x$.
- Additional terms do not change previous terms.
- Similar to Taylor approximation, an error estimate is given by the last term.
- The procedure generalizes directly for any degree of polynomial interpolation.
- A formal recurrence calculation is not difficult to derive.

**Example 5.2 Quadratic Interpolation**

Aqueous amine solutions are commonly used for the chemical absorption of acid gases such as hydrogen sulfide and carbon dioxide. A particular design problem for an absorption tower requires knowledge of the viscosity of a 43 wt. % solution of methyldiethanolamine (MDEA) in water at 45°C. Al-Ghawas *et al.* [1989] have measured the viscosity of aqueous solutions of MDEA over a range of temperatures and concentrations and have correlated their data according to

$$\ln \mu = B_1 + \frac{B_2}{T} + B_3 T$$

where $\mu$ is in poise and $T$ is in °K. The $B_i$ parameters are given in Table 5.2.

**TABLE 5.2** CORRELATION PARAMETERS FOR THE VISCOSITY OF AQUEOUS MDEA

| wt. fraction MDEA | $B_1$ | $B_2$ $\times 10^{-3}$ | $B_3$ $\times 10^{2}$ |
|---|---|---|---|
| 0 | −19.52 | 3.913 | 2.112 |
| 0.10 | −22.14 | 4.475 | 2.470 |
| 0.20 | −25.16 | 5.157 | 2.859 |
| 0.30 | −28.38 | 5.908 | 3.255 |
| 0.40 | −31.52 | 6.678 | 3.634 |
| 0.50 | −34.51 | 7.417 | 3.972 |

For this problem we wish to employ quadratic interpolation to determine values of $B_1$, $B_2$, and $B_3$ that correspond to a 43 wt. % MDEA solution; thus, we use program NEWTINT. Since the desired weight fraction of 0.43 is closest to the table entries for a weight fraction of 0.40, we take $x(0) = 0.40$, $x(1) = 0.30$, and $x(2) = 0.50$. Implementation of the program for each of the $B_i$'s at 0.43 weight fraction gives

$$B_1 = -32.44 \qquad \text{Est. error} = -1.7 \times 10^{-2}$$
$$B_2 = 6903 \qquad \text{Est. error} = 3.3$$
$$B_3 = 0.03740 \qquad \text{Est. error} = 4.3 \times 10^{-5}$$

Substitution of these values into the correlation formula for $\mu$ at $T = 45°\text{C}$ (318°K) gives $\mu = 3.200$ poise. Since the estimated errors for the $B_i$'s are very small, we can be quite confident in this interpolated value of the viscosity.

## 5.3 MISCELLANEOUS NOTES ON POLYNOMIAL INTERPOLATION

Ordinary polynomial interpolation, as discussed above, is useful for problems where low-degree approximation, as evidenced by error estimates, is appropriate. Such a situation often occurs when linear or quadratic, or possibly cubic, interpolation is seen to suffice. In other situations—for example, where much data is to be used or where there are steep variations in the function—ordinary polynomial interpolation may not be suitable, and we may need to use spline or rational approximation. These methods will be discussed subsequently. It should be stressed that polynomial interpolation is generally *not* useful for extrapolation outside the range of the data. If extrapolation is necessary, rational approximation is usually desirable, since experience shows that it is generally applicable over a relatively broad range.

Another point that should be borne in mind when pursuing approximation problems is that it is often useful to employ a *change of variables*. Thus, for example, if the function $y(x)$ being approximated behaves nearly as an exponential, linear or quadratic approximation will not be appropriate over a wide range of $x$. However, if we introduce $z = \ln y$ as the new dependent variable, it may well happen that $z(x)$ will be well approximated by a low-degree polynomial. This idea is valuable regardless of the type of approximation we may be using. We also mention that interpolation at the so-called "Chebyshev" points of an interval is sometimes useful [Kahaner, 1989].

### 5.3.1 Inverse Interpolation

It occasionally happens that it is useful to interchange the roles of the dependent ($y$) and independent ($x$) variables in an interpolation problem. This would be the case, for example, if we wished to predict $x$ for a given $y$. To accomplish this task, we simply trade $x$ for $y$ and $y$ for $x$ in our approximation formulas. It is important to note, however, that such an interchange usually changes the mathematical model which is the basis of the calculation. Thus, as an example, if we originally express $y(x)$ as the quadratic $y(x) = A + Bx + Cx^2$, then, if we solve this equation for $x(y)$, the result involves a square-root function of $y$. But an interchange of the roles of $x$ and $y$ in the original equation would show $x$ to be a *quadratic* in $y$, not a square-root function. Normally this is not a problem, but it is important to be aware of the change in the model.

## 5.4 SPLINE INTERPOLATION

It has been mentioned earlier that higher-degree (e.g., higher than third-degree) polynomial interpolation of discrete data often shows undesirable behavior, such as exhibiting wiggles between data points (see later in Figure 5.3). This is because in most engineering and scientific applications, the mathematical models that seem to work best are generally *not* of the polynomial type. Thus, for example, we seldom see in nature processes that evolve in time toward some *unbounded* asymptote, which polynomial behavior would predict. Low-degree polynomial approximation is often very useful *over a relatively short range of the variables,* and this is where it should be used. When the range of variables is too wide for ordinary polynomial approximation, other procedures have been found to be useful. Two of these are so-called spline approximation and rational approximation. These techniques both require considerably more computation than ordinary polynomial approximation, so they are ideally suited for computer work. Cubic spline approximation will be discussed in this section, while rational approximation will be put off until the next section.

The name "spline" comes from a draftsman's tool in the form of a flexible rod which is used to produce a "smooth" curve. The idea of spline approximation is to use a different relatively low-degree approximation to the function between each pair of data points, and systematically match function values and derivatives up to some order at each data point. In this way an approximation is generated that is continuous and has (some) continuous derivatives, while not oscillating much between data points. The simplest example of such a spline approximation corresponds to using linear interpolation between each data point. This produces a continuous approximation which does *not* have a continuous first derivative (see Figure 5.2).

To obtain a spline approximation which is useful in practice, it has been found that using cubic interpolation on each subinterval is particularly convenient. Such a procedure leads to the *cubic spline*, for which the approximation has both continuous first and second derivatives. A computer program that implements this procedure,

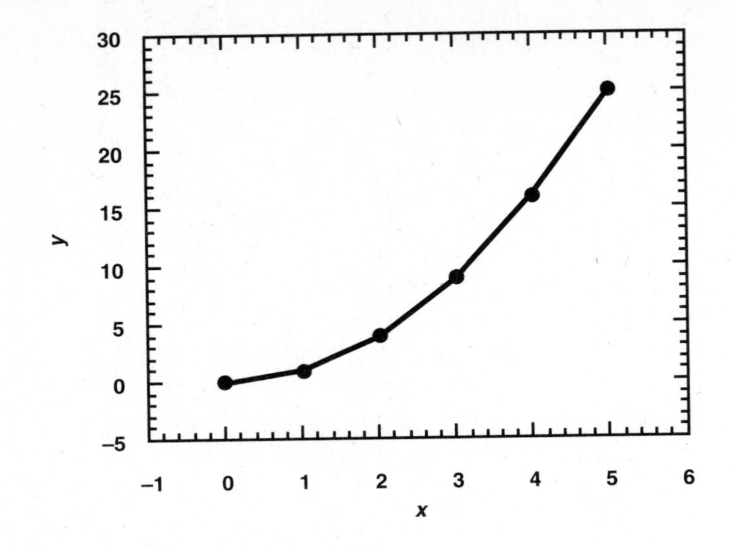


**Figure 5.2.** Linear Spline Interpolation

CUBSPL, is discussed at the end of the chapter. Some advantages of cubic spline approximation include:

- Good for non-smooth or rapidly varying functions
- Good when there are many data points to be used
- Good for calculating the derivative of the function
- Good for calculating the integral of the function

The disadvantages of cubic spline approximation are that it does not produce a single formula for approximation in the desired interval, providing required end conditions is a little vague, and it is generally not very useful for extrapolation beyond the range of the data.

We now give a derivation of formulas for cubic spline approximation, generally following the very clear exposition by Gerald [1978]. Another useful reference is Johnson and Riess [1982]. The procedure to be followed is straightforward, but the development is somewhat lengthy and tedious. Suppose that we have $(m + 1)$ data points, which divide the region of interest into $m$ parts having $(m-1)$ *interior* points. We first wish to explore the degrees of freedom of the problem, namely the relationship between the number of constants and the number of equations. On each of the $m$ subintervals, we will use one cubic polynomial, which gives a total of $4m$ constants to be determined. (For 11 data points, we have $10(4) = 40$ constants; obviously, a computer is necessary!) The number of equations is as follows. We need $2(m-1) + 2 = 2m$ equations to specify the known $y$ values at the interior points, plus the end points; the factor two is because we have a $y$ equation for both the left and right sides of each interior point. We also need $(m-1)$ equations for the equal (but unknown) first derivative specifications at the interior node points, plus this many again for the second derivative specifications. Thus, the degrees of freedom of the calculation are (number of variables) − (number of equations) = $4m - [2m + 2(m-1)] = 2$. Therefore, we need two more equations, and these can be specified in various ways. First, the best procedure would involve specifying the first derivatives of the function at the end points, if these happen to be known (usually they are not known). Lacking this information, other procedures generally specify some kind of relationships for $y''$ at the end points. These include using $y'' = 0$ at the two end points (linear or "natural" spline), using the equations $y_1'' = y_2''$, $y_n'' = y_{n-1}''$ at the end points, or using a linear extrapolation of $y''$ from the two closest interior values at the end points. This "vagueness" regarding end conditions to complete specification of the cubic spline approximation is annoying, but it does not seem to cause many problems in practice.

We now write down the actual equations to be used in our cubic spline approximation. For convenient reference, we call the region between points $(x_i, y_i)$ and $(x_{i+1}, y_{i+1})$ interval "$i$". On interval $i$, we write our cubic polynomial, together with its derivatives, as

$$\begin{aligned} y &= a_i(x-x_i)^3 + b_i(x-x_i)^2 + c_i(x-x_i) + d_i \\ y' &= 3a_i(x-x_i)^2 + 2b_i(x-x_i) + c_i \\ y'' &= 6a_i(x-x_i) + 2b_i \end{aligned} \tag{5-11}$$

We now let $h_i \equiv (x_{i+1} - x_i)$ and then evaluate the above relationships at the two end points of interval $i$. This gives

$$y \qquad \begin{aligned} y_i &= d_i \\ y_{i+1} &= a_i h_i^3 + b_i h_i^2 + c_i h_i + d_i \end{aligned} \tag{5-12}$$

$$y' \qquad \begin{aligned} y_i' &= c_i \\ y_{i+1}' &= 3a_i h_i^2 + 2b_i h_i + c_i \end{aligned} \tag{5-13}$$

$$y'' \qquad \begin{aligned} y_i'' &= 2b_i \\ y_{i+1}'' &= 6a_i h_i + 2b_i \end{aligned} \tag{5-14}$$

We now express the unknowns $a_i$, $b_i$ in terms of $y_i''$ and $y_{i+1}''$ and also express $c_i$ in terms of these same second derivatives as well as $y_i$ and $y_{i+1}$. In doing this, it is convenient to introduce the notation $S_i \equiv y_i''$. Using Equations (5-12) and (5-14), we get

$$\begin{aligned} b_i &= S_i/2 \\ a_i &= \frac{S_{i+1} - S_i}{6h_i} \\ c_i &= \frac{y_{i+1} - y_i}{h_i} - \left(\frac{2h_i S_i + h_i S_{i+1}}{6}\right) \end{aligned} \tag{5-15}$$

The slopes $y_i'$ and $y_{i+1}'$ are now equated using Equations (5-13) to give, after considerable algebraic simplification,

$$h_{i-1}S_{i-1} + (2h_{i-1} + 2h_i)S_i + h_i S_{i+1} = 6\left[\left(\frac{y_{i+1} - y_i}{h_i}\right) - \left(\frac{y_i - y_{i-1}}{h_{i-1}}\right)\right] \tag{5-16}$$

**TABLE 5.3** DATA TO BE INTERPOLATED

| $x$ | $y$ |
|---|---|
| 0.0 | 0.0 |
| 0.5 | 0.9330 |
| 1.0 | 1.0000 |
| 2.0 | 1.0718 |
| 3.0 | 1.1161 |
| 4.0 | 1.1487 |

Equation (5-16) represents a tridiagonal linear algebraic system for the unknowns $S_i$ which holds for $i = 2, 3, \ldots, (n-1)$ where there are $n$ data pairs and $(n-1)$ separate cubic equations. As mentioned earlier, several different end conditions can be used to establish an algebraically determinate system. The simplest of these corresponds to using $S_1 = S_n = 0$, which is known as a "linear" condition (also as "natural" boundary conditions). A second set of boundary conditions, known as "parabolic" conditions, involves setting $S_1 = S_2$ and $S_n = S_{n-1}$. These two options are employed in our computer program CUBSPL, which is discussed at the end of the chapter. There are various other possibilities, but in most situations the different end conditions do not produce widely different results. Usually, it is hard to decide if one set of conditions makes more sense than another. One case where a decision would be clearcut is if the *derivatives* of the unknown function are known at the two end points; these values could then be used to specify the spline. Unfortunately, this information is rarely available.

Practical error estimation is very difficult for spline approximation. Depending on the particular boundary conditions, results vary somewhat, but the essential idea is that the error for cubic spline approximation is proportional to both max $|f^{iv}|$, and the maximum data increment $|x_i - x_{i+1}|^4$. This is also the same general result applicable to ordinary cubic interpolation. A crude way to get some idea of the error is to delete one of the data points, recalculate the cubic spline for this case, and take the difference between these two spline approximations as reflective of the error. In some applications it is desirable to have a more systematic way to deal with spline functions. So-called B-splines can be used for this purpose [Kahaner, 1989].

Program CUBSPL, discussed at the end of the chapter, permits a cubic spline fit to either a set of data or an analytically defined function. It also allows for evaluation of the fitted function, both its first and second derivatives, and its integral between any two limits. An option to plot the data and the function is also available. The analytically defined function facility is useful for problems where a plot of an analytical function is desired (e.g., root-finding, etc.), or where it is desired to use spline approximation to calculate either the derivative or the integral of an analytical function.

**Example 5.3 Cubic Spline Interpolation**

We wish to use cubic spline approximation for interpolation of the highly nonlinear data given in Table 5.3.

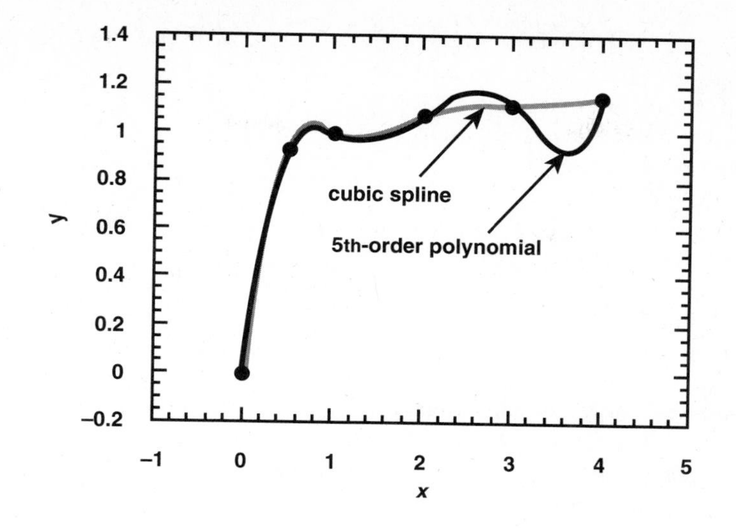


**Figure 5.3.** Interpolation Using Cubic Spline

For this problem, we can use the program CUBSPL to interpolate between any pair of $x$-values. Figure 5.3 shows the data of Table 5.3 and an interpolation curve generated from CUBSPL. For this problem, it is found that there is little difference between the "linear" or "parabolic" boundary conditions. We find that, for this particular problem, the fitted curves between $x = 0.5$ and 1.0 and between $x = 1.0$ and 2.0 show some oscillatory behavior. However, interpolation for $x > 2.0$ appears to be quite regular. Also shown in Figure 5.3 is the interpolation curve generated from a fifth-order polynomial. It is seen that cubic spline interpolation is superior to polynomial interpolation in this case.

## 5.5 RATIONAL INTERPOLATION

The interpolation approximations we have dealt with so far, namely ordinary polynomial and cubic spline, are often useful, but another procedure using polynomials is sometimes superior. This process, called *rational* interpolation, utilizes a *ratio* of polynomials to express an approximation to the data. Before deriving the appropriate equations for implementing rational approximation, we pause to give some justification for this approach. We know that ordinary polynomials are useful for approximation in a broad context (e.g., Taylor's theorem, polynomial interpolation, orthogonal polynomials). If we consider the use of the basic arithmetic operations to generalize polynomial approximation, we see that the operations of addition, subtraction, and multiplication of polynomials all lead to new polynomials, so that nothing different is produced. On the other hand, *division* of polynomials does *not* produce a polynomial, so that this operation does indeed give a new kind of function. Of course, rational functions include polynomials as a special case, corresponding to the situation where the denominator polynomial is just unity. An important reason why a *ratio* of polynomials (rather than just a polynomial itself) is often important in practice is that in engineering and science, it often happens that physical phenomena (and their mathematical models) have an asymptotic behavior as an independent variable becomes large. Common asymptotic behavior corresponds to the dependent variable approaching zero, a constant, or a bounded oscillation. Such behavior cannot be well-described by an ordinary polynomial, which necessarily has an *unbounded* asymptotic behavior. However, rational functions can mimic such behavior quite well. Functions that have a zero denominator (pole) at certain points can also be well represented by a rational function, but not by a polynomial. Finally, we note that a rational approximation version of a Taylor expansion (called a Padé approximation) is known, for many of the reasons given above, to provide in many cases a much better and wider-range approximation than the original Taylor series. Thus, we can also view rational interpolation as having the same relationship to Padé approximation that ordinary polynomial interpolation has to the Taylor expansion.

We now consider the details of rational interpolation. A computer program that implements the algorithm we discuss, called RATINTG, is discussed at the end of the chapter. The main advantages of rational approximation are that it often affords a better approximation than we can achieve otherwise, including estimation of derivatives and integrals; it gives a single relatively simple equation for the approximation; and it is often more useful than other approximations for *extrapolation* beyond the range of the data. A disadvantage is that the rational interpolation problem does not *always* have a solution; sometimes a solution can be effected by deleting a particular data point or by changing the model formulation to the form $y(x) = F(x) + G(x) \cdot \text{Rational}(x)$, where $\text{Rational}(x)$ is a rational function of $x$.

The standard notation we shall use for a rational approximation is as follows:

$$y(x) = \frac{p_0 + p_1 x + p_2 x^2 + \cdots + p_k x^k}{1 + q_1 x + q_2 x^2 + \cdots + q_m x^m} \tag{5-17}$$

By convention, we take the leading term of the denominator to be unity. To get equations for the unknowns $p_i, q_i$, we cross-multiply in Equation (5-17) and require that the resulting equation be satisfied for each of the data pairs $(x_i, y_i)$. This process produces the following system of linear algebraic equations for $(p_0, \ldots, p_k; q_1, \ldots, q_m)$:

$$p_0 + x_i p_1 + x_i^2 p_2 + \cdots + x_i^k p_k - x_i y_i q_1 - x_i^2 y_i q_2 - \cdots - x_i^m y_i q_m = y_i \tag{5-18}$$

Since for $n$ data pairs we determine the $n$ unknowns, $p_0, p_1, \ldots, p_k$; $q_1, q_2, \ldots q_m$, we must in general have $n = k + 1 + m$. Thus, as we take more terms in the denominator, we must have fewer in the numerator, and vice versa. When we use different $(k, m)$ combinations for a fixed $n$, the $p_i, q_i$ coefficients will change. As a simple, concrete illustration of the above algorithm, consider the case of three data points that are to be fitted to the rational function $y = (p_0 + p_1 x)/(1 + q_1 x)$. By cross-multiplication we get $y + q_1 y x = p_0 + p_1 x$. By slightly rearranging this equation, using a subscript $i$ to denote the $i$th data point, we obtain $p_0 + p_1 x_i - q_1 y_i x_i = y_i$ $(i = 1, 2, 3)$. This represents a system of three equations in three unknowns for $p_0$, $p_1$, and $q_1$.

It has been found empirically that *the best rational approximations usually correspond to $k \approx m$*; that is, either $k = m$ or $k$ differs from $m$ by no more than one.

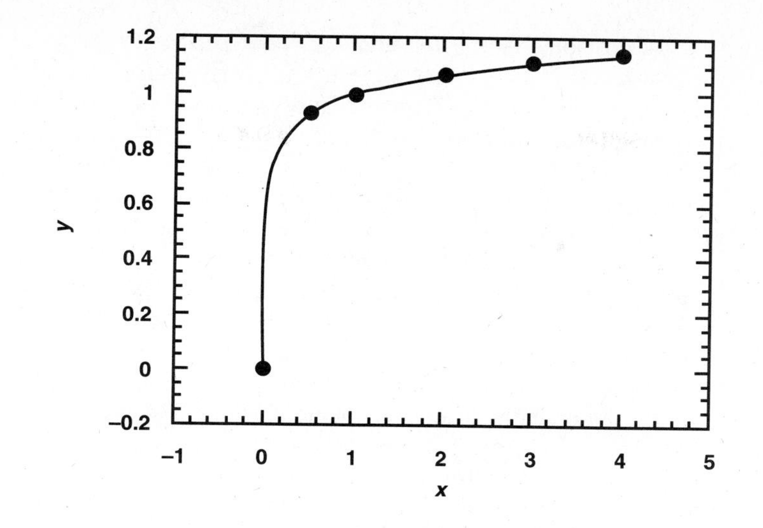


**Figure 5.4.** Rational Interpolation with $k = 3, m = 2$

To estimate error, we can compare two successive rational approximations with $n$ and $(n + 1)$ data points. Computer program RATINTG takes the specified data and gives the option to calculate $y_{k,m}(x)$ together with a plot, the derivative $y'(x)$ using a variable-increment numerical process, and the integral of $y(x)$ over any range by error-monitored numerical quadrature.

**Example 5.4 Rational Interpolation**

We wish to use rational interpolation to interpolate the data given in Table 5.3. We may use the program RATINTG with $k = 3$ and $m = 2$ to generate the interpolation curve shown in Figure 5.4. Note that other $k$ and $m$ combinations could be used so long as $(k + 1 + m)$ is equal to the number of data points $n$. We choose $k = 3$ and $m = 2$ since this keeps the degrees of the numerator and denominator nearly the same, which has been found to be useful by experience. Rational interpolation is seen to do an excellent job of interpolating these data. In particular, it provides a smooth, nonoscillatory fit in this problem. The equation of the interpolation curve for these data is

$$y = \frac{37.80x + 28.06x^2 + 0.2742x^3}{1 + 42.12x + 23.02x^2} \qquad \text{(a)}$$

## 5.6 RATIONAL LEAST SQUARES

In the authors' experience, perhaps the most powerful, general technique for obtaining an analytical approximation to a function or data set is that of *rational least squares.* We give the derivation of the equations for this technique, along with further discussion and applications, in Chapter 6 on least squares. But it is important to emphasize that this technique is just as important for purely deterministic problems (where the data are known exactly) as it is for stochastic problems (where the data may be inexact, due to some contamination such as experimental error). The rational least-squares method requires only known numerical values of the function to be approximated, and it is broadly applicable and easy to implement using program RATLSGR, which is discussed in the chapter on least squares.

The idea is based on the following observations and experience. It is known that a (rational) Padé approximation is often *vastly* superior to a Taylor approximation involving the same number of terms (see the chapter on acceleration and summation). To obtain a Padé approximation we must know the Taylor approximation, and in many difficult problems this is virtually impossible to obtain. Moreover, the Padé approximation has an asymptotic character, so it will have reduced accuracy for larger values of $x$. In addition, it is generally impossible to bound the error of Padé approximation, except by numerical comparisons. To get a rational approximation to a data set we could, of course, use rational interpolation, as discussed previously. However, rational interpolation has the great disadvantage that it is extremely *inflexible*; that is, for a given number of data points, the number of parameters is fixed. Thus, rational interpolation is not very useful for large data sets or for analytical functions (for which we require many data points).

To get around the drawbacks of Padé approximation and rational interpolation, and yet retain the important advantages of rational approximation, we instead use *rational least squares*. This procedure offers a number of advantages for fitting a set of data approximately with a rational function. First, no Taylor approximation need be computed, and only numerical data are required. Second, by using rational *least squares* (rather than rational interpolation), a large data set can be approximated using a small number of parameters. As we increase the number of parameters, the error generally decreases. Least-square approximation produces errors that are *distributed* over the interval of interest, so we generally get fairly close to a minimum maximum-error (or *minimax*) result. Also, if we have a fairly large number of data points, say 20 or more, the maximum of the discrete errors (at the known data points) can normally be expected to be close to the unknown *continuous* maximum error (at the infinity of points where there is no data). It is also easy to set up a *hybrid* rational least-squares approximation, which is also asymptotic at some point.

The procedure is to start with a reasonable number of data points (say 20 to 30, if available) for the function to be approximated, which can be either a known analytical function or a set of numerical data. Generating this information is normally not a problem since these data points are produced by some computer program. Then we simply use program RATLSGR to fit this data to the desired accuracy *using the smallest number of parameters possible*. This is accomplished by computer experiments as follows. We know from experience that best rational approximations generally correspond to cases where the degrees of the numerator and denominator differ by no more than unity. Thus, we start, for example, using a first-degree numerator ($k = 1$) and a first-degree denominator ($m = 1$). The program computes (iteratively) the values of the rational function coefficients, together with the maximum error, maximum relative error, and least-squares error sum-of-squares. For deterministic fitting problems, the user is generally interested in having a maximum error that is less than some specified value. If the ($k = 1, m = 1$) result

**TABLE 5.4** RATIONAL LEAST-SQUARES APPROXIMATION FOR $e^{-x}$

| Number of Parameters | $k$ | $m$ | Maximum \|Error\| Rational | Maximum \|Error\| Taylor |
|---|---|---|---|---|
| 3 | 2 | 0 | $2 \times 10^{-3}$ | $5 \times 10^{-2}$ |
| 3 | 1 | 1 | $9 \times 10^{-4}$ | $5 \times 10^{-2}$ |
| 5 | 4 | 0 | $3 \times 10^{-6}$ | $2 \times 10^{-3}$ |
| 5 | 2 | 2 | $5 \times 10^{-7}$ | $2 \times 10^{-3}$ |

is inadequate (which is usually the case), one progresses by adding in one or two more parameters at a time, and so on, until the desired error tolerance has been achieved.

The broad generality of this technique for deterministic problems can hardly be overemphasized. It is useful for the development of parsimonious (small number of parameters) approximations with prescribed error tolerances for known functions, and to analytically represent numerical solutions of algebraic, differential, or any other kind of equations. Rational least squares, as implemented in our program RATLSGR, represents a powerful approximation tool for both deterministic and stochastic (noisy) problems.

Program RATLSGR requires from the user only the data points and a specification of the desired degrees $(k, m)$ of the numerator and denominator, respectively. *No initial parameter estimates are required.* The program has a facility to compute the required data from an analytically specified function. The program also provides options for transforming variables for both $x$ and $y$; evaluation, numerical differentiation, or numerical integration of the fitted function; and a plot of the fitted function.

**Example 5.5 A Rational Least Squares Fit of $e^{-x}$**

To illustrate the use of rational least squares and program RATLSGR for a deterministic analytical function, we consider the approximation of $e^{-x}$ on the interval (0 to $\ln 2$). Similar polynomial approximations for this case are given in Abramowitz and Stegun [1964, p. 71]. Here we show, in Table 5.4, the results of implementing program RATLSGR for the fitting model

$$y(x) = \frac{p_0 + p_1 x + p_2 x^2 + \cdots + p_k x^k}{1 + q_1 x + q_2 x^2 + \cdots + q_m x^m} \tag{a}$$

where $y(x) = e^{-x}$. The results show that for this problem, rational least-squares approximation is more efficient that polynomial least-squares approximation (which is just a special case). Also, in terms of maximum error, the least-squares approximations are considerably better than Taylor approximations with the same number of terms.

The maximum errors for the rational approximations shown in the table correspond to using 30 data points to represent $e^{-x}$. If we consider the case $(k = 1, m = 1)$ and use 60, 100, and 200 points to represent $e^{-x}$, the maximum error at the data points increases from $8.9 \times 10^{-4}$ with 30 points, up to $9.6 \times 10^{-4}$ with 60 points, $9.9 \times 10^{-4}$ with 100 points, and $1.0 \times 10^{-3}$ with 200 points. In this example, this maximum error increases only slightly as we go from only 30 points up to 200 points.

## 5.7 GENERAL ONE-DIMENSIONAL INTERPOLATION

We have considered up to now interpolation approximation using very specific functions, either polynomials, cubic splines, or rational functions. While these are the most important in chemical engineering applications, there are situations where it is desired to do interpolation with some other functions. In principle, one can, for any approximating function, set up algebraic relations corresponding to requiring that the function and the $n$ data points agree at $n$ locations. This would generally lead to a system of either nonlinear or linear algebraic equations. These equations could then be solved, for example, by program SIMLE in the case of linear equations or by program NEWTNLE in the case of nonlinear equations.

A more convenient way to deal with such a general problem is to use our general least-squares computer program, LSTSQGR. This program is set up to handle the general interpolation problem in the event that (in contrast to least squares) the number of data points is equal to the number of parameters to be determined. The program will work directly for an approximation that is *linear* in the parameters. For models that are nonlinear in the parameters, the user must furnish initial parameter estimates. For further information, see the chapter on least squares.

## 5.8 MULTIDIMENSIONAL INTERPOLATION

It is possible to carry out multidimensional interpolation by using a sequence of one-dimensional interpolations. This is done by setting up an appropriate gridwork in each coordinate and using any of the one-dimensional interpolation procedures discussed previously. The process is best explained by an example, which follows. We should also mention that a very convenient way to carry out multidimensional interpolation (and to obtain an interpolation equation, if desired) is to use our multidimensional least-squares computer program, LSTSQMR, for the case where the number of data points equals the number of parameters to be determined. For further information on this topic, see the chapter on least squares.

**Example 5.6 Multidimensional Interpolation**

Consider the data in Table 5.5 which represents a subset of the diffusion coefficient data for nitrous oxide in aqueous methyldiethanolamine (MDEA) as measured by Al-Ghawas *et al.* [1989]. We wish to interpolate these data to estimate the diffusion coefficient of nitrous oxide in aqueous MDEA at a temperature of 27°C and an MDEA weight fraction of 0.45.

For this problem we can use repeated linear interpolation using the data at 25 and 30°C for 0.40 and 0.50 wt. fraction MDEA. We can use the program NEWTINT to carry out these calculations. For instance, at 25°C we use linear interpolation to get $D = 0.6285 \times 10^{-5}\ \text{cm}^2/\text{s}$ at an MDEA wt. fraction of 0.45. Then we repeat this process at 30°C to get $D = 0.6925 \times 10^{-5}\ \text{cm}^2/\text{s}$ at an MDEA wt. fraction of 0.45. Finally, for an MDEA wt. fraction of 0.45, using the previously calculated values for $T = 25°$ and $T = 30°$, we linearly interpolate to get $D = 0.654 \times 10^{-5}\ \text{cm}^2/\text{s}$ at 27°C.

An alternative approach for this problem would be to fit the data to an equation of the form

$$D = C_0 + C_1 x_1 + C_2 x_2 \tag{a}$$

**TABLE 5.5** DIFFUSIVITY DATA FOR $N_2O$ IN AQUEOUS MDEA

| $T$ (°C) | Weight Fraction MDEA | Diffusion Coefficient $\times 10^5$(cm$^2$/s) |
|---|---|---|
| 20 | 0.30 | 0.823 |
| 20 | 0.40 | 0.639 |
| 20 | 0.50 | 0.430 |
| 25 | 0.30 | 0.973 |
| 25 | 0.40 | 0.751 |
| 25 | 0.50 | 0.506 |
| 30 | 0.30 | 1.032 |
| 30 | 0.40 | 0.824 |
| 30 | 0.50 | 0.561 |

where $x_1$ is the temperature, $x_2$ is the wt. fraction, and the $C$s are constants to be fitted by the data. In this case we would need three sets of data to solve for the three constants; program SIMLE could be used to effect the solution. Alternatively, it is often more convenient to carry out these calculations using the capabilities of program LSTSQMR (which is described in the chapter on least squares). Using the data for 25°C and 0.40 and 0.50 wt. fraction MDEA and the data point for 30°C and 0.40 wt. fraction MDEA, we find

$$C_0 = 1.366$$
$$C_1 = 0.0146$$
$$C_2 = -2.45$$

With these values of the constants, we calculate $D = 0.658 \times 10^{-5}$ cm$^2$/s. The $D$ values interpolated in these two ways are seen to show good agreement. It should be pointed out that the latter method has the advantage that there are no restrictions on the location of the data. The linear interpolation method, however, requires that there be data entries for the diffusivity (the dependent variable) at the same two values of the temperature for two different values of the wt. fraction of MDEA.

## 5.9 SUMMARY

When it is applicable, simple linear or quadratic polynomial interpolation is very useful. To test for applicability, we use the error estimate of the Newton interpolation algorithm. When ordinary polynomial approximation is not adequate, such as when we have many data points or a rapidly changing function, or when we need the derivative, it is often best to employ cubic spline approximation. In many cases, particularly when we want an equation for the approximation or need to extrapolate beyond the range of the data, it is often best to use rational interpolation (or rational least squares). For one-dimensional interpolation involving models more general than those above, it may be convenient to use program LSTSQGR, with the number of data points equal to the number of parameters. For multidimensional interpolation, it may be convenient to use program LSTSQMR.

## 5.10 COMPUTER PROGRAMS

In this section, we discuss the computer programs that are applicable to various problems in the chapter. For each program we give a brief discussion and a concise listing, which includes (i) a program description, (ii) how to run, (iii) user-specified subroutines including a sample problem, and (iv) selected output. Detailed program listings can be obtained from the diskette.

It is important to note that in the interpolation programs CUBSPL and RAT-INTG, as well as in least-squares programs LSTSQGR and RATLSG, the data are specified in exactly the same format in the user data subroutine. This is very convenient when it is desired to use the same data set in any of these different programs. This data transfer is easily accomplished by either the Keep/Include True BASIC commands or through use of the clipboard on the Apple Macintosh.

### 5.10.1 Linear/Quadratic Interpolation Program NEWTINT

Instructions for running are given in the concise listing that follows. The program prompts for whether the data are to be specified by input prompt or in subroutine Data at the end of the program. After the data has been specified, the program prints the number of data points and the actual data pairs. The program then calculates and prints the Newton polynomial coefficients, $c(i)$, the second derivative $(d2ydx2)$, and the maximum or minimum $y$ with corresponding $x$ value. The second and third items are given only if $n = 3$ (quadratic interpolation). The program then gives the option of calculating the integral of the interpolation function. If this is desired, the program asks for the integration limits $a$ and $b$. Finally, an infinite loop is exercised to prompt the user for the calculation of $y$ and the first derivative $dy/dx$ at any desired $x$ values. To get out of this loop, one must use the stop command. The data given in the present program data subroutine correspond to the function $y(x) = x$, which is useful for checking purposes. Both the linear and quadratic options of the program return appropriate results for this particular test case.

**Program NEWTINT.**

```
!This is program NEWTINT.tru (Newton linear or quadratic interpolation)
!by O.T.Hanna. n is number of data pairs; linear(n=2),quadratic(n=3)
!After calculation of Newton interpolation coefficients c(k) in
!y = c(0) + c(1)*(x-x(0)) + c(2)*(x-x(0))*(x-x(1)), program returns
!c() coefficients,and x and y optimum,y" if n = 3. Program has options
!to calculate integral a to b,as well as y, y' values at a given x
!with estimated errors. Estimated error is from last term of Newton
!interpolation approx.
!************************************************************
!*To run, specify no. of data pts n and actual x, y data either *
!*via prompt or in Sub Data. Respond to integration prompt, and *
!*finally specify x values where y,y' are desired               *
!************************************************************
```

```
SUB Data(n,x(),y())          !Subroutine option for data input
   LET n = 3
   LET x(0) = 0
   LET y(0) = 0
   LET x(1) = 1
   LET y(1) = 1
   LET x(2) = 2
   LET y(2) = 2
END SUB
```

### *Cubic Spline Interpolation Program CUBSPL*

Instructions for running the program are given in the concise listing that follows. First, we must specify the number and values of data pairs in Sub Data; the program prints this original data. *It is important to note that the data values must be given in increasing order of* x. Next, the program asks if data is to be transformed; if desired, specification by the user of a transformation of either $y$ or $x$ (or both) is provided for in Sub Data. If the data are transformed, this is also printed. At this point, the user is asked to select either the linear or quadratic boundary conditions, as described earlier. Next, the user is asked if a value of the integral of the spline is desired; if so, a prompt is given for specification of the integration's limits, which may be arbitrary. Finally, the user is asked to choose the evaluation of $x$ either by a prompt (to get specific $y, y', y''$ values at a given $x$) or in a loop (for graphics). Both the integral and derivatives are calculated using the appropriate cubic spline formulas.

If a graphical display of $y$ versus $x$ is desired, the user must specify, prior to running the program, three parameters in subroutine Plot_info at the end of the program. These parameters are n1, the number of plot increments to be used, x1, the $x$ starting value for the plot, and x2, the $x$ ending value for the plot. When the plot routine is executed, automatic scaling of the $x$ and $y$ variables is performed. The user is asked to approve this scaling; in the event that it is not acceptable, the user is provided the option of specifying the $x$, $y$ scales and the axis tick-mark increments.

There are two possible problems shown in the concise listing below. If the function option is selected, the specified function is $3x^4 - x^3 + 10x$, and the $x$ range of calculation is from $x0 = -1$ to $x1 = 0$. To get a proper plot of this function, we must use these same values of $x0$ and $x1$ in Sub Plot_info. Alternatively, if we select the data option, we will use the $n = 6$ data points shown; the present form of Sub Plot_info is set up for this case ($x0 = 0, x1 = 1$) using $n1 = 30$ plot increments.

#### Program CUBSPL.

```
!This is program CUBSPL.tru by R.A.Davis & O.T.Hanna.
!Data or a specifed function are fitted to a cubic spline.
!There must be at least three data pairs. The program allows
!an option for calculation of the integral between any limits a,b.
```

```
!The user has option to plot results or interpolate for y, y' and y".
!The plot module for graphics is by John D. Rickey. If y info does
!not show completely, we may need to increase value of
!l_margin at line 4220.
!*******************************************************************
!*To run, specify n data pairs x(i),y(i) or else an analytically    *
!*defined function [x(i) must be increasing, in either case] in     *
!*Sub Data at end of program. Case = 1 for natural(linear) end      *
!*conditions (y" = 0 at ends). Case = 2 for parabolic end           *
!*conditions (adjacent y" values equal at each end). Program        *
!*prompts for evaluation of integral (and limits a,b), as well as   *
!*for calculation of y, y', y" at any x. If a plot is desired, user *
!*must specify parameters in Sub Plot_info at end of program        *
!*******************************************************************
SUB Data(n,x(),y())
  !Here we generate x(i),y(i) either by read statements if data
  !is specified or by user-defined function f(x) below.
  !The marks *** designate where the user must modify program.

  INPUT prompt"Supply data(0) or specify function(1) ?":ques
  IF ques = 1 THEN
    !Option for spline fit to user-defined function y = f(x) on
    !the interval x0 < x < x1
    !Function is y(i) = f(z), where z replaces x
    !This allows cubic spline approx., integral, derivative,
    !plot, for any analytically defined function.
    LET x0 = -1        !***initial x; use same in Sub Plot_info
    LET x1 = 0         !***final x
    INPUT prompt"No. of parts into which interval is divided?":m
    LET n = m + 1      !no. of 'data' pts. for function evaluation
    MAT redim x(n),y(n)
    FOR i = 1 to n
       LET delta_x = (x1 - x0)/m !x increment for function
       LET x(i),z = x0 + (i-1)*delta_x
       !y(i)=f(z) below is user-defined, z is dummy for x
       LET y(i) = 3*z^4 - z^3 +10*z          !***y(i)=f(z)
    NEXT i
  ELSE IF ques = 0 THEN
     !Option for user-specified data
     !This allows cubic spline approx.,integral,derivative,
     !plot, of the specified data.
     LET n = 6             !*** - no. of data pairs specified
     MAT redim x(n),y(n)
     PRINT"Number of data pairs = ";n
     PRINT
     PRINT"Original Data"
     FOR i = 1 to n
       READ y(i),x(i)
       PRINT"y(";i;")= ";y(i);Tab(25);"x(";i;")= ";x(i)
     NEXT i
```

```
  END IF

  PRINT
  INPUT prompt"Is data to be transformed? (No-0,Yes-1) ":q
  IF q = 1 then
     PRINT"Transformed data Log(y), 1/x" !***user specified
     FOR i = 1 to n
       LET y(i) = Log( y(i) )           !***
       LET x(i) = 1/x(i)                !***
       PRINT"y(";i;")= ";y(i);Tab(25);"x(";i;")= ";x(i)
     NEXT i
  END IF
  PRINT

  !Specify data pairs as DATA y(i), x(i)  !***
  DATA   1,          0.0
  DATA   .368,       .2
  DATA   .135,       .4
  DATA   .0498,      .6
  DATA   .0183,      .8
  DATA   .00674,     1.0

END SUB

SUB Plot_info (n1,x1,dx1)
   !Usually convenient to use from 20 to 50 plot points
   !*** denotes required user specification
   LET n1 = 30              !*** no. of plot increments to be used
   LET x1 = 0               !*** xstart for plot of curve
   LET x2 = 1               !*** xend for plot of curve
   LET dx1 = (x2 - x1)/n1 !x increment for plot
   LET n1 = n1 + 1          !no. x(i),y(i) to be calculated
END SUB
```

### ***Rational Interpolation Program RATINTG***

Instructions for running the program are given in the concise listing that follows. We note that the general organization for this program (data format, graphics) is virtually the same as for program CUBSPL discussed above. Rather than repeat everything, we note what is *different* about using program RATINTG. First, the data values can be given in any order. The degrees of the polynomials in the numerator ($k$) and denominator ($m$) of the desired rational approximation must be specified subject to the algebraic requirement $n = (k + 1 + m)$. No end conditions are called for. The integral and derivative of the rational approximation are computed numerically, the integral by subroutine TRMPQUAD and the derivative by an automatic variable-increment calculation. For the integral, the user must respond to the prompt for the number of intervals to use; the program returns the value of the integral together with a conservative estimate of the relative error (provided the value of the calculated

Ratio is between 0.8 and 1.2). For information on the Ratio, see the chapter on differential equations.

### Program RATINTG.

```
!This is program RATINTG.tru for rational interpolation by O.T.Hanna
!Program includes J.Rickey automatic-scaling plot module.
!Given n data pairs x(i),y(i),program calculates rational polynomial
!approx. of the form y = (p0 + p1*x ..+ pk*x^k)/(1 + q1*x ..+ qm*x^m).
!Option is given for evaluation of integral from a to b, as well as
!for derivative of rational function.

!*********************************************************************
!*To run,specify n data pairs x(i),y(i) in Sub Data at end of program    *
!*The degrees of the numerator (k) and denominator (m) must be          *
!*specified; we must have n = (k+1+m). Data can be transformed if       *
!*desired. Program prompts for calculation of integral from a to b      *
!*and calculation of y, y', y" at any x. If a plot is desired, user must *
!*specify parameters in Sub Plot_info at end of program                 *
!*********************************************************************


SUB Data(n,x(),y(),k,m)
   !Here we generate x(i),y(i) by read statements.
   !The marks *** designate where the user must modify program.
   LET k = 5                          !***degree of numerator
   LET m = 0                          !***degree of denominator
   PRINT"k = ";k;Tab(20);"m = ";m
   LET n = 6                          !*** - no. of data pairs
   PRINT"Number of data pairs = ";n
   PRINT
   PRINT"Original Data"
   FOR i = 1 to n
      READ y(i),x(i)
      PRINT"y(";i;")= ";y(i);Tab(25);"x(";i;")= ";x(i)
   NEXT i
   PRINT
   INPUT prompt"Is data to be transformed?(No-0,Yes-1) ":q
   IF q = 1 then
      PRINT"The following is transformed data Log(y), 1/x"  !***
      FOR i = 1 to n
         LET y(i) = Log( y(i) )       !***
         LET x(i) = 1/x(i)            !***
         PRINT"y(";i;")= ";y(i);Tab(25);"x(";i;")= ";x(i)
      NEXT i
   END IF
   PRINT

   !Specify data pairs as DATA y(i), x(i)           !***
   DATA    1,      0.0
   DATA    .368,   .2
```

```
    DATA     .135,    .4
    DATA     .0498,   .6
    DATA     .0183,   .8
    DATA     .00674, 1.0

END SUB

SUB Plot_info (n1,x1,dx1)
    !Usually convenient to use from 20 to 50 plot points
    !*** denotes required user specification
    LET n1 = 30             !*** no. of plot increments to be used
    LET x1 = 0              !*** xstart for plot
    LET x2 = 1              !*** xend for plot
    LET dx1 = (x2 - x1)/n1  !x increment for plot
    LET n1 = n1 + 1         !no. x(i),y(i) to be calculated
END SUB
```

## PROBLEMS

**5.1.** The following data for the thermal conductivity of acetone vapor are taken from Bennett and Myers [1984]:

| $k$(Btu/hr-ft-°F) | T(°F) |
|---|---|
| 0.0057 | 32 |
| 0.0074 | 115 |
| 0.0099 | 212 |
| 0.0147 | 363 |

It is suggested that $\ln k$ varies approximately linearly with $\ln T$, where $T$ is expressed in °R(°R =° F + 460).

It is desired to estimate the value of $k$(300°F) by using interpolation methods in the suggested coordinates. Estimate the accuracy of your results. To accomplish this, use quadratic interpolation by means of the Newton algorithm. Solve this without using the computer (use a calculator as if this were an exam), and also use program NEWTINT. Compare the results.

**5.2.** Using the data from problem 5.1, it is necessary to estimate the temperature (°F) at which $k = 0.008$. Do this using program NEWTINT and (i) a trial process together with the direct procedure from problem 5.1 and (ii) inverse quadratic interpolation. Are the results the same?

**5.3.** Fit the data points $x = 1$, $y = 0.4$ and $x = 2$, $y = 0.1$ with the following two models:
(i) $y = c_1 \operatorname{Exp}(c_2 * x)$ and
(ii) $y = (2 + p_1 * x)/(3 + q_2 * x^2)$. Use program SIMLE.

**5.4.** Use the data of Table 5.5 to estimate the value of the diffusion coefficient $D$ at $T$ = 22°C and wt. fraction MDEA = 0.36. Do this by fitting an equation of the form $D = c_0 + c_1 * X_1 + c_2 * X_2$, where $X_1$ is $T$ and $X_2$ is wt. fraction MDEA. Use program SIMLE.

**5.5.** Use the data of Table 5.5 together with program LSTSQMR to determine an appropriate two-dimensional quadratic interpolation approximation for $D(T, X)$ of the form $D = c_1 + c_2T + c_3X + c_4T^2 + c_5TX + c_6X^2$. For $T = 22°C$ and $X = 0.36$, compare the results of this quadratic prediction for $D$ with that of linear interpolation in problem 5.4. The difference may be reflective of the error as (i) a realistic estimate of the error in the linear approximation or (ii) a conservative estimate of the error in the quadratic approximation.

**5.6.** Measurements on a gas have yielded the following experimental values of $y$ versus $x$.

| $x$ | $y$ |
|---|---|
| 25 | 0.148 |
| 50 | 0.149 |
| 100 | 0.156 |
| 150 | 0.160 |
| 175 | 0.167 |

It is required to estimate, as accurately as possible, the value of $y$ at $x = 0$. To do this requires *extrapolation* of the data, which is always a hazardous process. There are many possible procedures, including polynomial, rational, and spline interpolation, as well as least-squares approximation. Also, it may not be clear how much of the data should be used in the process. Since we generally cannot know in advance which techniques may be most appropriate, an experimental approach is often desirable. Consistencies (as well as inconsistencies!) are useful as a guide to the best estimate. Graphical assessments of the various procedures are also helpful.

Use polynomial interpolation, together with error estimates and graphical interpretation, to analyze this problem and determine $y(0)$.

**5.7.** Solve problem 5.6 using rational interpolation, together with error estimates and graphical interpretation, to determine $y(0)$.

**5.8.** Solve problem 5.6 using cubic spline interpolation, together with error estimates and graphical interpretation, to determine $y(0)$.

## REFERENCES

ABRAMOWITZ, M. and STEGUN, I. (eds.) (1964), *Handbook of Mathematical Functions*, Washington, DC: National Bureau of Standards.

AL-GHAWAS, H. A., HAGEWIESCHE, D. P., RUIZ-IBANEZ, G., and SANDALL, O. C., (1989), *J. Chem. Eng. Data,* 34, p. 385.

DAHLQUIST, G., BJORCK, A., and ANDERSON, N. (1974), *Numerical Methods*, Prentice-Hall, Englewood Cliffs, NJ

GERALD, C. (1978), *Applied Numerical Analysis,* 2nd ed., Addison-Wesley, Reading, Mass.

JOHNSON, L. W., and RIESS, R. D. (1982), *Numerical Analysis,* 2nd ed., Addison-Wesley, Reading, Mass.

KAHANER, D., MOLER, C., and NASH, S. (1989), *Numerical Methods and Software*, Prentice-Hall, Englewood Cliffs, NJ

PRESS, W. H., FLANNERY, B. P., TEUKOLSKY, S. A., and VETTERLING, W. T. (1986), *Numerical Recipes*, Cambridge Univ. Press, New York

# 6

# Least-Squares Approximation

## 6.1 LINEAR LEAST SQUARES

The so-called method of least squares is a very important approximation method in engineering. This method is perhaps the best known "distributed error" approximation method. Distributed error methods are characterized by a decrease of some global error measure with respect to the *whole approximation interval* as the order of approximation increases. This is in contrast to asymptotic methods, such as Taylor approximation, which are characterized by a decrease of error at some special point as the order of approximation increases.

Least-squares approximation is valuable in problems such as fitting equations to discrete data points and in analyzing measurement errors. The subject of least-squares analysis also plays a central role in the application of the theory of statistics, which treats problems involving random quantities such as experimental errors. The subject of random variables and statistics is beyond the scope of our presentation; we will therefore use least squares only in a deterministic sense. Least squares is also useful for continuous approximations, such as developing simple approximations to known functions. There are several variations on the theme of least-squares approximation, such as discrete versus continuous, constrained versus unconstrained, and linear versus nonlinear.

### 6.1.1 Simple "Linear" Least Squares

Since unconstrained, discrete, linear least squares is used most frequently in applications, this situation will now be developed in detail. Discrete least-squares approximation is very valuable because of its relationship to statistical analysis of process or measurement "noise." Here, we discuss the elementary mechanics of implementation. However, *interpretation* of the results is the most important part, and this requires statistical analysis, which is beyond the scope of our treatment. To illustrate the ideas, suppose the data points in Figure 6.1 represent some calibration data for air flow as measured by a rotameter. If the data points were considered to be "exact," then clearly we must be observing a *peculiar* x-y function which would necessarily somehow wiggle through these data points. Another, perhaps more plausible, view might be that the actual x-y relationship is not "peculiar," but rather, the data points are inexact and contaminated by some sort of errors. Suppose, for example, that there is a basis for believing that the x-y relationship is actually linear, of the form $y_p(x_i) = B_1 + B_2 x_i$, where the notation $y_p$ refers to $y$ "predicted." The least-squares procedure would be to form the sum of the squares of the differences between the observed values $y_i$ and the predicted values $y_p(x_i)$ and minimize this with respect to the unknown parameters $B_1$ and $B_2$. Thus, we would form a normalized "Error" quantity

$$\text{"Error"} = \frac{\sum_{i=1}^{N} [y_i - (B_1 + B_2 x_i)]^2}{N - 2} \tag{6-1}$$

Here, $N$ is the total number of data points, and the factor $(N - 2)$ is used rather than $N$ since there are two parameters to be determined. To determine least-squares values of $B_1$ and $B_2$, we minimize "Error" with respect to $B_1$ and $B_2$ by using $\partial\text{Error}/\partial B_1 = 0$, $\partial\text{Error}/\partial B_2 = 0$. This then yields the two equations

$$\begin{aligned} 0 &= \sum_{i=1}^{N} [y_i - (B_1 + B_2 x_i)] \\ 0 &= \sum_{i=1}^{N} [y_i - (B_1 + B_2 x_i)] x_i \end{aligned} \tag{6-2}$$

These two equations can be represented in the convenient compact form

$$0 = \sum_{i=1}^{N} [y_i - (B_1 + B_2 x_i)] x_i^k \qquad (k = 0, 1) \tag{6-3}$$

By expanding the sums, we generate two linear algebraic equations for the coefficients $B_1$ and $B_2$. When $B_1$ and $B_2$ are known, the quantity "Error" can be determined. The computed value of "Error" (using the optimally determined coefficients) gives an indication of the goodness of fit of the data to the model. To make this statement more precise requires statistical analysis.

### 6.1.2 General Linear Regression

We now proceed to develop general weighted, discrete, *linear* least squares using general approximating functions. The term "linearity" refers to the dependence of the prediction function on the *parameters*. The least squares approximation process is often called "regression," which has its origin in the subject of statistics. Suppose we wish to fit $N$ data pairs $(x_i, y_i)$ with an approximation function that has $p$

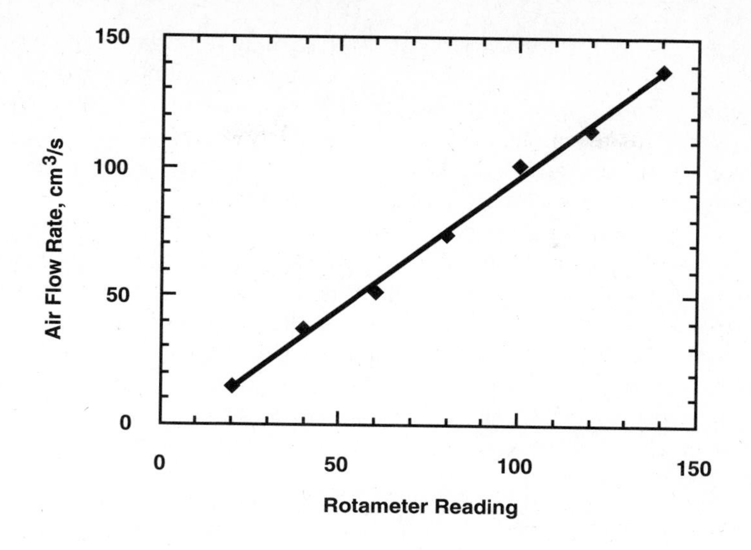


**Figure 6.1.** Rotameter Calibration Data for Air Flow at 22 C and 1 Atm

parameters, of the form $y_i \approx y_p \equiv g_0(x_i) + B_1 g_1(x_i) \cdots + \cdots + B_p g_p(x_i)$. We then define "Error" as

$$\text{"Error"} = \frac{\sum_{i=1}^{N} w_i [y_i - (g_0(x_i) + B_1 g_1(x_i) + \cdots + B_p g_p(x_i))]^2}{N - p} \tag{6-4}$$

Here, Equation (6-4) includes the "weight" factors $w_i$ which are sometimes used; for our purposes, the $w_i$ will always be taken as unity. To determine the coefficients $B_1, \ldots, B_p$, we form the necessary conditions for a minimum, $0 = \partial \text{Error}/\partial B_k$, $k = 1, \ldots, p$. These $k$ conditions yield the so-called normal equations of least squares, which are shown below and are merely linear algebraic equations for the $B_i$.

$$0 = \sum_{i=1}^{N} w_i [y_i - (g_0(x_i) + B_1 g_1(x_i) + \cdots B_p g_p(x_i))] g_k(x_i) \quad (k = 1, \ldots, p) \tag{6-5}$$

The $g_k(x)$ functions can be chosen arbitrarily; power functions are often used. The weights, $w_i$, are normally positive and reflective of the relative importance of the data points. However, in the absence of other information, they are usually chosen to be unity.

Once the $B_i$ coefficients are known, "Error" can be calculated and the goodness of fit assessed. One important question concerns the degree of approximation to be used. If this is known and fixed, then Error is fixed. But if the degree of approximation is allowed to vary by using more and more $g_k(x)$ functions in the approximation, we need to know where to stop. In general, for any number of data points, we can get exact agreement (and the *numerator* of Error = 0) if we choose $p = N$. This is usually undesirable; we ordinarily wish to choose $p$ so that Error has its smallest (or nearly smallest) value corresponding to use of a small number of parameters. That is, if Error is just a bit smaller for a larger $p$, the additional degree of approximation may not be worth a small reduction of error.

It should be mentioned that least squares problems are sometimes algebraically *ill-conditioned.* This means that, in some cases, the calculated parameter values may be sensitive to small changes in the data. This tends to happen especially when there are a large number of parameters in the problem. To guard against possible ill-conditioning, it is generally a good idea to work in double-precision arithmetic on the computer. This is done automatically for all calculations in True BASIC.

Computer program LSTSQGR can be used to carry out general linear (or non-linear) regression analysis. In addition to calculating the least squares coefficients, it has facilities for evaluation, numerical differentiation, and numerical integration of the prediction function. A plot of the data and fitted function is also an option. See the discussion under "Computer Programs" at the end of the chapter.

In the case of *continuous* linear least-squares approximation, we minimize the following integral with respect to the $B_i$:

$$\text{Error} = \int_a^b w(x)[y(x) - (g_0(x) + B_1 g_1(x) + \cdots B_p g_p(x))]^2 dx \tag{6-6}$$

In this case, the normal linear equations for the $B_i$ are the same as Equations 6-5 except that the summations are replaced by the corresponding integrals.

$$0 = \int_a^b w(x)[y(x) - (g_0(x) + B_1 g_1(x) + \cdots B_p g_p(x))]g_k(x)dx \tag{6-7}$$

Here, $w(x)$ is a positive weight function (usually unity), and the $g_k(x)$ can be chosen arbitrarily (often as power functions). Once the $B_i$ are calculated, the optimal least-squares approximation is known. It turns out that, as would be expected, continuous least-squares problems can often be well approximated by the corresponding *discrete* cases, provided that a reasonably large number of data points (20 or more) from the continuous function are used. An example of this type, using the function $e^{-x}$ and rational least squares, was considered in the chapter on interpolation.

**Example 6.1 Fit of Heat Capacity Data to Quadratic Equation in Temperature**

It is desired to fit the heat capacity data for methylcyclohexane given by Vargaftik [1975] to a quadratic function of temperature, $C_p = a + bT + cT^2$. Here, $C_p$ is the heat capacity, $T$ is absolute temperature, and $a$, $b$, and $c$ are parameters to be determined by linear least squares. The data are shown in Table 6.1.

For convenience in notation, we redefine the original variables as follows: $C_p = y_i$; $T = x_i$; $a = B_1$, $b = B_2$, and $c = B_3$. In these new variables, the prediction function is $y_p(x) = B_1 + B_2x + B_3x^2$. We therefore minimize

$$\text{Error} = \frac{\sum_{i=1}^{N}[y_i - (B_1 + B_2x_i + B_3x_i^2)]^2}{N-3} \tag{a}$$

with respect to $B_1$, $B_2$, and $B_3$. This gives the normal equations

$$\begin{aligned}
&(1)\quad B_1N + B_2\sum x_i + B_3\sum x_i^2 = \sum y_i \\
&(2)\quad B_1\sum x_i + B_2\sum x_i^2 + B_3\sum x_i^3 = \sum x_iy_i \\
&(3)\quad B_1\sum x_i^2 + B_2\sum x_i^3 + B_3\sum x_i^4 = \sum x_i^2y_i
\end{aligned} \tag{b}$$

**TABLE 6.1** HEAT CAPACITY OF LIQUID METHYLCYCLOHEXANE ($C_7H_{14}$)

| $T$, °K | $C_p$, kJ/kg°K |
|---|---|
| 150 | 1.426 |
| 160 | 1.447 |
| 170 | 1.469 |
| 180 | 1.492 |
| 190 | 1.516 |
| 200 | 1.541 |
| 210 | 1.567 |
| 220 | 1.596 |
| 230 | 1.627 |
| 240 | 1.661 |
| 250 | 1.696 |
| 260 | 1.732 |
| 270 | 1.770 |
| 280 | 1.808 |
| 290 | 1.848 |
| 300 | 1.888 |

Thus, we calculate the coefficients defined by the data pairs $(x_i, y_i)$ and then solve three linear equations for $B_1$, $B_2$, and $B_3$. Knowing $B_1$, $B_2$, and $B_3$ we can calculate Error, which is a measure of the goodness of fit. Then $y_p(x) = B_1 + B_2x + B_3x^2$ represents the least-squares approximation. Note that, for example, the above set of equations is also valid for calculating a *linear* least-squares approximation of the form $y_p = B_1 + B_2x$. To do this, we simply delete equation (3) and set $B_3 = 0$ in equations (1) and (2). This leaves two equations in two unknowns for $B_1$ and $B_2$, which, it should be noted, will generally have *different* values depending on whether we use a linear or quadratic prediction function. When we run program LSTSQGR for this problem, we get $a = B_1 = 1.32803$, $b = B_2 = -5.50567 \times 10^{-4}$, $c = B_3 = 8.07773 \times 10^{-6}$, and Error $= 1.82 \times 10^{-6}$. A plot of the data, shown in Figure 6.2, indicates the fit to be very good.

### 6.1.3 Numerical Differentiation and Integration of Noisy Data

An important application of least squares concerns the mathematical manipulation of so-called noisy data, that is, data which are somehow contaminated by experimental error. The process of numerical differentiation is particularly susceptible to errors in the data. We can think of fitting a function with least squares as interpolation with noisy data. In a similar way, if we need to perform differentiation or integration of noisy data, it is generally not useful to do this with the original data. Instead, the data can be fit with an appropriate prediction function using least squares, and then the *prediction formula* can be used to get any desired derivatives or integrals, either analytically or numerically. This procedure usually diminishes the effects of random errors.

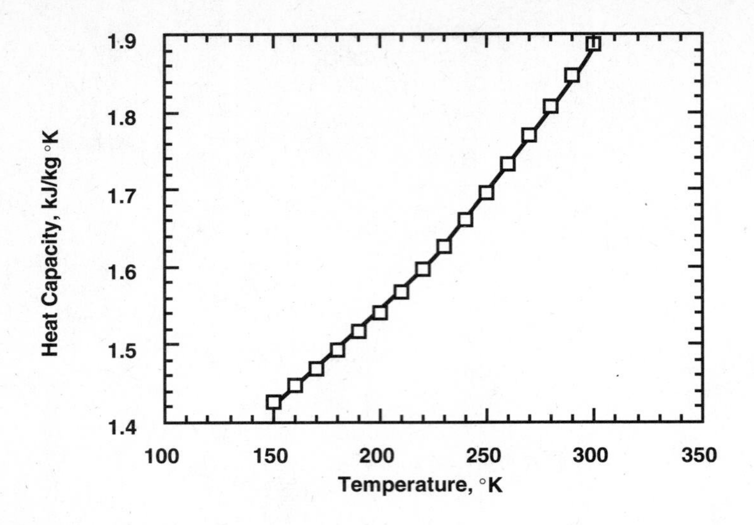


**Figure 6.2.** Heat Capacity vs. Temperature for Methylcyclohexane

### *6.1.4 Effects of Linear Constraints*

It occasionally happens that the prediction equation is required to satisfy certain constraints exclusive of the least-squares process. This means that the parameters in the prediction equation must satisfy particular constraints, and therefore, the parameters cannot all be independent, as is assumed in the usual least-squares algorithm. In such a situation the least-squares procedure must be modified appropriately.

In the case where the constraints take the form of one or more linear equations involving the parameters, the approach is as follows. In principle, any $k$ equation constraints could generally be solved for $k$ of the parameters in terms of those remaining, and the $k$ solved-for parameters could be eliminated from the prediction expression. Then the *modified* prediction expression would be used in the usual unconstrained least-squares procedure described above. Alternatively, the Lagrange Multiplier technique could be used to incorporate constraints. The ideas are best illustrated by means of an example.

**Example 6.2** $y_p = B_1 + B_2x + B_3x^2$ **Subject to** $y(1) = 2$

In this problem, the given constraint, when substituted into the prediction equation, gives $2 = B_1 + B_2(1) + B_3(1^2)$. Thus, $B_1$, $B_2$, and $B_3$ cannot all be independent. We have a choice of eliminating any one of these parameters in terms of the other two (of course, the final predictions will be the same regardless of which parameter is eliminated). Here we *arbitrarily* eliminate $B_3$ in favor of $B_1$ and $B_2$ according to $B_3 = 2 - B_1 - B_2$. This causes the new *unconstrained* prediction equation to become $y_p = 2x^2 + B_1(1 - x^2) + B_2(x - x^2)$.

This is just a particular case of the general prediction function $y_p = g_0(x) + B_1g_1(x) + B_2g_2(x)$, where $g_0(x) = 2x^2$, $g_1(x) = 1 - x^2$, and $g_2(x) = x - x^2$. We now minimize, as usual, $\sum(y_i - y_p)^2$ with respect to the *independent* parameters $B_1$ and $B_2$ of the new prediction function as follows:

$$\min_{B_1, B_2} \sum_{i=1}^{N} [y_i - (2x_i^2 + B_1(1 - x_i^2) + B_2(x_i - x_i^2))]^2 \quad \text{(a)}$$

By differentiating with respect to $B_1$ and $B_2$, and setting equal to zero in the usual way, we get the following linear algebraic equations for $B_1$ and $B_2$:

$$0 = \sum_{i=1}^{N} [(y_i - 2x_i^2) - B_1(1 - x_i^2) - B_2(x_i - x_i^2)](1 - x_i^2) \quad \text{(b)}$$

$$0 = \sum_{i=1}^{N} [(y_i - 2x_i^2) - B_1(1 - x_i^2) - B_2(x_i - x_i^2)](x_i - x_i^2) \quad \text{(c)}$$

After $B_1$ and $B_2$ have been calculated from Equations (b) and (c), we substitute them back into the modified prediction equation. This equation automatically satisfies the original constraint.

### 6.1.5 Comparison of Different Models

In some problems it is necessary or desirable to discriminate between more than one possible prediction equation. One common example of this corresponds to the fitting of data to different degree polynomials. A simple and useful decision criterion is use of the "Error" quantity, defined in general as $\sum(y_i - y_p)^2/(N - p)$, where $N$ is the number of data points and $p$ is the number of parameters in the prediction equation. For an equal number of parameters between two models, we would generally accept the one having the smallest value of Error.

Notice, however, that if we consider a succession of possible models having an increasing number of parameters (such as a sequence of polynomials of increasing degree), the numerator of Error will generally decrease as more parameters are added, while the denominator will also decrease. As more parameters are added, we finally reach the point where, when $N = p$, the numerator and denominator both equal zero. This is the case of *interpolation*, where the prediction equation goes exactly through all of the data points. Thus, as we increase the number of parameters in a model, whether or not Error decreases depends on whether the numerator or denominator decreases most rapidly. If we start with a small number of parameters, Error usually decreases fairly rapidly and eventually achieves a sort of "bouncing around" minimum. The "best" model is usually taken to be one with a *small* number of parameters which also has a small value of Error. Obviously, some judgment is required in this discrimination. In general, it is good practice to keep the number of parameters as small as is reasonably possible. This also helps avoid ill-conditioning problems.

## 6.2 NONLINEAR LEAST SQUARES

In many situations, the mathematical model describing the variation of $y$ with $x$ is a nonlinear function of the *parameters* (for example, $y = B_1x^3 + B_2\exp(-2x)$ is linear in the parameters $B_1$, $B_2$; on the other hand, $y = B_1x^2 + \exp(-B_2x)$ is *nonlinear*). In such a case, the least-squares procedure generally involves the solution of nonlinear algebraic equations. To illustrate the basic ideas, suppose the prediction equation is of the form $y_p(x) = f(B_1, B_2, x)$ in the case of two unknown parameters $B_1$ and $B_2$. Then the least squares procedure would involve minimizing the quantity $\sum[y_i - f(B_1, B_2, x_i)]^2$ with respect to $B_1$ and $B_2$. The necessary conditions for a minimum would be expressed as

$$\sum_{i=1}^{N}[y_i - f(B_1, B_2, x_i)]\frac{\partial f}{\partial B_j} = 0 \quad (j = 1, 2) \tag{6-8}$$

This condition corresponds to two nonlinear algebraic equations for the unknowns $B_1$ and $B_2$. More general problems could be approached in an analogous way.

### 6.2.1 The Gauss-Newton Method

Although it is possible to attempt to solve nonlinear least-squares problems by dealing directly with the standard nonlinear "normal" equations for the parameters, in practice it is usually much better to proceed in a slightly different way, due to the special structure of the least-squares minimization problem. One useful approach, which is relatively simple, is often referred to as the Gauss-Newton Method. This method is a successive-approximation procedure which is based on linearizing the prediction function with respect to the *parameters* about estimated values, prior to the least-squares minimization. This leads to a *sequence* of *linear* least-squares problems whose solutions (hopefully) will approach the desired solution of the original nonlinear problem. To develop the method for a two-parameter situation, consider the nonlinear least-square minimization

$$\min_{B_1, B_2} \sum_{i=1}^{N} \frac{[y_i - f(B_1, B_2, x_i)]^2}{N-2} \tag{6-9}$$

Now we use Taylor's theorem as follows to linearize the prediction function with respect to the parameters $B_1$ and $B_2$ about some known values $B_{1a}$, $B_{2a}$:

$$f(B_1, B_2, x_i) = f(B_{1a}, B_{2a}, x_i) + \frac{\partial f}{\partial B_{1a}}(B_1 - B_{1a}) + \frac{\partial f}{\partial B_{2a}}(B_2 - B_{2a}) \tag{6-10}$$

Here, the partial derivatives are evaluated at the known values $B_{1a}$, $B_{2a}$, $x_i$. By substituting this approximate relation into the original least-squares minimization, we generate the following *linear* least-squares problem for the unknown values $B_1$ and $B_2$:

$$\min_{B_1, B_2} \sum_{i=1}^{N} \frac{[y_i - (f_a + \frac{\partial f}{\partial B_{1a}}(B_1 - B_{1a}) + \frac{\partial f}{\partial B_{2a}}(B_2 - B_{2a}))]^2}{N-2} \tag{6-11}$$

The "normal" linear equations for $B_1$ and $B_2$ are therefore

$$0 = \sum_{i=1}^{N}\left[y_i - (f_a + \frac{\partial f}{\partial B_{1a}}(B_1 - B_{1a}) + \frac{\partial f}{\partial B_{2a}}(B_2 - B_{2a}))\right]\frac{\partial f}{\partial B_{j_a}} \quad (j = 1, 2) \tag{6-12}$$

Starting with initial approximate values $B_{1a}$, $B_{2a}$, we compute $B_1$ and $B_2$, the (hopefully) improved estimates of these quantities. To continue, the newly calculated values $B_1$ and $B_2$ are taken to be the approximate values $B_{1a}$ and $B_{2a}$ for the next iteration, and so on. The process is terminated when successively calculated values of the parameters agree to the desired accuracy. Note that if $B_1$ and $B_2$ are the *exact* values, the Gauss-Newton normal equation reduces to the standard normal equation. It is useful to observe that in terms of algebra, each computational cycle of Gauss-Newton is the same as a *linear* regression problem if $g_0$ is replaced by $f(B_1, \ldots, B_p, x_i)$, and $g_i$ by $\partial f/\partial B_i$. Thus, it is convenient to have a computer program that will do either linear or nonlinear regression, according to the proper definition of the $g_i$ functions and whether or not iteration is required. The programs LSTSQGR and LSTSQMR are set up in this way; they are discussed in the "Computer Programs" section.

The convergence of the Gauss-Newton method is dependent on determining, at some point in the process, parameter values that are sufficiently close to the true values. The necessity of "good" starting values is the essential difficulty of nonlinear least squares, as opposed to linear least squares, where *no* starting values are required. If divergence of Gauss-Newton is encountered, different approximate parameter values can be tried. One procedure to get appropriate initial starting values of the parameters is to fit various data subsets exactly to the approximating function through a nonlinear interpolation process (see the section on "Miscellaneous Topics in Least Squares"). In some problems we can initially break up the model equation and the data into a small $x$ region and a large $x$ region and treat each region separately in order to get initial starting values. Alternatively, we can sometimes proceed by assuming "trial" values for the nonlinear parameters and directly solve the remaining linear problem in order to move toward the nonlinear solution. Another possible approach is to attempt to transform variables in order to achieve a linear problem. This is discussed in the next section.

**Example 6.3 Fitting Vapor-Liquid Equilibrium Data to the Wilson Equation**

Vapor-liquid equilibrium data, shown in Table 6.2, were obtained for the heptane-toluene binary system at one atmosphere pressure using a Myers Equilibrium Still. These data were collected in the UCSB undergraduate laboratory. It is desired to fit the experimental activity-coefficient data to the Wilson equation.

**Solution.** The activity coefficients as predicted by the Wilson equation [Prausnitz, 1969] are

$$\ln \gamma_H = -\ln(x_H + Ax_T) + x_T\left(\frac{A}{x_H + Ax_T} - \frac{B}{Bx_H + x_T}\right) \tag{a}$$

and

$$\ln \gamma_T = -\ln(x_T + Bx_H) - x_H\left(\frac{A}{x_H + Ax_T} - \frac{B}{Bx_H + x_T}\right) \tag{b}$$

In Equations (a) and (b) the $\gamma$'s are activity coefficients, the $x$'s are mole fractions, $A$ and $B$ are the parameters to be determined, and the subscripts $H$ and $T$ refer to hexane and toluene, respectively. The experimental data could be fitted to either Equation (a) or (b); but, because of experimental errors, the values of $A$ and $B$ determined from the two data sets would be different. In this problem it is better to use all of the data and work in terms of the excess Gibbs free energy, which leads to a standard least-squares problem. The Wilson equation prediction of the excess Gibbs free energy is

**TABLE 6.2** ACTIVITY-COEFFICIENT DATA FOR THE HEPTANE-TOLUENE SYSTEM

| $x_H$ | $\gamma_H$ | $x_T = 1 - x_H$ | $\gamma_T$ | $g^E/RT$ |
|---|---|---|---|---|
| 1.000 | 1.000 | 0.000 | | 0.0000 |
| 0.790 | 1.058 | 0.210 | 1.473 | 0.1259 |
| 0.596 | 1.160 | 0.404 | 1.167 | 0.1509 |
| 0.480 | 1.189 | 0.520 | 1.114 | 0.1392 |
| 0.390 | 1.234 | 0.610 | 1.073 | 0.1250 |
| 0.293 | 1.184 | 0.707 | 1.091 | 0.1111 |
| 0.220 | 1.313 | 0.780 | 1.046 | 0.0950 |
| 0.150 | 1.215 | 0.850 | 1.050 | 0.0707 |
| 0.065 | 1.335 | 0.935 | 1.011 | 0.0290 |
| 0.000 | | 1.000 | 1.000 | 0.0000 |

$$y_p = \left(\frac{g^E}{RT}\right)_{\text{theory}} = -x_H \ln\{x_H + A(1 - x_H)\} - (1 - x_H)\ln\{1 - x_H + Bx_H\} \tag{c}$$

The quantity $g^E/RT$ is determined from the experimental data as

$$y_i = \left(\frac{g^E}{RT}\right)_{\text{expt.}} = x_H \ln \gamma_H + (1 - x_H)\ln \gamma_T \tag{d}$$

To solve this nonlinear least-squares problem using program LSTSQGR, we must first calculate the required partial derivatives of the prediction function, Equation (c), with respect to the parameters $A$ and $B$. These are

$$\frac{\partial y_p}{\partial A} = \frac{-x_H(1 - x_H)}{x_H + A(1 - x_H)}, \quad \frac{\partial y_p}{\partial B} = \frac{-(1 - x_H)x_H}{1 - x_H + Bx_H} \tag{e}$$

From past experience with this problem, it is known that reasonable starting values for the Gauss-Newton iteration in program LSTSQGR are $A = 1.8$ and $B = 0.16$. When these starting values are used, we obtain convergence to the parameter estimates $A = 1.238$, $B = 0.3285$ with Error $= 3.3 \times 10^{-5}$. Figure 6.3 shows a comparison of the experimental data with the least-squares fit.

### 6.2.2 Transformation of Problems to Linear Form

In some cases it is possible to transform a nonlinear least-squares problem into an alternative form which is linear in the parameters. This has the great advantage of leading to a straightforward calculation which does not require starting values and can usually be solved routinely on the computer. However, it must be borne in mind that the parameter values calculated by least squares from different forms of the same original equation will not be *exactly* the same, but often close. Either the dependent or independent variables (or both) may be transformed in order to achieve linearity in the new variables. Even if the desired model is nonlinear, solution of some linearized version may be useful in providing starting values for the nonlinear calculation. Some pertinent references on this subject include Himmelblau [1970] and Ratkowsky [1990].

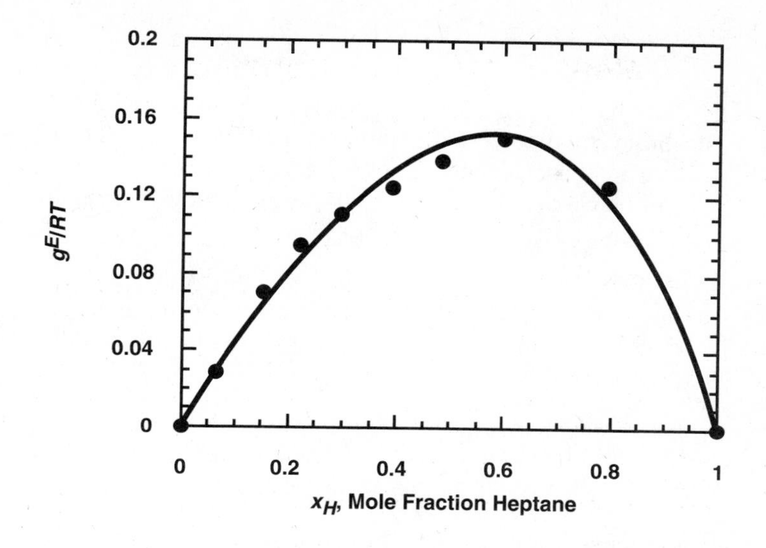


**Figure 6.3.** Vapor/Liquid Equilibrium Data Fitted to the Wilson Equation

**6.2.2.1 Some Linearizing Transformations.** We consider here several simple but typical examples of nonlinear models that can easily be converted to linear form by a suitable change of variables. First consider the common exponential decay model $y_p = B_1 \exp(-B_2 x)$. This nonlinear model can be linearized by taking logarithms of both sides to give $\ln y_p = \ln B_1 - B_2 x$. With the new dependent variable $\ln y$, the new *linear* prediction equation corresponds to $g_0(x) = 0, g_1(x) = 1$, and $g_2(x) = -x$ with parameters $\ln B_1$ and $B_2$. Various other nonlinear models are linearized by taking logarithms.

Another example is the nonlinear prediction function $y_p = B_1 x/(B_2 + x)$. In this case taking logarithms does not achieve linearity, but instead, if we cross-multiply, we get $1/y_p = (B_2 + x)/B_1 x = (B_2/B_1)/x + (1/B_1)$, a linear expression in the new parameters. Thus, by taking $1/y$ as the new dependent variable, we obtain a linear prediction equation with $g_0(x) = 0$, $g_1(x) = 1$, and $g_2(x) = 1/x$ having parameters $(1/B_1)$ and $(B_2/B_1)$. It is also interesting to observe that the original nonlinear prediction equation has the form of a *rational* function (see "Rational Least-Squares Approximation"), which can be used to advantage.

## 6.3 GENERAL MULTIPLE LINEAR REGRESSION

The problem of multiple linear regression (the name comes from statistics) refers to the situation where the dependent variable $y$ depends on more than one (say $k$) independent variables $x_1, x_2, \ldots, x_k$. Thus, each data "point" corresponds to a value of $(x_1, x_2, \ldots, x_k, y)$. The corresponding linear prediction equation is the same as before except that now, in general, all the $g_i$ functions may depend on any or all of the $k$ independent variables. Thus, the linear prediction function is written as

$$y_p(x_1, x_2, \dots, x_k) = g_0(x_1, \dots, x_k) + B_1 g_1(x_1, \dots, x_k) + B_2 g_2(x_1, \dots, x_k) + \dots + B_p g_p(x_1, \dots, x_k) \tag{6-13}$$

Now exactly the same minimization of Error is used to determine the so-called normal linear algebraic equations for the unknown parameters $B_j$. The calculations are virtually identical to those of one-dimensional general linear regression considered earlier. They can be carried out conveniently using computer program LSTSQMR. An example problem illustrates the details.

**Example 6.4 Multiple Linear Regression**

Gilliland and Sherwood [1934] obtained mass transfer data for the evaporation of several liquids falling down the inside of a vertical wetted-wall column into a turbulent air stream flowing countercurrent to the liquid. Data were obtained for the evaporation of nine different liquids at several air flow rates. The data in Table 6.3 represent a relatively small sample of the data reported by these authors. It is desired to obtain the least-squares fit of these data to an equation of the form

$$Sh = B_1 Re^{B_2} Sc^{B_3} \tag{a}$$

This is an example of a problem in which the equations defining the $B_i$ parameters will be nonlinear. However, this problem can be converted into a linear problem by taking logarithms of both sides.

$$\log Sh = \log B_1 + B_2 \log Re + B_3 \log Sc \tag{b}$$

To make this correspondence we take $y = \log \text{Sh}$, $x_1 = \log \text{Re}$, and $x_2 = \log \text{Sc}$. Using program LSTSQMR by fitting the logarithms, and then returning to the original variables, we find that the best correlation is

$$Sh = 0.0370 Re^{0.777} Sc^{0.455} \tag{c}$$

This correlation is quite similar to that presented by Gilliland and Sherwood based on their complete set of approximately 360 data points ($B_1 = 0.023, B_2 = 0.83, B_3 = 0.44$). Figure 6.4 shows a plot of the data compared to the correlation equation.

## 6.4 GENERAL MULTIPLE NONLINEAR REGRESSION

Multiple nonlinear regression is merely a multidimensional form of the nonlinear least-squares problem treated earlier, and the Gauss-Newton approach to this problem is virtually identical to the one-dimensional version. The only difference is that the nonlinear prediction function now may depend on any or all of the $k$ independent variables $x_1, x_2, \dots, x_k$ in addition to the $p$ unknown parameters. As in the one-dimensional case, the $g_i$ functions in computer program LSTSQMR must be replaced by $g_0 = f(B_1, \dots, B_p, x_1, \dots, x_k)$, $g_i = \partial f / \partial B_i (i = 1, \dots, p)$. With this correspondence, program LSTSQMR will generally converge to the desired nonlinear solution if the starting estimates of the parameters are sufficiently close to their true unknown values. The following example serves to illustrate the solution of a nonlinear multiple regression problem.

**TABLE 6.3** MASS TRANSFER DATA OF GILLILAND AND SHERWOOD

| Sh | Re | Sc |
|---|---|---|
| 43.7 | 10,800 | 0.60 |
| 21.5 | 5,290 | 0.60 |
| 24.2 | 3,120 | 1.80 |
| 88.0 | 14,400 | 1.80 |
| 51.6 | 6,620 | 1.875 |
| 50.7 | 8,700 | 1.875 |
| 32.3 | 4,250 | 1.86 |
| 56.0 | 8,570 | 1.86 |
| 26.1 | 2,900 | 2.16 |
| 41.3 | 4,950 | 2.16 |
| 92.8 | 14,800 | 2.17 |
| 54.2 | 7,480 | 2.17 |
| 65.5 | 9,170 | 2.26 |
| 38.2 | 4,720 | 2.26 |
| 93.0 | 16,300 | 1.83 |
| 70.6 | 13,000 | 1.83 |
| 42.9 | 7,700 | 1.61 |
| 19.8 | 2,330 | 1.61 |

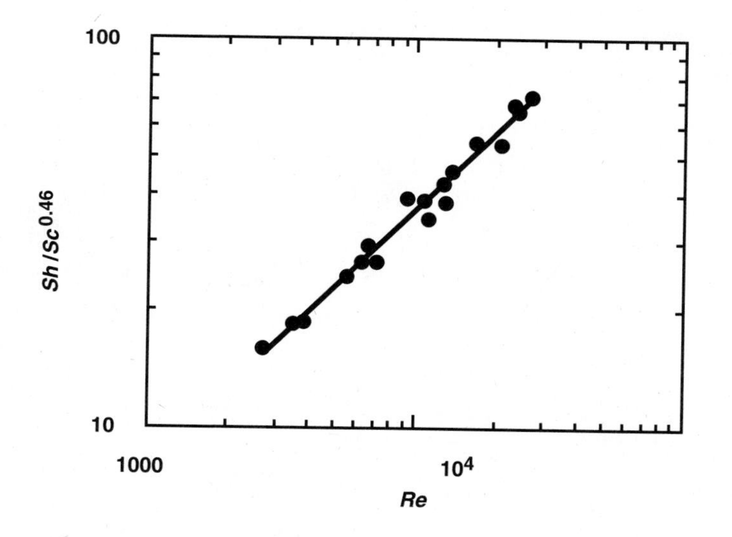


**Figure 6.4.** Evaporation Data of Gilliland and Sherwood

**Example 6.5 Multiple Nonlinear Regression**

The Gilliland-Sherwood data in Table 6.3 may be fitted directly to the nonlinear model

$$Sh = B_1 Re^{B_2} Sc^{B_3} \tag{a}$$

In this case, we use the values found in Example 6.4 as the initial guesses for the $B$ parameters and use the nonlinear least-squares capabilities of LSTSQMR to get

$$\text{Sh} = 0.0336\text{Re}^{0.789}\text{Sc}^{0.436} \tag{b}$$

It is interesting to compare how well these correlations fit the data. If the standard deviation is defined as

$$\text{std. dev.} = \left\{ \frac{\sum_{i=1}^{n} [\text{Sh}_{\text{exp}} - \text{Sh}_{\text{pred}}]^2}{N-3} \right\}^{1/2} \tag{c}$$

we find that the nonlinear regression gives a standard deviation of 3.36; whereas the linear regression analysis of Example 6.4, in which the logarithm of the Sherwood numbers was fitted, gives a standard deviation of 3.59. Thus, for this example, the goodness of the fit is comparable for both cases with the nonlinear regression giving a slightly better fit of the data. This is expected since the nonlinear regression fits the actual Sh and the linear regression finds the best fit for log Sh.

## 6.5 MISCELLANEOUS TOPICS IN LEAST SQUARES

There are several other topics in least-squares analysis that deserve attention. These include rational least squares, hybrid least squares, and interpolation by least squares.

### 6.5.1 Rational Least-Squares Approximation

Rational least squares refers to using a prediction function which is a *ratio* of polynomials. Such a prediction formula is typically of the following form:

$$y_p(x) = \frac{p_0 + p_1 x + p_2 x^2}{1 + q_1 x + q_2 x^2} \tag{6-14}$$

In general, we could have any number of power terms in either the numerator, the denominator, or both. This type of function is special in two ways. First, it is known that, in many approximation problems, rational functions are more efficient than ordinary polynomials (recall Padé approximation, rational interpolation, and see the discussion of rational least squares in Chapter 5). This makes them ideal candidates as fitting functions in cases where there is no known "model" function. Second, while the rational prediction function Equation (6-14) is obviously nonlinear in the parameters $p_i$, $q_i$, there is a simple way to render it linear in an approximate way. This has the advantage of yielding a linear problem that *requires no starting guesses of the parameters*. The hope (usually realized) is that these approximate parameter values would be good starting guesses for a fully nonlinear solution using Gauss-Newton. The linear approximation is developed as follows. By cross-multiplying in Equation (6-14) and then transposing, we obtain

$$y_p = p_0 + p_1 x + p_2 x^2 - y_p(q_1 x + q_2 x^2) \tag{6-15}$$

In this equation, if we *replace* $\text{y}_\text{p}$ *on the right-hand side by* $\text{y}_\text{i}$ (the corresponding observation rather than the prediction), then the prediction formula is linear in the parameters and can be handled in the usual way. It should be noted that some prediction equations which are not rational functions can be made so approximately through appropriate Taylor series approximations.

Program RATLSGR is constructed to carry out the entire calculation, so that *the user need not furnish any starting values*. This program fits a rational least-squares approximation to either a given data set or a specified analytical function. (An example of the fit of the function $e^{-x}$ was given in the chapter on interpolation.) The program allows easy change of the degrees of both the numerator and denominator, if desired. The parameters of the rational approximation are computed, along with options for evaluation, numerical differentiation, or integration of the fitted function. There is also an option for a plot of the data and the fitted function.

**Example 6.6 The Prediction Function $y_p = B_1x/(B_2 + x)$**

In Section 6.2 we considered this prediction function and showed how to linearize it by using the new dependent variable $1/y$, the new independent variable $1/x$, and a reinterpretation of the parameters. Here, we note that this model is actually a rational approximation and can therefore be approximately "linearized" by the procedure discussed in the present section. It is also useful to note that if we divide both sides by $x$ and divide the right-hand side of both the numerator and denominator by $B_2$, we get the prediction equation in the form

$$\frac{y}{x} = \frac{B_1/B_2}{1 + x/B_2} = \frac{p_0}{1 + q_1 x} \tag{a}$$

The advantage of this form is that it can be used directly in program RATLSGR, with $B_0/B_1 = p_0$, $1/B_2 = q_1$, and the dependent variable being $y/x$, whereas the program does not accommodate the original form.

**Example 6.7 Rational Least Squares**

It is desired to fit the vapor pressure data for toluene as given in the *Chemical Engineers' Handbook* [Perry *et al.*, 1984] and shown in Table 6.4 to the Antoine Equation. The Antoine Equation is expressed as

$$\log_{10} P^* = B_1 - \frac{B_2}{T + B_3} \tag{a}$$

$P^*$ is the vapor pressure in mm Hg and $T$ is the temperature in °C. $B_1$, $B_2$, and $B_3$ are the empirical constants to be fitted.

To solve this problem, we can use program RATLSGR. However, in order to use this program we need to rearrange the Antoine Equation to get it in the form

$$\log_{10} P^* = \frac{p_0 + p_1 T + \cdots}{1 + q_1 T + \cdots} \tag{b}$$

If we get a common denominator for the Antoine Equation and carry out the algebra to get the proper form, we obtain

$$\log_{10} P^* = \frac{(B_1 B_3 - B_2)/B_3 + B_1 T/B_3}{1 + T/B_3} \tag{c}$$

Thus, we find

$$p_0 = \frac{B_1 B_3 - B_2}{B_3}; \quad p_1 = \frac{B_1}{B_3}; \quad q_1 = \frac{1}{B_3} \tag{d}$$

**TABLE 6.4** VAPOR PRESSURE DATA FOR TOLUENE

| $P^*$ (mm Hg) | $T$ (°C) |
|---|---|
| 5 | −4.4 |
| 10 | 6.4 |
| 20 | 18.4 |
| 40 | 31.8 |
| 60 | 40.3 |
| 100 | 51.9 |
| 200 | 69.5 |
| 400 | 89.5 |
| 760 | 110.6 |
| 1520 | 136.5 |

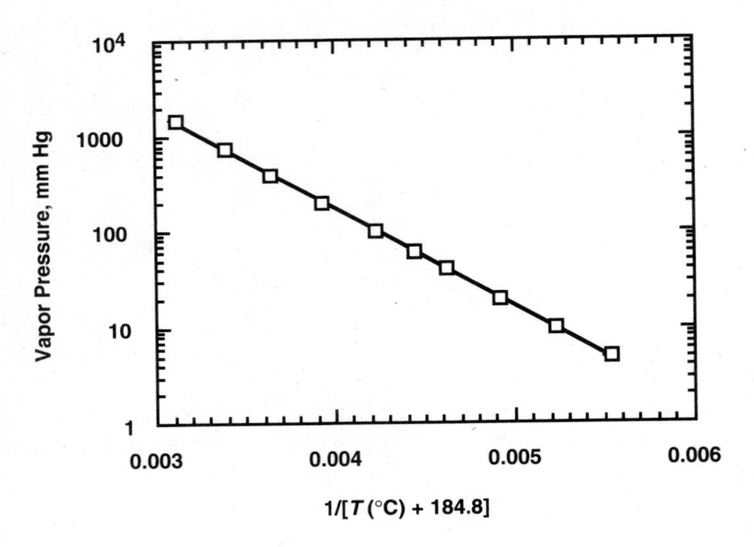

**Figure 6.5.** Toluene Vapor Pressure Data Fitted to the Antoine Equation

From program RATLSGR we obtain the following values of the parameters:

$$p_0 = 0.7734; \quad p_1 = 3.468 \times 10^{-2}; \quad q_0 = 5.412 \times 10^{-4} \tag{e}$$

In terms of the $B$ parameters, we find

$$B_3 = \frac{1}{q_1} = 184.8; B_1 = B_3 * p_1 = 6.408; B_2 = B_3(B_1 - p_0) = 1041 \tag{f}$$

The data are shown plotted according to this correlation in Figure 6.5.

**Example 6.8 Fitting the Numerical Solution of a Differential Equation Using Rational Least Squares**

We consider now an important general approximation situation, that of representing the numerical solution of some problem by a rational function having a known maximum

**TABLE 6.5** RATIONAL LEAST SQUARES FIT TO NUMERICAL SOLUTION OF $y' = -y/(1 + 0.1y)$; $y(0) = 1$ ON INTERVAL (0,5) (26 DATA POINTS)

| Number of Parameters | $k$ | $m$ | Maximum \|Error\| |
|---|---|---|---|
| 3 | 2 | 0 | $1.3 \times 10^{-1}$ |
| 3 | 1 | 1 | $3.6 \times 10^{-2}$ |
| 5 | 4 | 0 | $4.6 \times 10^{-3}$ |
| 5 | 2 | 2 | $1.1 \times 10^{-3}$ |
| 7 | 6 | 0 | $9.9 \times 10^{-5}$ |
| 7 | 3 | 3 | $9.8 \times 10^{-6}$ |
| 8 | 3 | 4 | $1.0 \times 10^{-6}$ |

error. For concreteness, we select the numerical solution of the following nonlinear differential equation on the $x$ interval $(0, 5)$.

$$\frac{dy}{dx} = \frac{-y}{1 + 0.1y}, y(0) = 1 \tag{a}$$

This equation describes a simple chemical reactor with so-called Langmuir-Hinshelwood kinetics. The numerical solution (which will not be tabulated here) is obtained using the IEX method (Improved Euler/Extrapolation), which is discussed in Chapter 10 on numerical solution of initial value problems. The accuracy has been monitored so that we are confident that the numerical values are accurate to at least six significant figures. We use the 26 data points corresponding to increments of 0.2 over the interval from $x = 0$ to $x = 5$.

To proceed, we input the data into program RATLSGR and then select various combinations of $(k, m)$, which represent the degrees of the numerator and denominator, respectively. This produces the results shown in Table 6.5 and Figure 6.6. The maximum error of approximation at the data points is given by the program. The total number of parameters in a standard rational approximation is always equal to $(k + 1 + p)$. The results of Table 6.5 show typical behavior. For a given number of parameters, rational approximation is generally better than polynomial approximation ($m = 0$). As the number of parameters increases, the maximum error at the data points decreases. We see that in this problem, rational approximation represents the data very well. The maximum error is about $10^{-3}$ with five parameters, and it can be reduced to about $10^{-6}$ using eight parameters. A graph of the fit of the five-parameter rational function corresponding to $k = 2$, $m = 2$ is shown in Figure 6.7.

### 6.5.2 Hybrid Least-Squares Approximation

There are occasionally situations where it is very useful to *combine* least-squares and other types of approximation, such as Taylor series or interpolation, in the same problem. We have already discussed an implicit example of this, least squares with equality constraints; this is just interpolation at certain points followed by least squares. Another common scenario, primarily of interest in continuous problems, is where it is desired for the prediction function to have proper *asymptotic* behavior near some point (often $x = 0$). In this case, we could form a "hybrid" least-squares approximation by directly using the first few terms of the Taylor approximation, followed by polynomial terms having unknown parameters to be determined by

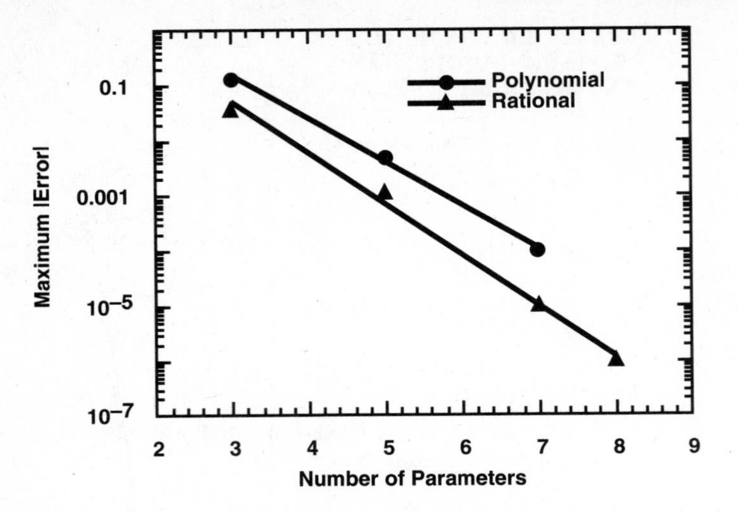


**Figure 6.6.** Comparison of the Maximum Error for a Polynomial and a Rational Least-Squares Fit of the Numerical Solution of a Differential Equation

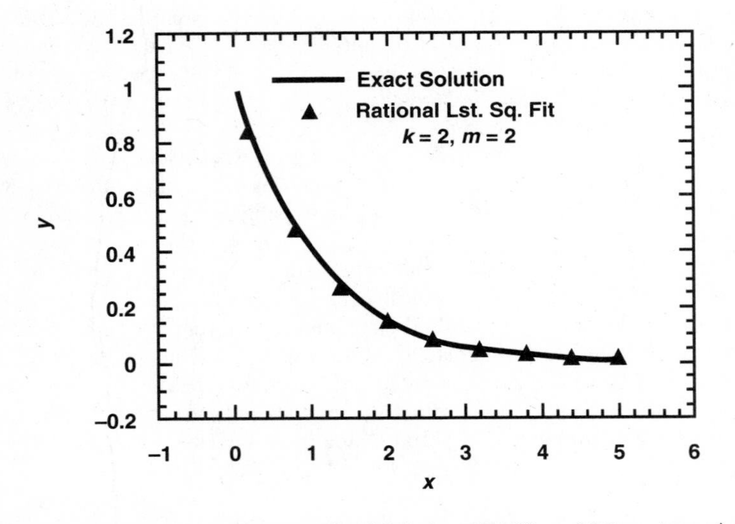


**Figure 6.7.** Rational Least Squares Fit of Solution of Differential Equation: $y' = y/(1 + 0.1y)$, $y(0) = 1$

least squares. This idea can be pursued in many directions. As a simple illustration, suppose we wish to develop an approximation to $\sin x$ on the interval $(0, \pi/2)$. The first two terms of the (asymptotic) Taylor approximation are $(x - x^3/3!)$. In view of this, to retain the proper limiting behavior near $x = 0$, we would retain the leading term $x$ and perhaps use least squares to determine an additional term of the form $B_1 x^3$. Thus, we might reasonably try to approximate $\sin x$ on the desired interval by fitting a least-squares function of the form $y_p = x + B_1 x^3$.

### 6.5.3 Interpolation via Least Squares

As we noted earlier, the least-squares calculation reduces to an exact fit to the data points, otherwise known as interpolation, in the case where the number of parameters $p$ is equal to the number of data points $n$. Although it is a little unorthodox to do interpolation by means of a least-squares calculation, it is very convenient to do so when we have *very general* least-squares computer programs such as LSTSQGR and LSTSQMR. Thus, these programs can be used to accomplish one- or multidimensional interpolation for arbitrary prediction functions, including those nonlinear in the parameters.

It is convenient to use the interpolation capability of program LSTSQGR to evaluate, numerically differentiate or integrate, and plot a known analytical function $y = f(x)$. This is accomplished by using case = 1, one parameter, and one data point that corresponds to $y_0 = f(x_0)$. To implement, use $g0 = 0$ and $g(1) = f(x)$. This causes $c_1$ in the prediction function $y_p = c_1 \cdot f(x)$ to be unity, so that the prediction function is just the desired function $f(x)$.

One application of nonlinear interpolation is in the problem of determining starting guesses for nonlinear least squares. Sometimes this can be successfully accomplished by *interpolating* the prediction function at selected values of $(x_i, y_i)$. Because of the possibility of ill-conditioning in larger problems, it is always a good idea to check the results of such an interpolation calculation to see that the data values are correctly predicted.

## 6.6 COMPUTER PROGRAMS

In this section we discuss the computer programs that are applicable to various problems in the chapter. For each program we give a brief discussion and a concise listing, which includes (i) a program description, (ii) how to run, (iii) user-specified subroutines, including a sample problem, and (iv) selected output. Detailed program listings can be obtained from the diskette.

### General One-Dimensional Regression Program LSTSQGR

Instructions for running the program are given in the concise listing. The program prompts for case (linear or nonlinear *in the parameters*) and the number of parameters $p$. Examples are shown in Sub Function of the proper specification of the $g_i(x)$ functions for both linear and nonlinear problems. The comment specifications (!) are used to conveniently deactivate portions of the program. In the case of a nonlinear problem, a starting guess for the unknown parameters must be given in response to the prompt. Program output includes parameter values and the "Error" value (estimate of variance), and allows prediction of $y$ at any $x$. Also, values of maximum error and maximum relative error are computed. For the nonlinear case, the program gives intermediate information, including "norm of corrections," which permits easy monitoring of the calculation toward convergence (hopefully!)

or divergence. The program also has an option for numerical differentiation or integration of the fitted function. In addition, a plot of the data and the fitted function is also available. The program (via interpolation) also allows evaluation, numerical differentiation and integration, and a plot of an analytically defined function.

The problem considered in Sub Function of the concise listing is for $y_p = c_1 + c_2 x$. We specify $p = 2$ parameters and $n = 3$ data pairs. The output shows that $c_1 = 1.07333$ and $c_2 = 1.5$. The Error value $\sum(y_i - y_p)^2/(n-2)$ is calculated to be $3.22667 \times 10^{-2}$. If we comment out the linear model functions and uncomment the nonlinear ones to make them active, we find that the best parameters are considerably different than in the linear case, and the Error value is now 0.266894, which is much larger than before. Based on comparison of the Error values, the linear fit seems substantially better. This is also intuitively reasonable when we look at the respective *predicted* $y_i$ values at the data values $x_i$. A graphical comparison (which is always very useful) would lead to the same conclusion.

### Program LSTSQGR.

```
!This is program LSTSQGR.tru by O.T.Hanna
!Program includes J.Rickey module plotting for graphics
!The program does linear or nonlinear least squares
!For linear probs. ypred = g0(x)+c1*g1(x)+c2*g2(x)...+cp*gp(x)
!For nonlinear probs. ypred = f(x,c1,c2..cp)
!For n = p, program does interpolation
!n = number of data pairs x,y
!p = number of parameters to be determined
!Case=1 for linear lst. sq., Case=2 for nonlinear lst. sq.
!Program gives option to calculate predicted function,
!derivative, or definite integral from a to b.
!Both error least-squares and error-max. are given.
!To get printed output, it is convenient to use (1)list function to
!print problem, then (2) give command run >> filename (eg. output).
!The file filename can be edited if desired and printed using
!either the list or print command.
!To print graph,use Shift-PrtSc when screen is correct(IBM)
!*********************************************************************
!*To run,specify x,y data,and p = no. of parameters in Sub Data: for    *
!*linear case define p+1 functions g0,g(1)..g(p) in Sub Function as in  *
!*linear model above; for nonlinear case define g0=f(x,c), g(1)=        *
!*df/dc(1)...g(p)=df/dc(p) For graphics specification see Sub Plot_info *
!*********************************************************************
!This program can also be used to evaluate,differentiate,integrate
!or plot any specified analytical function f(x). To do this use
!case=1, g0=0, g(1)=f(x), n=1, p=1 and choose 1 data pair (x0,y0)
!from y0 = f(x0). This should then produce c(1)=1. For integration
!option, use "min.no. parts interval divided into" = a value for
!which 0.8 < Ratio < 1.2 and Est. Rel. Error = small enough. For
!evaluation of function f(x) and its derivative, use Prompt(0)
!option. To plot f(x) use Loop(1) option and specify n1,x1,x2
!values in Sub Plot_info.
```

```
!This is program LSTSQGR.TRU
SUB Function(x,g0,g(),c())
    !*****Least Squares Solution of Problem Below*****
    !The marks *** designate user specified statements
    !Specify p = no. parameters in Sub Data; for least squares
    !specify g(1)..g(p) here.

    LET g0 = 0                          !***linear problem
    LET g(1) = 1                        !***
    LET g(2) = x                        !***

    !LET g0 = c(1)*Exp(c(2)*x)          !***nonlinear problem f(x,c1,c2)
    !LET g(1) = Exp(c(2)*x)             !***df/dc1
    !LET g(2) = c(1)*x*Exp(c(2)*x)      !***df/dc2
END SUB

SUB Data(n,x(),y(),p)
    !Here we generate x(i),y(i) by read statements.
    !The marks *** designate user-specified statements.

    LET p = 2           !*** - no. of parameters to be determined
    PRINT"Number of parameters = ";p
    LET n = 3           !*** - no. of data pairs
    PRINT"Number of data pairs = ";n
    PRINT
    PRINT"Original Data"
    FOR i = 1 to n
        READ y(i),x(i)
        PRINT"y(";i;")= ";y(i);Tab(25);"x(";i;")= ";x(i)
    NEXT i
    PRINT
    INPUT prompt"Is data to be transformed?(No-0,Yes-1) ":q
    IF q = 1 then
        PRINT"The following is transformed data Log(y), 1/x"   !***
        FOR i = 1 to n
            LET y(i) = Log( y(i))          !***
            LET x(i) = 1/x(i)              !***
            PRINT"y(";i;")= ";y(i);Tab(25);"x(";i;")= ";x(i)
        NEXT i
    END IF
    PRINT

    !Specify data pairs as DATA y(i), x(i)             !***
    DATA 1,    0
    DATA 2.72, 1
    DATA 4,    2

END SUB
```

```
SUB Plot_info (n1,x1,dx1,text$(),label$())
    !*** denotes required user specification
    !Usually convenient to use from 30 to 100 plot increments n1
    !Automatic scaling is done w.r.t. values of function

    LET n1 = 30                     !*** no. of plot increments to be used
    LET x1 = 0                      !*** xstart for plot
    LET x2 = 2                      !*** xend for plot
    LET dx1 = (x2 - x1)/n1          !x increment for plot
    LET n1 = n1 + 1                 !no. x(i),y(i) to be calculated for plot
END SUB
```

## General Multiple Regression Program LSTSQMR

The instructions for running are given in the concise listing below. This program operates in almost the same way as LSTSQGR, except that now the model equations may depend on $k$ different independent variables $x(k)$. Thus, each data "point" now involves the $(k+1)$ values $(y_i, x_i(1), \ldots, x_i(k))$, and this data must be specified in Sub Data. The example problem illustrated in the concise listing is for $y_p = c_1x_1 + c_2x_2$, which depends on two independent variables. For this problem, the program gives $c1 = -1.09143 \times 10^{-4}$ and $c2 = 1.26095 \times 10^{-2}$ together with Error $= 3.80 \times 10^{-4}$.

### Program LSTSQMR

```
!This is program LSTSQMR.tru by O.T.Hanna
!The program does linear or nonlinear least squares multiple regression
!for a single algebraic prediction function. Weights w(i) may be specified
!For linear probs. yp = g0(x(1),..x(k))+b(1)*g1(x(1),..x(k))..+b(p)*gp(..)
!For nonlinear probs. yp = f(x(1),x(2),..x(k),b(1),b(2),..b(p))
!n = number of data sets x(i,j),y(i)
!p = number of parameters to be determined
!k = number of independent variables
!Current dimensioning of x1(i,j) requires n <= 50, k <= 10
!Case = 1 for linear, 2 for nonlinear
!To get printed output use (1)list ,(2)Ctrl-PrtSc(or echo),(3)run
!*******************************************************************
!*To run,specify y(i),x1(i,j) data in Sub Data - i=data case,      *
!*j=indep't var. The weights w(i) are specified in Sub Data        *
!*For linear case define functions g0,g(1),...g(p)in Sub Function  *
!*For nonlinear case define g0=f(x,b), g(1)=df/db(1)..g(p)=df/db(p) *
!*******************************************************************

!*****Least Squares Solution of Problem Below*****
SUB Function(x(),g0,g(),b())
  !*** denotes user-specified quantities
    !linear model is ypred = b1*x1 + b2*x2
    LET g0 = 0
    LET g(1) = x(1)
```

```
    LET g(2) = x(2)

    !nonlinear model is ypred = b1*x1*Exp(b2*x2) = f
  !   LET c = Exp( b(2)*x(2))
  !   LET g0 = b(1)*x(1)*c        !f
  !   LET g(1) = x(1)*c           !df/db1
  !   LET g(2) = x(2)*g0          !df/db2

END SUB

SUB Data(n,k,x1(,),y(),w())
    LET k = 2        !***no. of indep't. variables x(1),x(2)..x(k)
    LET n = 4        !***no. of data sets; y(i),x(i,1),x(i,2)..x(i,k)
    PRINT "number of data sets = ";n
    PRINT "number of independent variables = ";k
    PRINT
    !x1(i,j) is ith data value for jth indep't variable
    FOR i = 1 to n
        READ y(i)
        FOR j = 1 to k
            READ x1(i,j)
        NEXT j
    NEXT i
    PRINT
    PRINT"Observations y(i) and xj(i)"
    PRINT"  y          x1         x2          x3          x4          ..."
    PRINT
    FOR i = 1 to n
        PRINT y(i);Tab(12);
        FOR j = 1 to k
            PRINT x1(i,j);Tab(12 +12*j);
        NEXT j
        PRINT
    NEXT i

    INPUT prompt"Is data to be transformed?(No-0,Yes-1) ":q
    IF q = 1 then
        !Example of transformation option
        PRINT"The following is transformed data Log(y),1/x1"     !***
        FOR i = 1 to n
            LET y(i) = Log( y(i))             !***
            LET x1(i,1) = 1/x1(i,1)           !***
            !Etc.
        NEXT i
        PRINT"Transformed Observations y(i) and xj(i)"
        PRINT"  y          x1         x2          x3          x4          ..."
        PRINT
        FOR i = 1 to n
            PRINT y(i);Tab(12);
            FOR j = 1 to k
```

```
                PRINT x1(i,j);Tab(12 +12*j);
            NEXT j
            PRINT
        NEXT i
    END IF

    !Option to read or calculate w(i) weights***
    FOR i = 1 to n
        ! READ w(i) !if READ then put in DATA statement below y(i) values
        LET w(i) = 1 !or,for example, w(i) = fn( x1(i,1))
    NEXT i

    !Put in n data pts. y(i),x1(i),x2(i),..xk(i) as shown below
    !x1(i,j) is symbol for xj(i) (ith data value for jth indep't variable)
    !Put in data as DATA y(i),x1(i,1),x1(i,2)..x1(i,k) by row    !***
    DATA .089, 500,  10
    DATA .071, 1500, 20
    DATA .054, 2500, 25
    DATA .047, 3000, 30

    !Next DATA statement is for option of reading w(i) values
!   DATA 1 , 1 , 1, 1, 1, 1, 1 , 1 !n w(i) values, if desired
!   PRINT
END SUB
```

### *Rational Least Squares Program RATLSGR*

The instructions for running are indicated in the concise listing below. Program RATLSGR is for the special one-dimensional general prediction function $y_p = (p_0 + p_1 x \dots)/(1 + q_1 x \dots)$. Thus, to use this program no prediction function is specified by the user, but rather the program prompts for the desired degrees of the numerator and denominator. Otherwise, the behavior of the program is quite similar to that of program LSTSQGR described above. There are options for evaluation, numerical differentiation, and numerical integration of the fitted function. Also, a plot of the data and the fitted function is available. The problem considered in the concise listing is used in Example 6.8.

#### Program RATLSGR

```
!This is program RATLSGR.tru by O.T.Hanna adapting prgm LSTSQGR.TRU.
!Program uses John Rickey plotting module.
!This is a subroutine for polynomial rational least squares approx.
!of the form y = (p0 + p1*x..+ pk*x^k)/(1 + q1*x..+ qm*x^m).
!Program allows use of data or an analytically defined function.
!n is no. of input data pairs x(j),y(j); k,m are degrees of
!numerator and denominator. param = (k+1+m) is number
!of parameters to be determined.
!Program has option for values of ypred, the integral of
!ypred from a to b with error estimate and derivatives of ypred via
```

```
!variable step numerical differentiation.
!Program gives least sq.error(est.of variance),max. (err/rel.err.).
!Program includes graphics capabilities for 1 data set,1 curve.
!To get printed output on MAC or IBM we can send output to a
!file via TB command run >> filename. Then edit filename and print.
!To print graph, use Shift-PrtSc when screen is correct(IBM).
!***************************************************************
!*To run,specify n sets of x,y data or an analytically defined *
!*function in Sub Data, along with desired degrees of numerator*
!*and denominator, k and m. Program has prompt to over-ride    *
!*k,m values for computational experiments.                    *
!*Supply plot information in Sub Plot_info.                    *
!***************************************************************
!For integration option, use "min.no. parts interval divided into"
!= a value for which 0.8 < Ratio < 1.2 and Est. Rel. Error = small
!enough. For evaluation of function f(x) and its derivative,
!use Prompt(0) option. To plot f(x) use Loop(1) option and
!specify n1,x1,x2 values in Sub Plot_info.

SUB Data(n,x(),y(),k,m)
    !Here we generate x(i),y(i) by read statements.
    !The marks *** designate where the user must modify program.
    LET k = 1           !***degree of numerator
    LET m = 1           !***degree of denominator
    INPUT prompt"Supply data(0) or specify function(1) ?":ques
    IF ques = 1 THEN
        !Option for rational least squares fit to user-defined
        !function y = f(x) on the interval x0 < x < x1
        !Function is y(i) = f(z), where z replaces x
        !This allows rational l.s. approx.,integral,derivative,
        !plot, for any analytically defined function.
        LET x0 = 0 !.001  !***initial x; use same in Sub Plot_info
        LET x1 = Log(2)   !***final x
        INPUT prompt"No.of parts into which interval is divided ?":m1
        LET n = m1 + 1    !no. of 'data' pts. for function evaluation
        MAT redim x(n),y(n)
        FOR i = 1 to n
            LET delta_x = (x1 - x0)/m1 !x increment for function
            LET x(i),z = x0 + (i-1)*delta_x
            !***y(i)=f(z) below is user-defined, z is dummy for x
            LET y(i) = Exp(-z)          !***y(i)=f(z)
        NEXT i
    ELSE IF ques = 0 THEN
        !Option for user-specified data
        !This allows rational l.s. approx.,integral,derivative,
        !plot, of the specified data.
        LET n = 26                      !*** - no. of data pairs
        PRINT"Number of data pairs = ";n
        PRINT
        PRINT"Original Data"
```

```
        FOR i = 1 to n
            READ y(i),x(i)
            PRINT"y(";i;")= ";y(i);Tab(25);"x(";i;")= ";x(i)
        NEXT i
    END IF
    PRINT
    INPUT prompt"Is data to be transformed?(No-0,Yes-1) ":q
    IF q = 1 then
        PRINT"The following is transformed data y/x " !***user specified
        FOR i = 1 to n
            LET y(i) = (y(i) - 1)/x(i)         !***
            LET x(i) = x(i)/(1 + x(i))
        NEXT i
    END IF
    PRINT

    !Specify data pairs as DATA y(i), x(i)     !***
    !Prob. is y'=-y/(1+.1*y), y(0)=1 ; 26 data points

    DATA 1.      , 0
    DATA .832555 , .2
    DATA .691333 , .4
    DATA .572767 , .6
    DATA .473615 , .8
    DATA .39098 , 1.
    DATA .322313 , 1.2
    DATA .265394 , 1.4
    DATA .218312 , 1.6
    DATA .179435 , 1.8
    DATA .14738 , 2.
    DATA .120984 , 2.2
    DATA 9.92685e-2 , 2.4
    DATA 8.14194e-2 , 2.6
    DATA 6.67584e-2 , 2.8
    DATA 5.47229e-2 , 3.
    DATA 4.48476e-2 , 3.2
    DATA 3.67479e-2 , 3.4
    DATA 3.01066e-2 , 3.6
    DATA 2.46626e-2 , 3.8
    DATA 2.02011e-2 , 4.
    DATA 1.65453e-2 , 4.2
    DATA 1.35502e-2 , 4.4
    DATA 1.10967e-2 , 4.6
    DATA 9.08702e-3 , 4.8
    DATA 7.44104e-3 , 5.

END SUB
```

```
SUB Plot_info (n1,x1,dx1)
    !Usually convenient to use from 20 to 50 plot points
    !*** denotes required user specification
    LET n1 = 20              !*** no. of plot increments to be used
    LET x1 = 0               !*** xstart for plot
    LET x2 = Log(2)          !*** xend for plot
    LET dx1 = (x2 - x1)/n1   !x increment for plot
    LET n1 = n1 + 1          !no. x(i),y(i) to be calculated
END SUB
```

## PROBLEMS

**6.1.** Consider the following set of data, which is to be fit by an equation of the form $y_p(x) = c_1 + c_2x + c_3x^2 + c_4x^3$. Use least-squares approximation (with weight function unity) to determine linear, quadratic, and cubic approximations to the data. That is, consider the cases $y_p(x) = c_1 + c_2x$, $y_p(x) = c_1 + c_2x + c_3x^2$, and $y_p(x) = c_1 + c_2x + c_3x^2 + c_4x^3$. Calculate the linear case by hand as well as with program LSTSQGR; use only the program to determine the quadratic and cubic approximations. In terms of the quantity "Error" = SSE/$(n-p)$, which representation seems to be most appropriate? Using only the four terms above in various combinations, how many different $y_p(x)$ functions could be constructed?

| $x$ | 0.1 | 0.3 | 0.4 | 0.5 | 0.7 | 0.9 | 1.0 |
|---|---|---|---|---|---|---|---|
| $y$ | 0.15 | 0.29 | 0.42 | 0.47 | 0.6 | 0.75 | 0.9 |

**6.2.** We wish to fit the experimental data in the table below with a second-degree polynomial of the form $a_0 + a_1x + a_2x^2$ using least squares. It is known from external information that the fitted equation should have a maximum value at $x = 0.6$, and it should pass through the point $x = 1$, $y = 0$. Determine the fitted equation that satisfies these requirements. Do this calculation (i) by hand and (ii) using program LSTSQGR.

| $x$ | 0.2 | 0.6 | 0.8 | 1.0 |
|---|---|---|---|---|
| $y$ | 0.7 | 2.0 | 1.0 | 0.5 |

**6.3.** A very common model for a dimensionless first-order chemical reaction is $dC/dt = -kC$, with $C(t = 0) = 1$. The integrated form of this model is $C = \exp(-kt)$, which is *nonlinear* in the parameter $k$. For the data given below, determine the best value for $k$. Do this using program LSTSQGR by first transforming the model to linear form in order to get a first estimate of $k$ to be used in calculating the nonlinear value. How much different are the linear and nonlinear values of $k$?

| $t$ | $C$ |
|---|---|
| 0.2 | 0.75 |
| 0.5 | 0.55 |
| 1.0 | 0.21 |
| 1.5 | 0.13 |
| 2.0 | 0.04 |

**TABLE 6.6**

| Rate | $C$ | $T$ |
|---|---|---|
| 0.0360 | 0.8 | 300 |
| 1.01 | 0.8 | 400 |
| 7.45 | 0.8 | 500 |
| 0.0231 | 0.4 | 300 |
| 0.649 | 0.4 | 400 |
| 4.79 | 0.4 | 500 |
| 0.0135 | 0.2 | 300 |
| 0.378 | 0.2 | 400 |
| 2.80 | 0.2 | 500 |

**6.4.** Reconsider problem 6.3. Since we do not have a measurement of the initial concentration at $t = 0$, it is possible that there is a systematic error (bias) in the concentrations, so that $C(0)$ is not equal to unity. To test for this possibility, extend the previous model to the form $C = A\exp(-kt)$, where now both $A$ and $k$ are to be determined. A simple test for bias involves recomputing to see if the two-parameter model causes the value of "Error" to be significantly reduced. Use program LSTSQGR as in the previous problem, by first transforming the original model to a linear form, and using these estimates of the parameters to solve the nonlinear model. Is there evidence of a significant bias?

**6.5.** Use all of the data of Table 5.5 to fit the following two-dimensional models for $D(T, X)$.
(i) $D = c_1 + c_2T + c_3X$
(ii) $D = c_1 + c_2T + c_3X + c_4T^2 + c_5TX + c_6X^2$
In each case, calculate $D(T = 22°\text{C}, X = 0.36$ wt. frac.); compare the two results.

**6.6.** This problem illustrates how program LSTSQGR can be used as a sophisticated calculator to evaluate a *known* (exact) function, as well as its derivative and definite integral. Consider the function $y(x) = (e^x + 2)/(1 + x^2)$. By direct calculation, $y(0) = 3$. To use program LSTSQGR, specify $g0 = 0$, $g(1) = y(x)$ (above), $p = 1$, and $n = 1$. Use case $= 1$ and give the single data pair $y = 3$, $x = 0$. (i) Determine $dy/dx(x = 1.5)$, and check this result by analytical differentiation. (ii) Evaluate $\int y(x)dx$ between the limits of 0 and 2 correct to six figures. Use the "Est. Rel Error" and "Ratio" values to validate the results.

**6.7.** Consider the function of problem 6.6. Use the graphical representation (in program LSTSQGR.) to estimate the approximate $x$, $y$ values where this function has a relative maximum or minimum. Then refine these estimates by brute force, using the program to calculate values of $y(x)$ and $dy/dx$.

**6.8.** It is desired to fit the data in Table 6.6 to the *nonlinear* model equation

$$\text{rate} = K\frac{C}{1 + 0.3C}e^{-a/T}$$

in order to determine $K$ and $a$ by least squares. (i) First linearize the above equation with respect to the parameters by taking logarithms. (ii) Use the estimates from the linear problem as the first guesses for the nonlinear problem. (iii) How much different are the linear and nonlinear parameter estimates?

**6.9.** The data below are to be fit with a "generic" model having relatively few parameters. It is not known whether the data may contain experimental errors. As possible generic models, use various least-squares polynomial and rational approximations (use only rational approximations whose numerator and denominator differ by no more than one degree). To determine a good model, use program RATLSGR to select one that has a relatively small number of parameters, a relatively small "error variance" (as given by the program), and a reasonable graph.

| $x$ | $y$ |
|---|---|
| 0.0 | 0.0 |
| 0.1 | 0.074 |
| 0.2 | 0.11 |
| 0.3 | 0.12 |
| 0.4 | 0.12 |
| 0.6 | 0.093 |
| 0.8 | 0.065 |
| 1.0 | 0.042 |
| 1.2 | 0.025 |
| 1.4 | 0.015 |
| 1.6 | 0.0082 |
| 1.8 | 0.0044 |
| 2.0 | 0.0023 |

## REFERENCES

BEREZIN, I. S. and ZHIDKOV, N. P. (1965), *Computing Methods*, Addison-Wesley, Menlo Park, CA

GILLILAND, E. R. and SHERWOOD, T. K. (1934), *Ind. & Engr. Chem.*, **26,** No. 5, p. 516

HIMMELBLAU, D. M. (1970), *Process Analysis by Statistical Methods*, Wiley, New York

KAHANER, D., MOLER, C., and NASH, S. (1989), *Numerical Methods and Software*, Prentice-Hall, Englewood Cliffs, NJ

PERRY, R. H., GREEN, D. W., and MALONEY, J. O.(1984), *Chemical Engineers Handbook*, 6th ed., McGraw-Hill, New York

PRAUSNITZ, J. M. (1969), *Molecular Thermodynamics of Fluid-Phase Equilibria*, Prentice-Hall, Englewood Cliffs, NJ

PRESS, W. H., FLANNERY, B. P., TEUKOLSKY, S. A., and VETTERLING, W. T. (1986), *Numerical Recipes*, Cambridge Univ. Press, New York

RATKOWSKY, D. A. (1990), *Handbook of Nonlinear Regression Models*, Marcel Dekker, New York

VARGOFTIK, N. B. (1975), *Tables of the Thermophysical Properties of Liquids and Gases*, 2nd ed., Hemisphere, Washington D.C.

# 7

# Nonlinear Algebraic Equations

## 7.1 SOLUTION OF A SINGLE EQUATION IN ONE UNKNOWN

A nonlinear algebraic equation is, by definition, an equation that is not linear. Or, in other words, we may say that a nonlinear algebraic equation in one variable is one which *cannot* be written in the form $ax + b = 0$. We know how to get a simple, explicit solution to a linear equation, and we can also easily identify situations where a linear equation has no solution. In the case of nonlinear equations, however, the state of affairs is quite different.

In general, only a few nonlinear equations can be solved explicitly, and even when this is possible, it is often not practical to do so. For example, there exists an algebraic formula for the solution of a cubic equation, but it is so complicated that it is seldom used. Instead, it is usually more practical to employ a numerical approximation procedure. In regard to questions of the existence and uniqueness of solutions of nonlinear algebraic equations, matters are again complicated. It is easy to construct simple examples of nonlinear equations that have either zero, any finite number, or infinitely many roots. In view of the previous discussion, it is apparent that dealing with nonlinear algebraic equations is fraught with hazards. These difficulties must be faced, however, since so many problems in chemical engineering involve nonlinear equations.

### 7.1.1 The Simple Newton Method

In most problems involving nonlinear equations, it is either necessary or desirable to use numerical methods of solution. A general procedure that is useful for this purpose is the method of successive approximations. The nature of the method is to somehow successively refine an initial approximation to a desired root to the point where the root is known to the desired accuracy.

Before undertaking a detailed study of this general approach, it is instructive to begin by considering the simple Newton method. We will see later, in our more general considerations, how the Newton method can be interpreted to be one special case of the method of successive approximations. The Newton method is developed most easily through the use of Taylor's formula applied to the nonlinear equation

$$f(x^*) = 0 \tag{7-1}$$

We desire to compute the unknown value $x^*$ for which $f(x)$ is zero (of course, there may be more than one such value). For the moment, we assume the existence of a single desired root $x^*$. If we have some approximation to $x^*$, say $\overline{x}$, then by Taylor's formula we can write

$$f(x^*) = 0 \cong f(\overline{x}) + f'(\overline{x})(x^* - \overline{x}) \tag{7-2}$$

The value of $x^*$ computed from this expression, namely, $\overline{x} - f(\overline{x})/f'(\overline{x})$, should be close to the true value if $\overline{x}$ is close to $x^*$. The essence of the Newton method is to note that if the calculated $x^*$ is an improvement over $\overline{x}$, then we should use this $x^*$ value as a new approximation $\overline{x}$, from which a further improvement can be obtained, and so on. This idea can be implemented neatly by writing a recurrence relation based on replacing $\overline{x}$ by $x^{[n]}$ and $x^*$ by $x^{[n+1]}$. This process gives

$$x^{[n+1]} = x^{[n]} - \frac{f(x^{[n]})}{f'(x^{[n]})} \tag{7-3}$$

To start the iteration, some initial approximation $x^{[0]}$ must be chosen. Then with $n = 0$, Equation (7-1) determines $x^{[1]}$. Taking $n = 1$ in Equation (7-3) then leads to $x^{[2]}$, and so on. The hope is that this process will eventually lead to a very accurate approximation to the desired root. Figure 7.1 shows an example of a Newton iteration calculation.

The simple Newton method generally works very well in practice. Its only real disadvantage is that it requires computation of the derivative $f'(x)$. For very complicated $f(x)$ functions, the calculation of $f'(x)$ may be difficult. Otherwise, however, the Newton method normally converges very rapidly if it converges at all, and it always converges to a particular root if the initial approximation $x^{[0]}$ is sufficiently close to the root.

**Example 7.1 The Newton Method**

Suppose we wish to find a root of $x^3 = x^2 + 2$ which is known to lie between $x = 1$ and $x = 2$. In this case, the Newton formula (7-3) becomes

$$x^{[n+1]} = x^{[n]} - \frac{(x^{[n]})^3 - (x^{[n]})^2 - 2}{3(x^{[n]})^2 - 2x^{[n]}} \tag{a}$$

We first start the iteration with $x^{[0]} = 1$; then we restart the iteration with $x^{[0]} = 2$ in order to assess the effect of the starting value on the convergence of the method. The results are seen in Table 7.1. We see that for $x^0 = 2$, the iterations converge monotonically to the true value 1.6956 in three cycles. On the other hand, for $x^0 = 1$, the iteration

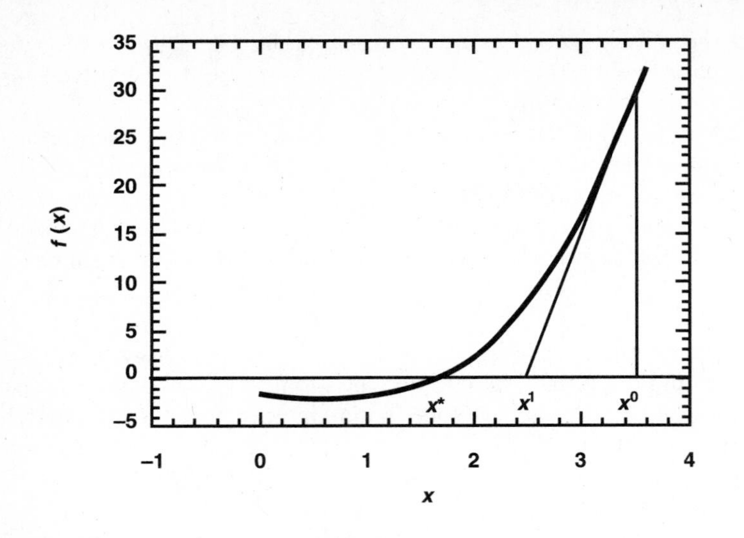


Figure 7.1. Example of Newton Iteration

**TABLE 7.1** USE OF THE NEWTON METHOD TO FIND A ROOT OF $x^3 = x^2 + 2$ BETWEEN 1 AND 2

$$x^{[n+1]} = x^{[n]} - \frac{(x^{[n]})^3 - (x^{[n]})^2 - 2}{3(x^{[n]})^2 - 2x^{[n]}}$$

$$(x^* = 1.6956)$$

| $n$ | $x^n$ | $x^n$ |
|---|---|---|
| 0 | 2.0000 | 1.0000 |
| 1 | 1.7500 | 3.0000 |
| 2 | 1.6978 | 2.2380 |
| 3 | 1.6956 | 1.8400 |
| 4 | 1.6956 | 1.7097 |
| 5 | | 1.6958 |
| 6 | | 1.6956 |
| 7 | | 1.6956 |

starts to diverge by jumping over the root, but quickly recovers and approaches the true value monotonically in a total of six cycles. The present example represents a typical performance of the Newton method.

### 7.1.2 The Secant Method

Another useful technique for solving nonlinear equations is known as the secant method. This procedure has the advantage that no derivative of the function $f(x)$ is required. The method has its primary application in situations where solution of a nonlinear equation is only *part* of the problem being solved (for example, in the "shooting" approach for solving a boundary value problem). The secant method is closely related to the Newton method. It requires using one previous value, so it is a two-step iteration procedure. To derive the secant method, we can start from the Newton formula and replace the derivative by a difference quotient as follows. For the Newton method, the new approximation $x^{[n+1]}$ is given by $x^{[n+1]} = x^{[n]} - f(x^{[n]})/f'(x^{[n]})$. To get the secant method, we replace $f'(x^{[n]})$ by $(f(x^{[n]}) - f(x^{[n-1]}))/(x^{[n]} - x^{[n-1]})$, which is a numerical differentiation approximation based on the consecutive iterates $x^{[n-1]}$ and $x^{[n]}$. This then gives the following equation for the secant method:

$$x^{[n+1]} = x^{[n]} - \frac{f(x^{[n]})(x^{[n]} - x^{[n+1]})}{f(x^{[n]}) - f(x^{[n+1]})} \tag{7-4}$$

To start the calculation, two different values of $x$ are needed, say $x^{[0]}$ and $x^{[1]}$. It is customary to choose these values close to each other, for example, $x^{[1]} = 1.05x^{[0]}$. Note that the secant method is *not* quite the same thing as the Newton method implemented with numerical evaluation of the derivative. This is because the secant iterates themselves are used to form the difference quotient, rather than $x$ values needed to give the best approximation to the derivative. This causes the convergence properties of the two methods to be a little different, with Newton being somewhat better. The secant method can also be interpreted more precisely as inverse linear interpolation applied to the equation $f(x) = 0$.

### 7.1.3 The Method of Successive Approximation

We now consider the general method of successive approximation as applied to the solution of nonlinear algebraic equations. This technique is often called the method of *ordinary iteration.* The general idea, which has many uses in applied mathematics, is to effect some kind of partial solution to the equation being considered, leaving behind the relatively unimportant parts of the equation. Thus we wish to partially solve the equation $f(x) = 0$ by expressing it as $x = g(x)$. Here, the "solved" portion corresponds to the $x$ on the left-hand side and the "unsolved" portion of $x$ remains in the $g(x)$ on the right-hand side. We might consider the magnitude of $dg/dx$ from our partial solution to be indicative of the degree to which the partial solution is reasonable. This is because in the limiting case where the partial solution becomes exact, the right-hand side no longer varies with $x$, and $g'(x) \to 0$. On the other hand, the value of the partial solution would seem to be less for relatively larger magnitudes of $g'(x)$. A further study of this matter will show that it is crucial whether $|g'(x)|$ is less than or greater than unity. Since there is normally a number of different partial solutions to a nonlinear equation, it is best, if possible, to choose one that corresponds to $|g'(x)| < 1$.

We now proceed to analyze the method of successive approximation as applied to the solution of a nonlinear algebraic equation. Suppose we wish to solve $f(x^*) = 0$ for a root $x^*$. We then form some kind of partial solution to this equation, which can be written in the form $x^* = g(x^*)$. The $g(x)$ function is often selected by rearranging $f(x) = 0$ but the only requirement is that the roots $x^*$ of both representations be the same. The successive approximations are then defined by the recurrence relation

$x^{[n+1]} = g(x^{[n]})$. This is implemented by selecting a starting value $x^{[0]}$, then choosing $n = 0$ to get $x^{[1]}$; next, choose $n = 1$ to get $x^{[2]}$, and so on. The calculations are very easy to carry out on a computer or calculator. The hope is that this process will converge to an unknown desired root $x^*$.

To examine the convergence question, we *define* the error at the $n$th stage of this process to be $e^{[n]} = x^{[n]} - x^*$. If we subtract the equations $x^{[n+1]} = g(x^{[n]})$, $x^* = g(x^*)$ and use Taylor's formula as $g(x^{[n]}) - g(x^*) = g'(\xi_1)(x^{[n]} - x^*) = g'(\xi_1)e^{[n]}$, and so on, we get

$$x^{[n+1]} - x^* = e^{[n+1]} = g'(\xi_1)e^{[n]} = g'(x^*)e^{[n]} + \frac{g''(\xi_2)(e^{[n]})^2}{2!} \tag{7-5}$$

The above expression utilizes two different Taylor approximations, each of which has importance. First, we observe that if $|g'(x)| < K < 1$ in the region including the root and the iteration values, then $|e^{[n+1]}| < K|e^{[n]}|$ and the iterations converge to the root, with the error decreasing by roughly the factor $K$ at each stage. Thus, if $|g'(x)| \approx K$ is small, the error diminishes rapidly and convergence is fast. On the other hand, if $|g'(x)|$ is greater than unity, the errors increase in magnitude at each stage and the process diverges. It is also useful to note from Equation (7-5) that for either convergence or divergence, the errors remain of the same sign if $g'(x)$ is positive, while they oscillate in sign at every stage if $g'(x)$ is negative. This means, in the former case, that the iterations will stay on the same side of the root for so-called monotonic convergence or divergence, while in the latter case, the iterations will be alternately above and below the root. To determine the error in approximation $x^{[n+1]}$, we may write

$$x^{[n+1]} - x^{[n]} = (x^{[n+1]} - x^*) - (x^{[n]} - x^*) = (x^{[n+1]} - x^*)\left(1 - \frac{1}{g'(\xi_1)}\right) \tag{7-6}$$

Here we have replaced $e^{[n]} = (x^{[n]} - x^*)$ by $(x^{[n+1]} - x^*)/g'(\xi_1)$ from Equation (7-5). This expression is rearranged to give

$$x^{[n+1]} - x^* = e^{[n+1]} = \frac{g'(\xi_1)}{(g'(\xi_1) - 1)}(x^{[n+1]} - x^{[n]}) \tag{7-7}$$

If $g'$ is uniformly negative, $|e^{[n+1]}| < |x^{[n+1]} - x^{[n]}|$. If $g'$ is positive and less than 1 in magnitude, with $g' < K < 1$, then $|e^{[n+1]}| < (K/(1-K)) \cdot |x^{[n+1]} - x^{[n]}|$. If $g'$ is positive and greater than 1 with $1 < k < g' < K$, then $|e^{[n+1]}| < (K/(k-1)) \cdot |x^{[n+1]} - x^{[n]}|$. Obviously, if $|K| = 1$, the iterations simply repeat and do not either converge or diverge.

If we happen to be close to the root $x^*$, we see that $e^{[n+1]} \approx -(g'(x)/(1-g'(x)) \cdot (x^{[n+1]} - x^{[n]})$. But in this case, $g'(x)$ can be estimated well by the numerical differentiation approximation $g'(x) = (g(x^{[1]}) - g(x^{[0]}))/(x^{[1]} - x^{[0]}) = (x^{[2]} - x^{[1]})/(x^{[1]} - x^{[0]})$, so that two consecutive iterations furnish an estimate of $g'(x)$ and hence of the error $e^{[n+1]}$. This is the basis of the delta-square acceleration procedure, to be discussed later.

In the selection of an iteration function $g(x)$, an important special case arises if it happens that $g'(x^*) = 0$. Reference to Equation (7-5) shows that in this case, $e^{[n+1]} = (g''(\xi_2)/2)(e^{[n]})^2$. Whenever $e^{[n+1]}$ is proportional to the square of $e^{[n]}$, we say that the convergence is quadratic, as opposed to linear convergence if $g'(x^*) \neq 0$. With quadratic convergence, where $e^{[n+1]} \approx C(e^{[n]})^2$, we have $e^{[n+1]} = (Ce^{[n]})e^{[n]}$, so for a sufficiently good initial guess, $e^{[n]}$ will be small enough that the errors will decrease in magnitude regardless of the value of $C$. Also, when such a process converges, the speed of convergence will be enhanced, since the error at each stage is proportional to the *square* of the error at the previous stage. A simple example of the above results corresponds to the interpretation of the Newton method in terms of ordinary iteration. Since the Newton method is defined by $x^{[n+1]} = x^{[n]} - f(x^{[n]})/f'(x^{[n]})$, this can be viewed as ordinary iteration applied to the function $g(x) = x - f(x)/f'(x)$. By direct calculation, $g'(x) = 1 - (f'(x)^2 - f(x)f''(x))/(f'(x))^2 = f(x)f''(x)/(f'(x))^2$ Thus, since $f(x^*) = 0$ at the root, $g'(x^*) = 0$, and we have quadratic convergence for the Newton method.

We should mention a generalization of the above behavior known as *superlinear* convergence. This corresponds to the general iterative situation where $e^{[n+1]} = K(e^{[n]})^{1+\alpha}$, if $\alpha > 0$. This situation includes the Newton method, as well as the secant method, the delta-square method, and some others. In this case, we always get convergence to a root if we start close enough to the root, since then $e^{[n+1]}$ will always be smaller than $e^{[n]}$. Also, for such superlinear behavior, when we are close enough to the root, the approximation error will be less than the difference between two successive iterates. This can be seen from the general relation $d^{[n+1]} \equiv x^{[n+1]} - x^{[n]} = e^{[n+1]} - e^{[n]}$, which shows for superlinear convergence that when $|e^{[n]}|$ is small enough, $|e^{[n+1]}| < |d^{[n+1]}|$.

**Example 7.2 Ordinary Iteration**

Again consider the equation $x^3 - x^2 - 2 = 0$. We have seen that there is one real root in the region $1 < x^* < 2$. To form an iteration function $g(x)$, there are several possible obvious rearrangements of the original equation. Thus, for example, we could solve for the $x$ in the $x^3$ term to give $x = g_1(x) = (x^2 + 2)^{1/3}$. Alternatively, we could solve for the $x$ in the $x^2$ term to get $x = g_2(x) = (x^3 - 2)^{1/2}$. Since we know $x^*$ is larger than one, it would seem that the $x^3$ term is more important than the $x^2$ term, so we might expect $g_1(x)$ to yield better results than $g_2(x)$.

The results of performing iterations with $g_1(x)$ and $g_2(x)$ are shown in Table 7.2. The function $g_1(x)$ is seen to yield monotonic convergence to the desired root $x^* = 1.6956$ for starting values of both $x = 1$ and $x = 2$.

It is easy to verify, in fact, that the iterations based on $g_1(x)$ will converge for any finite (positive or negative) starting value. The iterations based on $g_2(x)$, by contrast, do not converge for *any* starting value, including $x^{[0]} = 1.7$, which is initially in error only by 0.004. Computation of the $g'(x)$ values shows that they are less than one in magnitude for $g_1(x)$ and greater than one in magnitude for $g_2(x)$. Note that the use of three consecutive iterations to estimate $g'(x)$, according to $g'(x) = (x^{[2]} - x^{[1]})/(x^{[1]} - x^{[0]})$, produces reasonable results. This rather typical example illustrates the hazards of using the ordinary iteration method. When it works, it is fine, except perhaps that the rate of convergence is rather slow compared to the Newton method. When it does not work, however, it appears useless. In cases where a "good" $g(x)$ function is not rather obvious (such as in the present case), the effort required to find an appropriate $g(x)$ might not be worth the return. For this reason, it is best to employ ordinary iterations in conjunction with the delta-square ($\Delta^2$) process, to be discussed next.

**TABLE 7.2** ITERATIVE SOLUTION OF $x^3 = x^2 + 2$ BY MEANS OF $x^{[n+1]} = g(x^{[n]})$ FOR VARIOUS STARTING VALUES

(a) $g_1(x) = (x^2 + 2)^{1/3}$ (b) $g_2(x) = (x^3 - 2)^{1/2}$

($x^* = 1.6956$)

| $g_1 = (x^2 + 2)^{1/3}$ | | | $g_2 = (x^3 - 2)^{1/2}$ | | |
|---|---|---|---|---|---|
| $x$ | $x$ | $n$ | $x$ | $x$ | $x$ |
| 2. | 1. | 0 | 2. | 1.7 | 1.6 |
| 1.82 | 1.44 | 1 | 2.45 | 1.71 | 1.45 |
| 1.74 | 1.60 | 2 | 3.56 | 1.72 | 1.02 |
| 1.71 | 1.66 | 3 | 6.58 | 1.77 | # |
| 1.703 | 1.681 | 4 | 16.80 | 1.88 | |
| 1.699 | 1.690 | 5 | 68.87 | 2.14 | |
| 1.697 | 1.693 | 6 | 571. | 2.81 | |
| 1.6961 | 1.6947 | 7 | * | * | |
| 1.6958 | 1.6953 | 8 | | | |
| 1.6957 | 1.6955 | 9 | | | |
| 1.6956 | 1.6956 | 10 | | | |

* Diverges

\# Negative square root

### 7.1.4 The Delta-Square ($\Delta^2$) Process

To avoid the problems of divergence and slow convergence associated with ordinary iteration, while at the same time retaining the advantage of not requiring derivatives, we introduce the $\Delta^2$ process. This important acceleration technique for nonlinear algebraic equations is known under various names, including those of Aitken, Shanks, and Steffensen. The method can be derived in various ways, but for our purposes it is easiest to base the derivation on Equation (7-7). If we are fairly close to a root, then two ordinary iterations starting with some value $x^{[0]}$ will produce $x^{[1]}$ and $x^{[2]}$. As mentioned earlier, these three values can then be used to estimate $g'(x)$ as $(x^{[2]} - x^{[1]})/(x^{[1]} - x^{[0]})$. Then, replacing $g'(\xi_1)$ in Equation (7-7) with this estimate for $g'(x)$, we can solve (7-7) for $x^*$; this is by definition the $\Delta^2$ approximation. It can be written as

$$x^{\Delta^2} = x^{[2]} + \frac{g'}{1 - g'}(x^{[2]} - x^{[1]}); \quad g' \equiv \frac{x^{[2]} - x^{[1]}}{x^{[1]} - x^{[0]}} \tag{7-8}$$

The procedure is to start with $x^{[0]}$, do two ordinary iterations to get $x^{[1]}$ and $x^{[2]}$, then calculate $x^{\Delta^2}$. This completes one $\Delta^2$ cycle; to continue, take the previous $\Delta^2$ value as $x^{[0]}$ in the new cycle, do two ordinary iterations to get the new values $x^{[1]}$ and $x^{[2]}$, then calculate a new $\Delta^2$ value from Equation (7-8), and so forth.

It can be shown that the $\Delta^2$ process is quadratically convergent; that is, that $e^{[n+1]}$ is proportional to the square of $e^{[n]}$. Thus, the $\Delta^2$ process will converge rapidly (comparable to Newton) for any $g(x)$, provided that the starting value is close enough to the unknown root. Also, because of the superlinear convergence, the error in the

**TABLE 7.3** $\Delta^2$ ACCELERATION OF THE ITERATIVE METHOD FOR SOLVING $x^3 = x^2 + 2$

| | (a) $x^{[n+1]} = [(x^{[n]})^2 + 2)]^{1/3}$ | | |
|---|---|---|---|
| $x^0 =$ | 2. | **1.6945** | **1.6956** |
| | 1.871 | 1.6952 | 1.6956 |
| | 1.744 | 1.6954 | 1.6956 |
| $x^0 =$ | 1. | **1.6825** | **1.6956** |
| | 1.442 | 1.6905 | 1.6956 |
| | 1.598 | 1.6936 | 1.6956 |
| | (b) $x^{[n+1]} = [(x^{[n]})^3 - 2)]^{1/2}$ | | |
| $x^0 =$ | 2. | **1.6959** | **1.6956** |
| | 2.449 | 1.6952 | 1.6956 |
| | 3.563 | 1.6971 | 1.6956 |

$x^* = 1.6956$

delta-square values, $e^{[n+1]}$, will be less in magnitude than the difference in successive iterates, $d^{[n+1]} (= x^{[n+1]} - x^{[n]})$, when the iterations are close enough to the root.

**Example 7.3 The $\Delta^2$ Method**

We now apply the $\Delta^2$ method to the equation $x^3 - x^2 - 2 = 0$. This has been done for both of the iteration functions considered previously, and the results are shown in Table 7.3. A column of numerical values represents one $\Delta^2$ cycle. To elaborate, the value at the top of each column is the $\Delta^2$ value (printed in **bold)** calculated from the previous cycle of two ordinary iterations. The second and third entries in a column represent the result of two successive ordinary iterations starting with the previous $\Delta^2$ value. The first "$\Delta^2$" value (which is necessarily unknown) is the user-specified starting value $x^0$.

In case (a), we see that the $\Delta^2$ method converges rapidly for both $x^{[0]} = 2$ and $x^{[0]} = 1$. The convergence is significantly faster than that of the corresponding ordinary iteration, as can be seen from Table 7.2.

The behavior of case (b) is particularly interesting. The subiterations here are *diverging*, as was seen in Table 7.2. However, the $\Delta^2$ process operates on diverging subiterations and produces rapidly convergent $\Delta^2$ iterations. Thus, in this case, the $\Delta^2$ process turns a disastrous, divergent ordinary iteration into a successful calculation.

The preceding results are rather typical of what is found in practice. It must again be emphasized that success in both the Newton and $\Delta^2$ methods depends strongly on a good initial estimate of the root. The $\Delta^2$ process is comparable in efficiency to the Newton method, as can be seen from Tables 7.1 and 7.3. The computer costs are similar for the $\Delta^2$ and Newton methods, since $\Delta^2$ requires two $g(x)$ values per cycle and Newton requires both $f(x)$ and $f'(x)$ in each cycle. However, the work for the programmer is much less for the $\Delta^2$ process, since the derivative does not have to be calculated or programmed.

### 7.1.5 The Extended Newton Method

A simple extension of the standard Newton method for a single equation is sometimes useful, especially in asymptotic analysis. This extended Newton method is based on an approximate extension of the Taylor expansion used in deriving the standard method. The extended Taylor expansion is

$$f(x) = 0 \cong f(\bar{x}) + f'(\bar{x})(x - \bar{x}) + \frac{f''(\bar{x})}{2!}(x - \bar{x})^2 \tag{7-9}$$

In the *standard* Newton method, we delete the final term of Equation (7-9) to get $(x - \bar{x}) \cong -f(\bar{x})/f'(\bar{x})$. In the extended Newton procedure, we replace one of the $(x - \bar{x})$ factors in the last term of Equation (7-9) with the regular Newton expression just referred to. This yields

$$(x - \bar{x}) = \left[\frac{-f}{\left(f' - \dfrac{f''f}{2f'}\right)}\right]_{\bar{x}} \tag{7-10}$$

Equation (7-10) has the advantage of avoiding the quadratic equation required for the exact solution of Equation (7-9), but still contains a significant contribution from the quadratic term. This approximation is obviously more accurate near the root, so this extended Newton procedure is especially useful for a one-step improvement of an asymptotic approximation.

### 7.1.6 Existence and Uniqueness of Roots

Nonlinear algebraic equations do not necessarily have any solution at all. It is therefore very desirable to have some means of assessing when the equation of interest either cannot have or may not have any roots. It is usually simplest to ask these and similar questions separately for positive and negative roots. One important elementary result is that $f(x)$ must have at least one (or an *odd* number) of real roots in regions separating $f(x)$ values of *opposite* sign. This is easily seen by means of a simple graphing exercise. Thus, for example, when looking for positive roots it is usually easy to find the sign of $f(0)$ and $f(\infty)$. If these are opposite, there must be at least one, or three, and so on roots in this region. If $f(0)$ and $f(\infty)$ have the *same* sign, there may still be roots in this range, but there must be an *even* number of them, such as zero, two, and so on. It is also often convenient to quickly determine the sign of $f(x)$ at some simple integer values, such as 1 or 2. Sometimes it is useful to examine $f'(x)$ in connection with the existence of roots. Thus, for example, if $f'(x)$ is positive for $x > a$ and if $f(a)$ is positive, then $f(x)$ must remain positive for $x > a$, and there can be no root for $x > a$.

We now consider the matter of uniqueness of roots to a single nonlinear equation. As mentioned earlier, it often happens that a given nonlinear algebraic equation has more than one solution. An important condition related to the uniqueness of solutions of $f(x^*) = 0$ is the sign of $f'(x)$. *In any region where* f′(x) *has a uniform sign, there can be no more than one real root of the equation* f(x) = 0. This is easy to see geometrically, and it can be shown analytically with Taylor's theorem as follows. Suppose there are two roots $x_1^*$ and $x_2^*$ of $f(x) = 0$ in a region where $f'$ has a uniform sign. Then, by Taylor's formula, $f(x_2^*) = f(x_1^*) + f'(\xi)(x_2^* - x_1^*)$. But if $x_1^* \neq x_2^*$ and both $f(x_1^*)$ and $f(x_2^*)$ are zero, then $f'(\xi)$ must be zero, in contradiction to the assumption that $f'(x)$ is either uniformly positive or uniformly negative in the region under consideration. Thus, to consider uniqueness it is always useful to see what signs $f'(x)$ may have. Of course, if $f'(x)$ changes sign, this does not *necessarily* mean that there will be multiple roots, but rather that there *might* be. In searching for multiple roots, a practical procedure is to implement a numerical method using different starting values; if possible, it is desirable to monitor $f'(x)$ during the computation to look for sign changes.

### 7.1.7 Bracketing Real Solutions of a Polynomial Equation

To solve nonlinear algebraic equations using an iterative method such as delta-square or Newton, it is important wherever possible to provide good initial estimates. We present a simple systematic procedure for accomplishing this in the case of real solutions of a polynomial equation. In some instances, the ideas can be applied to more general nonlinear equations. In particular, if the non-polynomial terms in an equation are represented by rational approximations (such as Padé or rational interpolation), then the original equation could be approximated by a new polynomial equation.

The bracketing procedure considered here, which might be called "region elimination," is based on determining regions where there *cannot* be a solution and using this to establish upper and lower bounds on any possible solutions. It must be emphasized that our calculation of bounds between which a solution must lie does *not* imply that there actually *is* a root in that region. The method proceeds by first considering positive, real roots and determining an upper bound for them. Then we calculate lower bounds for positive, real roots. Finally, we do a simple change of variable and repeat the process to get upper and lower bounds for the negative real roots.

To illustrate the technique, we use an equation due to Berezin and Zhidkov [1965]. We always write the equation so that the highest-degree term has a positive coefficient. The polynomial equation of fourth degree examined here is

$$x^4 - 35x^3 + 380x^2 - 1350x + 1000 = 0 \tag{7-11}$$

We first consider possible positive roots of this equation. It is noted that if there are any positive roots of this (or any other) polynomial equation, at least one coefficient in the equation must have a negative sign; for if this were not true (if all signs were positive), we would have the sum of positive terms being equal to zero, which is impossible. Thus, if all signs of the equation are positive, there cannot be any positive real roots, so we needn't consider the region $0 \leq x \leq \infty$ any further. However, in our example, we observe that there are two negative terms. The procedure is now to factor out the highest degree term and collect with it all of the negative terms in group I as follows:

$$0 = \underbrace{x^4\left(1 - \frac{35}{x} - \frac{1350}{x^3}\right)}_{\text{group I}} + \underbrace{(380x^2 + 1000)}_{\text{group II}} \tag{7-12}$$

The remaining positive terms are lumped together in group II at the end of the equation. Now, the group II terms always have a positive value when $x$ is positive. However, if there is a root to the equation, it must correspond to the group I terms being *negative*, for otherwise we would have two positive terms summing to zero. The way the group I terms have been constructed, it is easily seen that if we find a value of $x$ for which group I is positive, for all larger $x$ values group I will remain positive. This is because the value of group I increases monotonically from $-\infty$ at $x = 0$ to $+1$ at $x = \infty$. Therefore, if we simply find any value of $x$ for which group I is positive, that $x$ represents an upper bound to any positive root of the equation. We simply try a few rather obvious $x$ values to accomplish this. Naturally, we wish to find a relatively small $x$ value for which group I is positive, since this would yield a relatively good upper bound. The best we could do using this procedure would correspond to finding the $x$ value at which group I just becomes zero. For the problem at hand, it is clear that the most important term in group I is $-35/x$. For example, if we choose an $x$ which is a little larger than 35, say 40, we find that group I is positive. Thus, $x = 40$ is an upper bound to any positive root of the equation, although our analysis does *not* show that there is necessarily a root for $0 \leq x \leq 40$. A slightly refined calculation shows that $x < 36.1$.

Having found an *upper* bound for positive real roots, we now turn our attention to calculation of a lower bound for positive real roots. To do this, we introduce the transformation $x = 1/y$ and re-express the original equation as

$$0 = 1000y^4 - 1350y^3 + 380y^2 - 35y + 1 \tag{7-13}$$

Lower bounds for $x$ correspond to upper bounds for $y$. We therefore, as before, write the new equation by taking the (positive) highest-degree term and putting it with all negative terms in group I, and we put all remaining positive terms in group II. This gives

$$0 = \underbrace{1000y^4\left(1 - \frac{1.35}{y} - \frac{0.035}{y^3}\right)}_{\text{group I}} + \underbrace{1000(0.38y^2 + 0.001)}_{\text{group II}} \tag{7-14}$$

As before, we now simply find a relatively small $y$ value for which group I is positive; this will give an upper bound to any positive $y$ roots. By inspection, $y$ must be a bit larger than 1.35. A simple trial calculation shows that for $y = 1.4$, group I is barely positive, so this is a good value. Since $y = 1/x$, we have that $1/x = y < 1.4$, or $1/1.4 = 0.714 < x$, for a positive real root $x$. Thus, our calculations, when put together, show that any positive real roots of the original equation must lie in the region $0.714 < x < 36.1$. It would be pointless to take starting values outside this range when looking for positive real roots.

To bracket *negative* roots, it is convenient to introduce the change of variable $x = -z$ so that for $x$ negative, $z$ will be positive. Then exactly the same ideas that were used above are now applicable to determine upper and lower bounds for any $z$ roots. In the present example, the original equation in $x$ now becomes

$$0 = z^4 + 35z^3 + 380z^2 + 1350z + 1000 \tag{7-15}$$

However, we immediately see that the equation in $z$ has *no negative coefficients*, and therefore, no positive $z$ value can satisfy it. So in this example there are no negative roots.

The previous example should clarify the general real-root bracketing procedure that can be followed in any polynomial (or power) equation. The steps of the general recipe are:

1. Put equation in a form where highest-degree term is positive.
2. Group terms according to (a) group I—highest-degree term and all negative terms, (b) group II—all remaining positive terms.
3. Find (nearly) smallest $x$ for which group I is positive; this $x$ is an upper bound to any positive real roots.
4. To get lower bound for positive roots, introduce $x = 1/y$ and repeat.
5. To get bounds for negative real roots, use $x = -z$ and repeat as for positive roots.

Finally, it should be noted that if, for some case, the computed bounds *underlap*, that is, if the upper bound is *less than* the lower bound, then because of this inconsistency there can exist no roots in this case.

**Example 7.4 The Peng-Robinson Equation of State**

Suppose we wish to use the Peng-Robinson equation of state to calculate the density of supercritical $CO_2$ at a pressure of $1 \times 10^4$ kPa and temperature 340°K. The Peng-Robinson equation is a pressure-explicit, two-constant equation of state which is given as

$$P = \frac{RT}{V - b} - \frac{a}{V(V + b) + b(V - b)} \tag{a}$$

where $a$ and $b$ are constants. For the conditions of this problem, and using the critical constants and acentric factor given in Modell and Reid [1983] for $CO_2$, these parameters are $a = 364.61$ m$^6$kPa/(kg mole)$^2$ and $b = 0.02664$ m$^3$/kg mole.

**Solution.** The density of $CO_2$ is simply the reciprocal of the specific volume ($\rho = 1/V$) Thus, we need to calculate the specific volume $V$ from the equation of state. Several approaches are now possible and we comment on them; all lead, when successful, to the answer $V = 0.05513$ m$^3$/kg mole. For any iteration procedure, it is apparent that the ideal gas value of $V$, $V = RT/P$, might be a good starting value.

First, we may note that the above equation could be used directly in the Newton method if we simply transpose the $P$ to get $f(V) = 0$. Alternatively, we could rationalize the denominator in the above equation to get a cubic equation in $V$. For this particular problem, the specific form of the cubic is

$$V^3 - 0.06760V^2 + 0.005004V - 0.0002380 = 0 \tag{b}$$

In addition, if desired, we could use a natural ordinary iteration procedure based on solving the original equation for the $V$ in the term $RT/(V - b)$ to get

$$V = b + \frac{RT}{P} - \frac{a(V - b)}{P[V(V + b) + b(V - b)]} = g(V) \tag{c}$$

Although it does require an algebraic rearrangement, there are some advantages that accrue to using the cubic equation (b) given above. The cubic equation formulation allows use of the polynomial bracketing procedure described earlier. Applied to the above cubic, this technique yields the upper bound $V < 0.07$. A second advantage of the cubic is that it allows use of the program POLYRTS (to be discussed later in this chapter) to determine *all* the roots of this third-degree polynomial equation. This is often useful to verify that the programmed equation of state does not have any more than one *real* root. For this problem, POLYRTS yields the desired real root plus a pair of conjugate complex roots, therefore accounting for all possible roots. It also happens in this problem (and it would be difficult to predict in advance) that the Newton method applied to the cubic equation (b) converges for a much wider range of initial guesses than it does when applied to the original non-polynomial form, Equation (a). For example, in this problem the Newton method converges for the cubic equation when $V$ is as large as 100, whereas for the original form it diverges when $V$ is as large as 0.09.

It is also noteworthy to observe that the $\Delta^2$ method applied to the iteration function $g(V)$ given above in Equation (c) here performs even better than the Newton method applied to the cubic equation. The range of convergence seems about the same, but for remote starting values (e.g., $V = 100$), $\Delta^2$ converges considerably faster.

## 7.2 ALL ROOTS OF A POLYNOMIAL EQUATION

In certain important computational problems it is necessary to find *all* of the roots of a polynomial equation. Examples include algebraic eigenvalue problems and inverting rational Laplace transforms. For real roots, this could be accomplished using the methods we have described by selecting various appropriate starting values. In such a situation the systematic bracketing of both positive and negative roots, which we discussed earlier, might be very useful. However, determination of any *complex* roots that may be present is another matter. We consider here the case where the given polynomial has real coefficients, so that any complex roots must occur in conjugate pairs. The procedures to be developed to determine all of the roots, whether real or complex, are employed in the computer program POLYRTS.

### 7.2.1 Real Roots

To deal with real roots, we consider the so-called synthetic division recurrence algorithm for finding a root using Newton's method and then deflate the original polynomial to form a new one whose degree is one less. This process is then repeated until all the real roots are found. However, it is up to the user to determine the number of real roots, as well as to furnish starting values. Suppose we are given the polynomial whose roots are to be found, as

$$p(x) = a_0x^n + a_1x^{n-1} + \cdots + a_{n-1}x + a_n \tag{7-15}$$

For *any* value of $x$, say $x = \alpha$, we can factor the original polynomial as

$$p(x) = (x - \alpha)(b_0x^{n-1} + b_1x^{n-2} + \cdots + b_{n-2}x + b_{n-1}) + r_0(\alpha) \tag{7-16}$$

In Equation (7-16), $r_0$ is a constant (with respect to $x$) whose value depends on $\alpha$. By directly multiplying and equating coefficients of powers in Equations (7-15) and (7-16), we can define the $b_i$ and $r_0$ in terms of the original $a_i$. This relationship turns out to be extremely neat and simple when expressed in recurrence form. By equating powers, we get

$$\begin{aligned}
b_0 &= a_0 \\
b_1 &= \alpha b_0 + a_1 \\
b_2 &= \alpha b_1 + a_2 \\
&\vdots \\
b_{n-1} &= \alpha b_{n-2} + a_{n-1} \\
r_0 &= \alpha b_{n-1} + a_n
\end{aligned}$$

Thus, if we *define*, for convenience, $b_n = r_0$ and $b_{-1} = 0$, then the simple recurrence relation $b_k = \alpha b_{k-1} + a_k$ is valid for $k = 0, 1, \ldots, n$. If $x = \alpha^*$ is a root of $p(x) = 0$, then we must have $r_0(\alpha^*) = 0$. In order to find a real root $\alpha^*$, we want to effectively solve the equation $r_0(\alpha) = 0$. When this root is found, we will also know the coefficients $b_k$ of the deflated equation, so we can repeat the process to look for a new root and so forth. To solve the equation $r_0(\alpha) = 0$ we use Newton's method together with the above recurrence relation. If we express Equation (7-16) as $p(x) = (x - \alpha)q_{n-1}(x) + r_0$, we can get $p'(\alpha) = q_{n-1}(\alpha) = r_1$, where $q_{n-1}(x) = (x - \alpha)q_{n-2}(x) + r_1$. Thus, using the same recurrence relation *twice* in the same loop gives the $p(\alpha)$, $p'(\alpha)$ values needed for one Newton iteration. To illustrate this in detail, we show the following computational loop, which evaluates both $p(\alpha)$ and $p'(\alpha)$:

```
b(-1) = 0
c(-1) = 0
For i = 0 to n
     b(i) = αb(i-1) + a(i)
     c(i) = αc(i-1) + b(i)
Next i
p(α) = b(n)
p'(α) = c(n-1)
```

Program POLYRTS starts with $x = 0$ and uses the above algorithm to attempt to find a real root. If it succeeds, it uses the current root as the starting value to find a new real root of the deflated polynomial, and so on. If the maximum number of iterations (user-specified) does not yield a root (as measured by the user specified-error tolerance), the user is allowed to either (i) continue with a new initial guess, or (ii) go on to the complex-root part of the program.

### 7.2.2 Complex Roots

To determine complex roots, the Bairstow method [Henrici, 1964] may be used. This calculation is based on the idea that if the coefficients of the polynomial are real (which we take to be the case), then, since the complex roots are complex conjugates, they necessarily appear in *real* quadratic factors. The Bairstow process, analogous to synthetic division, seeks to find a quadratic factor containing two complex conjugate roots, which then deflates the original polynomial by two degrees. This process is continued until all of the complex roots are found. It should be noted that the Bairstow process can also find a pair of real roots, if they are present. To derive the method, we start again with the original given polynomial of Equation (7-15), which we write as

$$p(x) = (x^2 - sx - t)(b_0 x^{n-2} + b_1 x^{n-3} + \cdots + b_{n-2}) + b_{n-1}(x - s) + b_n \tag{7-17}$$

The reason for writing the second-last term of Equation (7-17) as $b_{n-1}(x - s)$, rather than as the more obvious $b_{n-1}x$, is that as written in Equation (7-17), multiplication and comparison with Equation (7-15) leads to the following very simple and convenient recurrence relation for the $b_k$ values:

$$b_k = a_k + sb_{k-1} + tb_{k-2}; b_{-2} = b_{-1} = 0 \quad (k = 0, 1, \ldots, n) \tag{7-18}$$

For a root to the polynomial to be contained in the quadratic factor $(x^2 - sx - t)$, we must have $b_{n-1} = b_n = 0$, and these are the two equations in the two unknowns $s$ and $t$ that are solved by the Newton method. Fortunately, it turns out that the use of the recurrence relation in Equation (7-18) yields very similar relationships for the required derivatives $\partial b_n/\partial s$, $\partial b_n/\partial t$, $\partial b_{n-1}/\partial s$ and $\partial b_{n-1}/\partial t$. With $c_k = \partial b_{k+1}/\partial s$, and $d_k = \partial b_{k+2}/\partial t$ we find that

$$c_k = b_k + sc_{k-1} + tc_{k-2}; \quad c_{-2} = c_{-1} = 0 \quad (k = 0, 1, \ldots, n)$$

and

$$d_k = c_k \quad (k = 0, 1, \ldots, (n-2)) \tag{7-19}$$

These relationships are used in program POLYRTS.

Program POLYRTS, as shown in the "Computer Programs" section, requires that the user specify the degree of the polynomial together with its coefficients. It is generally best to determine all real roots first; to carry this out, it is very useful (but not absolutely necessary if luck is on your side!) to know the number and approximate location of the real roots. The program first tries to find a real root using synthetic division, and, if successful, deflates the polynomial and looks for a new root, and so on, until it no longer succeeds in finding a real root. Its failure could be because all real roots have been found *or* because the current starting value is not close enough to any remaining root. The user is given the choice of specifying a new starting value or proceeding to the Bairstow calculation to seek complex conjugate roots. This latter part of the program is constructed to work only with an *even*-degree polynomial; if the polynomial at this point is of odd degree, an error message will appear. The Bairstow procedure may by itself succeed in finding all of the complex roots. If not, it will prompt the user for new initial guesses for $s$ and $t$.

**Example 7.6 Find All Roots of $x^4 - 1.1x^3 + 2.3x^2 + 0.5x + 3.3 = 0$**

This problem is considered by Gerald [1978]. As the reader can verify, use of the root bracketing technique shows that there can be *no real roots* to this equation. This is because the bracketing bounds produce no interval containing either positive or negative real roots; that is, the *inconsistent* upper and lower bounds cannot be satisfied by any real root. Application of program POLYRTS initially produces no real roots. When the program is permitted to stop searching for real roots, it then produces the following two pairs of complex conjugate roots, which are the same as those given by Gerald [1978]:

$$z_1 = -0.45 \pm 0.947365i \tag{a}$$

$$z_2 = 1 \pm 1.41421i \tag{b}$$

## 7.3 SYSTEMS OF NONLINEAR ALGEBRAIC EQUATIONS

Systems of nonlinear algebraic equations arise frequently in chemical engineering, especially in connection with problems in thermodynamics, fluid mechanics, reaction kinetics, and optimization. In mathematical form we have, in general, $f_i(x_1{}^*, x_2{}^*, \ldots, x_k{}^*) = 0$. For $(i = 1, 2, \ldots, k)$, this represents a system of $k$ equations in $k$ unknowns involving $k$ different functions $f_1, f_2, \ldots, f_k$. We will discuss, for systems of equations, extensions of both the ordinary-iteration on $\Delta^2$ and the Newton methods, as well as the method of continuation.

### 7.3.1 The Newton Method

We consider briefly, for illustration, the case where we have two equations in two unknowns. In this situation, the analog of the one-dimensional Newton method is just

$$\begin{aligned}
0 &= f_1(x_1^{[n]}, x_2^{[n]}) + \frac{\partial f_1}{\partial x_1}(x_1^{[n+1]} - x_1^{[n]}) + \frac{\partial f_1}{\partial x_2}(x_2^{[n+1]} - x_2^{[n]}) \\
0 &= f_2(x_1^{[n]}, x_2^{[n]}) + \frac{\partial f_2}{\partial x_1}(x_1^{[n+1]} - x_1^{[n]}) + \frac{\partial f_2}{\partial x_2}(x_2^{[n+1]} - x_2^{[n]})
\end{aligned} \tag{7-20}$$

The above expression is derived using two Taylor series expansions, one for each $f_i$ function. In these equations the partial derivatives are all evaluated at $(x_1^{[n]}, x_2^{[n]})$. Thus, for *each Newton iteration,* two linear algebraic equations in the two unknowns $x_1$ and $x_2$ for the $n$ + 1st stage must be solved. In addition, four partial derivatives must be calculated. For the general $k$th order system, the Newton method is represented by the following system of $k$ linear algebraic equations for the $k$ unknowns $x_i^{[n+1]}$, $(i = 1, 2, \ldots, k)$:

$$0 = f_i(x_1^{[n]}, x_2^{[n]}, \ldots, x_k^{[n]}) + \sum_{j=1}^{k} \frac{\partial f_i}{\partial x_j}(x_j^{[n+1]} - x_j^{[n]}) \tag{7-21}$$

In this equation, the $k^2$ partial derivatives are all evaluated at $x_i^{[n]}$. *Every step* of the standard $k$-dimensional Newton method requires evaluation of $k^2$ partial derivatives and the solution of a system of $k$ linear algebraic equations. Thus, the method is very computationally intensive; for *large* systems this can be a serious limitation. A great advantage of the method, however, is its quadratic convergence. If the initial guess is close enough to a solution, we always get convergence. Moreover, when convergence occurs, it is (at least eventually) quite rapid. It must be emphasized that determination of a suitable starting value is generally much more difficult in the case of systems of equations. To reduce the cost of the standard Newton method, sometimes the partial derivatives are evaluated only periodically; in this case, the quadratic convergence is lost.

Two Newton programs for nonlinear algebraic systems are given here. The first is NEWTNLE (Newton nonlinear equations), and it requires specification of the $n^2$ partial derivatives, along with the equations to be solved $f_i = 0$ $(i = 1, \dots, n)$. The second program is NEWTONVS (Newton variable step). This program calculates the required partial derivatives numerically using a variable-step procedure (hence the suffix VS in the name of the program). Each program solves the requisite linear algebraic system at each step. For program NEWTONVS, the user need only specify the number of equations, the functions $f_i(x_1, \dots, x_n)$, and the starting guesses.

**Example 7.7 Solution of Necessary Conditions for an Optimum**

Consider the problem of finding the minimum with respect to $x_1, x_2, \dots, x_n$ of the function of $n$ variables $\phi(x_1, x_2, \dots, x_n)$. One direct approach to this problem involves solving the necessary conditions for a minimum, namely $\partial\phi/\partial x_1 = 0, \dots, \partial\phi/\partial x_n = 0$. These conditions correspond to the usual first derivative requirements for the minimum of a function of one variable, except that one such equation applies for each independent variable, and this yields a system of $n$ (generally nonlinear) equations in $n$ unknowns. As a particular example we present a two-dimensional problem in chemical reactor design due to Rinker [1994]. The function to be minimized in the range $(0 < x_1 < 1, 0 < x_2 < 1)$ is

$$\phi(x_1, x_2) = \frac{1}{x_1} + \frac{1}{x_2} + \frac{1 - x_2}{x_2(1 - x_1)} + \frac{1}{(1 - x_1)(1 - x_2)} \tag{a}$$

The two necessary conditions corresponding to $\partial\phi/\partial x_1 = 0$, $\partial\phi/\partial x_2 = 0$ are expressed as follows:

$$f_1 = 0 = -\frac{1}{x_1^2} + \left(\frac{1 - x_2}{x_2} + \frac{1}{1 - x_2}\right)\frac{1}{(1 - x_1)^2} \tag{b}$$

$$f_2 = 0 = -1 + \frac{1}{(1 - x_1)}\left(\frac{2x_2 - 1}{(1 - x^2)^2}\right) \tag{c}$$

If we now solve Equations (b) and (c) using the Newton method (with either program NEWTNLE or NEWTONVS), we obtain, with the starting values $x_1 = x_2 = 0.5$, rapid convergence to the final values $x_1 = 0.363696$ and $x_2 = 0.561246$. We can note that, even in this relatively simple problem, computation of the required four partial derivatives is rather tedious. Program NEWTONVS carries out this calculation numerically so that the user need not furnish these derivatives.

### 7.3.2 Successive Approximation and the Delta-Square Method

It requires much less effort to use ordinary iteration, as opposed to Newton, for systems of equations, but this method is subject to the very serious hazard of possible divergence. For a second-order system the iteration procedure would be, for example,

$$x_1^{[n+1]} = g_1(x_1^{[n]}, x_2^{[n]}) \tag{7-22}$$

$$x_2^{[n+1]} = g_2(x_1^{[n+1]}, x_2^{[n]}) \tag{7-23}$$

Notice that the new value of $x_1$ is used in the second equation as soon as it is available; this idea is usually beneficial. If the ordinary iteration method is used, linear equations need not be solved; nor is it necessary to compute partial derivatives. Thus, this method has great computational advantages, so it may be worth some effort to find an iteration procedure that works. In the case of systems, one problem is that there are *many* possible $g_i$ functions, some of which may be useful, others not. The question is how to find a "good" (convergent) set of $g_i$ functions. In general, analogous to the one-dimensional case, we would like to choose $g_i$ functions for which the partial derivatives $|\partial g_i/\partial x_j|$ are "small." From a purely mathematical standpoint this is usually not possible, but sometimes the physical problem itself suggests a natural iteration method in terms of solving which equation for which variable. Also, it is often useful to employ a variant of the one-dimensional $\Delta^2$ process, known as the "modified" $\Delta^2$ method, to attempt to accelerate the convergence of successive approximation. This is accomplished by simply applying the one-dimensional algorithm component-wise, as carried out in program DELSQ. In this case, while the great advantage of computational ease is retained, the modified $\Delta^2$ procedure does *not* converge quadratically, so there is no guarantee of convergence even if we start right at the root. Experience has shown that in many problems (but *not* all, as in the one-dimensional case), the modified $\Delta^2$ procedure does indeed improve convergence, and sometimes causes convergence even when the underlying ordinary iteration diverges. So in the authors' experience, it is always worth a try if ordinary iteration is being attempted. It must be kept in mind for *systems*, however, that there are instances where $\Delta^2$ diverges while ordinary iteration converges. Program DELSQ is set up to handle ordinary iteration or modified $\Delta^2$ for a system of $n$ nonlinear algebraic equations. The user need only specify the number of equations $n$, the $g_i$ functions, and the starting values. To employ modified $\Delta^2$, it is generally desirable, if possible, to arrange the numbering of the equations to have the dominant $|\partial g_i/\partial x_i|$ (diagonal) value correspond to the $i$th equation.

**Example 7.8 Solution of the Problem of Example 7.7 by Ordinary-Iteration/$\Delta^2$**

We reconsider the problem of Example 7.7 and use ordinary-iteration/$\Delta^2$ and program DELSQ to effect the solution. For convenience we repeat the equations for the two necessary conditions corresponding to $\partial\phi/\partial x_1 = 0$, $\partial\phi/\partial x_2 = 0$. These equations are

$$f_1 = 0 = -\frac{1}{x_1^2} + \left(\frac{1-x_2}{x_2} + \frac{1}{1-x_2}\right)\frac{1}{(1-x_1)^2} \tag{a}$$

$$f_2 = 0 = -1 + \frac{1}{(1-x_1)}\left(\frac{2x_2-1}{(1-x_2)^2}\right) \tag{b}$$

In this case, a good choice of the $g(x)$ functions corresponds to solving the $f_1$ equation for $x_1$ and the $f_2$ equation for $x_2$ as follows:

$$x_1 = (1-x_1)\left(\frac{1-x_2}{x_2} + \frac{1}{1-x_2}\right)^{-\frac{1}{2}} = g_1(x_1, x_2) \tag{c}$$

$$x_2 = \frac{(1-x_1)(1-x_2)^2 + 1}{2} = g_2(x_1, x_2) \tag{d}$$

Use of either ordinary iteration or $\Delta^2$ by means of program DELSQ, using initial values $x_1 = x_2 = 0.5$, produces the same final results given previously for Newton. In this case the $\Delta^2$ method converges in four iterations (same as Newton), whereas unaided ordinary iteration takes somewhat longer. Thus, in this problem it may seem that $\Delta^2$ is more powerful than Newton, since it does not require partial derivatives. However, it must be emphasized that while the indicated $g(x)$ functions did well, some other rearrangements of the original equations failed to produce convergence. The modified $\Delta^2$ method is wonderful when it works, but in many instances for *systems* it does not converge, even for good starting values.

### 7.3.3 Miscellaneous Procedures

A few comments are in order regarding other ideas and approaches that are sometimes useful in the solution of nonlinear systems. We can observe the equivalence between solving a nonlinear system and certain optimization problems. For example, the problem of solving the system $f_1(x_1, x_2) = 0$, $f_2(x_1, x_2) = 0$ has for its solution(s) the same $(x_1, x_2)$ values as the minimum of the function $\phi = f_1^2 + f_2^2$. An advantage of this correspondence which has sometimes been employed is that a direct minimization calculation is often insensitive to the starting values.

The idea of finding an efficient "organization" of nonlinear system calculations can sometimes be very important, especially in large problems. This concept, which is sometimes called "precedence ordering," involves a systematic analysis of such things as ordering the equations and deciding which variables to iterate on, and so forth.

In some large problems it is very useful to employ the so-called "modified" Newton method, which is a close relative of the modified $\Delta^2$ approach described earlier. This procedure involves retaining only the *diagonal* elements $(\partial f_i/\partial x_i)$ in the Newton equations. The resulting simplified algorithm requires only $k$ (as opposed to $k^2$) partial derivatives per step, and *no* algebraic equations to be solved. An option for this procedure is contained in program NEWTONVS. Because of the computational savings, this method is very good *if it works*. The key to success is the presence and identification of an ordering of the equations for which the diagonal partials dominate the nondiagonal ones. This often performs better than one might expect, so if computer storage or time is a problem, this method may be useful. Another variant which is only slightly more complicated involves retaining a tridiagonal block of partials (of size $3n$) and using the tridiagonal algorithm to solve the linear equations at every step. Finally, we should mention that more sophisticated procedures for the solutions of nonlinear algebraic systems are discussed, for example, in Dennis and Schnabel [1983].

## 7.4 THE METHOD OF CONTINUATION (HOMOTOPY)

The method of continuation, also known as the "homotopy" method, is an important procedure for solving systems of nonlinear algebraic equations. It generally requires more sophisticated calculations, but it has the great advantage of being able to find roots even when the starting values are poor (where, for example, the Newton method might not converge). We preface the discussion of the method of continuation with a consideration of the related important problem of the numerical solution of a single nonlinear equation containing a parameter.

Suppose we wish to find the values of $x$ which are the solutions of $f(x, B) = 0$ for certain $B$ values. For any given fixed $B$, the problem is merely the solution of a single nonlinear equation, which can be handled by the methods discussed earlier. However, we can see that $x$ for the above equation depends on $B$, and this relationship can be exploited to advantage. For example, if we solve the equation for a particular $B$ in the desired $B$ range, we expect that for a "nearby" $B$ value, the solution won't change too much. Thus, we might use the calculated $x$ for a given $B$ as the starting guess for the unknown $x$ at a nearby $B$. This process can be continued, and we can therefore creep along from the first solution to other nearby ones until we have calculated the $x$ values for the entire $B$ range of interest. In this way, the only difficult starting value is the first one. Obviously, the success of this procedure requires that we keep moving from some solution to another one which is sufficiently close, and this clearly depends on the particular equation being solved. Another way to move from one solution to the next is to note that since $x$ depends on $B$, we can form a differential equation for $dx/dB$ by differentiating the equation $f(x, B) = 0$. This gives

$$\frac{\partial f}{\partial x}\frac{dx}{dB} + \frac{\partial f}{\partial B} = 0 \tag{7-24}$$

By solving this differential equation for $x$ as a function of $B$, using a calculated initial condition corresponding to one $(x, B)$ solution pair, we can determine, in just one integration run, the whole set of $x$ solutions in the desired $B$ range. In the same way, if we had a *system* of nonlinear equations depending on parameter $B$, we could generate a system of ordinary differential equations for the solution vector $dx_i/dB$. In this case, the system of differential equations would have exactly the same form as the equation above, except that now the matrix $\partial f_i/\partial x_j$ replaces $\partial f/\partial x$, the vector $dx_i/dB$ replaces $dx/dB$, and the vector $\partial f_i/\partial B$ replaces $\partial f/\partial B$.

Having seen the rather natural way to simplify the calculation of roots for functions depending on a parameter, we now proceed to see how we can use this idea for systems in a general way. Suppose we wish to solve the *system* of equations $\mathbf{f}(\mathbf{x}^*) = 0$. We use boldface here as a convenient notation to indicate a vector quantity (e.g., $\mathbf{f}$ refers to $(f_1, \ldots, f_n)^T$). Let us introduce into the equations a parameter $\lambda$, so that $\mathbf{x}$ depends on $\lambda$. We wish to do this in such a way that for some arbitrary initial value $\mathbf{x}_0$, $\lambda = 0$ and for $\mathbf{x} = \mathbf{x}^*$ (the desired root), $\lambda = 1$. This can be accomplished, for example, by the relationship

$$\mathbf{f}[\mathbf{x}(\lambda)] = (1 - \lambda)\mathbf{f}(\mathbf{x}_0) \tag{7-25}$$

Here, we see that for $\lambda = 0$, $\mathbf{f}(\mathbf{x}) = \mathbf{f}(\mathbf{x}_0)$, and for $\lambda = 1$, we have $\mathbf{f}(\mathbf{x}) = 0 = \mathbf{f}(\mathbf{x}^*)$ so that at $\lambda = 1$, $\mathbf{x} = \mathbf{x}^*$. By differentiation of Equation (7-25), we get

$$\sum_{j=1}^{n} \frac{\partial f_i}{\partial x_j}\frac{dx_j}{d\lambda} = -f_i(x_0), \quad (i = 1, \ldots, n) \tag{7-26}$$

Thus, the solution of the differential equation system Equation (7-26), starting with $\mathbf{x} = \mathbf{x}_0$ at $\lambda = 0$, should yield $\mathbf{x}^*$ at $\lambda = 1$ This procedure is utilized in Example 7.9 below (where difficulties are encountered with the Newton method). The differential continuation method described here obviously requires some rather sophisticated calculations, but it does pay off when serious problems occur with starting values. A computer program utilizing this procedure, called METHCON, is described at the end of the chapter.

**Example 7.9 Chemical Equilibrium**

A mixture of 1 mole of CO and 3 moles of $H_2$ is allowed to reach equilibrium at 500°C and 1 atm total pressure. It is desired to estimate the composition of the equilibrium mixture. The significant reactions (along with their equilibrium constants $K_y$) for this system are thought to be

| | | |
|---|---|---|
| (1) | $3H_2 + CO = CH_4 + H_2O$ | $K_y = 69.18$ |
| (2) | $CO + H_2O = CO_2 + H_2$ | $K_y = 4.68$ |
| (3) | $CO_2 + C = 2CO$ | $K_y = 0.0056$ |
| (4) | $5H_2 + 2CO = C_2H_6 + 2H_2O$ | $K_y = 0.141$ |

**Solution.** All compounds appearing in the reactions, with the exception of $C$, will appear in the gas phase. From standard chemical equilibrium relationships, assuming perfect gas behavior, we can write

$$\frac{y_{CH_4} y_{H_2O}}{y_{H_2}^3 y_{CO}} = 69.18 \tag{a}$$

$$\frac{y_{CO_2} y_{H_2}}{y_{CO} y_{H_2O}} = 4.68 \tag{b}$$

$$\frac{y_{CO}^2}{y_{CO_2}} = 0.0056 \tag{c}$$

$$\frac{y_{C_2H_6} y_{H_2O}^2}{y_{H_2}^5 y_{CO}^2} = 0.141 \tag{d}$$

As written, Equations (a) through (d) involve four equations in the six unknown mole fractions. We use the "reaction coordinate" formulation of the chemical equilibrium equations, as described by Smith and Van Ness [1987]. We choose the species labels [1-$H_2$, 2-CO, 3-$CH_4$, 4-$H_2O$, 5-$CO_2$, 6-$C_2H_6$] and the reaction coordinates $x1$, $x2$, $x3$, $x4$ for reactions 1 to 4, respectively. With this formulation, the equilibrium mole fractions expressed in terms of the $x_i$ become

$$\begin{aligned}
y_1 &= (3 - 3x_1 + x_2 - 5x_4)/D \\
y_2 &= (1 - x_1 - x_2 + 2x_3 - 2x_4)/D \\
y_3 &= x_1/D \\
y_4 &= (x_1 - x_2 + 2x_4)/D \\
y_5 &= (x_2 - x_3)/D \\
y_6 &= x_4/D
\end{aligned} \tag{e}$$

Here, $D = (4 - 2x_1 + x_3 - 4x_4)$ represents the total moles present at equilibrium. When these expressions are substituted into Equations (a)–(d), we obtain the following nonlinear system of four equations in the four $x_i$ unknowns:

$$
\begin{aligned}
69.18 &= \frac{x_1(x_1 - x_2 + 2x_4)D^2}{(3 - 3x_1 + x_2 - 5x_4)^3(1 - x_1 - x_2 + 2x_3 - 2x_4)} \\
4.68 &= \frac{(x_2 - x_3)(3 - 3x_1 + x_2 - 5x_4)}{(1 - x_1 - x_2 + 2x_3 - 2x_4)(x_1 - x_2 + 2x_4)} \\
0.0056 &= \frac{(1 - x_1 - x_2 + 2x_3 - 2x_4)^2}{(x_2 - x_3)D} \\
0.141 &= \frac{x_4(x_1 - x_2 + 2x_4)^2 D^4}{(3 - 3x_1 + x_2 - 5x_4)^5(1 - x_1 - x_2 + 2x_3 - 2x_4)^2}
\end{aligned}
\tag{f}
$$

We could now try to solve these equations using the the Newton method. If we use program NEWTNLE, we must calculate 16 ($4^2$) partial derivatives. To avoid the necessity of calculating and programming the partial derivatives, we could instead use program NEWTONVS. As we know from earlier discussions, *the main difficulty is to find starting values that result in convergence.* The only obvious physical information available is that each $y_i$ (not $x_i$) must lie between 0 and 1. Also, in this particular (rather complicated) problem, since *solid* carbon is indicated in reaction 3 but is not present in the feed, a carbon balance on the equilibrium mixture shows that we must have $x_3 < 0$. This information could be used in conjunction with Equations (e) to try to determine some necessary bounds for the $x_i$. Instead, we *arbitrarily* choose as initial values $x_1 = 0.1$, $x_2 = 0.2$, $x_3 = 0.3$, and $x_4 = 0.4$. These initial values do *not* produce convergence with the Newton method. As we know, the method of continuation is designed to be of assistance for problems where starting values are a source of difficulty. When we employ this method *carefully*, using program METHCON, we do achieve convergence to the desired results

$$
\begin{aligned}
x_1 &= 0.681604 \\
x_2 &= 1.58962 \times 10^{-2} \\
x_3 &= -0.128703 \\
x_4 &= 1.40960 \times 10^{-5}
\end{aligned}
\tag{g}
$$

From the error monitoring in program METHCON, we are virtually certain that the results in Equations (g) are easily accurate to the six figures shown. Note that the value of $x_3$ is negative, as required by the equilibrium carbon balance.

To obtain these results for the initial values given above, we needed to proceed with care (to understand what follows, the reader is encouraged to see material on program IEXVS in the chapter on numerical solution of initial value problems in ordinary differential equations). Since we don't have good initial values, it is generally best to proceed by using the automatic error control option. In this problem, we have used the rather large "tol" (local relative error tolerance) value of 0.1. If we use one integration trajectory to start, together with an initial integration step size of 0.01, we do achieve a result (but not a very good one, judging by the function check values) in the first cycle. We then continue the problem, repeating the integration procedure of the first cycle. This time the local relative errors are smaller, and the function check values are smaller. So now we continue using two trajectories and error control; this produces even better values. For the mathematical *coup de grace*, we then continue and employ three trajectories, which yields the maximum global error information. The result of this is the values shown in Equations (g), along with estimated errors of about $10^{-14}$ and good error ratios. Then just for fun, one can continue and use one trajectory with a step size of unity. This just repeats the previous values, along with the local error estimate (which is global for the case of only one step) of zero.

On the other hand, if we try to use either three or two trajectories from the beginning, we encounter difficulties. Surprisingly, we also experience difficulties if we start with an error tolerance much smaller than 0.1. Thus, it is best in this problem to start with a crude approach and refine it later. In general (but in this problem we have an exception), when there is a problem with the integration, we can either (a) decrease the step size (fixed-step operation) or (b) decrease the local error tolerance parameter "tol" (variable step operation). If we can't get continuation to work at all with given starting values, we must generally have recourse to altering the starting values. To give a better indication of the difficulty of the above problem, we note that the Newton method does *not* converge, even for the initial values ($x_1$ = 0.5, $x_2$ = 0, $x_3$ = −0.1, $x_4$ = 0), which seem intuitively very "close" to the known true values.

When we substitute the $x_i$ values from Equations (g) into Equations (e), we finally obtain the desired equilibrium mole fractions as

$$y_1 = 0.387162$$
$$y_2 = 0.0179676$$
$$y_3 = 0.271769$$
$$y_4 = 0.265442$$
$$y_5 = 0.0576549$$
$$y_6 = 5.62034 \times 10^{-6}$$

## 7.5 ASYMPTOTIC SOLUTIONS OF AN EQUATION WITH A PARAMETER

This section involves material which is at a higher mathematical level than that given previously in the chapter, so it has deliberately been placed at the end of this chapter. Frequently, the roots of a nonlinear equation depend on one or more parameters. Usually in such a case, it is very desirable to express the solution as a function of the parameters. Since the implicit character of the original equation does not allow any analytical representation of the root even for specific parameter values, generally it is not possible to express the root explicitly as a function of parameters. Of course, one approach is to carry out many numerical solutions and correlate the results by some type of interpolation scheme. Rational least squares is naturally suggested. But another attractive procedure, which is often feasible, is to find a simple, approximate analytical solution that becomes more valid as a parameter approaches some limiting value (often either zero or infinity). There are basically two general approaches for developing asymptotic solutions, one of which is very simple and direct (giving less accuracy), while the other is more sophisticated and requires more computation (which can be done on the computer) but may yield more accuracy. Which of these procedures is more appropriate in given circumstances depends on the problem requirements. Both methods are now discussed.

The first asymptotic procedure involves identifying an appropriate "asymptotic balance" among the terms of the equation, to see if under the limiting conditions of interest only certain terms contribute heavily in the algebraic balance, while others are essentially negligible. This is attempted by trying the balance of various pairs of terms in the equation to see what the corresponding simplified root would be.

Any "trial" balance is checked for consistency by using the result of the balance to substitute back into the assumed "small" terms to see if, for the trial balance, they are indeed small. If one of these balances is consistent for a case of interest, then the approximate expression for the root found in this way can be improved by one step of either the Newton or extended Newton method. We illustrate this technique now with an example.

**Example 7.10 Solution of $1/f^{1/2} = 1.763\ln(\text{Re}\, f^{1/2}) - 0.6$**

This implicit equation represents the friction coefficient $f$ as a function of Reynolds number Re for developed flow in smooth circular ducts over a large range of Reynolds numbers ($\text{Re} > 10^4$). We wish to find a simple but accurate approximation that expresses $f$ explicitly as a function of Re for $\text{Re} > 10^4$, the practical range of application. An asymptotic balance is sought among the terms of the equation for the limit of $\text{Re} \to \infty$; the first approximation obtained in this way is refined by either the Newton or extended Newton method.

The character of the physical problem represented by the equation suggests that there should be one positive root $f$ for any physically realistic value of Re. For convenience of mathematical manipulation, the new variables $S = 1/f^{1/2}$ and $B = (1.763\ln \text{Re} - 0.6)$ are introduced. Then the new equation to be solved becomes

$$S = B - 1.763\ln S \tag{a}$$

where now $B \to \infty$ for $\text{Re} \to \infty$ and $S \to \infty$ for $f \to 0$. If we write

$$F(S) = S + 1.763\ln S - B = 0 \tag{b}$$

we see that $F'(S) = 1 + 1.763/S$, which is positive for all positive $S$. Thus, there can be no more than one positive root of this equation. Also, for $S \to \infty$, $F \to \infty$, while for $S = 1$, $F(1)$ is negative if $B > 1$, showing that there must be exactly one positive root of the equation for this case. To balance terms in the equation when $B$ is large, the quantity $S + 1.763\ln S$ must become large. Since $\ln S \ll S$ for large $S$, we must have approximately that $S \approx B$ for $B \to \infty$. This is the desired first approximation for large $B$. In terms of the original variables,

$$f \cong 1/(1.763 \ln \text{Re} - 0.6)^2 \tag{c}$$

To substantially improve this result in a simple way, we use one step of the regular or extended Newton process. Using one step of regular Newton with the initial approximation $S = B$ gives

$$\frac{1}{f^{1/2}} = S \cong B - \frac{CB\ln B}{C+B} \quad \text{(Regular Newton)} \tag{d}$$

where $C = 1.763$. By using the extended Newton method, Equation (7-10), with $S^0 = B$, we get the slightly more complicated, but also more accurate, formula

$$\frac{1}{f^{1/2}} = S \cong B - \frac{CB\ln B}{C + B + \dfrac{C^2}{2}\dfrac{\ln B}{C+B}} \quad \text{(Extended Newton)} \tag{e}$$

A comparison of the regular and extended Newton asymptotic solutions to this problem is shown in Table 7.4. For the practical range of interest, $\text{Re} \geq 10^4$, we see that regular Newton has a maximum error of less than 1.3%, while the maximum error for

**TABLE 7.4** ASYMPTOTIC SOLUTION OF $1/f^{1/2} = 1.763 \ln(\text{Re} f^{1/2}) - 0.6$ FOR $\text{Re} \to \infty$

| Re | $f \times 10^3$ Regular Newton | $f \times 10^3$ Extended Newton | $f_{\text{exact}} \times 10^3$ |
|---|---|---|---|
| $10^4$ | 7.86 | 7.77 | 7.76 |
| $10^5$ | 4.52 | 4.49 | 4.49 |
| $10^6$ | 2.90 | 2.89 | 2.89 |
| $10^7$ | 2.010 | 2.005 | 2.004 |
| $10^9$ | 1.117 | 1.116 | 1.115 |

extended Newton is about ten times smaller, or less than 0.13%. Depending on accuracy requirements, either of these formulas is quite useful.

A second, more sophisticated technique for developing asymptotic analytical solutions to nonlinear equations involves the following steps. As before, we use the idea of "dominant balance" in the equation to identify the first approximation to a more complete limiting solution. After this first approximation has been found, we systematically refine it so as to make it useful for a large range of the parameter. To accomplish this, we wish to express the parameter as a Maclaurin series in terms of the variable. Depending on the circumstances, this may require a change of variables in the equation so that the transformed parameter has such an expansion in terms of the transformed variable in the vicinity of the desired root. Once the appropriate variables are defined to yield such an expansion, the remaining steps are mechanical and can be accomplished with the help of the appropriate computer programs. The steps required can be listed as follows:

1. Obtain a Maclaurin expansion for the *parameter* in terms of the variable.
2. *Revert* this series to get the variable as a Maclaurin series in terms of the parameter.
3. Attempt acceleration of this series by means of Euler's transformation, Padé approximation or Shanks transformation.
4. Compare the asymptotic analytical approximation with selected numerical results to determine the region of validity.

The technique is now illustrated with an example.

**Example 7.11 Expansion of the Peng-Robinson Equation of State for Small Pressure**

As an example of the series asymptotic method, we consider the problem of expressing the volume in terms of the pressure using the Peng-Robinson equation of state for the case where the pressure is "small." The temperature is fixed. It is convenient to write the equation in the form

$$\frac{P}{RT} = \frac{1}{V-b} - \frac{a(t)/RT}{V(V+b)+b(V-b)} \tag{a}$$

Since $b$ is a constant independent of temperature, by using the variable $P/RT$ we have left only one quantity depending on temperature, the ratio $a(T)/RT$. For $P \to 0$, $V \to \infty$ and we have ideal gas behavior. Therefore, instead of $V$ we use the variable $y = 1/V$ and we also, for convenience, define $\phi = b \cdot y$, $A = a(T)/RT$, and $x = P/RT$. Then Equation (a) becomes

$$x = \frac{y}{1-\phi} - \frac{Ay^2}{1+2\phi-\phi^2} \tag{b}$$

With these convenient definitions, the rest is straightforward. First, we expand the *RHS* of Equation (b) as a power series in $y$ (keeping in mind that $\phi = b \cdot y$). This is accomplished using the geometric expansion for the first term and the division algorithm (see Appendix D and program SERIES) for the second term. This gives

$$x = y(1+\phi+\phi^2+\phi^3 \ldots) - Ay^2(1-2\phi+5\phi^2-12\phi^3 \ldots) \tag{c}$$

Putting in $\phi = b \cdot y$ then yields

$$x = y + y^2(b-A) + y^3(b^2+2Ab) + y^4(b^3-5Ab^2) \ldots \tag{d}$$

For a given gas and temperature, the coefficients in Equation (d) are known. This series for $x$ in terms of $y$ is now reverted to give the series for $y$ in terms of $x$ (see Appendix D and program SERIES). This expresses the density $y(=1/V)$ as a series in powers of $P/RT$. If the convergence of the series for the desired range of $P/RT$ is satisfactory, the problem is solved. If not, then acceleration can be attempted using Euler's transformation, Padé approximation, or Shanks transformation, as discussed in Appendix E.

## 7.6 RECOMMENDATIONS AND COMMENTS

Since we have discussed a number of methods for solving nonlinear algebraic equations, it seems appropriate to make some remarks as to when the various methods are most useful. The secant method is often useful when we need a simple, one-dimensional method to be used as part of another computer program, such as in the shooting method for boundary value problems in ordinary differential equations.

The ordinary iteration or successive approximation (SA) method used in the DELSQ program is useful for problems where a "natural" iteration function is apparent, regardless of the number of variables, and where simplicity and low-cost computation are desired.

The $\Delta^2$ acceleration of SA, also featured in program DELSQ, can always be considered as a possible option. It is normally the method of choice for a *one-dimensional* problem, since for this case it has quadratic convergence and does not require a derivative. Note that for *any* equation $f(x) = 0$ we can write $x^{n+1} = x^n + Cf(x^n) = g(x^n)$. Here, $C$ can be any number, but $C \approx -1/f'(x^0)$ is usually a good value.

The Newton method, utilized in program NEWTONVS, is usually best for systems that are not *too* large, where no ordinary iteration function $\mathbf{g}(\mathbf{x})$ is apparent, and when reasonable starting values are known. The powerful continuation method is especially useful when reasonable starting values are not known and when various "guesses" don't work.

The asymptotic approach is most useful when the problem has parameters and it is desired to determine a solution formula as a function of parameters. Also, it can be useful in finding approximate root locations.

The synthetic division/Bairstow method, as developed in program POLYRTS, is useful for *polynomial* problems where it is desired to find all roots. The algorithm is especially valuable for problems having some *complex* roots. One general and important application is in finding eigenvalues of matrices through determination of the roots of the characteristic equation (see program MATXEIG).

## 7.7 COMPUTER PROGRAMS

In this section we discuss the computer programs that are applicable to various problems in the chapter. For each program we give a brief discussion and a concise listing, which includes (i) a program description, (ii) how to run, and (iii) user-specified subroutines, including a sample problem. Detailed program listings can be obtained from the diskette.

### Ordinary Iteration/$\Delta^2$ Nonlinear Equation Program DELSQ

Instructions for running the program are given in the concise listing that follows. The program prompts for the method (ordinary iteration or delta-square), initial estimates of all the unknowns, and whether intermediate results are to be printed. This print control option is useful for problems involving a number of variables and slow convergence; printing all intermediate results in such a case slows down the operation. If intermediate results are *not* printed, the user can conveniently monitor the "norm" value, which is printed for each step, to see if convergence is occurring. The function check option merely evaluates the iteration functions using the starting values. These $g(x^0)$ values can then be checked independently (by a calculator, for example) to verify that the proper $g(x)$ functions are indeed contained in the program.

The program prints the successive iterates, the norm (convergence criterion), the number of iterations, and, in the case of the delta-square option, the "estimated relative error" (est.rel.err), which is just the delta-square correction divided by the current estimate of the unknown. When a math coprocessor is contained in the system, True BASIC calculates internally to 14 figures; if no coprocessor is present, nine figures are used. Thus, since the output is normally given to six figures, the norm_tol value in Sub Function is appropriately taken in the vicinity of $10^{-7}$.

The problem shown in Sub Function is for the solution of the two nonlinear equations shown in the concise listing, which are just the equations used in Example 7.8. For many problems, a maximum number of iterations equal to 100 is adequate. In some difficult problems, this may need to be increased. In solving nonlinear equations by some iterative method, we often obtain either convergence or divergence rather quickly. In other cases, where the iterations seem to "wander around," it sometimes pays off to be patient and let the iterations continue for a while, in hopes that some iterate will eventually bounce into a region of attraction.

### Program DELSQ

```
!This is program DELSQ.tru, the ordinary iteration/delta-square
!solution of a system of n nonlinear equations by O.T.Hanna
!Equations must be in form x(i) = gi(x1,x2,...xn), i = 1,2,...n
!mx is maximum no. of iterations allowed
!norm_tol is convergence criterion
!method is 1 for ordinary iteration, 2 for delta-square

!****************************************************************
!*To run define parameters, equations in Sub Function at the    *
!*end of the program; system prompts for method, initial values, *
!*print control. To check Function use mx = 1, method = 1. Stop  *
!*program via (i)Ctrl-Break(IBM) or Stop(MAC),(ii)maximum        *
!*number of iterations(mx), or (iii)norm_tol.                    *
!****************************************************************

!This is DELSQ.tru program for problem below
SUB Function(mx,norm_tol,x())      !***denotes user specified quantity
    LET mx = 100                   !***maximum number of iterations
    LET norm_tol = 1e-7            !***norm tolerance stop control
    !Eqns x = g(x)                 !***
    LET a = 1 - x(2)
    LET x(1) = (1-x(1))*( a/x(2) + 1/a )^(-0.5)
    LET x(2) = ( (1-x(1))*a^2 + 1 )/2
END SUB
```

### ***Newton Nonlinear Equation Systems Program NEWTNLE***

Instructions and information for running the program are contained in the concise listing given below. Program NEWTNLE prompts for the option of overriding the starting values for the unknowns. The user is also given options for verifying the functions $f(i)$ by computing them using the starting values, and verifying the partials $\partial f_i/\partial x_j$ based on the starting values. The problem considered in Sub Func involves four equations in four unknowns. The parameters mx and norm_tol are set as discussed above for the DELSQ program.

### Program NEWTNLE

```
!This is program NEWTNLE.tru for a system of nonlinear eqns by
!O.T.Hanna. Program solves a system of nonlinear algebraic
!equations by the Newton method.
!****************************************************************
!*To run we must specify the n equations f(i) as a function     *
!*of x(j) in Sub Func, as well as partial derivatives           *
!*df(i)/dx(j) = f(i,j), max. number of iterations(mx) and       *
!*convergence tolerance parameter norm_tol. Specify initial     *
!*values of x(j) in Sub Init. Program prompts for option to     *
```

```
!*over-ride initial values or verify f(i),df(i)/dx(j).            *
!*Program is stopped by either convergence or max. iterations.    *
!*****************************************************************
!This is NewtNLE.tru of problem below
SUB Init(n,x())
    !***denotes user-specified statement
    LET n = 4                        !***number of equations
    MAT redim x(n)
    LET x(1) = 5                     !***initial values of variables
    LET x(2) = 5
    LET x(3) = 5                     !***initial values of variables
    LET x(4) = 5
END SUB

SUB Func(x(),f(),f1(,),mx,norm_tol)
    !***denotes user-specified quantity
    LET mx = 100                     !***max allowed iterations
    LET norm_tol = 1e-7              !***convergence tolerance

    !***Specify Equations
    LET f(1)=x(3)*x(1)-5+x(4)
    LET f(2)=x(3)*x(2)-7+x(4)
    LET f(3)=x(3)-12+x(4)
    LET f(4)=x(4)-60*x(1)*x(2)^0.6

    !***Specify n x n partial derivatives as df(i)/dx(j) = f(i,j)
    LET f1(1,1) = x(3)
    LET f1(1,2) = 0
    LET f1(1,3) = x(1)
    LET f1(1,4) = 1
    LET f1(2,1) = 0
    LET f1(2,2) = x(3)
    LET f1(2,3) = x(2)
    LET f1(2,4) = 1
    LET f1(3,1) = 0
    LET f1(3,2) = 0
    LET f1(3,3) = 1
    LET f1(3,4) = 1
    LET f1(4,1) = -60*x(2)^0.6
    LET f1(4,2) = -60*x(1)*(.6)*x(2)^(-.4)
    LET f1(4,3) = 0
    LET f1(4,4) = 1
END SUB
```

### ***Newton Nonlinear Equation Systems Program NEWTONVS***

Instructions and information for running the program are contained in the concise listing given below. The CASE statements contained in Sub Func defining the functions $f(i)$ are needed as part of the logic for the program to individually calculate the required partial derivatives $\partial f_i/\partial x_j$ by a numerical variable increment algorithm in Sub Parder contained within the program. Program NEWTONVS prompts for starting values for the unknowns. The user is also given options for verifying the functions $f(i)$ by computing them using the starting values, and displaying the partials based on the starting values.

The program also gives the option of using either the full Newton method or the so-called modified Newton method based on retention of the *diagonal* partials $(\partial f_i/\partial x_i)$ only. This latter procedure, *if it works*, has the considerable advantage of using only $n$ (as opposed to $n^2$ for full Newton) partial derivatives per cycle, and it does not require solution of algebraic equations. However, unlike Newton (but the same as modified Delta-Square for systems), it does *not* guarantee convergence given sufficiently good starting values.

The problem considered in Sub Func is the four-variable problem used in Example 7.9. The parameters mx and norm_tol are set as discussed above for the DELSQ program.

### Program NEWTONVS

```
!This is program NEWTONVS.tru by O. T. Hanna
!Program solves a system of nonlinear algebraic equations by either the
!modified Newton or full Newton method. The partial derivatives are
!computed by a variable-step algorithm in Sub Pardervs.
!************************************************************************
!*To run we must specify the n equations f(i) as a function of x(j) in   *
!*Sub Func, as well as max. number of iterations(mx) and norm_tol.        *
!*Program prompts for x(j), option to verify f(i), option to cumpute     *
!*df(i)/dx(j), selection of method as full or modified Newton            *
!************************************************************************

SUB Init(n,x())
    !***denotes user-specified statement
    LET n = 4                          !***number of equations
    MAT redim x(n)
    LET x(1) = 0.1
    LET x(2) = 0.2
    LET x(3) = 0.3
    LET x(4) = 0.4
END SUB

SUB Func(i,x(),f(),mx,norm_tol,n)
    !Function subroutine
    LET mx = 100                       !max allowed iterations
    LET norm_tol = 1e-6                !convergence tolerance

    LET D = 4 - 2*x(1) + x(3) - 4*x(4)
    LET y1 = (3 - 3*x(1) + x(2) - 5*x(4))/D
    LET y2 = (1 - x(1) - x(2) + 2*x(3) - 2*x(4))/D
    LET y3 = x(1)/D
    LET y4 = (x(1) - x(2) + 2*x(4))/D
```

```
    LET y5 = (x(2) - x(3))/D
    LET y6 = x(4)/D

    SELECT CASE i
    CASE 1
         LET f(1) = y3*y4/( (y1^3)*y2 ) - 69.18
    CASE 2
         LET f(2) = y5*y1/( y2*y4 ) - 4.68
    CASE 3
         LET f(3) = y2^2/y5 - .0056
    CASE 4
         LET f(4) = y6*y4^2/( (y1^5)*(y2^2) ) - .141
    CASE else
    END SELECT
END SUB
```

### *Polynomial All-Root Solver Program POLYRTS*

Instructions for running the program are given in the concise listing below. The program starts without any prompts and attempts to find the real roots one by one. If it cannot converge to a real root, the user is given the option of trying a new starting value. When all real roots have been found (the user must decide this), answer no to the prompt for a new starting value. The program then commences to find pairs of complex conjugate roots. If convergence should fail, the program prompts the user for new starting values for coefficients of a quadratic factor.

The problem considered in Sub Data is the same one used in Example 7.6. The parameters used are similar to those in the above two programs.

#### Program POLYRTS

```
!This is program POLYRTS.tru by O.T.Hanna
!Program solves for all roots of a real polynomial eqn. by first using
!synthetic division/Newton to get real roots and then using
!Bairstow/Newton for the complex roots. The polynomial eqn. is
!a0*x^n + a1*x^n-1 +...+ an-1*x + an.
!a1(i) are working values of a(i)

!*******************************************************************
!*To run we must specify in Sub Data polynomial coefficients a(0)    *
!*to a(n) and degree n of polynomial,along with an error tolerance   *
!*and maximum no. of iterations                                      *
!*******************************************************************

Sub Data(n,a(),norm_tol,mx)
!***denotes user specified values
!n is degree of polynomial(dim requires n <= 10)***
!a(0 to n) are coefficients of a0*x^n + a1*x^n-1 +..+an***
```

```
!tol is convergence tolerance,maxiter is max. no. of iterations***

Let norm_tol = 1e-9
Let mx = 50

!Problem from Gerald[1978] p 29
Let n = 4
Let a(0) = 1
Let a(1) = -1.1
Let a(2) = 2.3
Let a(3) = .5
Let a(4) = 3.3

End Sub
```

### *Continuation Program METHCON for Solving a System of Nonlinear Equations*

Instructions for running the program are given in the concise listing below. This is our most powerful program for solving nonlinear algebraic systems, because it is relatively insensitive to poor initial values and it also allows error estimation. The required partial derivatives are computed with a variable-increment numerical procedure. The program converts the given algebraic system to a differential continuation system, which is then solved through any option available in the ODE solver IEXVS, which is also part of the program. The integration process may be continued from the current values after each full continuation cycle (this is a very useful feature). Detailed discussion of the integration options, including number of trajectories, global extrapolation, and automatic local error control, is given in the chapter on numerical solution of initial-value problems in ordinary differential equations.

#### Program METHCON

```
!This is program METHCON.tru (meth. of continuation) by O.T.Hanna
!for a system of nonlinear algebraic equations f[x(i)] = 0.
!Solution is via ODE system [df(i)/dx(j)] dx(j)/dlambda = -f(x0),
!where lambda is continuation parameter which goes from 0 to 1.
!Integration is by improved Euler based on program IEXVS.tru.
!Program IEXVS does 2nd order improved Euler integration, with
!options for 1 (3rd order) or 2 (4th order) global Richardson
!extrapolations. Global error information is provided, along with
!the option for automatic variable-step integration in each case.
!Partial derivatives are calculated numerically in Sub Pardervs.
!n=number of algebraic (also differential) equations
!x,y(i) are independent and dependent variables in ODE algorithm
!h=local step size; hp is print increment; pr is print parameter for
!fixed step. j1 is number of trajectories;j1=1 for h,j1=2 for h,h/2;
!j1=3 for h,h/2,h/4
```

```
!ext is extrapol. paramter(with j=2 only); ext=0,j1=2 gives direct
!brute-force error calc.
!To print every step,comment out (!DO !LOOP) in Integration section.
!Present dimensioning is for n <= 20.
!The following parameters are for the option of variable-step
!integration: error parameter tol is the desired local relative
!error, parameters eps,sf,are zero denominator and safety
!factor values.
!******************************************************************
!*To run,set n,initial estimates x(i) [as y(i)], any variable-step  *
!*parameters in Sub Init; specify system equations f(i) in terms    *
!*of x(i) in Sub Func; program prompts for (a) function check?,     *
!*(b) notraj,(ext),h,ec,(pr)? After each integration cycle program  *
!*prompts to "continue this problem?" for improved accuracy         *
!******************************************************************

SUB Init (n,x0,x1,y(),d(),hp,hmaxstep,tol,eps,sf,f0() )
    !***denotes user specified quantity
    DIM f2(1)
    LET n = 4                        !***number of equations
    MAT redim f0(n),f2(n)
    LET y(1) = .1                    !***set initial values of x(i)
    LET y(2) = .2                    !as y(i)
    LET y(3) = .3
    LET y(4) = .4

    LET x0 = 0                       !initial and final lambda values
    LET x1 = 1
    FOR i = 1 to n                   !initialize f(y_initial)
        CALL Func(i,y,f2)
        LET f0(i) = f2(i)
    NEXT i
    CALL Deriv (x0,y,d,f0,n)

    !The following is used only with error control(ec=1)
    LET hp = .2                      !***hprint
    LET hmaxstep = hp                !maximum user-specified step size

    LET tol = .1                     !***
    LET eps = .01
    LET sf = 2
END SUB

SUB Func(i,x(),f())
    !***user must specify f[x(i)] = 0 as shown below
    !Thermodynamic equilibrium problem
    LET D = 4 - 2*x(1) + x(3) - 4*x(4)
    LET y1 = (3 -3*x(1) +x(2) -5*x(4))/D
    LET y2 = (1 -x(1) -x(2) + 2*x(3) - 2*x(4))/D
    LET y3 = x(1)/D
```

```
    LET y4 = (x(1) - x(2) + 2*x(4))/D
    LET y5 = (x(2) - x(3))/D
    LET y6 = x(4)/D

    SELECT CASE i

    CASE 1
        LET f(1) = y3*y4/( (y1^3)*y2 ) - 69.18
    CASE 2
        LET f(2) = y5*y1/( y2*y4 ) - 4.68
    CASE 3
        LET f(3) = y2^2/y5 - .0056
    CASE 4
        LET f(4) = y6*y4^2/( (y1^5)*(y2^2) ) - .141
    CASE else
    END SELECT
END SUB
```

## PROBLEMS

**7.1.** Use the Newton method to find the root near $x = 3$ of the equation $f(x) = 0 = x^2 - 4e^{(-0.3x)} - 6$. Do this (i) by hand calculations and (ii) using program NEWTNLE.

**7.2.** Solve the equation of problem 7-1 using the ordinary iteration/delta-square method by means of program DELSQ. Use two different iteration functions $x = g(x)$ as follows: solve for the $x^2$ term and then divide by $x$; and solve for the $x^2$, term and then take the square-root of both sides. Compare the results of ordinary iteration and the delta-square method for both of these iteration functions.

**7.3.** Consider the nonlinear algebraic equation $x^3 - 4x^2 + 7x - 5 = 0$. Analyze this equation to see how many *real* roots it possesses (consider positive and negative roots separately). Use the bracketing technique outlined in the text to determine both upper and lower bounds to any positive roots. Use program DELSQ to calculate accurate values of all the real roots.

**7.4.** Use program POLYRTS to determine all of the roots (real and complex) of the polynomial equation in problem 7-3.

**7.5.** Consider the following nonlinear algebraic system of equations:

$$x_1^3 + \frac{2x_1x_2}{1 + e^{x_2}} = 43$$

$$x_1^{1/3} + x_1x_2^{1/2} - \frac{3}{x_2} = 12$$

We wish to find positive roots of these equations. Using appropriate computer programs, try this using (i) ordinary iteration, (ii) the delta-square method, and (iii) the Newton method. Only one of these methods is required to succeed.

**7.6.** This problem is due to Franks [1972]. A continuous-flow chemical reactor operating as a steady-state system can be described by the following set of equations:

$$Fx_A = FA - R$$
$$Fx_B = FB - R$$
$$F = FA + FB - R$$
$$R = 60x_A x_B^{0.6}$$

Solve this set of equations for $x_A$, $x_B$, $F$, and $R$, assuming $FA = 5$ and $FB = 7$. Physically, it is known that the reaction rate $R$ must lie in the interval $0 \le R \le FA$.

**7.7.** For steady turbulent flow in a circular pipe, the pressure drop $\Delta p$ is given by the expression $\Delta p = \rho f L u^2/2D$, where $\rho$ is fluid density, $L$ is pipe length, $D$ is pipe diameter, $u$ is flow velocity, and $f$ is the friction factor. The friction factor is approximately proportional to $1/\text{Re}^{0.25}$; Re is the Reynolds number ($\text{Re} = Du\rho/\mu$ where $\mu$ is fluid viscosity). In pipe network problems, it is often convenient to work in terms of $Q$, the volumetric flow rate, rather than velocity $[Q = (\pi D^2/4)u]$.

Consider a pipe network that starts at point 1, goes to point 2, and then splits into two branches, with one branch going to point 3 and the other to point 4. It is desired to calculate the pressure at intermediate point 2, and the volumetric flow rates in the line segments $Q_1$, $Q_2$, and $Q_3$. Expressed in terms of the volumetric flow rates and pressures, the flow network equations are

$$p_1 - p_2 = K_1 Q_1^{1.75}$$
$$p_2 - p_3 = K_2 Q_2^{1.75}$$
$$p_2 - p_4 = K_3 Q_3^{1.75}$$
$$Q_1 = Q_2 + Q_3$$

where pressures are in lb. force/sq. in., and the $Q$'s are in gallons/min. The values of the $K_i$ in these units are $K_1 = 2.35e-3$, $K_2 = 4.67e-3$, and $K_3 = 3.72e-2$. If the boundary pressures are $p_1 = 75$ psi, $p_3 = 20$ psi, and $p_4 = 15$ psi, determine the values of $p_2$, $Q_1$, $Q_2$, and $Q_3$. (Hint: Since each equation contains only a small number of variables, an ordinary iteration/$\Delta^2$ procedure may be useful.)

**7.8.** It is desired to calculate the dew point temperature and liquid composition for a mixture of benzene and toluene at a pressure $P$ of one atmosphere ($P = 760$ mm Hg). The mixture vapor composition is $y_1 = 0.77$ mole fraction benzene and $y_2 = 0.23$ mole fraction toluene. Assume ideal solution and perfect gas behavior for the liquid and gas mixtures, respectively. The vapor pressures as a function of temperature are given by the equation $\log_{10} P^* = A - B/(T + C)$ where $P^*$ is in mm Hg and $T$ is in °C.

| Benzene | Tolulene |
|---|---|
| A = 6.89745<br>B = 1206.35<br>C = 220.237 | A = 6.95334<br>B = 1343.94<br>C = 219.377 |

The equilibrium conditions to be satisfied (from Raoults law) are $y_1 P = P_1^* x_1$ and $y_2 P = P_2^* x_2$. These equations must be used along with the liquid phase mass balance equation $x_1 + x_2 = 0$. When $x_1$ and $x_2$ are eliminated by the equilibrium equations, we get the following final equation, which is implicit in temperature:

$$1 = P\left(\frac{y_1}{P_1^*} + \frac{y_2}{P_2^*}\right)$$

Solve this equation for $T$; then determine $x_1$ and $x_2$.

**7.9.** To perform a stability analysis for the Runge-Kutta Four integration method for differential equations, we need to find the smallest *real* root of the equations $|1 - A + A^2/2! - A^3/3! + A^4/4!| = 1$, where $||$ stands for absolute value. Use program POLYRTS to accomplish this task.

**7.10.** Consider the nonlinear algebraic equation $x^5 - Bx^2 + 12 = 0$, where $B$ is a positive parameter. Determine an approximate analytical expression for all of the *positive* roots of this equation for the cases of $B$ "large" and $B$ "small." Also give Newton improvements for these formulas.

**7.11.** For the equation in problem 7.10, determine numerical values for the positive roots in the cases $B = 100$ and $B = 0.01$.

**7.12.** A mass balance equation for a stirred-tank chemical reactor has the form

$$1 - x = \frac{Kx(0.5 + x)}{1 + 2x(0.5 + x)}$$

where $x$ and $K$ are dimensionless concentration and rate constant, respectively. It is desired to determine information about the behavior of $x$ for relatively large values of $K$.

(i) Compute the numerical solution for $x$ at $\epsilon = 1/K = 0.01, 0.05, 0.1, 0.15, 0.2, 0.25, 0.3, 0.4, 0.5, 0.6, 0.7, 0.8, 0.9, 1.0$.

(ii) Develop the simplest one-term approximate solution for $x(K)$ when $K \to \infty$.

(iii) Determine an improved asymptotic solution to $x$ for $K \to \infty$ by using $\epsilon = 1/K$, expand $\epsilon$ as a series in $x$ ($\epsilon = a_1 x + a_2 x^2 \ldots$) and then revert this series ($x = b_1 \epsilon + b_2 \epsilon^2 \ldots$). Use program SERIES and take $n = 8$. Compare the series computation with the numerical results to determine its region of usefulness.

**7.13.** Try to improve the $x(\epsilon)$ series from problem 7.12 by means of a Padé approximation (use program PADE). Use $n = 8$ and compare the results to the numerical solution.

**7.14.** For the numerical data of problem 7.12, fit the rational approximation $(x/\epsilon) = (p_0 + p_1\epsilon \ldots + p_k\epsilon^k)/(1 + q_1\epsilon \ldots q_m\epsilon^m)$ using rational least squares and program RATLSGR. Use the interval $(0 < \epsilon \leq 1)$. Select $(k, m)$ so that the maximum error is of the order $10^{-3}$. Compare with the numerical solution.

## REFERENCES


BEREZIN, I. S. and ZHIDKOV, N. P. (1965), *Computing Methods*, Addison-Wesley, Menlo Park, CA

DAHLQUIST, G., BJORCK, A., and ANDERSON, N. (1974), *Numerical Methods*, Prentice-Hall, Englewood Cliffs, NJ:

DENNIS, J. E., Jr., and SCHNABEL, R. B. (1983), *Numerical Methods for Unconstrained Optimization and Nonlinear Equations*, Prentice-Hall, Englewood Cliffs, NJ

FRANKS, R. G. E. (1972), *Modeling and Simulation in Chemical Engineering*, Wiley, New York

GERALD, C. (1978), *Applied Numerical Analysis,* 2nd ed., Addison-Wesley, Reading, MA:

HENRICI, P. (1964), *Elements of Numerical Analysis*, Wiley, New York

KAHANER, D., MOLER, C., and NASH, S. (1989), *Numerical Methods and Software*, Prentice-Hall, Englewood Cliffs, NJ

MODELL, M. and REID, R. C. (1983), *Thermodynamics and Its Applications*, Prentice-Hall, Englewood Cliffs, NJ

RINKER, R. G. (1994), personal communication.

SMITH, J. M. and VAN NESS, H. C. (1987), *Introduction to Chemical Engineering Thermodynamics*, McGraw-Hill, New York

# 8

# Optimization

Whenever there is some flexibility or indeterminacy in a mathematical system, it may be possible to take advantage of it in order to achieve some best or *optimum* result. We have had experience with this idea in the simple one-dimensional case of elementary calculus, where we "set the derivative equal to zero" in order to find a maximum or minimum of a function. In more complicated problems there will be a number of independent variables, and also, perhaps, various constraints these variables must satisfy.

Applications of optimization occur in virtually all fields of engineering. In particular, the fields of design, engineering economics, and operations research (resource allocation) are fundamentally concerned with optimization. Within chemical engineering, we can cite such well-known and important problems as optimum pipe diameters in fluid mechanics, optimum insulation thickness in heat transfer, chemical equilibrium in thermodynamics, and optimum yields in chemical reactors. The list of applications could go on and on.

The nature and difficulty of various optimization problems are such that a number of different approaches are necessary. The field is so large that there are many entire books on the subject. Here we can only introduce and implement some of the simpler approaches to these problems. For further detail and more advanced algorithms, the reader is referred to Edgar and Himmelblau [1988], Reklaitis *et al.* [1983], and Beveridge and Schecter [1970].

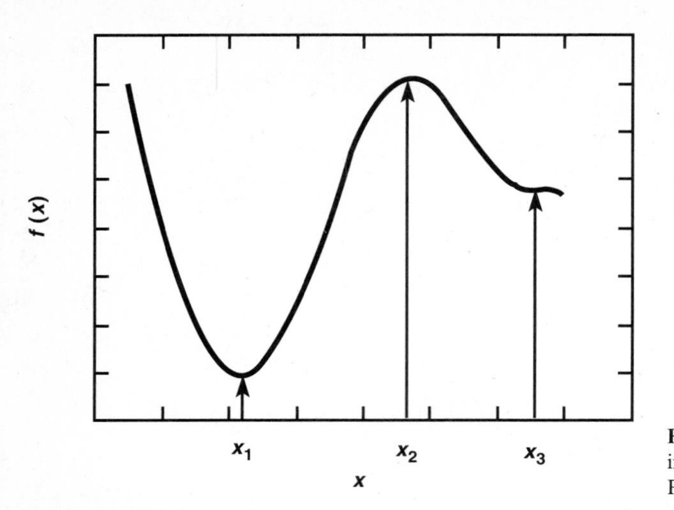


**Figure 8.1.** Illustration of Relative Maximum, Relative Minimum, and Saddle Point

## 8.1 RELATIONSHIPS FOR UNCONSTRAINED PROBLEMS

In general, we call the function that is to be optimized (maximized or minimized) the *objective* function. As an example of the types of behavior we may encounter, consider Figure 8.1. For this one-dimensional (one independent variable) objective function $f(x)$, we have a relative minimum at $x_1$, a relative maximum at $x_2$, and a so-called *saddle point* at $x_3$. A relative minimum (or maximum) corresponds to a point where, for all sufficiently small $\Delta x$, $\Delta f > 0$ (or $\Delta f < 0$). For the saddle point, neither of these definitions applies; for sufficiently small $|\Delta x|$, $\Delta f$ changes sign as $\Delta x$ does, so the saddle point is neither a relative maximum or minimum. However, as Figure 8.1 shows, at the saddle point, $df/dx = 0$. Thus, at this saddle point, if we look to the left of $x_3$, it appears we have a minimum; but if we look to the right of $x_3$, it appears we have a maximum. Point $x_3$ corresponding to the saddle point is *neither* a relative maximum nor minimum.

In trying to use elementary calculus to find a relative maximum or minimum for a function of one variable, we define the *stationary* points of the objective function to be those points where $df/dx = 0$. These are the points corresponding to a *possible* maximum or minimum. If we somehow determine the stationary points of a function, we still need to find out their character (maximum, minimum, saddle). To analyze this situation, we call upon Taylor's theorem. Here we proceed with algebraic arguments, since these must be used in higher-dimensional problems where we do not have the benefit of a simple geometrical picture.

For the simple one-dimensional problem, we proceed algebraically as follows. Suppose the objective function is $f(x)$, and suppose a stationary point (a point where $df/dx = 0$) exists at $x^*$. We define the increment $h$ by $h = (x - x^*)$. When $x$ is sufficiently close to $x^*$ (small $h$), we can use Taylor's theorem to write

$$f(x) - f(x^*) = \Delta f = \frac{df}{dx}(x^*)h + \frac{d^2f}{dx^2}(x^*)\frac{h^2}{2!} + \cdots \tag{8-1}$$

By using Equation (8-1), we can show that for either a relative maximum or minimum of $f$ at $x^*$, we must necessarily have $df/dx = 0$ at this point. For if this was *not* true, then for sufficiently small $h$ the sign of $\Delta f$ would be determined by the leading nonzero term of the Taylor expansion, $(df/dx)_{x^*}h$. In this case, for sufficiently small $|h|$, $\Delta f$ would not be uniformly either positive or negative, but rather would have the sign of $(df/dx)_{x^*}h$, which would vary according to whether $h$ was positive or negative.

The above *necessary* condition for an optimum is not, of itself, enough to ensure that a relative optimum actually exists at $x^*$. As we have seen geometrically, we could have a saddle point where $df/dx = 0$. To determine *sufficient* conditions to guarantee the existence of an optimum at point $x^*$, we examine the first *nonzero* term of the Taylor expansion in Equation (8-1). If the quantity $d^2f/dx^2$ at $x^*$ is not equal to zero, then we can guarantee that $x^*$ is a relative minimum if it is positive, while $x^*$ is a relative maximum if it is negative.

Most one-dimensional problems fall into the category where $d^2f/dx^2$ is not equal to zero at the stationary points. If, however, $d^2f/dx^2 = 0$ at a particular stationary point, the character of that point must be determined by the nature of the first nonzero term in the Taylor expansion of Equation (8-1). If the first nonzero term is of odd order (e.g., 3, 5, etc.), then the stationary point is a saddle, since the odd power of $h$ will change sign according to whether $h$ is positive or negative. If the first nonzero term is of even order, the stationary point is a relative minimum if this term is positive, while it is a relative maximum if this term is negative.

As simple examples of these principles, consider the functions $y = x^2$, $y = x^3$, and $y = x^4$, as shown in Figure 8.2. We immediately see that $dy/dx = 0$ at $x = 0$ for each of these functions. For $y = x^2$, the first nonzero derivative at $x = 0$ is $d^2y/dx^2 = 2$; thus, this function has a relative minimum at $x = 0$. For $y = x^3$, the first nonzero derivative at $x = 0$ is $d^3y/dx^3 = 6$; thus, this function has a saddle point at $x = 0$. For $y = x^4$, the first nonzero derivative at $x = 0$ is $d^4y/dx^4 = 24$; thus, this function has a relative minimum at $x = 0$.

It is important to realize that when we are seeking the optimum of a function in some particular range of the variables, the desired point may lie in the interior of the region (where we would find it by the usual calculus method), or it may lie instead at the boundary of the region. The calculus method would generally not identify a boundary optimum, so we need to look for it separately by evaluating the objective function at the boundary points and comparing to the calculus optima.

### 8.1.1 Functions of Two or More Variables

In the case of an unconstrained objective function that depends on two or more independent variables, it is more difficult to find the stationary points and also more difficult to determine their character. As in the case of a single variable, it is easy to see the *necessary* conditions for a relative optimum. These necessary conditions are simply that all of the partial derivatives $\partial f/\partial x_i$ $(i = 1, 2, \ldots, n)$ be equal to zero at a candidate point (stationary point) for an optimum.

**Necessary Conditions for an Optimum.**

$$\frac{\partial f}{\partial x_i} = 0 \ (i = 1, \ldots, n) \tag{8-2}$$

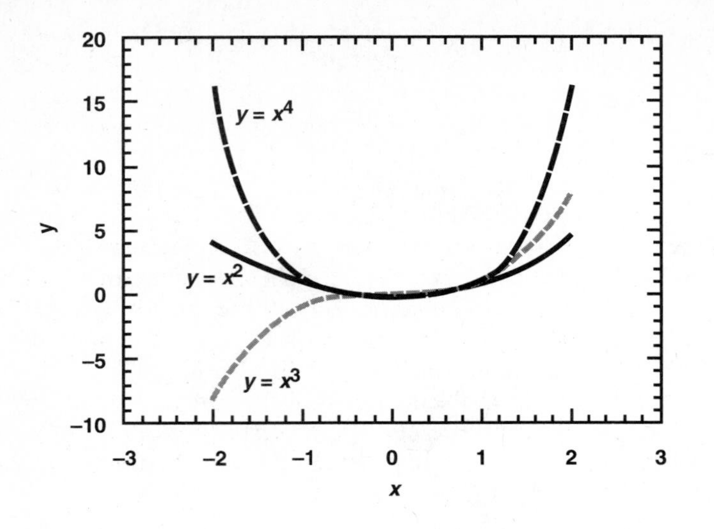


**Figure 8.2.** Stationary Points at $x = 0$

We can see this to be true in the following way. If we perturb one particular independent variable at an optimum point, while keeping all other variables at their optimal values, the problem reduces to a one-dimensional consideration of $\Delta f$ with respect to the perturbed variable. We have already seen, in the one-dimensional problem, that this then requires the derivative of the objective function with respect to the perturbed variable to be zero. By applying this reasoning with each independent variable being perturbed separately, we derive the above necessary conditions.

In the case where the objective function $f$ depends on $n$ independent variables, finding the stationary points generally requires the solution of the $n$ nonlinear algebraic equations $\partial f/\partial x_i = 0$. This could be attempted with any of the usual nonlinear equation solvers. The solution of these nonlinear equations may yield more than one set of roots, in which case we would need to test each stationary point (usually experimentally) to see if it is a maximum, minimum, or saddle point. To look for different roots, we try alternate starting points. Instead of using this traditional calculus approach, for larger problems it is generally easier to use one of the numerical methods designed for the direct determination of the optimum. This matter will be discussed later.

The determination of *theoretical* sufficient conditions for the optimum of a function of many variables is generally quite difficult, so it is usually not pursued in practice. However, the *concept* is important, so we discuss it briefly. To illustrate the analysis of the situation, we consider the two-dimensional case where we wish to optimize $f(x_1, x_2)$. Suppose that we have computed a stationary point $(x_1^*, x_2^*)$ by solving the equations $\partial f/\partial x_1 = 0$ and $\partial f/\partial x_2 = 0$. We can then write the Taylor expansion of $f(x_1, x_2)$ about the point $(x_1^*, x_2^*)$ as follows:

$$f(x_1, x_2) - f(x_1^*, x_2^*) = \Delta f = \frac{1}{2!} \cdot \left( h_1^2 \frac{\partial^2 f}{\partial x_1^2} + 2h_1 h_2 \frac{\partial^2 f}{\partial x_1 \partial x^2} + h_2^2 \frac{\partial^2 f}{\partial x_2^2} \right)_{x_1^*, x_2^*} + \cdots \quad (8\text{-}3)$$

**TABLE 8.1** CHARACTER OF STATIONARY POINTS FOR A TWO-DIMENSIONAL PROBLEM,

$$A = \frac{\partial^2 f}{\partial x_1^2} \quad B = \frac{\partial^2 f}{\partial x_1 \partial x_2} \quad C = \frac{\partial^2 f}{\partial x_2^2}$$

| $CA - B^2$ | C,A | Nature of Stationary Point |
|---|---|---|
| Positive | Positive | Minimum |
| Positive | Negative | Maximum |
| Negative | — | Saddle point |
| Zero | — | Indeterminate |

Here, $h_1 = x_1 - x_1^*$ and $h_2 = x_2 - x_2^*$. We also let $\partial^2 f/\partial x_1^2 = A, \partial^2 f/\partial x_1 \partial x_2 = B$ and $\partial^2 f/\partial x_2^2 = C$, with all these partials evaluated at the stationary point $(x_1^*, x_2^*)$. Using these definitions, Equation (8-3) can be written in the form

$$2\Delta f = h_1^2 A + 2h_1 h_2 B + h_2^2 C + \cdots = A h_2^2 \left[ \left( \frac{h_1}{h_2} \right)^2 + 2\frac{h_1}{h_2}\frac{B}{A} + \frac{C}{A} \right] + \cdots \tag{8-4}$$

A sufficient condition for an optimum at a stationary point is that the leading-order term of the Taylor expansion, Equation (8-3), have a uniform sign for *all* sufficiently small increments $h_i$ (both positive and negative). To see what this entails, we *complete the square* in Equation (8-4) by adding and subtracting the quantity $B^2/A^2$, so that with a little algebraic rearrangement, Equation (8-4) becomes

$$2\Delta f = A h_2^2 \left[ \left( \frac{h_1}{h_2} + \frac{B}{A} \right)^2 + \frac{CA - B^2}{A^2} \right] + \cdots \tag{8-5}$$

In this form, Equation (8-5) shows that the term in brackets will have a uniform sign (positive) for *any* $(h_1, h_2)$ pair which is not identically zero, provided that $(CA - B^2) > 0$. If $(CA - B^2) < 0$, then we have a saddle point, since for some $(h_1, h_2)$ pairs the term in brackets will be positive, while for other pairs it will be negative. If $(CA - B^2) = 0$, the character of the stationary point is indeterminate, since here there will be many "small" pairs $(h_1, h_2)$ for which the term in brackets will be zero. In this case, we would need to examine the next term in the Taylor expansion. We summarize these various results in Table 8.1. Note that if $(CA - B^2) > 0$, then $C$ and $A$ necessarily have the same sign.

It should be noted that it is sometimes useful to employ inequalities to bracket either the independent variables and/or the objective function. This can sometimes be accomplished by bracketing the necessary conditions, bracketing the constraints, or bracketing a systematically modified objective function.

**Example 8.1 Two-Variable Unconstrained Minimization**

We reconsider the following problem arising in chemical reaction analysis, which was introduced in the chapter on nonlinear algebraic equations. This problem is due to our colleague, Professor Robert G. Rinker. We wish to minimize, with respect to $x_1$ and $x_2$, the function

$$f(x_1, x_2) = \frac{1}{x_1} + \frac{1}{x_2} + \frac{1 - x_2}{x_2(1 - x_1)} + \frac{1}{(1 - x_1)(1 - x_2)} \tag{a}$$

We observe from Equation (a) that for the various cases $x_1 = 0, x_2 = 0, x_1 = 1$, and $x_2 = 1$, the denominator of $f$ becomes zero, and hence $f$ becomes infinite. This suggests the possibility that there may be a minimum for $0 < x_1 < 1$ and $0 < x_2 < 1$. The foregoing observation is useful in the selection of initial guesses for the solution of the necessary-condition algebraic equations. It is also consistent with the physical nature of the problem. By partial differentiation, together with some algebraic rearrangement, we obtain the following necessary conditions for a minimum:

$$\frac{\partial f}{\partial x_1} = 0 = \frac{-1}{x_1^2} + \left[\frac{1-x_2}{x_2} + \frac{1}{1-x_2}\right] \cdot \frac{1}{(1-x_1)^2}$$
$$\frac{\partial f}{\partial x_2} = 0 = \frac{-1}{x_2^2} + \left[\frac{-1}{x_2^2} + \frac{1}{(1-x_2)^2}\right] \cdot \frac{1}{1-x_1} \tag{b}$$

It is apparent from the first of Equations (b) that $x_2 > 1$ is impossible, since this would correspond to the sum of two negative terms adding to zero! To solve these equations, we use program NEWTONVS. By using the natural first guesses $x_1 = 0.5$, $x_2 = 0.5$, we quickly obtain the root $x_1 = 0.3637$, $x_2 = 0.5612$, with the corresponding value $f = 9.342$. Since this is a two-variable problem, the sufficient conditions of Table 8.1 can be used to test the nature of the stationary point. We can do this numerically by using Equations (b) as two input functions to program NEWTONVS, inputting the computed stationary points $x_1 = 0.3637$, $x_2 = 0.5612$ as the first guess and requesting the partial derivative option. This calculation yields $\partial^2 f/\partial x_1^2 = A = 65.33$, $\partial^2 f/\partial x_1 \partial x_2 = B = 4.99$, and $\partial^2 f/\partial x_2^2 = C = 66.30$. Thus, the quantity $(AC - B^2)$ is positive, and since $A$ and $C$ are both positive, the stationary point is a relative minimum. As expected, the analytical determination of the partial derivatives for $A, B$, and $C$ yielded the same results as those obtained from program NEWTONVS. Searching in the region $0 < x_1 < 1$, $0 < x_2 < 1$ produced no new roots. However, it is interesting to note that searching *outside* this region produced the root $x_1 = -1.20$, $x_2 = 0.64$, which apparently corresponds to a nonphysical saddle point.

## 8.2 RELATIONSHIPS FOR EQUALITY-CONSTRAINED PROBLEMS

We now look into how optimization is affected by side conditions, or *constraints*. In this section we will concentrate on constraints whose original form is an equation; these constraints are called equality constraints. Optimization problems involving both equality and inequality constraints come under the heading of *mathematical programming*; they will be discussed later. To keep things simple, we consider an objective function that depends on two independent variables, while the single constraint will have the form of an equation depending on these same two variables. This problem can be written as

$$\text{Opt } f(x_1, x_2) \quad \text{subject to} \quad g(x_1, x_2) = b \tag{8-6}$$

Here, Opt refers to either maximum or minimum. The quantity $b$ is a parameter that is constant for any particular situation. First, we observe that if it were possible to solve the constraint equation explicitly for either $x_1$ or $x_2$, [For example, as $x_1 = G(x_2)$], then we could substitute this expression into the original objective function to get the new objective function $F(x_2) = f(G(x_2), x_2)$, which would then be treated as a one-variable *unconstrained* problem in $x_2$. In such a case, the constraint actually makes the problem simpler by reducing the number of variables.

Unfortunately, the possibility of direct algebraic solution of the constraint equations does not happen often in practice, so that systematic procedures must be devised for the general problem. A more general approach for constrained problems involves the elimination of variable *increments*, rather than the variables themselves. This is because, through Taylor's theorem, we can write first-order expressions for $\Delta f$ and any $\Delta g$'s as linear expressions in the increments $\Delta x_i$. In the two-variable constrained problem described by Equation (8-6), the Taylor expansion for $\Delta f$ about an optimum point (analogous to Equation (8-3)) contains increments $\Delta x_1, \Delta x_2$, which are *not* independent. These increments are related by the first-order Taylor expansion for $\Delta g$ as follows:

$$\begin{aligned} f(x_1, x_2) - f(x_1^*, x_2^*) = \Delta f &= \frac{\partial f}{\partial x_1}\Delta x_1 + \frac{\partial f}{\partial x_2}\Delta x_2 + \cdots \\ 0 = \Delta g &= \frac{\partial g}{\partial x_1}\Delta x_1 + \frac{\partial g}{\partial x_2}\Delta x_2 + \cdots \end{aligned} \tag{8-7}$$

Let us now introduce a subscript notation for partial derivatives as $g_{x2} = \partial g/\partial x_2$, and so on. If we then solve for $\Delta x_1 = (-g_{x2}\Delta x_2/g_{x1})$ from the above expression for $\Delta g$, and substitute this for $\Delta x_1$ in $\Delta f$, we get

$$f(x_1, x_2) - f(x_1^*, x_2^*) = \Delta f = \left(\frac{-f_{x1}g_{x2}}{g_{x1}} + f_{x2}\right)\Delta x_2 + \cdots \tag{8-8}$$

In Equation (8-8) the $\Delta x_2$ increment is independent, so by the same reasoning as earlier, a necessary condition for the constrained optimum is that the coefficient of $\Delta x_2$ be zero. This equation depends on the two variables $(x_1, x_2)$ and it must be solved simultaneously with the constraint equation $g(x_1, x_2) = b$ to determine the stationary points where an optimum might exist.

### 8.2.1 Lagrange Multipliers

An alternative approach to a constrained optimum is the method of Lagrange multipliers. This approach is mathematically equivalent to that of eliminating increments, but it offers several computational advantages. One is that it treats all variables in a symmetrical way. For the two-variable problem of this section, we develop the method by returning to Equation (8-7). If we multiply the constraint equation by the undetermined *Lagrange multiplier* $\lambda$ and add this expression to that for $\Delta f$, we get

$$\Delta f = \left(\frac{\partial f}{\partial x_1} + \lambda\frac{\partial g}{\partial x_1}\right)_{x^*}\Delta x_1 + \left(\frac{\partial f}{\partial x_2} + \lambda\frac{\partial g}{\partial x_2}\right)_{x^*}\Delta x_2 + \cdots \tag{8-9}$$

To determine $\lambda$ and the necessary conditions for a constrained optimum, we reason as follows. We choose $\lambda$ to make the coefficient of $\Delta x_1$ equal to zero. When this is specified, $\Delta f$ depends only on the single independent increment $\Delta x_2$. In this event, as we saw earlier, the coefficients of any independent increments must necessarily be zero at any optimum. Thus, the necessary conditions for an optimum of Equation (8-9) are

$$\left(\frac{\partial f}{\partial x_1} + \lambda\frac{\partial g}{\partial x_1}\right)_{x^*} = 0, \quad \left(\frac{\partial f}{\partial x_2} + \lambda\frac{\partial g}{\partial x_2}\right)_{x^*} = 0 \tag{8-10}$$

To these two equations in the three variables $x_1$, $x_2$, $\lambda$, we must add the constraint equation $g(x_1, x_2) = b$. This then provides a set of three equations in three unknowns for the stationary points. In this particular problem, it is easy to see (by eliminating $\lambda$ between the equations of (8-10)) that the necessary condition of Equations (8-10) is exactly the same as that of Equation (8-8).

The above procedure for Lagrange multipliers extends naturally to the general problem of optimizing a function of $n$ variables subject to $m$ constraint equations (where $m < n$, so that there is some flexibility in the objective function to afford an optimum). In this more general case, there will be one multiplier $\lambda_k$ for each constraint equation ($k = 1, \ldots, m$), and, proceeding as above, a necessary condition for the constrained optimum is

$$\frac{\partial f}{\partial x_i} + \sum_{k=1}^{m} \lambda_k \frac{\partial g_k}{\partial x_i} = 0 \quad (i = 1, \ldots, n) \tag{8-11}$$

To these equations we must also add the constraint equations $g_k(x_1, \ldots, x_n) = b_k$ ($k = 1, \ldots, m$). This finally gives for the stationary points a set of $(n + m)$ equations, in the same number of unknowns, for the original $n$ variables $x_i$ plus the $m$ Lagrange multipliers. It is useful to note that these equations can be interpreted as the stationary points for the unconstrained Lagrangian function (which depends on the $n + m$ independent variables $x_i$, $\lambda_k$), defined as

$$L = f(x_1, \ldots, x_n) + \sum_{k=1}^{m} \lambda_k g_k \tag{8-12}$$

Another important characteristic of Lagrange multipliers is that they represent the change of the optimum with respect to the constraint parameter $b_k$. This can be shown as follows for the simple two-variable, one-constraint problem considered in this section. For a given $b$ value, the necessary conditions for an optimum are given by Equations (8-10), together with $g(x_1, x_2) = b$. If we now change $b$ by the amount $\Delta b$, both $f^*$ and $x^*$ will change. The optimum will change by the amount

$$f^*(b + \Delta b) - f^*(b) = \Delta f^* = \frac{\partial f^*}{\partial x_1} \Delta x_1^* + \frac{\partial f^*}{\partial x_2} \Delta x_2^* + \cdots \tag{8-13}$$

Here, the asterisk $*$ refers to the optimum condition for value $b$. From the constraint equation, $\Delta b$ and $\Delta x^*$ are related according to

$$\Delta b = \frac{\partial g}{\partial x_1} \Delta x_1^* + \frac{\partial g}{\partial x_2} \Delta x_2^* + \cdots \tag{8-14}$$

Eliminating the $\partial f^*/\partial x_i$ values using Equations (8-10), we obtain

$$\Delta f^* = -\lambda \frac{\partial g}{\partial x_1} \Delta x_1^* - \lambda \frac{\partial g}{\partial x_2} \Delta x_2^* + \cdots = -\lambda \Delta b + \cdots \tag{8-15}$$

By dividing by $\Delta b$ and passing to the limit as $\Delta b \to 0$, we obtain the desired result,

$$\frac{df^*}{db} = -\lambda \tag{8-16}$$

This result has some important applications. It shows that Lagrange multipliers are "sensitivity" coefficients for changes in a constraint parameter. This result will also be used in the later discussion of the Kuhn-Tucker necessary conditions for an *inequality*-constrained minimum. An analogous result can be derived in the same way for the general problem involving $n$ variables and $m$ equality constraints. In this more general situation, the above result becomes

$$\frac{\partial f^*}{\partial b_k} = -\lambda_k \tag{8-17}$$

To derive theoretical sufficient conditions for the equality-constrained problem, we must investigate the effects of the quadratic terms in both the objective function and the constraint equations at a stationary point. Except in very simple problems, this is not practical. In place of this, the behavior of these functions is often investigated with numerical experiments. This can often be accomplished using a direct search technique and computer program OPTSEARCH, to be discussed later.

**Example 8.2 A Two-Variable Equality Constrained Problem**

The following problem relates to chemical equilibrium and is adapted from Cooper and Steinberg [1970].

$$\begin{gathered} \text{Min} f(x_1, x_2) = 1 - x_1 - x_2 \\ \text{subject to} \\ g(x_1, x_2) = (1 - x_1 - x_2)^2 - \frac{16}{27} x_1 x_2^3 = 0 \end{gathered} \tag{a}$$

Note, first of all, that the objective function could not have a relative optimum *without the presence of the constraint*, since in this case the usual necessary conditions would not have a solution. A consideration of Equation (a) shows that it would be *possible* (but very messy) to solve explicitly for $x_1$ in terms of $x_2$ using the quadratic formula. Instead, we use Lagrange multipliers and the constraint equation to give

$$\left(\frac{\partial f}{\partial x_1} + \lambda \frac{\partial g}{\partial x_1}\right) = 0, \quad \left(\frac{\partial f}{\partial x_2} + \lambda \frac{\partial g}{\partial x_2}\right) = 0, \quad g(x_1, x_2) = 0 \tag{b}$$

By eliminating $\lambda$ in Equations (b) and putting in the partial derivatives, we obtain

$$\frac{f_{x1}}{f_{x2}} = \frac{g_{x1}}{g_{x2}} = \frac{-1}{-1} = \frac{-2(1 - x_1 - x_2) - (16/27)x_2^3}{-2(1 - x_1 - x_2) - 3(16/27)x_1 x_2^2} \tag{c}$$

By cross-multiplication in Equation (c) we find that $x_2 = 3x_1$. When this is substituted into the constraint equation to eliminate $x_2$, we get the equation

$$16x_1^4 = (1 - 4x_1)^2 \tag{d}$$

Equation (d) is actually a fourth-degree polynomial, but the factored form, which occurs naturally here, is more convenient for an analytical solution. By appropriately taking square-roots twice, we get the following roots, which represent stationary points:

$$x_1 = \frac{-1 \pm \sqrt{2}}{2} \textit{and} \frac{1}{2} \pm 0 \tag{e}$$

The four roots in Equation (e), to four figures, are $x_1 = 0.2071, -1.2071, 0.5$, and $0.5$ (0.5 is a double root). Note that since $f^* = 1 - x_1^* - x_2^* = 1 - x_1^* - 3x_1^* = 1 - 4x_1^*$, we see that $f^*$ will be *smallest* for the largest $x_1^*$ value, which is 0.5. However, this does *not* indicate that $f$ has a relative minimum at $x_1^* = 0.5$. To determine the character of the stationary points experimentally, we can choose small positive and negative increments in $x_1^*$ on each side of a stationary point. Then we must solve the constraint equation for the corresponding values of $x_2^*$ and then see if the $\Delta f$ values have a uniform sign. By proceeding in this way, we find that $x_1^* = 0.5$ is a saddle-point, $x_1^* = 0.2071$ is a relative maximum, and $x_1^* = -1.2071$ is the desired relative minimum. This behavior may seem a bit strange, since $f(-1.2071) > f(0.2071)$; however, there is a discontinuity in $f$ between $x_1 = -1.2071$ and $x_1 = 0.2071$. In order to visualize this, the reader might wish to make a simple graph of the above results and connect the stationary points qualitatively according to their character.

## 8.3 NUMERICAL METHODS FOR UNCONSTRAINED PROBLEMS

In optimization problems involving a number of variables, it is usually impossible to make much headway using analytical methods. In such cases various numerical methods can be used. This subject is very extensive, so here we merely introduce some concepts, as well as a relatively simple but useful approach. We discuss the problem in terms of minimization; for maximization, one can simply minimize the negative of the original objective function. One of the important ideas in multivariable optimization involves the definition and use of the so-called local steepest-descent directions (for minimization). We illustrate the derivation for the case of two independent variables; extension to higher dimensions follows the same ideas. Consider the objective function $f$ evaluated at the point $(x_0, y_0)$. We wish to find the "direction" in which $f(x_0, y_0)$ decreases most rapidly; this is called the direction of steepest descent. To properly define this direction, we proceed as follows. Consider the first-order Taylor expansion of $f$ about $(x_0, y_0)$.

$$f(x_0 + h, y_0 + k) = f(x_0, y_0) + (f_x h + f_y k)_{x_0, y_0} + \cdots \tag{8-18}$$

Assuming we are *not* at a stationary point (where both $f_x$ and $f_y = 0$), the linear terms of the Taylor expansion determine how $f$ changes for small-enough $h$ and $k$. The appropriate relative values of $h$ and $k$ determine the steepest-descent direction. To obtain this, we wish to make the linear Taylor term $(f_x h + f_y k)$ in Equation (8-18) as small as possible with respect to $h$ and $k$, subject to the normalizing condition ($h^2 + k^2 = \Delta s^2 =$ constant). This normalizing condition is that the "arc length" of the change in $x$ and $y$, which is just $\Delta s^2 = \Delta x^2 + \Delta y^2 = h^2 + k^2$, be equal to an arbitrary constant during the minimization process. Thus, to find the steepest-descent directions we solve the optimization problem

$$\min_{h,k}(f_x h + f_y k) \quad \text{subject to} \quad h^2 + k^2 = \Delta s^2 \tag{8-19}$$

To accomplish this, we introduce Lagrange multipliers according to

$$\psi = (f_x h + f_y k) + \lambda(h^2 + k^2 - \Delta s^2) \tag{8-20}$$

The necessary conditions for the constrained optimum of Equation (8-19) are that the partial derivatives of $\psi$ in Equation (8-20) with respect to $h, k$, and $\lambda$ be equal to zero. This gives the three equations

$$\begin{aligned} \psi_h &= 0 = f_x + \lambda \cdot 2h \\ \psi_k &= 0 = f_y + \lambda \cdot 2k \\ \psi_\lambda &= 0 = h^2 + k^2 - \Delta s^2 \end{aligned} \tag{8-21}$$

By solving Equations (8-21) for $h, k$, and $\lambda$, and then eliminating $\lambda$, we find

$$h = -\frac{f_x \Delta s}{T} \qquad k = -\frac{f_y \Delta s}{T} \qquad T = (f_x^2 + f_y^2)^{1/2} \tag{8-22}$$

Here, the minus sign is selected in Equations (8-22) to give the *descent* direction; if the sign was positive, we would have the steepest-ascent direction. Since $\Delta s$ is arbitrary, Equations (8-22) show the relative changes in $x$ and $y$ that correspond to the steepest-descent direction.

Equations (8-22) yield the *instantaneous* direction of steepest descent for the function $f$ at the point $(x_0, y_0)$. Thus, if we move in this direction, we are decreasing $f$ at the *maximum possible rate* in the vicinity of $(x_0, y_0)$. However, since the partial derivatives $f_x$ and $f_y$ depend on the particular $x, y$ values, it is clear that the steepest-descent direction is, in general, *constantly changing*. As soon as we move a finite distance in the given steepest-descent direction at any point, the function will no longer be decreasing at the maximum possible rate, since the new steepest-descent direction will generally be different from the original one. Thus, the steepest-descent direction is a *local* quantity that is constantly changing as we move toward the optimum.

### 8.3.1 A Variation of Steepest Descent

One way to utilize the steepest descent concept is as follows. One can move in a given steepest-descent direction until the function has reached a relative mininum. At that point, a new steepest-descent direction is computed, movement continued until a new relative minimum has been reached, and so on. This process is often referred to as the "method of steepest descent" for unconstrained optimization. This procedure is conceptually simple and has the advantage of generally avoiding saddle points. However, a major drawback is that while the method is effective far away from the optimum, *its convergence becomes very slow near the optimum.* For this reason, other, more complicated, methods (particularly conjugate direction and quasi-Newton methods) are generally more useful in practice, particularly in larger problems. For information on these procedures, see Bunday [1984], Edgar and Himmelblau [1988], and Reklaitis *et al.* [1983].

### 8.3.2 The Method of Conjugate Directions

A very useful and *computationally* simple (but not conceptually simple) procedure for unconstrained minimization is known as the method of conjugate directions. This method is a considerable improvement over that of classical steepest descent. The basic idea is to suppose that the objective function is *quadratic* in its variables (it necessarily will be *in the vicinity of the minimum*). Then a new set of coordinates is effectively found which represent a kind of "orthogonalization" of the search directions. This permits minimization of the quadratic form in $n$ steps. The procedure is also found to work well for non-quadratic functions. The original algorithm is due to Fletcher and Reeves, with further improvements given by Polak and Ribiere. The theory is beyond the scope of our discussion, but it may be found in Bunday [1984] or Reklaitis *et al.* [1983]. As discussed in these references, the method has performed well in practice and is attractive because of its computational simplicity and small storage requirements. The computational aspects are discussed in Press *et al.* [1986]. At the end of this chapter, we give a computer program for the method, called OPTFRPR (Fletcher-Reeves-Polak-Ribiere), which is similar to that given by Sprott [1991]. Our program uses a direct search for the required line minimization.

**Example 8.3 Minimization of $f(x_1, x_2) = (1 - x_1)^2 + 100(x_1^2 - x_2)^2$ Using the Conjugate Direction Method.**

This function has been designed to be difficult for "steepest-descent"-type algorithms. The minimum value of $f = 0$ clearly occurs at $x_1 = x_2 = 1$. We use program OPTFRPF to solve this problem. To use the program, we must specify the objective function $f(x_1, x_2)$ in Sub Function, evaluate the partial derivatives of $f(x_1, x_2)$, and then specify them in Sub Deriv. We use the program parameter values Tol $= 10^{-6}$, incr $= 1000$, and initial values $x_1 = x_2 = 2$, all specified in Sub Init. The results of this calculation are achieved in 0.47 min on the Macintosh SE 30 computer and give for the optimum conditions $x_1 = x_2 = 1.0000$, and $f_{\text{opt}} = 3.9 \times 10^{-10}$. In this problem, the conjugate gradient method (FRPR) is seen to perform quite well.

## 8.4 LINEAR PROGRAMMING

The problem of optimizing a linear objective function subject to linear inequality or equality constraints is called *linear programming* (LP). It is the simplest type of optimization problem that involves inequality constraints. This kind of problem is very important in the field of optimization, since it occurs in a variety of contexts, and it also serves as the prototype for more complicated nonlinear problems. Many applications of linear programming have been made in problems of engineering design, business, and economics. This is a very large subject, so we will concentrate here on the fundamentals of how to solve such problems numerically. This will be followed by a discussion of computer program OPTLP, which can be used to solve linear programming problems. We will then give references that readers can pursue for other aspects of this subject.

The linear programming problem can be represented in a mathematical form *convenient for solution purposes* as follows:

$$\text{Minimize} \qquad z = c_1 x_1 + c_2 x_2 + \cdots + c_n x_n$$

subject to (in matrix notation)

$$\begin{aligned}
&A_1 \mathbf{x} \geq \mathbf{b}_1 && \text{(GT, number of greater-than constraints)} \\
&A_2 \mathbf{x} = \mathbf{b}_2 && \text{(EQ, number of equal-to constraints)} \\
&A_3 \mathbf{x} \leq \mathbf{b}_3 && \text{(LT, number of less-than constraints)} \\
&\text{All } \mathbf{x}, \mathbf{b}_i \geq 0
\end{aligned} \tag{8-23}$$

Here, the $A_i, \mathbf{b}_i$ quantities represent appropriate matrices and vectors for the three types of systems of inequalities or equations. The objective function to be minimized is $z$, and it depends on $n$ original variables. The number of constraints can be less than, equal to, or greater than the number of variables $n$. The constraints, expressed above in matrix form, are divided into three groups according to whether they are

**TABLE 8.2** MAKEUP COAL CHARACTERISTICS FOR EXAMPLE 8.4

| Coal # | % Sulfur | % Phosphorus | % Ash | Cost, $/ton |
|---|---|---|---|---|
| 1 | 0.2 | 0.05 | 2 | 30 |
| 2 | 1.0 | 0.06 | 3 | 15 |
| 3 | 0.5 | 0.03 | 1 | 25 |
| 4 | 0.7 | 0.03 | 2 | 20 |

greater than (GT), equal to (EQ), or less than (LT) the positive right-hand-side values $\mathbf{b}_i$. For *convenience*, we have specified that all $x_i$ and all $\mathbf{b}_i$ must be $\geq 0$. If some $\mathbf{b}_i$ are originally $< 0$, we can multiply through those expressions by $-1$ to achieve a positive value; this will then merely change the sense of those inequalities. If some variable $x_k$ is known to be negative, we can replace it by means of $x_k' = -x_k$. On the other hand, if some variable $x_m$ has an unrestricted sign (that is, it is not known whether it is positive or negative), then we introduce *two* new variables in its place according to $x_m = x_n - x_p$, where both $x_n$ and $x_p$ are positive. Also note that if it is desired to *maximize* the objective function $z$ rather than minimize it, one need only minimize the function $(-z)$.

### 8.4.1 Problem Formulation

The formulation of linear programming problems is often quite similar to that of a material balance problem in chemical engineering. That is, the formulation involves defining appropriate variables for the problem and using them in conservation equations or resource-limiting inequalities, which constrain the objective function to be minimized. We give an example to illustrate the ideas.

**Example 8.4 A Blending Problem**

A utility requires that the blend of coal it burns contain no more than 0.6% sulfur, 0.05% phosphorus, and 2% ash. Four different types of makeup coal are available, with their characteristics and costs shown in Table 8.2. Because of a restricted supply of coal #2, it can constitute no more than one-half of the blend. How should the makeup coals be mixed in order to meet the above constraints and also achieve a minimum cost for the utility?

To proceed, we take a basis of one ton of the final mixed coal. The variables will be $x_1, x_2, x_3$, and $x_4$, the amounts in tons of makeup coals #1, 2, 3, and 4. This yields the overall material balance equation

$$x_1 + x_2 + x_3 + x_4 = 1 \tag{a}$$

The material balances for sulfur, phosphorus, and ash are in terms of *inequality restrictions*, since these quantities have certain percentages they cannot exceed. These inequalities take the respective forms

$$\begin{aligned} 0.2x_1 + x_2 + 0.5x_3 + 0.7x_4 &\leq 0.6 \\ 0.05x_1 + 0.06x_2 + 0.03x_3 + 0.03x_4 &\leq 0.05 \\ 2x_1 + 3x_2 + x_3 + 2x_4 &\leq 2 \end{aligned} \tag{b}$$

We also must have $x_2 \leq 0.5$, since the amount of coal #2 cannot exceed this due to supply limitations. In addition, we have the requirements that $x_1$, $x_2$, $x_3$, and $x_4$ must all be greater than or equal to zero, since these amounts cannot be negative. This latter, rather obvious, *non-negativity* requirement for the variables is quite important as regards use of the well-known simplex method (to be described later) for the solution of the linear programming problem. *Usually*, the variables in a linear programming problem are non-negative from the nature of the physical problem, but they need not be. However, use of the simplex solution process *requires* that the variables be non-negative. Thus, if any variables in the original problem are either negative or unrestricted in sign, appropriate new variables must be used in the simplex algorithm.

The total cost per ton of the blended coal, using the data from Table 8.2, is

$$\text{Cost} = 30x_1 + 15x_2 + 25x_3 + 20x_4 \tag{c}$$

Taking the cost function to be $z$, the full linear programming problem for this example is written as

$$\begin{gathered} \text{Minimize } z = 30x_1 + 15x_2 + 25x_3 + 20x_4 \\ \text{subject to} \\ x_1 + x_2 + x_3 + x_4 = 1 \\ 0.2x_1 + x_2 + 0.5x_3 + 0.7x_4 \leq 0.6 \\ 0.05x_1 + 0.06x_2 + 0.03x_3 + 0.03x_4 \leq 0.05 \\ 2x_1 + 3x_2 + x_3 + 2x_4 \leq 2 \\ x_2 \leq 0.5 \\ x_1, x_2, x_3, x_4 \geq 0 \end{gathered} \tag{d}$$

This problem involves four variables, one equality constraint, and four less-than inequality constraints, along with the nominal non-negative constraints on the variables. When program OPTLP is used to solve this problem by the simplex method, we obtain the unique solution $x_1 = 0.2$, $x_2 = 0$, $x_3 = 0$, $x_4 = 0.8$, and the minimum cost is $22/ton. Thus, in this particular problem, it happens that the minimum cost strategy is to use *none* of the #2 and #3 coals. It is easily verified that the $x_i$ solution values satisfy the constraints given above.

As an illustration of how the solution can change when the problem changes somewhat, suppose that the cost of coal #4 increases from $20/ton to $25/ton, all other conditions remaining the same. In this case, we simply change $c(4)$ in program OPTLP from 20 to 25. We find that the new solution gives a minimum cost of $22.73/ton with $x_1 = 0.272727$, $x_2 = x_3 = 0.363636$, and $x_4 = 0$. In this case, the minimum cost strategy uses none of the #4 coal.

Problem formulation is very important in linear programming. In some cases, an incorrect formulation will lead to either inconsistent constraints or to an unbounded solution. But sometimes, a correctly formulated problem can lead to these same results. For example, truly inconsistent constraints in a problem could cause the formulator to see that the problem should be altered so that some constraint should become, instead, the new objective function. At the end of any linear programming problem, it is wise to do an independent verification that the solution does indeed satisfy the constraints.

### 8.4.2 The Geometrical Approach for Two Variables

It happens that linear programming problems involving only *two* variables (but with any number of constraints) can be illustrated geometrically. This is very useful as an aid in understanding the nature of linear programming problems. As we shall discuss later, there are four noteworthy possible outcomes associated with any linear programming problem. These are:

1. A successful unique solution
2. A successful *non-unique* solution
3. An unbounded solution
4. No solution due to inconsistent constraints

It is easy to illustrate all of these behaviors graphically for a two-variable problem. This is accomplished by first using the constraints to define the *feasible region* within which a solution may exist. Following this, a set of lines can be drawn which represent various values of the objective function. If a feasible region exists and is finite, the solution is obtained where the objective function line with the smallest possible value is just about to leave the feasible region. This always occurs at a vertex or "corner point" where two (or more) constraints meet. Possible behaviors are shown in Figures 8.3–8.6.

For the first case above, we see a successful, unique solution where the minimum objective function line intersects the feasible region at a constraint boundary (see Figure 8.3).

For the second case, we have a successful *non-unique* solution (see Figure 8.4). This means that we have a finite feasible region, but the objective function line at the minimum coincides with a constraint line. Thus, we have a unique minimum that corresponds to many different (non-unique) combinations of the $x_i$ values.

For the third case, the feasible region is unbounded, and in the case shown, the minimum objective function value is also unbounded (see Figure 8.5).

The fourth case, illustrates what can happen if the constraints are inconsistent (see Figure 8.6). In such a situation there is *no feasible region*, and there can thus be no solution to the LP problem.

It is useful to note that the two-variable behaviors noted here also describe, in a qualitative way, what happens in more general problems. We can appreciate this by observing that for *any* size of problem, we can imagine that all but two variables have their optimal values (assuming that a solution exists). Then the situation for the remaining two variables can be depicted graphically, as we have done here.

### 8.4.3 The Simplex Algorithm

We now take up a discussion of the simplex method for the solution of linear programming problems, which is attributed to G. Dantzig. The simplex method represents a neat and systematic algebraic approach for LP problems of any size which allows appropriate recognition of all of the possible outcomes discussed earlier. Moreover, the method lends itself to a convenient computer implementation.

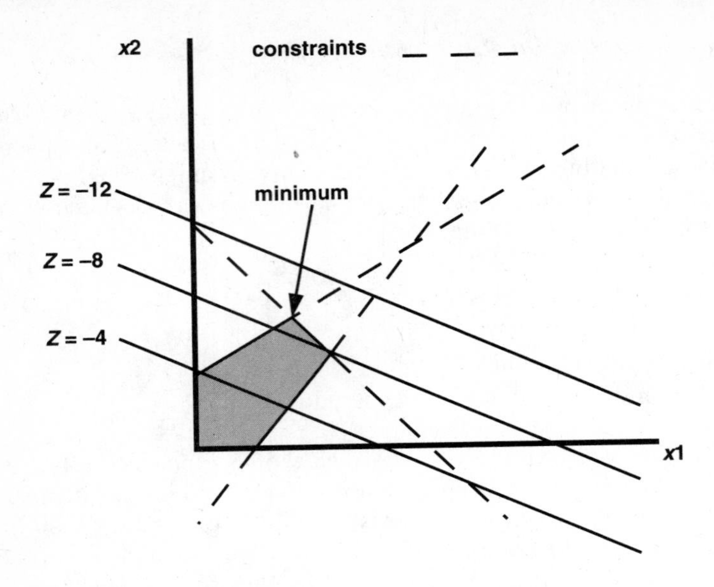


Figure 8.3. Linear Programming: Case (1)—Unique Solution

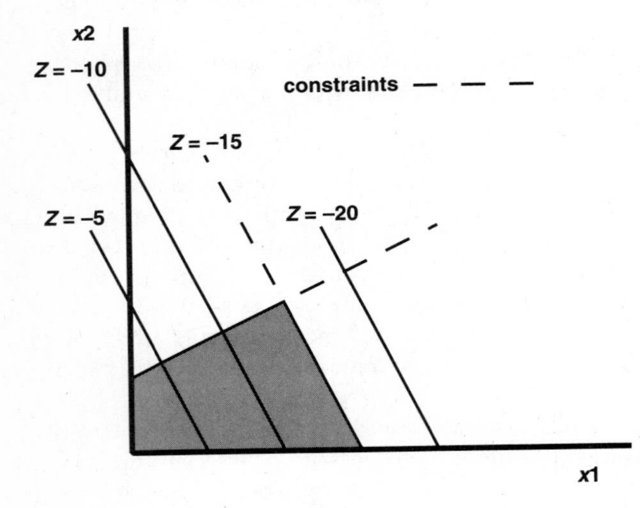


Figure 8.4. Linear Programming: Case (2)—Non-Unique Solution; Objective Function Parallel to Constraint Function

Program OPTLP (OPTimization, Linear Programming) employs the simplex method and is discussed in detail at the end of the chapter.

The fundamental simplex method is a procedure for moving from an initial feasible point, through appropriate vertex points of the constraints, toward the optimum. Each successive simplex iteration is designed to improve the objective function. If a solution exists, the simplex method will find it in a finite number of steps. The simplex method utilizes *only positive variables* $x_i$ and *positive right-hand-side values* $b_i$. If any variable is either negative or unrestricted in sign, a new variable must be introduced, as discussed earlier.

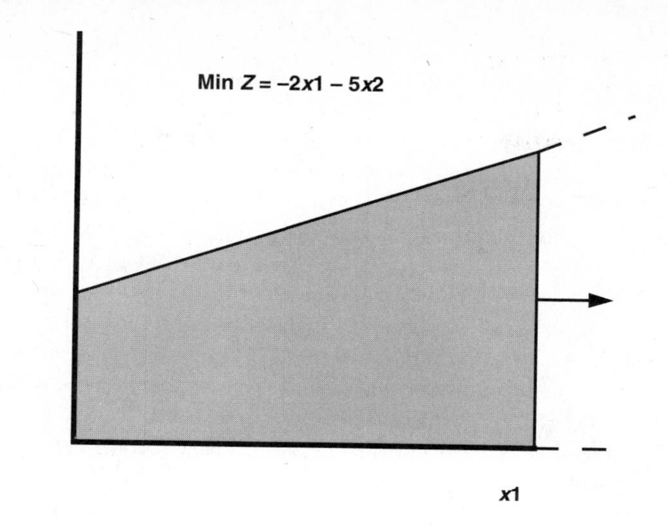


**Figure 8.5.** Linear Programming: Case (3)—Unbounded Solution

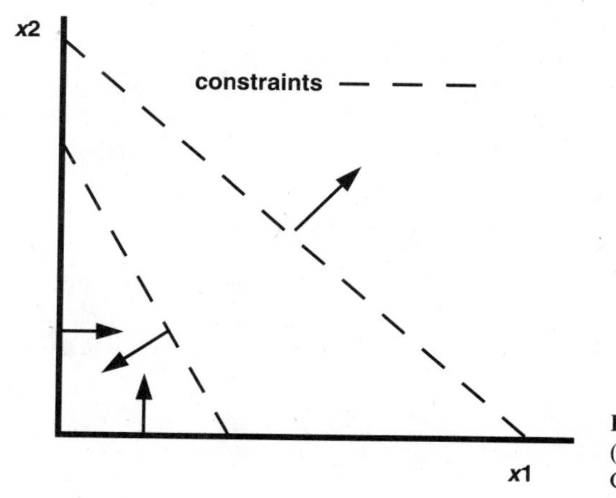


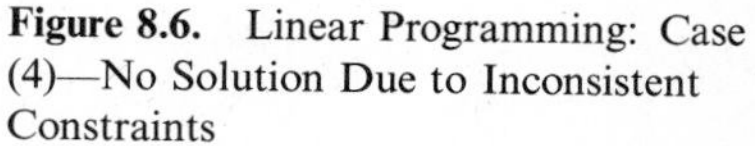
**Figure 8.6.** Linear Programming: Case (4)—No Solution Due to Inconsistent Constraints

The simplex algorithm in general proceeds in two stages. In phase I (if it is necessary), the algorithm is used to find an initial feasible solution. Following the successful completion of phase I, the method moves in phase II toward minimizing the objective function.

As it happens, *it is simpler to discuss the phase II calculations first*; a slight modification of these is used in phase I (if necessary). Suppose that our problem has $n$ original variables and $m$ linear constraint expressions (any combination of EQ equalities and $(GT + LT)$ inequalities). Refer back to Equations (8-23). The simplex method first involves the introduction of additional, so-called slack variables (which must be positive) into any inequality constraints, in order to convert them to equalities. There will thus be $(GT + LT)$ new slack variables in the equations. Now *assume* that we can solve the new set of equations for some "basic" (solved-for) set of variables, all of which are $\geq 0$, *leaving all the other (non-basic) variables equal to zero*. As we will show, if a solution to the original problem exists, this can always be done. Thus, there will be $m$ basic variables, one for each constraint equation. Such a solution of the constraint equations is called a basic feasible solution, or basic feasible point. To reiterate on this important point, a basic feasible solution to the constraints is a point at which each constraint equation $k$ is satisfied by one particular variable (the basic variable of that equation) being equal to $b_k$, *while all other variables in that equation are temporarily equal to zero*. Each equation has its own distinct basic variable, which is different from that in the other equations.

Once an initial basic feasible solution is known, we first test to see if we are at the desired minimum, by seeing if any coefficients of the non-basic variables in the expression for $z$ are negative. If there are *not* any negative coefficents, $z$ cannot be decreased by increasing any non-basic variable, so we are at the desired minimum. If we are not yet at the minimum, the problem is to move to a *new* basic feasible point for which the objective function is improved. This is accomplished by eliminating the current basic variables from the objective function using the constraint equations. Then we examine the objective function to see which variable would produce the greatest improvement if it was increased by a unit amount. We then introduce this currently non-basic variable (which presently has a value of zero) into the basis by "exchanging" it with one of the current basic variables. The choice of which basic variable to exchange is made by considering the constraint equations, to see which one would allow the greatest increase of the new basic variable. Following this exchange of one basic and non-basic variable, we re-express the objective function in terms of the *new* non-basic variables and repeat the process until no further improvement can be made in the objective function.

**Example 8.5 Illustration of a Phase II LP Problem**

The general situation where we do not need phase I to find an initial basic feasible solution of the constraints corresponds to the case where the constraints are only of the less-than (LT) type. In this case, the slack variables are used as the initial basic variables. Consider the problem

$$\begin{gathered}\text{Minimize } z = 4x_1 - 2x_2 - x_3 \\ \text{Subject to} \\ 3x_1 + 4x_2 - x_3 \le 1 \\ -2x_1 + 10x_2 + 7x_3 \le 3\end{gathered} \tag{a}$$

together with non-negativity of $x_1, x_2$, and $x_3$. We first introduce the positive slack variables $x_4$ and $x_5$ into the two constraint inequalities to produce the constraint *equations* shown next.

$$\begin{gathered}3x_1 + 4x_2 - x_3 + x_4 = 1 \\ -2x_1 + 10x_2 + 7x_3 + x_5 = 3\end{gathered} \tag{b}$$

In this problem, which involves only less-than (LT) constraints, it is easy to see that we can take the slack variables ($x_4$ in the first equation, $x_5$ in the second equation) as the basic variables. Their values will be $x_4 = 1$ and $x_5 = 3$. The non-basic variables $x_1$, $x_2$, and $x_3$ have the values zero, and therefore, the initial value of the objective function $z$ is zero. This solution, while feasible, is *not* optimal. We can see this from the objective function, whose coefficients of both $x_2$ and $x_3$ are negative. Thus, by increasing either $x_2$ or $x_3$ from their values of zero, we could further decrease the objective function $z$.

Since the coefficient of $x_2$ in the objective function is the most negative, we choose to increase $x_2$. We wish to let $x_2$ become a basic variable in one of the constraint equations, so we need to exchange it with one of the current basic variables, which would then become non-basic and have a value of zero. To decide in which equation this exchange of variables will be conducted, we consider the exchange in all possible equations and choose that corresponding to *the maximum increase of* $x_2$ *which will not violate any of the constraint equations.* To do this in the above two equations, we see that $x_2$ can become as large as 0.25 in the first equation of (b) when $x_4$ goes to zero. In the second equation of (b), $x_2$ can become as large as 0.3 when $x_5$ goes to zero. Thus, $x_2$ can become as large as the *minimum* of (0.25,0.3) without violating *any* of the constraints. We therefore make the exchange in the first equation of (b), where $x_2$ becomes basic and has the value 0.25, while $x_4$ becomes non-basic and has the value zero. Following this exchange of variables, we use the first equation to eliminate $x_2$ from the objective function and the other constraint equations (of which there is only one more in this example). After this first simplex step, our problem can be rewritten as

$$
\begin{gathered}
z = -0.5 + 5.5x_1 - 1.5x_3 + 0.5x_4 \\
0.75x_1 + x_2 - 0.25x_3 + 0.25x_4 = 0.25 \\
-9.5x_1 + 9.5x_3 - 2.5x_4 + x_5 = 0.5
\end{gathered} \tag{c}
$$

Several items are worthy of note in Equations (c). First, $x_1$, $x_3$, and $x_4$ are now the non-basic variables, all having the value zero. We see that the objective function $z$ is expressed solely in terms of these non-basic variables, and $z$ therefore now has the value $-0.5$. The basic variables $x_2$ and $x_5$ have the values 0.25 and 0.5, respectively. Thus, the first simplex step has decreased the objective function from its original value of zero to $-0.5$, with the corresponding $x_i$ values becoming $x_1 = 0, x_2 = 0.25, x_3 = 0, x_4 = 0$, and $x_5 = 0.5$. The simplex algorithm has therefore found a new "vertex" point (intersection of constraints) which has an improved value for the objective function.

To continue further simplex steps, we merely repeat this process with equations (c), finally stopping when either (i) a non-basic variable that decreases $z$ can be increased indefinitely without violating the constraints—this is the "unbounded solution"; or (ii) all coefficients of the non-basic variables in the objective function become non-negative. In this latter case, there is no increase of a non-basic variable that would decrease $z$, so we have reached the desired minimum.

While it is crucial to the understanding of the simplex method to carry out detailed calculations of the type considered above, when solving a linear programming problem we would usually wish to use a computer program to carry out all the arithmetic and do the necessary "bookkeeping." Program OPTLP at the end of the chapter is available to do this, and it is useful to now discuss the interpretation of its output as we continue this example. The problem *input* required in this example is illustrated in subroutine INIT, in the concise listing, at the points specified by !***. This includes the original number of variables $n = 3$, the total number of constraints $m = 2$, the number of greater-than constraints $GT = 0$, the number of equal-to constraints $EQ = 0$, and the number of less-than constraints $LT = 2$. Also required is the specification of the coefficients in the objective function and the constraints. The program does the rest, including putting in the slack variables, checking for an unbounded solution, and continuing until the optimum is reached.

When program OPTLP is run on this example, the initial information regarding the basic variables, the constraint equations, and objective function $z$ is printed as follows:

```
Basic  Var.  Value
        4      1
        5      3
        z      0

Var. j = 1    2    3    4    5
Eqns
A(1,j)    3    4   -1    *    0
A(2,j)   -2   10    7    0    *
z              4   -2   -1    *    *
```

This output first shows that originally, $x_4$ and $x_5$ are the basic variables having the values of 1 and 3, respectively, and the objective function $z$ is zero. The notation $A(i,j)$ under the label Equations refers to the coefficients of $x_j$ in constraint equations $i$. The numbers following $z$ refer to these same coefficients in the objective equation. The current *value* of the objective function is set off by itself and does *not* occur again in the row of numbers following z. The asterisks (*) refer to a basic variable; thus, each $A(i,j)$ row (equation) will contain exactly one asterisk, while the objective function, which is always expressed in terms of non-basic variables only, will always contain exactly $m$ asterisks. The output given above should be compared in detail to the objective function $z$ in Equations (a) and the constraint equations in Equations (b) to be sure the notation is clear.

If we now continue the simplex calculation using program OPTLP, the next step yields

```
Basic Var. Value
       2      0.25
       5      0.5
       z     -0.5

Var. j =  1      2      3        4     5
Eqns
A(1,j)    0.75   *   -0.25    0.25   0
A(2,j)   -9.5    0    9.5    -2.5    *
z         5.5    *   -1.5     0.5    *
```

This computer output should be compared to Equations (c) above; it represents the same information. It turns out that only one more simplex step is required to reach a unique minimum for $z$. The computer output for this final step is

```
Basic  Var.  Value
        2      0.263158
        3      0.0526316
        z     -0.578947

Var. j = 1    2  3  4              5
Eqns
A(1,j)    0.5   *  0    0.184211   0.0263158
A(2,j)   -1     0  *   -0.263158   0.105263
z         4     *  *    0.105263   0.157895
```

The final "tableau" above shows that we do indeed have a unique optimum, since the coefficients of the non-basic variables in the expression for $z$ are all positive. We see that the final solution corresponds to a minimum of $z = -0.578947$ (to six figures) with the $x_i$ values $x_1 = 0$, $x_2 = 0.263158$, $x_3 = 0.0526316$, $x_4 = 0$, and $x_5 = 0$. At the optimum, both constraints are tight (that is, both inequalities are actually *equalities*) since the slack variables $x_4$ and $x_5$ are both equal to zero.

### 8.4.4 A Phase I LP Problem

We now discuss how to handle the phase I calculation in linear programming, for the situation where an initial basic feasible solution *cannot* be obtained by merely using the slack variables as basic (as in Example 8.5). This happens any time that the constraints (after adjustment to make all $b_i$ positive) involve some combination of greater-than (GT) or equal-to (EQ) constraints. In such a situation the procedure followed in program OPTLP is as follows. Appropriate positive slack variables are introduced into the (GT) and (LT) inequality constraints, to make them equalities. Then one positive so-called artificial variable is introduced into each (GT) and (EQ) constraint. We can then take as initial basic variables in the constraints both the artificial variables in the GT and EQ constraints, and the slack variables in the LT constraints. The phase I problem is then to minimize the *artificial* objective function $w$, which is the sum of the artificial variables or $w = \sum$(artificial variables). This is accomplished by applying the simplex algorithm, as described above, to the objective function $w$. Assuming that a solution exists, when $w$ has been driven to zero, the artificial variables will all be zero, and we will have achieved a basic feasible solution to the original constraints. At that point, we merely drop the artificial variables and begin the phase II calculation of minimizing the original objective function $z$. This again involves the simplex procedure discussed in detail above. If it happens that the minimum of the artificial objective function $w$ is *greater than zero*, then the original constraints *do not have a solution*, and the original LP problem does not have a solution.

It should be mentioned that computer program OPTLP takes care of all these contingencies *automatically*, including the introduction of all necessary slack and artificial variables. The user need only define the original problem as described in the concise listing of the program. The program finds whether the constraints are consistent, finds a basic feasible solution, and then finds either an unbounded solution or a successful solution, as appropriate.

### 8.4.5 Further Notes on Linear Programming

There are several additional topics in linear programming that are of interest for further study. These include sensitivity studies, duality, and integer LP. These topics can be pursued in references such as Dantzig [1963], Bunday [1984], and Hillier and Liebermann [1967]. The Karmakar LP algorithm is discussed by Edgar and Himmelblau [1988].

## 8.5 NONLINEAR PROGRAMMING (NLP)

The general nonlinear programming problem may be defined as follows.

$$\begin{array}{lll} \text{Optimize} & f(\mathbf{x}) & \\ \text{Subject to} & \mathbf{h}(\mathbf{x}) = 0 & \qquad (8\text{-}24) \\ \text{and} & \mathbf{g}(\mathbf{x}) \le 0 & \end{array}$$

Here, the objective function $f(\mathbf{x})$ and the constraint functions $\mathbf{h}(\mathbf{x})$, $\mathbf{g}(\mathbf{x})$ all may depend *nonlinearly* on the $n$ variables. A distinction is made between the equality constraints $\mathbf{h}(\mathbf{x})$, and the inequality constraints $\mathbf{g}(\mathbf{x})$. We can see that the general NLP problem is an extension of the various problems we have considered previously, all of which may be regarded as special cases of Equation (8-24). Thus, unconstrained optimization corresponds to the absence of any constraint functions $\mathbf{h}(\mathbf{x})$ or $\mathbf{g}(\mathbf{x})$; equality-constrained problems have some $\mathbf{h}(\mathbf{x})$ constraint functions but no $\mathbf{g}(\mathbf{x})$ constraints; and linear programming problems possibly have both $\mathbf{h}(\mathbf{x})$ and $\mathbf{g}(\mathbf{x})$ constraints, all of which must be of *linear* form.

In general, NLP problems are much more difficult to solve than those we have considered earlier, in large measure because the optimum can be either on the boundary or in the interior of the feasible constraint region. Our discussion of nonlinear programming will necessarily be at an introductory level, but this should give the reader a basis for pursuing the subject later at a more advanced level. To this end, a number of pertinent references will be cited at the completion of our discussion.

To begin, we note that when nonlinear inequality constraints are present, the constraint feasible region will generally be curvilinear. For example, given the nonlinear constraints $g_1(x_1, x_2) = x_1 + x_2^2 \le 4$, $g_2(x_1, x_2) = x_1^2 + x_2 \le 4$, and $x_1, x_2 \ge 0$, we obtain the constraint feasible region shown in Figure 8.7. Moreover if the objective function $f(\mathbf{x})$ is not linear, it is possible for the partial derivatives of $f$ to be zero. Thus, for the general NLP problem, it is possible for the optimum to be located in a region interior to the feasible region, where the inequality constraints will be inactive or "loose." In such a case, the optimum could be obtained by merely ignoring the loose constraints. Unfortunately, we never know in advance which of the inequality constraints will be loose and which tight at the unknown optimum. In very small problems, we could consider all possible combinations of tight and loose constraints, solving each case using Lagrange multipliers and selecting as the optimum the case that has the best value of the objective function and also satisfies all constraints. But it is obvious that this strategy is not practical except in very small problems.

### 8.5.1 Direct Search

Before proceeding to discuss the calculus necessary conditions for an NLP optimum, we pause to mention a very elementary but practical technique that can be helpful in solving relatively small NLP problems (here, the requirement is that there be relatively few variables). This technique is nothing more than a systematic direct evaluation of *discrete* feasible solutions within a given range of the problem variables. This can be accomplished very easily on a computer; the principal limitation is simply the time required to accomplish this evaluation and comparison. Often, the range of the problem variables can be determined from the constraints of the problem.

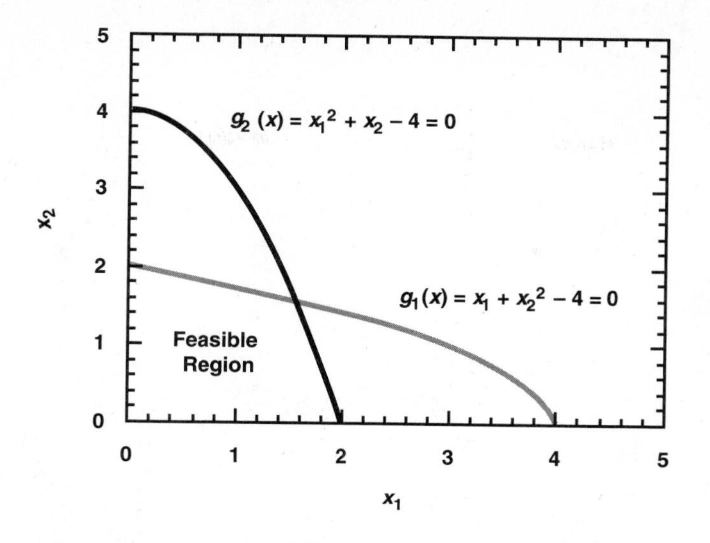


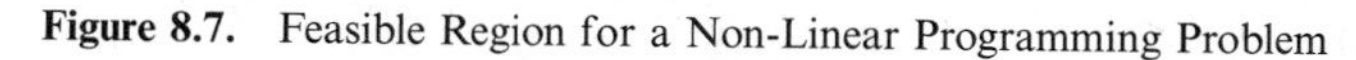

**Figure 8.7.** Feasible Region for a Non-Linear Programming Problem

As computers become faster, the direct search approach becomes more practical. A detailed discussion together with many examples is given by Conley [1984], where extension to larger problems through random sampling of feasible solutions is also discussed. To illustrate the procedure, we consider the following NLP problem:

$$\begin{gathered} \text{Min } Z = 2x_1^3 + 3x_2^2 \\ \text{Subject to} \\ x_1 + x_2 = 10 \\ x_1 + \frac{2}{x_2} \le 50 \\ x_1, x_2 \ge 0 \end{gathered} \tag{8-25}$$

We will first solve this problem for only *integer* values of the variables. This so-called integer programming type of problem is important in itself, and it is for this case that the search method is entirely rigorous. Since $x_1$ and $x_2$ are both positive, the equality constraint $x_1 + x_2 = 10$ shows that both $x_1$ and $x_2$ must be less than 10. Thus, we can consider all integer combinations of $x_1, x_2$ corresponding to $0 \le x_1 \le 10$ and $1 \le x_2 \le 10$. Naturally, we must exclude the possibility of $x_2 = 0$, since this would violate the inequality constraint. The following simple True BASIC computer program uses the 90 combinations of $x_1$ and $x_2$ to evaluate the constraints.

```
LET Zmin = 1e10                               !large value of Zmin
FOR i = 0 to 10
    LET x = i                                 !one loop for each variable
    FOR j = 1 to 10
        LET y = j
        LET Com = 1                           !initialize Com(pare)
        IF x + 2/y > 50 THEN LET Com = 0          !eval. of constraints
```

```
        IF x + y -10 <> 0 THEN LET Com = 0
        IF Com = 1 THEN                    !if constraints satisfied, compare
            LET Z = 2*x^3 + 3*y^2
            IF Z < Zmin THEN
                LET Zmin = Z               !if improvement,reset Z,x,y
                LET xmin = x
                LET ymin = y
            END IF
        END IF
    NEXT j
NEXT i
PRINT"Zmin = ";Zmin;"xmin = ";xmin;"ymin = ";ymin
END
```

For "feasible" points, the program evaluates the objective function $Z$, and if the new point represents an improvement, it records the new Zmin along with the corresponding values of the variables $x_1$ and $x_2$. Nonfeasible points (any that do not satisfy *all* constraints) are merely skipped. If we run this program, we get the results $Z\text{min} = 201$, $x\text{min} = 3$, and $y\text{min} = 7$. This represents the exact solution to the above *integer* programming problem.

Suppose now that we wish the answer to the above problem for *continuous* values of the variables. If we repeat the above calculation, now using discrete increments of 0.01 in the variables $x_1, x_2$ (using program OPTSEARCH at the end of the chapter), we get the result $Z\text{min} = 199.22360$, $x\text{min} = 2.7$, and $y\text{min} = 7.3$, which is an *approximation* to the exact continuous solution. By solving the continuous problem using the conventional calculus approach (eliminating a variable using the equality constraint), we get the exact solution as $Z\text{min} = 199.22359$, $x\text{min} = 2.70156$, and $y\text{min} = 7.29844$ (to this many significant figures). Thus, in this particular problem, the approximate simple discrete search produces results that are nearly the same as the calculus approach. Accuracy could generally be increased by using a finer discretization, at the expense of more computing time. An alternative procedure is to use the results of one particular discretization to sharpen the limits on the variables, so that in the next run the search will be conducted in a narrower region. For a calculation such as this, the approximate minimum will always necessarily be $\geq$ the true value.

The great advantage of the search technique is that it does not require partial derivatives and complicated calculations, but on the other hand, it produces only an approximate solution. In more complicated problems, the discrete search approach can be very valuable, since implementation of the calculus method becomes much more difficult. Even when the calculus method is to be employed to obtain an "exact" answer, the discrete search method can be very important in obtaining a good first approximation to the solution, or to see which constraints may be active. It is also valuable as a simple independent check of the complex calculation. It should be noted that the search technique is more suitable for problems involving only *inequality* constraints. This is because the search could easily miss many feasible points corresponding to equality constraints. If this seems to be happening in a given problem, it may be best to incorporate the equality constraints into a *penalty* objective function (see below) which is subject to the inequality constraints. Then one could consider a sequence of problems corresponding to the decrease of the penalty parameter, each of which could be solved using the search technique. As the penalty parameter goes to zero, the desired solution should be approached.

We will consider another example of this technique following discussion of the calculus approach to the NLP problem.

### 8.5.2 Penalty Functions for Problems with Equality Constraints

Another method for solving constrained optimization problems is known as the penalty function method. Here we merely introduce the ideas for an *equality*-constrained minimum and refer the reader to Reklaitis *et al.* [1983] for further information. Suppose we have an objective function $f$ that is to be minimized subject to the constraint equations $h_i = 0$ $(i = 1, 2, \ldots m)$, where $f$ and each $h_i$ may depend on the $n$ variables $x_i$. Then we can form a *new* objective function $z$, called a penalty function, by the equation

$$z = f + \frac{1}{r} \sum_{i=1}^{m} h_i^2 \tag{8-26}$$

The idea now is that if we carry out an *unconstrained* minimization of Equation (8-26), *using a very small positive value of* r, then the potential "penalty" of having the constraint functions not equal to (nearly) zero would cause the calculated minimum to be close to that of the original constrained problem. The advantage of this approach is that a given unconstrained minimization is generally much easier than that of an explicitly constrained problem. Unfortunately, one cannot merely prescribe some extremely small value of $r$ (such as $r = 10^{-6}$) and proceed directly to the desired result. As $r$ becomes smaller, each subsequent minimization generally becomes more ill-conditioned than the previous one, and therefore takes longer and is less accurate. Instead, a succession of decreasing small values of $r$ is chosen, and limiting optimal conditions are estimated. In this connection, it can sometimes be useful to extrapolate toward zero in $r$.

### 8.5.3 The Kuhn-Tucker Necessary Conditions for a Minimum

We now consider the calculus *necessary* conditions for the minimum of some objective function $Z = f(\mathbf{x})$ subject to general *inequality* (and perhaps also equality) constraints. The conditions to be discussed are usually referred to as the Kuhn-Tucker necessary conditions. To illustrate the ideas in a simple way, we shall use the problem of Equations (8-25). Following this, we will give the extension of the results to a more general problem. For convenience, we repeat the equations below.

$$\begin{gathered} \text{Min } Z = 2x_1^3 + 3x_2^2 \\ \text{Subject to} \\ x_1 + x_2 = 10 \\ x_1 + \frac{2}{x_2} \leq 50 \\ x_1, x_2 \geq 0 \end{gathered} \tag{8-27}$$

This problem involves two variables, one equality constraint, and one inequality constraint. The variables $x_1, x_2$ are to be non-negative. We first convert all inequalities to equalities by means of new "slack" variables. In our present problem, this leads to

$$x_1 + \frac{2}{x_2} + x_3^2 = 50 \tag{8-28}$$

Notice that by introducing $x_3^2$ instead of $x_3$, we automatically reduce the inequality to an equality regardless of the sign of $x_3$. We will ignore the non-negativity of $x_1, x_2$ at this point and seek solutions satisfying this requirement, rather than introduce two more slack variables. Thus, we now have a three-variable *equality*-constrained problem, so we can proceed with the Lagrange multiplier formalism that was applied to the equality-constrained problem earlier. To apply the Lagrange formalism, we define the Lagrangian function $L$ as

$$L = Z(x_1, x_2) + \lambda_1(x_1 + x_2 - 10) + \lambda_2\left(x_1 + \frac{2}{x_2} + x_3^2 - 50\right) \tag{8-29}$$

As in our earlier discussion of equality constraints, we form the partial derivatives of $L$ with respect to *all* of the variables $(x_1, x_2, x_3, x_4 \equiv \lambda_1, x_5 \equiv \lambda_2)$. This yields the following system of five nonlinear algebraic equations, which form a *necessary* condition for a minimum of the original inequality-constrained problem:

$$\begin{aligned}
&\frac{\partial L}{\partial x_1} = 0 = 6x_1^2 + \lambda_1 + \lambda_2 \\
&\frac{\partial L}{\partial x_2} = 0 = 6x_2 + \lambda_1 - \frac{2\lambda_2}{x_2^2} \\
&\frac{\partial L}{\partial x_3} = 0 = 2\lambda_2 x_3 \quad \text{(Complementary Slackness)} \\
&\frac{\partial L}{\partial x_4} = 0 = (x_1 + x_2 - 10) \\
&\frac{\partial L}{\partial x_5} = 0 = \left(x_1 + \frac{2}{x_2} + x_3^2 - 50\right) \\
&x_5 \equiv \lambda_2 \geq 0
\end{aligned} \tag{8-30}$$

There are several special points that need to be mentioned in regard to these necessary conditions. We observe that the first five equations of (8-30) nominally represent the result of applying Lagrange multipliers to the *transformed* equality-constrained problem. However, the slack variable $x_3$ did not appear in the original problem statement, but rather was introduced to make the inequality constraint an equality.

We now recall Equation (8-17), which showed that for an equality-constrained problem, $\partial Z/\partial b_i = -\lambda_i$, where $b_i$ refers to the right-hand-side of the $i$th constraint. In the case of an *inequality* constraint, this relationship shows that if the constraint is *inactive* at the minimum point, then changing $b_i$ slightly can have no effect on the minimum, and therefore, in this case, $\lambda_i = 0$. On the other hand, when the inequality is of the form $g_i(x) \leq b_i$, if the inequality constraint is active at the minimum point, then a small positive change in $b_i$ increases the feasible region, and therefore must correspond to an increment in $Z$ which is an *improvement* in the solution. For the case of a minimum, this corresponds to $\Delta Z$ being negative. Thus, another part of the necessary conditions is that *either* $x_3$ or $\lambda_2$ must generally be zero (complementary slackness, $0 = 2x_3\lambda_2$). If $\lambda_2 = 0$, the constraint is inactive and the value of $x_3$ is unimportant; if $x_3 = 0$, the constraint is active and, for a minimum, $\lambda_2$ *must be positive*. Thus, we can say in general, for this problem, that $\lambda_2 \geq 0$.

If we were attempting to maximize the objective function subject to the same constraints, then this condition would necessarily become $\lambda_2 \leq 0$. A further condition that rarely arises in applications is called the constraint qualification; it is discussed by Reklaitis *et al.* [1983].

In principle, we can try to solve Equations (8-30) for the values of the $x_i$ at the stationary point. One natural procedure for doing this is to form the *unconstrained* minimization problem

$$\text{Min} \sum_{i=1}^{5} \left(\frac{\partial L}{\partial x_i}\right)^2 \tag{8-31}$$

To accomplish this minimization, any unconstrained optimization algorithm could be used (such as program OPTFRPR) to drive the sum of squares to zero. As we mentioned earlier in our discussion of this rather simple example, it is much easier to anticipate that the inequality constraint may be inactive at the minimum, and simply use the equality constraint to eliminate one variable in the objective function so that we finally have to find the minimum of a function of only *one* variable.

In the more general problem where we wish to minimize $Z(\mathbf{x})$ subject to $p$ equality constraints of the form $h_i(\mathbf{x}) = 0$ and $q$ inequality constraints of the form $g_i(\mathbf{x}) \leq 0$, we introduce slack variables $x_k^2$ into each inequality constraint to form an equality and then form the Lagrangian, including one Lagrange multiplier for each constraint. Necessary conditions for a minimum are then obtained from $\partial L/\partial x_i = 0$, where $i$ includes the original variables, the slack variables, and the Lagrange multiplier variables. For each inequality constraint we will obtain one complementary slackness condition, and one non-negativity requirement for the Lagrange multiplier. In principle, these equations can be solved as above. We should note that there exist theoretical sufficient conditions for the NLP problem Reklaitis *et al.* [1983], but they are difficult to apply and are rarely used in practice. We now consider a more challenging example.

**Example 8.6 A Metal Cutting Problem Due to Reklaitis, et. al.**

For the problem formulation, see Reklaitis *et al.* [1983]. The mathematical optimization problem is expressed as

$$\begin{gathered}
\text{Min } Z = \frac{982}{x_1 x_2} + 8.1(10^{-9})x_2^{2.333}x_1^{1.222} \\
\text{Subject to} \\
1 \leq x_1 \leq 35, \quad 49.1 \leq x_2 \leq 982 \\
\frac{x_2 x_1^{0.78}}{4000} \leq 1, \quad \frac{3.8(10^5)}{x_1 x_2^2} \leq 1
\end{gathered} \tag{a}$$

To use the calculus method here requires six slack variables for the six inequality constraints, plus six Lagrange multiplier variables, along with the two original variables, for a total of *fourteen* variables. We could, in this way, form the Lagrangian and minimize as in Equation (8-30) (with 14 terms in the sum) using program OPTFRPR.

**TABLE 8.3** USE OF PROGRAM OPTSEARCH.tru FOR EXAMPLE 8.6

| $i$ max | $x_1$ | $x_2$ | $Z_{min}$ | Time (min.) |
|---|---|---|---|---|
| 10 | 35.0 | 142.39 | 0.2630 | 0.0086 |
| 100 | 35.0 | 151.7 | 0.2614 | 0.72 |
| Exact | 35.0 | 153.4 | 0.2613 | |

The exact solution given by Reklaitis *et al.* (using geometric programming, which is beyond the scope of our discussion) is $Z_{min} = 0.2613$ with $x_1 = 35.0$ and $x_2 = 153.4$. If we instead solve the original problem with the discrete search method using program OPTSEARCH, the search involves only two variables, and we get the results shown in Table 8.3. This example illustrates the potential power of the direct search method.

### 8.5.4 Other Methods for NLP Problems

The implementation of the Kuhn-Tucker conditions discussed above as a set of nonlinear algebraic equations involves a number of drawbacks, just as in the case of using the necessary conditions to find the minimum of an unconstrained problem. The usual difficulties of finding good starting values and the fact that any solution only represents necessary conditions are compounded in the case of inequality constraints by the fact that each inequality requires an additional slack variable.

Other important methods for the practical solution of larger NLP problems include successive linear programming, successive quadratic programming, the generalized reduced gradient method (GRG), and penalty function methods. These procedures are beyond the scope of our discussion; they are discussed in the references, particularly by Reklaitis *et al.* [1983] and Edgar and Himmelblau [1988].

## 8.6 COMPUTER PROGRAMS

In this section we discuss the computer programs that are applicable to various problems in the chapter. For each program we give a brief discussion and a concise listing, which includes (i) a program description, (ii) how to run, and (iii) user-specified subroutines, including a sample problem. Detailed program listings can be obtained from the diskette.

### General Unconstrained Optimization Program OPTFRPR

This program is for unconstrained minimization of a general function of $n$ variables using the conjugate directions method of Fletcher and Reeves, as modified by Polak and Riebiere. Instructions for running the program are given in the concise listing. A simple direct-search algorithm is used for the required line search. If necessary, this search can be "tuned" by increasing the parameter incr in Sub Init and by changing the (up,low) parameters in Sub Line_Search. The running value of 'Norm' shows the progress toward convergence, and the L value suggests the proper scale for (up, low). The automatic "restart" option following convergence is useful to verify that there are no further changes in the solution.

### Program OPTFRPR.

```
!This is program OPTFRPR.tru for unconstrained optimization
!of objective function by O.T.Hanna. Prgm uses conjugate gradient
!algorithm of Fletcher-Reeves/Polak-Ribiere(FRPR).
!A simple direct-search method is used for the line search.
!If initial run produces no change, try increasing incr in Sub Init.
!In difficult starting cases,decrease |up,low| in Sub Line_Search.
!Use optional restart at end to improve or verify solution if necessary.
!Prgm prints running 'Norm' value (which should converge toward
!zero) and L value which shows approximate correct scale for
!(up,low) in Sub Line_Search.

!*************************************************************
!*To run,specify no. of variables,initial x(i) and          *
!*parameters in Sub Init. Also prescribe objective          *
!*function z in Sub Function and partial derivatives        *
!*dz/dx(i)=g(i) in Sub Deriv.                               *
!*************************************************************

SUB Init(n,x(),ftol,itmax,incr )
    !***Denotes user specified statement
    LET itmax = 100              !***max. no. allowed iterations
    LET ftol = 1e-6              !***function convergence tolerance
    LET incr = 1000              !***no. increments in Sub Line_Search
    LET n = 2                    !***no. of variables
    MAT Redim x(n)
    LET x(1) = 2                 !***initial values of x(i)
    LET x(2) = 2
END SUB

SUB Function( x(),z )
    !***Denotes user specified statement
    !Rosenbrock prob. Bunday p 73
    LET z = (1-x(1))^2+100*(x(1)^2-x(2))^2      !***obj. fn. to be minimized
END SUB

SUB Deriv( n,x(),g() )
    !***Denotes user specified statement
    LET g(1) = -2*(1-x(1))+400*(x(1)^2-x(2))*x(1)    !***g(i) = dz/dx(i),
    LET g(2) = -200*(x(1)^2-x(2))                              etc.
END SUB
```

### *General Linear Programming Program OPTLP*

This program is used to minimize a linear objective function of $n$ variables subject to a total of $m$ linear equation or inequality constraints. Instructions for running the program are given in the concise listing. The program automatically introduces any slack or artificial variables that may be required. If a phase I calculation is needed to find an initial basic feasible solution, this is also done automatically. The program produces one of the three results: no solution due to inconsistent constraints, an unbounded solution, or the desired finite solution. The problem shown is from Example 8.4.

#### Program OPTLP.

```
!This is program OPTLP.tru (linear programming) by O.T.Hanna
!Program tries to minimize linear objective function Sum [c(i)*x(i)]
!subject to the linear constraints Ax >=,=,<= b using the
!Simplex algorithm of G. Dantzig. All x(i) are positive in algorithm.
!N is orignal no. of variables in objective.fn. and constraints.
!M is total no. of inequality/equation constraints. There are
!GT greater-than, EQ equal-to and LT less-than constraints, so
!that M = GT + EQ + LT.
!Constraints must have all b(i) positive and be arranged
!in the order (i) Ax >= b, (ii) Ax = b, and (iii) Ax <= b.
!Program puts in (GT+LT) proper slack variables and (GT + EQ)
!proper artificial variables as needed.
!bs(i) is label for basic variable in eqn i.
!If Phase I is required, program produces either (i)no solution,
!INCONSISTENT CONSTRAINTS or (ii)a successful initial basic
!feasible solution.If Phase I is successful, Phase II produces
!either (i)UNBOUNDED SOLUTION or (ii)desired solution.
!Optional printed Tableaux gives coefficients of constraints
![(A(1,j) to A(m,j)] and objective function(s) [A(m+1,j)-original,
!A(m+2,j)-artificial].

!*********************************************************
!*To Run,arrange constraints so that all b(i) are positive  *
!*Then specify the required information in Sub Init. This   *
!*includes GT=no.greater-than constraints,EQ=no.equal-to    *
!*constraints,LT=no.less-than constraints. Define left side *
!*of constraint rows A(i,j) in the order greater-than,      *
!*equal-to and less-than. Define b(i) values in above order *
!*(all b(i) must be positive). Define objective function    *
!*coefficients c(i).Prgm prompts for Tableux print option.  *
!*********************************************************

SUB Init(N,M,GT,EQ,LT,A(,),b(),c())
    !*** denotes user specified statement
    LET N = 4         !***original no. of variables
    LET M = 4         !***total no. of inequality/equation constraints
```

```
    LET GT = 0        !***no. of 'greater than' inequality constraints
    LET EQ = 1        !***no. of 'equal to' equation constraints
    LET LT = 3        !***no. of 'less than' inequality constraints
    !We must have M = GT + EQ + LT; if not we get error message

    MAT REDIM A(M+2,N+2*GT+LT+EQ),b(M+2),c(N)
    !Define system of constraints by row in the following order:
    !GT (greater than), EQ (equal to), LT (less than) as
    !Ax >= b, Ax = b, Ax <= b; ALL b(i) VALUES MUST BE POSITIVE!
    PRINT"SIMPLEX SOLUTION OF LINEAR PROGRAMMING PROBLEM"
    PRINT"Input data;Constraint Inequalites in the order Ax >=,=,<= b"
    PRINT"In output tableau, * denotes non-basic variable"
    PRINT
    FOR i = 1 to M
        FOR j = 1 to N
            READ A(i,j)
            PRINT"A(";Str$(i);",";Str$(j);")=";A(i,j);" ";
        NEXT j
        PRINT
    NEXT i
    PRINT
    !Data for A(i,j):GT rows of >= EQ rows of =, LT rows of <=.
    DATA 1,1,1,1                  !***,etc.
    DATA .2,1,.5,.7
    DATA .05,.06,.03,.03
    DATA 2,3,1,2

    FOR i = 1 to M
        READ b(i)
        IF b(i) < 0 THEN PRINT"ERROR-b(i) must be positive!"
        PRINT"b(";Str$(i);")=";b(i);" ";
    NEXT i
    PRINT
    PRINT
    !Data for b(i)
    DATA 1,.6,.05,2               !***
    !Objective function is z = sum [c(j)*x(j)]
    PRINT"c(i) are original coefficients in objective function"
    FOR j = 1 to N
        READ c(j)
        PRINT"c(";Str$(j);")=";c(j);" ";
    NEXT j
    PRINT
    PRINT
    !Data for c(j) objective function
    DATA 30,15,25,20              !***
END SUB
```

## *General Direct-Search Optimization Program OPTSEARCH*

This program conducts a systematic search in a given coordinate region to find the minimum of a given function subject to various inequality or equation constraints. Instructions for running the program are given in the concise listing. The program produces *exact* results for an integer or discrete range of the independent variables, and approximate results for a continuous range. This program is very unusual in that it requires the user to specify some statements in the main program, as discussed in the concise listing. The problem shown is from Equations (8-25).

### Program OPTSEARCH.

```
!This is program OPTSEARCH by O.T.Hanna for the approximate
!solution of general constrained minimum problems by a direct
!discrete search procedure described by Conley.
!n is the number of variables x(i), m is the number of
!(inequality or equality) constraints of the form g(i) (<,=,or >) 0.
!XL(i),XU(i) are the lower and upper bounds for the variables X(i).
!Z is the objective function which is to be minimized.
!max(i) is the number of parts into which interval for
!variable (i) is divided.
!***Denotes user-specified statement
!****************************************************************
!*To Run: In main program, set up FOR/NEXT loops for the n      *
!*variables; set up infeasible constraint conditions for the    *
!*m constraints; use generic forms shown.At end of program,user *
!*must give information to Subroutines Init, Constraints        *
!*and Function. In Sub Init specify the initial Zmin,           *
!*the no. of variables n, the no. of constraints m, the         *
!*upper and lower limits XU(i), XL(i) for variables and loop    *
!*limits imax(i) for variable (i). In Sub Function specify the  *
!*objective function to be minimized (Z). In Sub Constraints    *
!*specify m constraint functions of the type g(i) (<,>,=) 0.    *
!*Specification of the sense of the constraints is indicated    *
!*by user in main program.                                      *
!****************************************************************

!Start of Main Program-user must specify statements where !***
!Here we show the parts of main program user must specify.

FOR i1 = 0 to imax(1)                  !***one loop for each of n variables
    LET X(1) = XL(1) + incr(1)*i1      !imax(i) are positive integers
FOR i2 = 0 to imax(n)
    LET X(n) = XL(n) + incr(n)*i2

!Put in infeasible constraint conditions
IF g(1) > 0 THEN LET Com = 0           !***make proper choice of <,> or =
IF g(m) <> 0 THEN LET Com = 0          !***one for each of m constraints
```

```
NEXT i2                                      !***close loops on x(i) variables
NEXT i1                                      !in reverse order

!End of Main Program

SUB Init( Zmin_init,m,n,XL(),XU(),imax() )
!***Denotes user-specified statement
LET Zmin_init = 1e10          !***Zmin initial
LET m = 2                     !***No. of constraints
LET n = 2                     !***No. of variables
!For constant XL(i),XU(i),or imax(i) use loop
MAT REDIM XL(n),XU(n),imax(n)
!Now define limits on variables and imax
LET imax(1) = 10              !***max. loop index for variable X(i)
LET imax(2) = 9
LET XL(1) = 0                 !***lower limit on variable X(i)
LET XL(2) = 1
LET XU(1) = 10                !***upper limit on variable X(i)
LET XU(2) = 10
END SUB

SUB Constraints( X(),g(),m )
!g(i) are m constraint functions ( g(i) <,>, ) = 0
!***Denotes user-specified statement
LET g(1) = X(1) + 2/X(2) - 50       !***constraints,
LET g(m) = X(1) + X(2) - 10
END SUB

SUB Function( X(),Z )
!Z is objective function. to be minimized
!***Denotes user-specified statement
LET Z = 2*X(1)^3 + 3*X(2)^2         !***objective function
END SUB
```

## PROBLEMS

**8.1.** Find the stationary points of the function $f(x) = 3x^4 - x^3 + 10x$ and determine if they correspond to maxima, minima, or saddle points. Calculate accurately any maximum or minimum values of $f(x)$, along with the values of $x$ where they occur.

**8.2.** To gain some experience using the simple search technique in optimization, use program OPTSEARCH to verify *directly* that the optimum found in Example 8.1 is indeed a minimum. To do this, conduct the search in the vicinity of the known stationary point (and at the stationary point) to see if the results do indeed stay where the minimum is thought to be located.

**8.3.** It is desired to find the optimal dimensions of a cylindrical container of fixed volume $V$ which corresponds to the minimum cost of production. The container has no top, and the production costs include a materials cost $M$ (\$/area) and a "welding" cost $W$

**TABLE 8.4**

| Petroleum Component | Max. quantity available (barrels per day) | Selling Price (per barrel) |
|---|---|---|
| 1 | 1000 | $10 |
| 2 | 2000 | $15 |

**TABLE 8.5**

| Grade of Gasoline | Product Specification | Price (per barrel) |
|---|---|---|
| A | $\leq$ 35% of 1<br>$\geq$ 70% of 2 | $16 |
| B | $\leq$ 60% of 1<br>$\geq$ 30% of 2 | $14 |

($/length of weld) which is incurred in fastening the circular bottom to the cylindrical sides. Express the solution in terms of the problem parameters.

(i) Solve by using the constraint to eliminate a variable.

(ii) Solve by using the Lagrange multiplier method.

(iii) If $V = 10\ \text{in}^3, M = \$1/\text{in}^2$, and $W = \$0.30/\text{in}$, determine the optimal ratio of height to radius.

**8.4.** Consider the following three-dimensional function $Z(x_1, x_2, x_3)$ for positive values of $x_1, x_2$, and $x_3$.

$$Z = 4x_1^2 x_2 + \frac{2}{x_1 x_2 x_3} + 3x_1 x_3^2 + 5x_2$$

We wish to minimize this unconstrained function. To accomplish this, proceed as follows: First use program OPTSEARCH (several times, if necessary) to attempt to find reasonable starting values, as well as an estimate of the minimum; then use program OPTFRPR to refine these values.

**8.5.** Solve for $(x_1, x_2, x_3)$ at the stationary points in problem 8.4 corresponding to $\partial Z/\partial x_i = 0$. To do this, use program NEWTONVS and transfer the required partial derivatives directly from program OPTFRPR in problem 8.3. Look for other stationary points (including negative ones) by using different starting values.

**8.6.** Look for the *maximum* of the unconstrained function $Z$ in problem 8.4 by minimizing $Y = -Z$ using program OPTFRPR.

**8.7.** Find the solution of the following linear programming problem:

$$\text{Max}\, Z = 3x_1 + 6x_2 + 2x_3 \qquad \text{subject to}$$

$$3x_1 + 4x_2 + x_3 \leq 20$$

$$x_1 + 3x_2 + 2x_3 \leq 10$$

$$x_1, x_2, x_3 \geq 0$$

**8.8.** A refinery is blending two petroleum components (1 and 2) to form two different gasoline products, A and B. The maximum available quantities of components 1 and 2, along with their cost, is shown in Table 8.4.

The product specifications and prices charged for products A and B are shown in Table 8.5.

(i) By taking as variables

$$y_1 = \text{amount of 1 in product A}$$
$$y_2 = \text{amount of 2 in product A}$$
$$y_3 = \text{amount of 1 in product B}$$
$$y_4 = \text{amount of 2 in product B}$$

show that the problem formulation for maximum profit leads to the standard linear program

$$\text{Max } [6y_1 + y_2 + 4y_3 - y_4] \quad \text{subject to}$$
$$y_1 + y_3 \leq 1000$$
$$y_2 + y_4 \leq 2000$$
$$0.65y_1 - 0.35y_2 \leq 0$$
$$0.3y_2 - 0.7y_1 \geq 0$$
$$0.4y_3 - 0.6y_4 \leq 0$$
$$0.7y_4 - 0.3y_3 \geq 0$$
$$y_1, y_2, y_3, y_4 \geq 0$$

(ii) Solve this linear programming problem for the maximum profit and the amounts of 1 and 2 in products A and B.

**8.9.** The nonlinear objective function of two variables $f = x_1^2 + 4x_2^2$ is to be maximized subject to the constraints $x_1 + x_2^3 \leq 20$ with $x_1, x_2 \geq 0$. Find the optimal values of $f, x_1$, and $x_2$. Is the constraint binding at the optimum? Formulate the necessary conditions for the optimum using both the direct algebraic elimination and the Lagrange multiplier method.

(i) Obtain an approximate solution to the problem using program OPTSEARCH (several times, if necessary).

(ii) Use the necessary conditions to solve for an accurate solution. What are the advantages and disadvantages of the various approaches?

**8.10.** Consider the function $Z(x_1, x_2, x_3)$ defined in problem 8.4. We now impose the inequality constraint

$$h(x_1, x_2, x_3) = x_1^2 + x_2^2 + x_3^2 - 0.5 \leq 0$$

Find the minimum of this constrained problem. To accomplish this, first use program OPTSEARCH to find approximate values of the minimum and the corresponding $x_i$. Then use the Lagrange multiplier method together with the NEWTONVS program to get a more accurate solution.

## REFERENCES

BEVERDIGE, G. S. G. and SCHECHTER, R. S. (1970), *Optimization: Theory and Practice*, McGraw-Hill, New York

BUNDAY, B. D. (1984), *Basic Linear Programming*, Arnold, Baltimore

BUNDAY, B. D. (1984), *Basic Optimisation Methods,* Arnold, Baltimore

CONLEY, W. (1984), *Computer Optimization Techniques*, Petrocelli, New York

COOPER, D. and STEINBERG, D. (1970), *Introduction to Methods of Optimization*, Saunders, Philadelphia

DANTZIG, G. B. (1963), *Linear Programming and Extensions,* Princeton Univ. Press, Princeton, NJ

DENN, M. M. (1969), *Optimization by Variational Methods*, McGraw-Hill, New York

DENNIS, J. E., Jr., and SCHNABEL, R. B. (1983), *Numerical Methods for Unconstrained Optimization and Nonlinear Equations*, Prentice-Hall, Englewood Cliffs, NJ

EDGAR, T. F. and HIMMELBLAU, D. M. (1988), *Optimization of Chemical Processes,* McGraw-Hill, New York

GOTTFRIED, B. S. and WEISMAN, J. (1973), *Introduction to Optimization Theory*, Prentice-Hall, Englewood Cliffs, NJ

HILLIER, F. S., and LIEBERMAN, G. J. (1967), *Introduction to Operations Research*, Holden-Day, San Francisco

HIMMELBLAU, D. M. (1972), *Applied Nonlinear Programming*, McGraw-Hill, New York

JELEN, F. C. (1970), *Cost and Optimization Engineering*, McGraw-Hill, New York

PRESS, W. H., FLANNERY, B. P., TEUKOLSKY, S. A., and VETTERLING, W. T. (1986), *Numerical Recipes*, Cambridge Univ. Press, New York

REKLAITIS, G. V., RAVINDRAN, A., and RAGSDELL, K. M. (1983), *Engineering Optimization*, Wiley, New York

SPROTT, J. C. (1991), *Numerical Recipes Routines and Examples in BASIC*, Cambridge Univ. Press, New York

WILDE, D. J., and BEIGHTLER, C. S. (1967), *Foundations of Optimization*, Prentice-Hall, Englewood Cliffs, NJ

# 9

# Evaluation of Integrals

The mathematical problem of the evaluation of integrals occurs very often in engineering applications. Applications are especially important in the areas of ordinary and partial differential equations, as well as statistics. The problem of evaluation of integrals is a very fundamental one, since there is no chain rule for integration. Thus, for example, even the integral of the simple function $f(x) = \exp(-x^2)$ cannot be expressed as a finite combination of elementary functions. For this reason, it is necessary to devise approximate analytical and numerical processes for the evaluation of integrals that are efficient and for which the accuracy can be assessed. The term "quadrature" we will use in our discussion refers to the evaluation of a definite integral, as compared to the term "integration," which usually refers to solution of a differential equation. In some cases this is redundant, but usually a differential equation cannot be solved by quadrature.

The subject of quadrature is very large and many different techniques, both analytical and numerical, have been proposed to solve this problem. Here we will present various procedures that we consider to have a combination of simplicity, efficiency, and practical value. We begin by first discussing numerical techniques, and then exploring approximate analytical methods.

## 9.1 NUMERICAL QUADRATURE: THE TRAPEZOIDAL RULE WITH RICHARDSON EXTRAPOLATION

The trapezoidal rule for the evaluation of an integral involves replacing the integrand by a straight-line segment between the end points of the interval, and using the

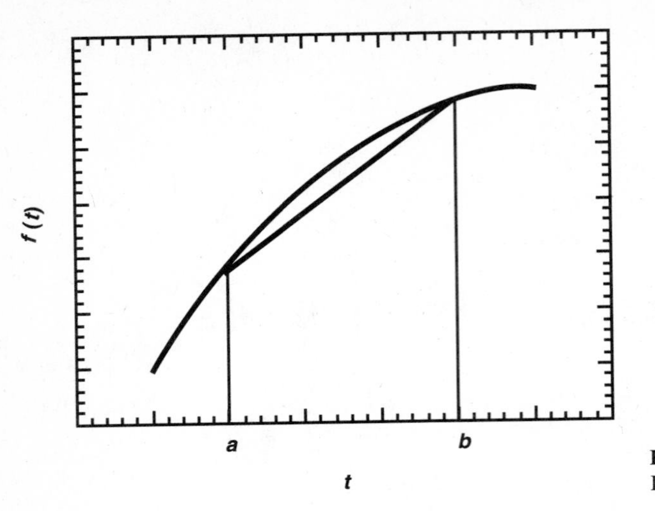


**Figure 9.1.** Crude Trapezoidal Rule Integration

area of the trapezoid formed in this way as an approximation to the true area (see Figure 9.1). For a given integral, the accuracy of this process can generally be increased by dividing the interval of integration into a larger number of trapezoids, each having the same width (see Figure 9.2). For example, if a single trapezoid is used, we would have

$$\int_a^b f(t)\,dt \cong [(b-a)/2] \cdot (f(a)+f(b))$$

The foregoing procedure would appear to be too crude to be useful in practice. However, when modified by an appropriate *Richardson extrapolation* process, the trapezoidal rule is probably the most practical all-purpose numerical quadrature method. The extrapolation improvement is the key to both efficiency and error estimation. We should mention that the result of using trapezoidal integration with a single extrapolation gives the well-known *Simpson's rule,* which corresponds to fitting the integrand with a quadratic function.

The algorithm for trapezoidal quadrature with Richardson extrapolation is based on the Euler-Maclaurin summation formula, which is derived in detail in Appendix F. The final result is repeated here for convenience.

$$\int_a^b f(x)\,dx = h\left[\frac{f(a)}{2} + f(a+h) + f(a+2h) \cdots + f(b-h) + \frac{f(b)}{2}\right] - \frac{h^2[f'(b)-f'(a)]}{12} + \frac{h^4[f'''(b)-f'''(a)]}{720} - \frac{h^6[f^v(b)-f^v(a)]}{30240} + R \tag{9-1}$$

In Equation (9-1), the trapezoidal approximation is on the right-hand side of the top line. The remainder $R$ in this equation is $O(h^8)$, which would generally be *very* small for small $h$ (except in some unusual cases, which we will discuss). Thus, the Euler-Maclaurin formula is generally extremely useful for the evaluation

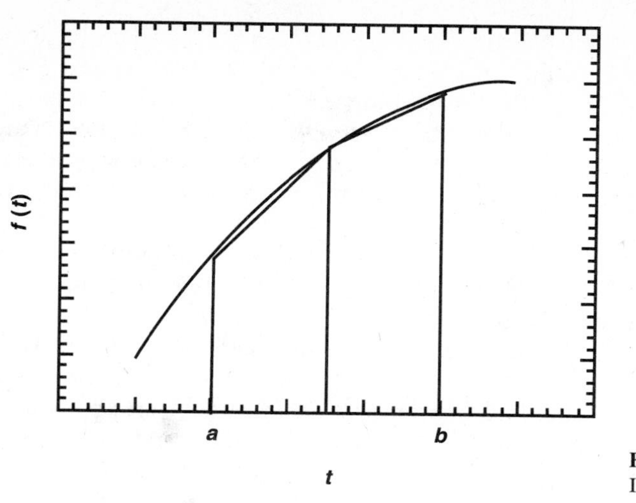


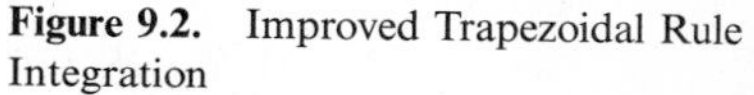

**Figure 9.2.** Improved Trapezoidal Rule Integration

of integrals, when at least some of the correction terms (second line of Equation (9-1)) are included. However, the correction terms depend on various derivatives of the function, which would be awkward (or nearly impossible for a data-defined function) to calculate. The saving feature here is that the correction terms above can be written in the form $I = A(h) + K_1 h^p + K_2 h^q + O(h^r)$, where $I$ is the unknown integral, $A(h)$ is the trapezoidal approximation for spatial increment $h$, $p = 2$, $q = 4$, and $r = 6$. The (unknown) constants $K_1$ and $K_2$ are *independent of* h. This formula is precisely of the type that is appropriate for repeated Richardson extrapolation, as discussed in detail in Appendix E. Thus, for example, in the case of a single trapezoidal quadrature, we would directly determine the two values $A(h/2)$ and $A(h)$ and use these values to calculate an estimate of the first correction term $K_1 h^2$. This would yield the result

$$I = A(h/2) + \frac{A(h/2) - A(h)}{2^2 - 1} + O(h^4) \tag{9-2}$$

Equation (9-2) shows that the once-extrapolated approximation is in error by $O(h^4)$, which is generally much more accurate than the $O(h^2)$ accuracy of the unextrapolated trapezoidal rule. Also, the quantity $[A(h/2) - A(h)]/3$, which is an estimate of the quantity $K_1 h^2$, can be taken as a conservative estimate of the unknown error, provided that $h$ is sufficiently small. This is because the $O(h^2)$ estimate will be large compared to the actual error, which is $O(h^4)$ under these circumstances.

As shown in Appendix E, repeated Richardson extrapolation can be accomplished for an equation like Equation (9-1). The result of this is to generally produce even more accuracy, with corresponding error estimates. In connection with this matter, we have found that a very good algorithm involves just two extrapolations, using the trapezoidal approximations $A(h/4)$, $A(h/2)$, and $A(h)$. This allows a simple calculation that gives quadrature results with error $O(h^6)$, and a conservative error estimate, which is $O(h^4)$. Moreover this is the simplest calculation that allows determination of the normalized "error ratio," which can be shown to be a useful reliability indicator for the extrapolation process. This ratio is defined as Ratio = $E(h/2, h)/[4E(h/4, h/2)]$. When the error ratio (which normally approaches unity as $h \rightarrow 0$) is between 0.8 and 1.2, we can be reasonably confident that the unknown actual error is smaller than the estimated error. For further information on the error ratio, see the chapter on numerical integration of differential equations (initial value problems).

The procedure for evaluating integrals as described above is employed in computer programs IEX and TRMPQUAD. Program IEX is set up to integrate ordinary differential equations and is discussed in that chapter. However, it contains a very convenient option for *evaluating integrals of analytically defined functions,* which is discussed at the end of this chapter. Program TRMPQUAD, which is also described at the end of this chapter, is employed to determine the integral of the approximating function in the data-fitting programs RATINTG, LSTSQGR, and RATLSGR. It should also be mentioned that, for convenience, a facility for the numerical evaluation of integrals of analytically defined functions is contained in programs CUBSPL and RATLSGR. Program LSTSQGR can also be used for this purpose

The trapezoidal rule together with Richardson extrapolation represents a very simple, effective procedure for increasing the accuracy and estimating errors of numerical quadrature. This method allows us to get good results at a relatively low cost, since we can normally use a larger step size $h$ and fewer function evaluations. This characteristic also helps avoid roundoff error, which could be incurred if $h$ is too small. A few words of caution must be mentioned regarding Richardson extrapolation. The procedure, as it is implemented here, assumes that $f(t)$ has bounded derivatives within the region of integration. If $f(t)$ has *unbounded* derivatives at some point, difficulties can appear. In particular, for functions involving fractional powers in a range including zero, problems are often encountered. In such a case, it is usually best to change variables so as to eliminate the fractional power. We will illustrate this behavior in an example. Another situation where extrapolation does not help is when the term being estimated happens to be exactly zero.

**Example 9.1 Evaluation of $I = \int_0^1 \frac{x^{1/3}}{1+x}\,dx$**

This integral is a typical example of an integrand containing a fractional power in a range including $x = 0$. Since the function being integrated will therefore have some very large derivatives, we anticipate that the trapezoidal rule with extrapolation will encounter some difficulties. We therefore choose to eliminate the fractional power through the change of variable $u = x^{1/3}$. This yields the new form of the integral

$$I = \int_{u=0}^{1} \frac{3u^3}{1+u^3}\,du \tag{a}$$

The integrand in Equation (a) does *not* contain any fractional powers, and its derivatives vary smoothly with $u$. We do not expect this form of the integral to cause any problems. To see the performance of the trapezoidal rule with extrapolation on the

**TABLE 9.1** EVALUATION OF TWO FORMS OF AN INTEGRAL BY TRAPEZOIDAL RULE WITH EXTRAPOLATION

$$I = \int_{u=0}^{1} \frac{3u^3}{1+u^3}\, du$$

| $h$ | $I$ | Estimated Relative Error | Ratio |
|---|---|---|---|
| 0.20 | 0.493053 | $4.1 \times 10^{-7}$ | 1.007 |
| 0.10 | 0.493053 | $1.6 \times 10^{-8}$ | 1.002 |
| 0.01 | 0.493053 | $2.6 \times 10^{-12}$ | 1.00002 |

$$I = \int_{0}^{1} \frac{x^{1/3}}{1+x}\, dx$$

| $h$ | $I$ | Estimated Relative Error | Ratio |
|---|---|---|---|
| 0.1000 | 0.492153 | $5.2 \times 10^{-5}$ | 0.625 |
| 0.0100 | 0.493012 | $2.4 \times 10^{-6}$ | 0.630 |
| 0.0010 | 0.493052 | $1.1 \times 10^{-7}$ | 0.630 |
| 0.0001 | 0.493053 | $5.1 \times 10^{-9}$ | 0.630 |

two different forms of the integral, we use program IEX to obtain the results shown in Table 9.1.

In the first part of Table 9.1, we see excellent behavior as evidenced by *ratios approaching unity* and very small estimated errors. Moreover, for *all* step sizes $h$ (including the very large one, $h = 0.2$), the estimated value of $I$ agrees to six significant figures. In the second part, by contrast, the ratios are *not* approaching unity, but rather 0.63, which is *not* in accordance with postulated extrapolation behavior. Also, in this second case, we see that we must go to very small increments $h$ (and hence high computing costs) before we finally seem to achieve a convergence to the answer obtained so easily by the first form.

Another important point to note relates to the estimated errors using the two different forms of the integrals. In the first form, the results suggest that the error estimates are conservative, as desired. But for the second form, since we know that $I = 0.493053$, correct to six figures, we can see in fact that the estimated errors are *not* conservative, and hence they could be very misleading if we only used the second form and did not compute the error ratio values. This example points up dramatically the value of the error ratio quantity in assessing the reliability of the extrapolation results. We also see the great value of removing the fractional power by means of the change of variable.

Further examples of numerical quadrature applied to more difficult problems will be given later in the chapter.

### 9.1.1 Dealing with Singularities

A fairly common problem in numerical quadrature involves integrands that contain an apparent singularity within the interval of integration. In such cases, after we have established that the integral does indeed exist, the problem is to find a convenient numerical procedure to evaluate the integral and monitor the error. The computer programs already mentioned will in general do the job nicely, *provided that the problem has been properly prepared.* The main tools available for problem preparation include changing variables and subtraction of the singularity. We illustrate the technique with an example.

**Example 9.2 Dealing with Singularities**

Consider numerical evaluation of the following integral:

$$J = \int_0^1 \frac{\sin t}{t^{3/2}}\, dt \tag{a}$$

The possible singular behavior occurs here at $t = 0$. Since $\sin t \approx t$ near $t = 0$, we see that the integrand behaves as $t^{-1/2}$ in this vicinity, so the integral does exist. It is clear that if we use, for example, program IEX to solve this problem, we would fail by virtue of the nonexistence of the integrand at $t = 0$. We first demonstrate the use of subtraction of the singularity to solve this problem. The idea is to extract the "singular" behavior of the integrand and deal with it analytically. This leaves behind a "less-singular" portion which can be handled better by a computer program. In some problems it may be desirable to continue this process for several steps. For the problem we consider, the first step is to write the integral $(a)$ as

$$J = \int_0^1 \frac{(\sin t - t) + t}{t^{3/2}}\, dt \tag{b}$$

Now we can directly evaluate $\int t^{-1/2} dt$ and determine the remaining integral numerically. The new integrand $(\sin t - t)/t^{3/2}$ is well-defined at $t = 0$ (where it is zero) and we can apply a computer program successfully. In fact, as the reader can verify, program IEX handles the new problem quite well. However, it must be noted that since the new integrand varies as $t^{3/2}$ near $t = 0$, trapezoidal integration with extrapolation will be a bit "queasy" because of the singular derivatives in the Euler-Maclaurin formula. Thus, if we subtract the next term of the series for $\sin t(-t^3/3!)$, the resulting performance of program IEX is even better.

The idea illustrated above is important in many applications where no other technique is applicable. However, in the particular problem we have discussed, it must be noted that a simple change of variable, $(t^{3/2} = u^3)$, actually represents a neater approach. In this case, the original integral $J$ in Equation (a) becomes

$$J = 2\int_0^1 \frac{\sin u^2}{u^2}\, du \tag{c}$$

Since this integrand is well-defined at $u = 0$, and since there are no fractional powers in the integrand, a direct application of a program such as IEX produces excellent results.

### 9.1.2 The Midpoint Rule

There is another useful method for numerical quadrature that is closely related to the trapezoidal method, but has some complementary features. This method is known as the midpoint rule, since it is based on approximation of the integral $\int_a^{a+h} f(t)\, dt$ by the area of a rectangle whose top coincides with $f(t)$ at the midpoint of the horizontal width of the rectangle. Thus, the midpoint approximation is $h \cdot f(a + h/2)$. The midpoint rule for integration is illustrated in Figure 9.3. For a function that has a second derivative of uniform sign on the integration interval,

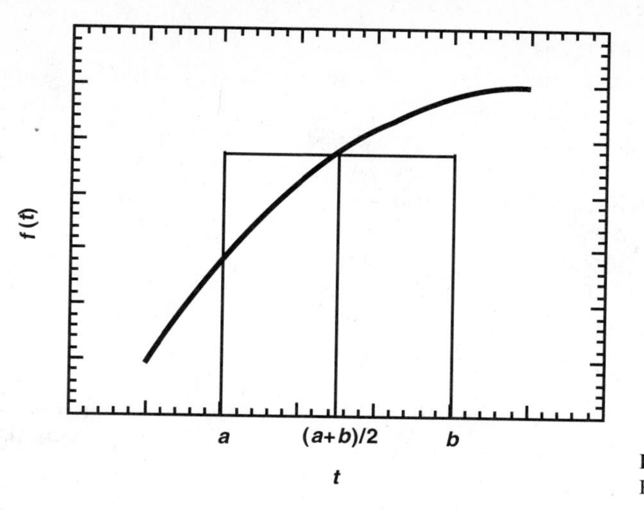


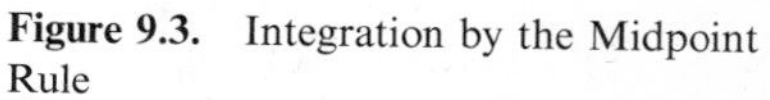
**Figure 9.3.** Integration by the Midpoint Rule

the midpoint and trapezoidal methods bracket the true unknown value. Also, there is an "Euler-Maclaurin"-type expansion that applies to midpoint quadrature which is very similar to that for trapezoidal quadrature, except that the correction terms have signs opposite to those for the trapezoidal case. Hence a Richardson extrapolation scheme of the same type is also applicable to the midpoint rule, and when $h$ is sufficiently small, corresponding approximations of the two schemes bracket the true result. A midpoint quadrature/extrapolation option is available in program TRMPQUAD.

The main practical benefit of the midpoint rule in quadrature is that it does *not* use function values from the ends of the interval. This may be convenient if one is dealing with a function that the computer has trouble evaluating at an endpoint (such as $\sin x/x$ on an interval including zero). However, one must be very *careful* in using the midpoint rule, since it will normally provide a finite answer for a problem having an actual singularity at an endpoint (even when the answer doesn't exist).

## 9.2 FUNCTIONS DEFINED BY DATA

It occasionally happens that it is necessary to evaluate the integral of a function which is not defined in terms of an analytical function, but rather by a set of discrete data. This could happen if the function comes from a discrete computation or perhaps from experimental data. In this situation, the approach to be taken depends on how much data are available, the spacing of the data, how much the function is varying, and whether or not it is assumed that the data are contaminated by experimental error.

We start with the most complicated case, namely that involving unequally spaced data contaminated with experimental error. In dealing with problems of experimental error contamination, it is useful to use least-squares approximation methods. We have already discussed this matter in the chapter on least squares, and we have used computer programs LSTSQGR and RATLSGR to fit data by least squares. Both of these programs contain the option of evaluating the integral of the approximating function; this is accomplished by the built-in subroutine TRMPQUAD referred to in the previous section. The spacing of the data can be arbitrary. In this kind of problem, we can't very well assess the error in the original least-square approximation, so this makes it difficult to assess error in the integral. Of course, it is easy to assess the error in the integral of the approximating function itself (and that feature is contained in the programs), but that is not the same thing as the error in the *unknown* function we are approximating.

Turning now to situations where there is assumed to be *no data noise,* we consider the case of unequally spaced data. In this situation, just as for interpolation, we could fit the data with a single polynomial (usually not so good), a cubic spline, or rational interpolation. We have discussed these techniques earlier and indicated computer programs for applying them (programs RATINTG for polynomials or rational approximation and CUBSPL for cubic splines). Of course, even though there is no data noise, it is sometimes desirable to fit the data using least squares (using program LSTSQGR or RATLSGR). All of these data-fitting programs contain a facility for calculating the integral of the approximation function. In the case of the cubic spline, the integration is exact; in the other cases, program TRMPQUAD is used, and this contains an error-monitoring capability which is explained in the concise listing. Thus, the above computer programs, all of which can be used for arbitrary spacing, allow estimation of the error of the integral of the *approximation,* and this can usually be made as small as necessary. However, what we really want in terms of error is an estimate in the error of the integral for the *unknown* function $y(x)$ when we approximate it with the interpolation or least-squares function $y_a(x)$. This error is expressed as

$$\text{Error} = \int_a^b y\,dx - \int_a^b y_a\,dx = \int_a^b (y - y_a)\,dx \tag{9-3}$$

Thus, by taking absolute values and using fundamental inequalities, we have

$$\left|\int_a^b (y - y_a)\,dx\right| \le \int_a^b |y - y_a|\,dx \le (b-a) \max_{a \le x \le b} |y(x) - y_a(x)| \tag{9-4}$$

This shows that the desired error is bounded by the length of the integration interval $(b-a)$ times the maximum difference between the unknown true function and the approximating function. As we mentioned in the chapter on interpolation, this difference can at least be estimated by the maximum absolute difference between two approximations differing by one data point. However, it is much easier for this purpose to use program RATLSGR, which automatically gives an approximate value of $\max |y(x) - y_a(x)|$ for the desired interval.

The final case of some interest corresponds to integration of *equally spaced* data. We could, of course, use the previous methods for this problem. However, in this situation, by using the trapezoidal rule with extrapolation, we can use the extrapolation procedure to estimate the error in the approximate value of the integral. Here, there are so many possibilities that there is not much point in developing a general computer program. The technique is best illustrated with an example.

**TABLE 9.2** THE DISCRETE FUNCTION $y(x)$

| $x$ | $y(x)$ |
|---|---|
| 0.00 | 1.000 |
| 0.25 | 0.667 |
| 0.50 | 0.586 |
| 0.75 | 0.536 |
| 1.00 | 0.500 |

**Example 9.3 Integration of Equally Spaced Data**

We wish to estimate the integral from zero to one of the discrete function $y(x)$ shown in Table 9.2.

We could, of course, fit this data by interpolation or least squares using programs RATINTG, CUBSPL, LSTSQGR, or RATLSGR, and compute an estimate of the integral directly from these programs. But with the possible exception of the case of rational least squares, this would not allow us to estimate the accuracy of the integral of the unknown function from which the data comes. Since we have equal spacing of the data, we can use trapezoidal quadrature with extrapolation to estimate the integral *and* the accuracy. We indicate below the details of such a hand calculation, where we carry only four significant figures. The notation $A(h/2, h)$ refers to a once-extrapolated value, and $A(h/4, h/2, h)$ is a twice-extrapolated value.

$$A(1) = 1(1/2 + 0.5/2) = 0.75$$

$$A(0.5) = 0.5(1/2 + 0.586 + 0.5/2) = 0.668$$

$$A(0.25) = 0.25(1/2 + 0.667 + 0.586 + 0.536 + 0.5/2) = 0.6347$$

$$A(0.5, 1) = 0.668 + (0.668 - 0.75)/3 = 0.668 - 0.0273 = 0.6407$$

$$A(0.25, 0.5) = 0.6347 + (0.6347 - 0.668)/3 = 0.6347 - 0.0111 = 0.6236$$

$$A(0.25, 0.5, 1) = 0.6236 + (0.6236 - 0.6407)/15 = 0.6236 - 0.00114 = 0.6225$$

The result of this calculation produces the estimate for the integral of $0.6225 \pm 0.00114$. The computed value of Ratio is $0.0273/[4(0.0111)] = 0.61$, which is not quite close enough to unity to be confident of the estimated accuracy. However, we should be very confident of at least two figures, that is, $I = 0.62$. This result is not bad at all, considering that we only had five data points to work with.

It is useful to note that we can produce the above results directly from computer program RATINTG *provided that in the integration option we specify the proper minimum number of intervals* (which in this example is one). This is because if we use exactly the same intervals in the computer calculation, the result is the same as using only the data. However, if we use more intervals than the data allows, we compute the integral based on some interpolated function values, which means we have assumed a "model" for the unknown function depending on which interpolation scheme we are using.

## 9.3 GAUSS QUADRATURE

A rather different numerical method that finds utility in certain applications is known as Gauss quadrature. This procedure has some deeper theoretical roots, but our purpose will be served if we present a simple discussion. We consider the generic case for the evaluation of $\int_{-1}^{1} F(x)\,dx$. Any integral with *finite* limits can be brought into this form by a suitable linear change of variable. Gauss methods of any order (in principle) can be developed for this integral. For illustration purposes, we will develop the two-point Gauss quadrature for this integral. For the simple algebraic approach we are taking, we express the integral as

$$\int_{-1}^{1} F(x)\,dx = w_1 F(x_1) + w_2 F(x_2) \tag{9-5}$$

The idea is to now require that Equation (9-5) be exact for the power functions $x^0 = 1$, $x$, $x^2$, and $x^3$. These requirements yield the following four equations for the Gauss points ($x_i$) and weights ($w_i$):

$$\begin{aligned} 2 &= w_1 + w_2 \\ 0 &= w_1 x_1 + w_2 x_2 \\ 2/3 &= w_1 x_1^2 + w_2 x_2^2 \\ 0 &= w_1 x_1^3 + w_2 x_2^3 \end{aligned} \tag{9-6}$$

These four nonlinear algebraic equations are particularly easy to solve. If we consider the second and fourth equations and subtract the first term on the right-hand side from each equation, and if we then divide the fourth equation by the second equation, we find $x_2 = -x_1$. Substitution into the second equation then gives $w_1 = w_2$. Then substitution into the first equation yields $w_1 = w_2 = 1$. Finally, substitution into the third equation gives $x_1 = -1/3^{1/2}$, and $x_2 = 1/3^{1/2}$. When we put these values of $w_i$ and $x_i$ back into Equation (9-5), we have the two-point Gauss quadrature formula for the integral. By virtue of the requirements it satisfies, this formula will be exact if $F(x)$ is any third-degree polynomial. Not surprisingly, the error of this approximation can be shown to be proportional to the maximum magnitude of the fourth derivative of $F(x)$ on the interval $(-1, 1)$. It is easy to verify for the integral considered above that *one-point* Gauss quadrature is exactly the same as the midpoint rule.

The calculations given above show the algebraic approach to Gauss quadrature. This procedure is, in a sense, the "most efficient" of any that can be devised for integrating polynomial functions, in the sense that it integrates any polynomial using the smallest possible number of quadrature points ($x_i$). It is thus well suited to applications where it is important to obtain high accuracy using relatively few points. Such an application occurs in the "method of weighted residuals" approach to the solution of differential equations, which we shall discuss later. One principal drawback to the Gauss procedure is that it is not easy to estimate the error of the approximation.

The illustration of two-point Gauss quadrature given above was especially simple, since we could get an explicit solution to the equations for the weights and quadrature points. It would be difficult to pursue the theory on this basis. Instead, the theory of Gauss quadrature is closely related to that of orthogonal polynomials, and this is discussed in Squire [1970] and Carnahan *et. al.* [1969]. The theory shows that the Gauss quadrature points, for integrals much more general than those considered above, are directly related to the zeroes of certain orthogonal polynomials. Tables of Gauss weights and quadrature points for various types of integrals are given in Abramowitz and Stegun [1965].

## 9.4 INFINITE INTEGRALS

Infinite integrals arise naturally as solutions to ordinary and partial differential equations, and they occur frequently in statistics and in the use of transform methods. They can often be artificially introduced into a problem where they do not occur naturally. In this chapter, we discuss some of the general properties of infinite integrals, together with certain basic computational procedures. Our main considerations include determination of whether an infinite integral converges, bracketing the value if it does exist, and finally, determining an accurate value together with an error estimate. We also introduce the gamma and beta functions, both of which are of utility in the evaluation of many infinite integrals.

If the limit $\lim_{N\to\infty} \int_a^N f(x)\,dx$ exists, then we call this quantity an infinite integral and denote it by the symbol $\int_a^\infty f(x)\,dx$. Similar considerations apply for cases where either the lower limit or both limits of integration approach unbounded values. When the limit exists, we say that the infinite integral in question converges. An infinite integral of the above type converges provided that the "tail" of the integral, $\int_N^\infty f(x)\,dx$, can be shown to be arbitrarily small for $N$ sufficiently large. In practical work, the procedures used to show convergence include knowledge of convergence behavior for certain simple functions, use of inequalities, and integration by parts. For example, if a function "tails off" for large $x$ as a constant divided by $x^{1+p}$, where $p > 0$, then the integral of the function converges at $\infty$. This is true, since we have

$$\int_N^\infty dx/x^{1+p} = 1/pN^p$$

and this goes to zero for $N \to \infty$ if $p > 0$.

**Example 9.4 Testing Convergence of Infinite Integrals**

We consider several illustrations. A very simple one is $\int_0^\infty 1/(1+x^3)\,dx$. Here, we can immediately see that this integral converges since the integrand is less than $1/x^3$ and the tail of this converges. A slightly more difficult case is

$$\int_0^\infty \frac{\sin x}{1+x^3}\,dx \tag{a}$$

We take absolute values and use $|\sin x| \le 1$ to show that the tail of the integral in Equation (a) is bounded by the integral of $1/x^3$, which converges. A more subtle situation is given by the integral

$$\int_0^\infty \sin x^2\,dx \tag{b}$$

In this case, if we proceed as in the previous example, we see that the tail is bounded by $\infty$, which is not a useful result. This example will show that the integrand does *not*

**TABLE 9.3** THE GAMMA AND BETA FUNCTIONS

| The Gamma Function |
|---|
| $\Gamma(x+1) \equiv \int_0^\infty t^x e^{-t}\,dt \quad (x > -1)$ <br> $\Gamma(x+1) = x\Gamma(x) \quad (x > 0)$ <br> $\Gamma(x+1) = (2\pi x)^{1/2} x^x e^{-x}[1+R]; \quad \lvert R\rvert < 1/12x$ |
| **The Beta Function** |
| $B(m,n) = \int_0^1 t^{m-1}(1-t)^{n-1}\,dt = 2\int_0^{\pi/2} \sin^{m-1}\theta\cos^{n-1}\theta\,d\theta \quad (m,n > 0)$ <br> $B(m,n) = \dfrac{\Gamma(m)\Gamma(n)}{\Gamma(m+n)}$ <br> $\int_0^\infty \dfrac{t^p}{(1+t^q)^r}\,dt = \dfrac{1}{q}B\left[\dfrac{rp-p-1}{q}, \dfrac{p+1}{q}\right]$ |

need to approach zero in order to have convergence. We need something more delicate to show convergence; first, we change variables according to $x^2 = u$, and then we use integration by parts (integrating $\sin u$) to get

$$\int_N^\infty \sin x^2\,dx = \int_{N^2}^\infty \frac{\sin u}{2u^{1/2}}\,du = -\frac{\cos u}{2u^{1/2}}\Bigg|_{u=N^2}^{\infty} - \int_{N^2}^\infty \frac{\cos u}{4u^{3/2}}\,du = \frac{\cos N^2}{2N} - \int_{N^2}^\infty \frac{\cos u}{4u^{3/2}}\,du \quad \text{(c)}$$

Since $\lvert\cos u\rvert \le 1$, the first term on the right of Equation (c) $\to 0$ for $N \to \infty$, and for the same reason we see that the integral on the right also goes to zero. Thus, the tail of the original integral goes to zero for $N \to \infty$, so the integral converges and has a finite value.

### *9.4.1 The Gamma and Beta Functions*

In dealing with infinite integrals two functions, the gamma and beta functions, are especially useful. We will indicate some of the most important properties of these functions and illustrate how they may be used. The gamma function in particular is very important in the topic of asymptotic expansion of integrals, which will be addressed later in the chapter. The main properties of these two functions (which are defined by definite integrals) are shown in Table 9.3.

The property $\Gamma(x+1) = x\Gamma(x)$ is obtained through integration by parts. When this is applied recursively with $x$ an integer $n$, we see that $\Gamma(n+1) = n!$, so that we may regard the gamma function as a generalization of the factorial function. The third relationship involving the gamma function in Table 9.3 is the famous Stirling formula; this is applicable to large factorial calculations when $x$ is an integer. The gamma function is tabulated in many places, including Abramowitz and Stegun [1965]. The beta function is very important and is related simply to the gamma function by means of $B(m,n) = \Gamma(m)\Gamma(n)/\Gamma(m+n)$.

**Example 9.5 Evaluation of $I = \int_0^\infty dx/(1+x^3)$**

This simple problem (which comes from a problem in convective transport) can be used to illustrate several approaches. First, regardless of what method we use to evaluate $I$, it is useful to *bracket* its value by determining simple upper and lower bounds. This is always a good idea, since one of the most common errors in using a working computer program is *programming the wrong function and not realizing it*. The simple bracketing helps uncover such errors. In this problem, it is tempting to simply get an upper bound to $I$ by making the denominator smaller through deleting one of the two terms. But if we try this, whichever term is deleted causes the remaining integral not to converge, either at 0 or $\infty$. Thus, what we need to do is to *split* the range of integration and employ appropriate inequalities in each part as follows:

$$I = \int_0^\infty \frac{dx}{1+x^3} = \int_0^1 \frac{dx}{1+x^3} + \int_1^\infty \frac{dx}{1+x^3} < \int_0^1 dx + \int_1^\infty \frac{dx}{x^3} = 1 + \frac{1}{2} = \frac{3}{2} \tag{a}$$

Here, we have deleted the $x^3$ term in the first part and unity in the second part. A lower bound can be calculated by splitting in the same way.

$$I = \int_0^\infty \frac{dx}{1+x^3} = \int_0^1 + \int_1^\infty > \int_0^1 \frac{dx}{1+1} + \int_1^\infty \frac{dx}{x^3(1+1/x^3)} >$$
$$\int_0^1 \frac{dx}{2} + \int_1^\infty \frac{dx}{x^3(1+1)} = \frac{1}{2} + \frac{1}{4} = \frac{3}{4} \tag{b}$$

We finally get $3/4 < I < 3/2$. Thus, if any method produces a value outside this range, we can be certain something is wrong, since it is easy to verify the correctness of our *simple* calculations. To evaluate $I$, we can observe from Table 9.2 that $I$ can be expressed in terms of a beta function with $p = 0$, $q = 3$, and $r = 1$. This gives $I = B(2/3, 1/3)/3 = \Gamma(2/3)\Gamma(1/3)/3$ (note that $\Gamma(1) = 1$). By using gamma values from Abramowitz and Stegun [1965], we obtain, to six figures, $I = 1.20920$. Note that this value is nearly halfway between the simple upper and lower bounds computed earlier.

It is also interesting to evaluate $I$ numerically by using the trapezoidal rule with extrapolation in program IEX. Since we can't specify in the program an *infinite* upper limit, we would often have to calculate an upper bound to the tail of the integral in order to see how far out we need to integrate. But fortunately, our present problem is amenable to a simple change of variable. We simply use the new variable $u = 1/x$ in the portion of the integral from 1 to $\infty$; this gives a new integral with the finite limits of 0 to 1. Then we can combine the two integrals into a single integral with limits from 0 to 1.

$$I = \int_0^\infty \frac{dx}{1+x^3} = \int_0^1 \frac{dx}{1+x^3} + \int_{u=1}^0 \frac{-u^{-2}\,du}{1+u^{-3}} = \int_0^1 \frac{(1+x)\,dx}{1+x^3} \tag{c}$$

When we calculate this finite integral by using program IEX, we get the value $I = 1.20920$ to six figures, with the error estimated to be less than $10^{-8}$. This is exactly the same as the value obtained using gamma functions.

We should also note that it is easy to obtain the value of the integral in this problem by using Taylor series methods, using two different series for the two parts of the interval from 0 to 1 and 1 to $\infty$. In our further discussions of evaluation of integrals, we shall have occasion to consider other infinite integrals.

## 9.5 APPROXIMATE ANALYTICAL EVALUATION

If an integral to be evaluated is simply an unknown number, numerical quadrature usually represents an entirely satisfactory approach. However, often the desired integral depends on one or more parameters (e.g., upper limit of integration or parameters of integrand). In such a case, the value of the integral is really an unknown *function,* and it is very beneficial to express it as an analytic function of the parameters. We have already considered previously two basic methods that can be used for approximation of integrals. These are Taylor series expansion, together with acceleration using Euler's transformation, Padé approximation, or Shanks transformation; and a combination of numerical quadrature followed by fitting the numerical results using interpolation methods or least squares. Rational least squares and program RATLSGR are especially useful for this purpose.

**Example 9.6 Approximate Evaluation of $K=\int_0^x t\,dt/(a^3+t^3)\,dt$**

We wish to obtain approximations that will be useful for $x \to 0$ or $\infty$ with fixed $a$ and for $a \to 0$ or $\infty$ with fixed $x$. This problem is cited in Goodwin [1961] as a typical example of an integral which can be expressed exactly, but may be easier to compute approximately for various values of $a$ and $x$. The integral involves two parameters, $x$ and $a$. By rescaling according to $t = au$, we find that the integral is expressed as

$$K = \frac{1}{a}\int_{u=0}^{x/a} \frac{u}{1+u^3}\,du \tag{a}$$

Thus, if we take $z = x/a$, the new representation of $K$ involves only the *single parameter* z in the integral, and we define for convenience $K = L/a$, where $L$ is the integral in Equation (a). For "small" values of $z$, expansion of the denominator as $1/(1+u^3) = 1 - u^3 + u^6 \ldots$, and term-by-term integration gives

$$L = \frac{z^2}{2} - \frac{z^5}{5} + \frac{z^8}{8} - \frac{z^{11}}{11} + \frac{z^{14}}{14} \cdots = z^2\left[\frac{1}{2} - \frac{z^3}{5} + \frac{z^6}{8} - \frac{z^9}{11} + \frac{z^{12}}{14} \cdots\right] \tag{b}$$

From the nature of the geometric expansion, the error in Equation (b) is less in magnitude than the first neglected term. It is clear that the convergence in Equation (b) will be rapid only provided $z$ is small (which corresponds to either $x$ small or $a$ large), and the series only converges for $z \le 1$. Thus, it may be useful to attempt to extend the accuracy and region of applicability of Equation (b) by using Padé approximation (see Chapter 4 on summation and acceleration), and this is the reason for factoring out $z^2$ on the right side. The term in brackets on the right-hand side of Equation (b) is now expressed in Padé form (retaining the five terms shown) by using computer program PADE, and $L$ is computed. The results of this calculation are shown below and in Table 9.4.

$$L = z^2\left[\frac{p_0 + p_1 x + p_2 x^2}{1 + q_1 x + q_2 x^2}\right] \quad (x = z^3) \tag{c}$$

$$p_0 = 0.5 \qquad p_1 = 0.371429 \qquad p_2 = 0.0262987$$
$$q_1 = 1.14286 \qquad q_2 = 0.259740$$

We are also interested in an approximation for $L$ in the case of large $z$. To pursue this, we see by reference to Equation (a) that the integral converges at $\infty$ (since the integrand tails off like $1/u^2$). Thus, we write for large $z$ that $L = \int_0^\infty - \int_z^\infty$, where the integrand

**TABLE 9.4** EVALUATION OF $L = \int_0^z \frac{u}{1+u^3}\,du$ FOR SMALL $z$

| $z$ | $L_{\text{numerical}}$ | $L_{\text{Padé}}$, Equation (c) | $L_{\text{Series}}$, Equation (b) |
|---|---|---|---|
| 0.0 | 0.00000 | 0.00000 | 0.00000 |
| 0.2 | 0.01994 | 0.01994 | 0.01994 |
| 0.5 | 0.11920 | 0.11920 | 0.11920 |
| 1.0 | 0.37360 | 0.37360 | 0.40550 |
| 1.5 | 0.58500 | 0.59110 | 1.58000 |

**TABLE 9.5** EVALUATION OF $L = \int_0^z \frac{u}{1+u^3}\,du$ FOR LARGE $z$

| $z$ | $L_{\text{numerical}}$ | $L_{\text{Padé}}$, Equation (f) | $L_{\text{Series}}$, Equation (e) |
|---|---|---|---|
| $\infty$ | 1.29020 | 1.29020 | 1.29020 |
| 10 | 1.109 | 1.109 | 1.109 |
| 5 | 1.010 | 1.010 | 1.010 |
| 2 | 0.7238 | 0.7238 | 0.7238 |
| 1 | 0.3736 | 0.3734 | 0.3394 |
| 0.7 | 0.2172 | 0.2120 | −5.312 |

is $u/(1+u^3)$. Using a new integration variable $y = 1/u$ and replacing $z$ according to $W = 1/z$, we find that

$$L = \int_0^\infty - \int_0^W (1 - y^3 + y^6 - y^9 + y^{12} \cdots)dy$$
$$= \int_0^\infty - \left[ W - \frac{W^4}{4} + \frac{W^7}{7} - \frac{W^{10}}{10} + \frac{W^{13}}{13} \cdots \right] \tag{d}$$

In Equation (d), the integrand of the infinite integral is again $u/(1+u^3)$. By using the new variable $u = 1/y$, this is seen to also be the integrand $1/(1+y^3)$, for which the integral is 1.20920. Equation (d) can be written, in preparation for Padé approximation, as

$$L = 1.20920 - W\left[1 + \frac{x}{4} - \frac{x^2}{7} + \frac{x^3}{10} - \frac{x^4}{13} \cdots \right] \quad (x = W^3) \tag{e}$$

We now use program PADE to get the following Padé approximation for the quantity in brackets in Equation (e) and then re-express $L$ as

$$L = 1.20920 - W\left[\frac{p_0 + p_1 x + p_2 x^2}{1 + q_1 x + q_2 x^2}\right] \quad x = W^3 = \frac{1}{z^3} \tag{f}$$

where

$$p_0 = 1.0 \qquad p_1 = 0.826923 \qquad p_2 = 0.0890110$$
$$q_1 = 1.07692 \qquad q_2 = 0.215385$$

The results of comparing Equations (e) and (f) with numerical values are shown in Table 9.5.

The previous example demonstrates how to get useful series approximations to the integral for both large and small $z$. But it is especially noteworthy to observe the efficacy of the Padé representations of the two Taylor series. The two Padé approximations provide near-four-figure accuracy over the whole range of $0 < z < \infty$ using only five terms of the respective Taylor series. To get comparable accuracy from the Taylor series alone (with no acceleration), we would need at least about 1670 terms from *each* series, and this includes using the known signs of the errors.

The preceding methods are useful as far as they go, but in many problems a more direct approach is necessary, since a simple series approximation may not be obvious. This will be especially true in the consideration of asymptotic expansion of integrals, which follows in the next section. The idea involved in the analytic approximation of integrals can be stated simply. A large contribution to the integral is extracted in a relatively simple way, leaving behind an unknown remainder which can be shown to be small in magnitude compared to the extracted approximation. This process is then repeated on the remainder, so that by means of "successive extractions," sufficient accuracy is finally achieved. In principle the objectives are clear, but in practice these ideas sometimes apply only with difficulty. Useful techniques for carrying out the preceding approach include: (a) bracketing by means of inequalities, (b) splitting the integration interval, (c) changes of variable, especially rescaling and translation of axes, (d) the judicious use of $\infty$, (e) approximating near the maximum of the integrand, and (f) judicious use of integration by parts. Difficult problems often yield to some combination of these techniques, performed in an appropriate order. Of course, it is always desirable to bound the errors of these approximations, either using inequalities or by comparison with selected numerical results.

The procedure described here is a sort of "brute-force" analytical approach, wherein the integral is forced to yield its behavior one term at a time. Although this approach does not have the elegance of producing the entire expansion at once, it is very useful for extracting the approximation one term at a time. In complicated problems where there is small hope of achieving the whole expansion, the one-term-at-a-time technique is extremely important. Under such circumstances, it can be appreciated that acceleration techniques (such as Padé, etc.) may be especially valuable.

**Example 9.7 Evaluation of I = $\int_0^\infty e^{-xt}dt/(1+t)$ for Small $x$**

The case of large $x$ falls into the category of asymptotic expansions, which is treated in the next section. The above integral is a form of what is known as the exponential integral, which has many applications in engineering and science. A brief inspection will reveal that *this is not a simple problem*; this example is for mathematically mature readers. If we expand the exponential for small $x$ as $\exp(-xt) = 1 - xt + (xt)^2/2! \ldots$, and integrate term by term, we find that *all of the integrals diverge at* $\infty$, so this approach is useless. We consider here an approach that is very specific for this particular problem, which at least shows some ideas that could be pursued in more difficult problems. First, we let $1 + t = z$ so that

$$I = e^x \int_1^\infty \frac{e^{-xz}}{z}\,dz \tag{a}$$

Now is it convenient to introduce $u = xz$ to get

$$I = e^x \int_x^\infty \frac{e^{-u}}{u}\,du \tag{b}$$

An integration by parts, differentiating $\exp(-u)$, gives

$$I = e^x \left( -e^{-x} \ln x + \int_x^\infty e^{-u} \ln u du \right) \tag{c}$$

But since the last integral in Equation (c) converges at zero, we can use $\int_x^\infty = \int_0^\infty - \int_0^x$, and this yields

$$I = -\ln x + e^x \int_0^\infty e^{-u} \ln u du + e^x \int_0^x e^{-u} \ln u du \tag{d}$$

The last integral in Equation (d) $\to 0$ for $x \to 0$, so an asymptotic approximation has been obtained for $x \to 0$. To further approximate the last integral, it is natural to use $\exp(-u) = 1 - u + u^2/2! \ldots$ and integrate term by term. This produces a rather awkward expansion in which logarithms continue to appear.

A neater method is to write

$$\begin{aligned} I &= e^x \int_x^\infty \frac{e^{-u}}{u} du = e^x \left( \int_x^1 \frac{e^{-u}}{u} du + \int_1^\infty \frac{e^{-u}}{u} du \right) \\ &= e^x \left( \int_x^1 \frac{du}{u} + \int_x^1 \frac{e^{-u} - 1}{u} du + \int_1^\infty \frac{e^{-u}}{u} du \right) \end{aligned} \tag{e}$$

Then use $\int_x^1 = \int_0^1 - \int_0^x$ for the middle integral of Equation (e), expand $[\mathrm{Exp}(-u) - 1]$, and integrate term by term. This process results in

$$I = e^x \left[ -\ln x + \int_1^\infty \frac{e^{-u}}{u} du + \int_0^1 \frac{e^{-u} - 1}{u} du + x - \frac{x^2}{2!2} + \frac{x^3}{3!3} \cdots \right] \tag{f}$$

Here the error is less in magnitude than the first neglected term, and it is easy to see that the series converges for all $x$, although it will not be useful for large $x$. Padé approximation could be used to increase the accuracy for a fixed number of terms. The constant term in Equation (f) (the sum of the two integrals) can be shown to be the famous Euler constant, whose value is $-0.57722$ to five figures.

### 9.5.1 Miscellaneous Techniques

Aside from the use of standard forms, change of variables, and integration by parts, there are few methods for evaluating integrals exactly in terms of elementary functions. The remaining common methods include partial fractions, differentiation or integration with respect to a parameter, and transform methods.

**Example 9.8 Evaluation of $F(t) = \int_0^1 (x^t - 1)\, dx/\ln x$ by Differentiation with Respect to a Parameter**

Here we use $x^t = \exp(t \ln x)$ to get $F(t) = \int_0^1 (e^{t \ln x} - 1)\, dx / \ln x$. Because of the occurrence of $\ln x$ in the denominator, it is natural to attempt differentiation with respect to $t$. Differentiation under the integral yields

$$F'(t) = \int_0^1 e^{t \ln x} dx = \int_0^1 x^t dx = \frac{1}{1+t} \tag{a}$$

Integrating this simple equation gives $F(t) = \ln(1 + t) + \text{constant}$. Since $F(0) = 0$, the integration constant is zero and the final result is $F(t) = \ln(1 + t)$.

## 9.6 ASYMPTOTIC EXPANSION OF INTEGRALS

We turn now to a further discussion of analytical approximation of integrals which is somewhat different from that considered earlier. As before, we seek asymptotic approximations, for which the relative error $\to 0$ as some parameter approaches a limiting value. But now we consider some specific but very important classes of problems where the approximate series generated often *diverge* for all values of the variables, but are nevertheless useful in computation. At first glance this seems very strange, but it turns out that such approximations can be very powerful for a wide variety of difficult applications. In particular, these methods apply to many problems of ordinary and partial differential equations, including many problems involving transport and/or chemical reaction.

### 9.6.1 Watson's Lemma

Probably the simplest, most familiar, and most important result in asymptotic expansion of integrals is known as Watson's lemma. In its simplest form, it can be written as

$$\int_0^\infty e^{-xt}F(t)dt \sim \frac{F(0)}{x} + \frac{F'(0)}{x^2} + \frac{F''(0)}{x^3} + \frac{F'''(0)}{x^4} + \cdots \tag{9-7}$$

The right-hand side of Equation (9-7) is, in general, *asymptotic* to the left-hand side in the sense that the relative error of approximation goes to zero for $x$ sufficiently large; the symbol $\sim$ is used to denote asymptotic equality. Some readers will recognize that the left-hand side of Equation (9-7) is just the Laplace transform of the function $F(t)$. The formula of Equation (9-7) is obtained formally by expanding $F(t)$ in a Taylor series about $t = 0$ and integrating term by term, using gamma functions. The gamma function relation needed to accomplish this is

$$\int_0^\infty t^p e^{-xt}dt = \frac{\Gamma(p+1)}{x^{p+1}} \tag{9-8}$$

Equation (9-8) is obtained by a simple change of variable from the definition of the gamma function given in the section on infinite integrals. We now stop to examine the derivation of Equation (9-7) and examine the restrictions on its use.

For concreteness, we consider evaluation of the integral $I = \int_0^\infty e^{-xt}\,dt/(1+t)$ for large values of $x$. This is the so-called exponential integral and is of the form of Equation (9-7) if $F(t) = 1/(1+t)$. We considered this problem for *small* values of $x$ in a previous section, and found that it was a difficult problem. Evaluation of $I$ for large $x$ is much simpler. The idea of Watson's lemma is that if $x$ is large, $\exp(-xt)$ decays very rapidly toward zero from its largest value of unity at $t = 0$, in which case the main contribution to the integral comes from near $t = 0$. If that is true, it makes sense to expand $F(t)$ about $t = 0$ and integrate term by term from $t = 0$ to $t = \infty$. Each of the individual terms involves an integral like that of Equation (9-8), which can be expressed neatly in terms of tabulated gamma functions. The only problem with this approach is that the technique usually involves (as in this example) integration of the series for $F(t)$ outside its radius of convergence. This in turn causes the series so produced to be *divergent for all* x, so that it does not

**TABLE 9.6** EXACT AND APPROXIMATE VALUES OF $I = \int_0^\infty e^{-xt}\,dt/(1+t)$

| $x$ | Exact | Series "best" | Padé (3,3) | Shanks III | Shanks Error Estimate |
|---|---|---|---|---|---|
| 10 | 0.09156 | 0.09152 | 0.09156 | 0.09156 | $3\times10^{-7}$ |
| 2 | 0.3613 | 0.375 | 0.3605 | 0.3612 | $9\times10^{-4}$ |
| 1 | 0.5963 | 0.5 | 0.5882 | 0.5949 | $1.3\times10^{-2}$ |

have any limit when interpreted as an infinite series. To circumvent the problem connected with development of a divergent series, we expand $F(t) = 1/(1+t)$ in a *finite* geometric series as in, for example, the following three-term plus remainder form:

$$I = \int_0^\infty \frac{e^{-xt}}{1+t}dt = \int_0^\infty e^{-xt}\left[1 - t + t^2 - \frac{t^3}{1+t}\right]dt = \frac{1}{x} - \frac{1}{x^2} + \frac{2}{x^3} - R \qquad (9\text{-}9)$$

Since $t^3/(1+t) < t^3$ for $t$ positive, we see that $|R|$ is bounded by $3!/x^4$, which is just the next term of the series. A similar argument for any *finite* number of terms of the series shows the same result; the difference between the integral and the finite series approximation is always bounded by the next term of the series. Since the terms of the finite series get uniformly smaller in magnitude for $x$ sufficiently large, we see that any finite series approximation will have an arbitrarily small error if $x$ is sufficiently large. Thus, the series in Equation (9-9) (which can be written as $(0!/x - 1!/x^2 + 2!/x^3 - 3!/x^4 \ldots)$ is *asymptotic* to the integral $I$. The asymptotic expansion in question is seen, for computational purposes, to behave *almost* as an ordinary alternating series with decreasing terms; the difference is that for a given "large" $x$, we *cannot* take as many terms as we please. We must not take terms beyond the point where they begin to increase in magnitude. Thus, whether or not our computation will be useful depends on our accuracy needs, as well as on the particular asymptotic series. In the series of Equation (9-9), for example, if $x = 10$, the remainder $R < 6/10^4$, and for many purposes this is useful. On the other hand, for $x = 1$, $R < 6$, with the approximation $= 1 - 1 + 2 = 2$, and the calculation is poor.

It is very important to note that it is often possible to obtain useful results from divergent approximations by using acceleration techniques such as Padé approximation or Shanks transformation (see Chapter 4). We have seen evidence of this earlier, where the so-called delta-square process (which is just Shanks transformation) was instrumental in developing convergent iterations from diverging subiterations in solving nonlinear algebraic equations. To illustrate how this works in the present example, we take seven terms in the asymptotic series, Equation (9-9), as follows:

$$I \cong 0 + \frac{1}{x} - \frac{1}{x^2} + \frac{2}{x^3} - \frac{6}{x^4} + \frac{24}{x^5} - \frac{120}{x^6} \qquad (9\text{-}10)$$

Using $1/x$ as the series variable, we can determine from the seven terms on the right of Equation (9-10) a (3,3) Padé approximation from program PADE and a triple-Shanks (III) transformation from program SHANKS. These results are compared to numerical values of $I$ in Table 9.6.

The Padé (3,3) approximation refers to polynomials of degree 3 in both numerator and denominator. The Shanks III value is the result of one triple-Shanks transformation, for which seven terms of the series is the minimum amount of information required. The series "best" values correspond to using the optimal number of terms ($\leq$ seven) and also using the known bracketing information based on the sign of the series error. For example, when $x = 1$, the first two terms of the series have the most information and give $0 < I < 1$, so that the best estimate for this case is $I = 0.5$, as shown in the table. For large values of $x$ (e.g., $x = 10$), the various approximations are almost the same. However, for smaller $x$, the Padé and Shanks values are *much* better. They are better still if more terms are taken; they are apparently converging towards the correct values, while the original asymptotic series is limited in accuracy for any given $x$. It is interesting to note that the Shanks error estimates are all conservative.

The preceding example was very straightforward due to the simple character of the geometric series. Fortunately, however, it is not difficult to show that the result of Equation (9-7) applies for a very broad class of functions $F(t)$. Suppose that $F(t)$ has a convergent Taylor expansion for some finite $t$ in the vicinity of zero. If it also satisfies $|F(t)| \leq M\,\text{Exp}(at)$, where $M$ and $a$ are positive constants, then $F(t)$ is said to be of exponential order. If both of these fairly mild restrictions apply, then the asymptotic expansion of Equation (9-7) applies. In this much more general case, it is not difficult to show that the error of the asymptotic series approximation to the integral can be expressed as (constant) $\cdot$ |first neglected term of series|. In fact, one can show that the following, even more general, result applies under the same circumstances:

$$\int_0^A t^\alpha e^{-xt} F(t)\,dt \sim \frac{\Gamma(\alpha+1)F(0)}{x^{\alpha+1}0!} + \frac{\Gamma(\alpha+2)F'(0)}{x^{\alpha+2}1!} + \frac{\Gamma(\alpha+3)F''(0)}{x^{\alpha+3}2!} + \cdots \tag{9-11}$$

Notice that Equation (9-11) has an upper limit $A$, which can be infinite, or *any finite value*. This is because the integrals of the individual terms from $A$ to $\infty$ can be shown to be exponentially small in $x$. Thus, for *large* $x$, there is no effect of the finite upper limit $A$ on Equation (9-11). For a reasonably large finite $x$, the effect of $A$ will be small. We now give two typical examples that illustrate the power of Watson's lemma.

**Example 9.9 Evaluation of $J = \int_0^x e^{-u^3}du$ for Large $x$**

This integral occurs in the Leveque solution for boundary-layer transport in the entrance region of a tube. Part of the power of Watson's lemma is that it can be applied to so many problems that do not originally appear to be in the proper form, as in this example. To develop a proper form, we first note that since the above integral converges at $\infty$ we can write $\int_0^x = \int_0^\infty - \int_x^\infty$. The infinite integral is just a constant, so to apply Watson's lemma, we now work with the remainder integral $K = \int_x^\infty e^{-u^3}\,du$. To proceed, introduce the new integration variable $t = u^3$, followed by the simultaneous translation and scaling transformation $x^3 w = t - x^3$. After some algebraic rearrangement, the result of this is

$$K = \int_x^\infty e^{-u^3}du = \int_{t=x^3}^\infty \frac{e^{-t}t^{-2/3}}{3}dt = \frac{e^{-x^3}x}{3}\int_{w=0}^\infty e^{-x^3 w}(1+w)^{-2/3}dw \tag{a}$$

We now see that the result of these transformations produces an integral which is of Watson's lemma form, if we take $x^3$ as the large parameter. Now we simply expand

**TABLE 9.7** ASYMPTOTIC EVALUATION OF $K \cdot 3x^2\text{Exp}(x^3)$ FOR LARGE $x$ (EQUATION c)

| $x$ | Exact | Asymptotic | Padé | Shanks II |
|---|---|---|---|---|
| 3.0 | 0.9767 | 0.9767 | 0.9767 | 0.9767 |
| 1.5 | 0.8605 | 0.9067 | 0.8613 | 0.8631 |
| 1.0 | 0.6970 | 9.3457 | 0.7113 | 0.7050 |
| 0.8 | 0.5832 | 4.1107 | 0.6254 | 0.6097 |

$F(w) = (1 + w)^{-2/3}$ in the Taylor expansion,

$$F(w) = (1 + w)^{-2/3} = 1 - \frac{2}{3}w + \frac{10}{9}\frac{w^2}{2!} - \frac{80}{27}\frac{w^3}{3!} + \frac{880}{81}\frac{w^4}{4!} \cdots \tag{b}$$

and integrate term by term, using gamma functions to get

$$K \sim \frac{e^{-x^3}}{3x^2}\left[1 - \frac{2}{3x^3} + \frac{10}{9x^6} - \frac{80}{27x^9} + \frac{880}{81x^{12}} \cdots\right] \tag{c}$$

In this expansion, because of the error character of Equation (b), the error is less in magnitude than the first neglected term and of the same sign. Although the asymptotic series in Equation (c) clearly diverges (since we have integrated Equation (b) beyond its radius of convergence of unity), it will be extremely valuable for large values of $x$. This is precisely the range where a simple Taylor expansion of the original integral would not be useful (even though this ascending series in $x$ converges for *all* values of $x$). Also, to attempt to make the divergent series, Equation (c), more useful for smaller values of $x$, we can try acceleration by means of Padé approximation or Shanks transformation. Numerical comparisons for some of these possibilities are shown in Table 9.7. The "Asymptotic" values given there are based on an evaluation of the five terms of Equation (c).

**Example 9.10 Dawson's Integral $D = \int_0^x e^{u^2}\,du$ for Large $x$**

This problem occurs in certain Laplace transform applications and is an example of a *growing* function that certainly does not initially appear to be of Watson's lemma type. If we translate axes to $u = x$, the new integrand will be of decaying type. To accomplish the several required transformations at once, we use $u^2 = A - Bt$, where $A$ and $B$ are to be chosen expeditiously. This transformation gives

$$D = \int_{t=A/B}^{(A-x^2)/B} e^{A-Bt}\frac{1}{2}(A - Bt)^{-1/2}(-B)\,dt \tag{a}$$

If we now choose $A = B = x^2$, the integral $D$ becomes

$$D = \frac{xe^{x^2}}{2}\int_{t=0}^{1}\frac{e^{-x^2t}}{(1-t)^{1/2}}dt \tag{b}$$

In this form, we see that Watson's lemma applies with large parameter $x^2$. By expanding $(1 - t)^{-1/2}$ as $1 + (2t) + (3/8)t^2$ and integrating term by term from zero to infinity (never mind that the denominator in Equation (b) $\rightarrow 0$ at $t = 1$), we get the following asymptotic expansion for $D$ when $x$ is large:

$$D = \frac{e^{x^2}}{2x}\left[1 + \frac{1}{2x^2} + \frac{3}{4x^4} \cdots\right] \tag{c}$$

It can be shown that Equation (c) is a valid asymptotic expansion for $D$; however, here the error is *not* less in magnitude than the first neglected term, but rather proportional to it, with unknown proportionality constant. An error bound can be determined through analysis (difficult), but as with Taylor series, we could determine an error bound *directly* with the use of selected numerical values and program SERIES.

### 9.6.2 The Laplace Method for $L = \int_a^b F(t)e^{xG(t)}\,dt$ with x Large

This integral is a generalization of the kind of large parameter problem we have been considering. If $G(t)$ has its maximum value in the interval $(a, b)$ at either *end point*, then by translating axes to the end where the maximum occurs and reverting the series for $G(t)$, we get a Watson's lemma form. However, in case the maximum of $G(t)$ occurs in the *interior* of the interval $(a, b)$, the situation is slightly different, and the approximation here is often called the Laplace method. We consider here the most common case of a single maximum located at the interior point $t = c$, where $a < c < b$. Since $G(t)$ has a maximum at $t = c$, we have $G'(c) = 0$ and $G''(c)$ is negative. When $x$ is large, the maximum contribution to the integral comes from near the maximum of $G(t)$, so we translate axes to the point $t = c$ by means of $t - c = u$. Near $u = 0$, $F(t) \approx F(c)$ and $G(u) \approx G(c) + G''(c)u^2/2$. Substitution of these approximations into the original integral gives

$$L \cong F(c)e^{xG(c)} \int_{a-c}^{b-c} e^{xG''(c)u^2/2}du \tag{9-12}$$

But now the integral of Equation (9-12) is, for large $x$, asymptotically just equal to twice the integral from 0 to $\infty$, so that $L$ is given asymptotically for large $x$ by

$$L = 2F(c)e^{xG(c)} \int_0^\infty e^{xG''(c)u^2/2}du = F(c)e^{xG(c)}\left[\frac{2\pi}{-G''(c)x}\right]^{1/2} \tag{9-13}$$

Equation (9-13) is the first approximation for $x \to \infty$. Higher-order terms can be determined, but this is rather tedious. See Bender and Orszag [1978] for further details.

**Example 9.11 Evaluation of J = $\int_0^\infty e^{-t^3+At}dt$ for Large $A$**

This integral occurs in the problem of the effect of surface mass transfer on boundary-layer transport. For $A$ small, we can simply expand $\exp(At)$ in powers of $A$ and integrate term by term using gamma functions. For $A$ *negative* and large in magnitude, we have a Watson's lemma problem. Here, we consider the case of $A$ positive and large in magnitude. First, we wish to introduce a change of variable to produce a Laplace method integral, if possible. Thus, we try letting $t = bu$, where we will choose $b$. By substitution into the integral, we see that we can factor out the $A$ dependence in the exponential if we choose $b^3 = Ab$, or $b = A^{1/2}$. This gives, for the new form of the integral,

$$J = A^{1/2} \int_0^\infty e^{A^{3/2}(-u^3+u)}du \tag{a}$$

Equation (a) shows that we now have an integral for which the Laplace method applies, with large parameter $A^{3/2}$, $F(u) = 1$, and $G(u) = (-u^3 + u)$. To proceed, we calculate

$G'(u) = -3u^2 + 1$, $G''(u) = -6u$. This implies that $G(u)$ has a maximum at $u^* = 1/3^{1/2}$ where $G(u^*) = 2/3^{3/2}$ and $G''(u^*) = -6/3^{1/2}$. Substitution into the formula for the Laplace method, Equation (9-13), gives the following large $A$ asymptotic formula for $J$:

$$J \sim e^{2(A/3)^{3/2}} \left[ \frac{\pi^{1/2}}{(3A)^{1/4}} \right] \tag{b}$$

This formula does not contain any error information. The simplest approach is to compare the formula with selected numerical values of the integral to determine the region of validity.

### Trigonometric Integrals of the Type $I = \int_0^\infty \sin kxF(x)\,dx$, $J = \int_0^\infty \cos kxF(x)\,dx$ for Large $k$

Integrals of this type are fundamental to the application of Fourier series and integrals. In Fourier series, the integrals have finite upper limits, but frequently it is useful to introduce $\infty$ into the upper limit as we have done before. We mention these problems here because they are important and because general asymptotic expansions exist that are quite similar to those for Watson's lemma, which we have already discussed. We consider only the *infinite* Fourier integrals indicated above; similar (but more awkward) results can also be derived for integrals with finite upper limits.

The procedure used to develop asymptotic expansions for Fourier integrals is recursive integration by parts. If we integrate each of the integrals $I$ and $J$ by parts *twice*, integrating the trigonometric function, we get the results

$$\begin{aligned} I &= \int_0^\infty \sin kxF(x)\,dx = \frac{F(0)}{k} - \frac{1}{k^2}\int_0^\infty \sin kxF''(x)\,dx \\ J &= \int_0^\infty \cos kxF(x)\,dx = -\frac{F'(0)}{k^2} - \frac{1}{k^2}\int_0^\infty \cos kxF''(x)\,dx \end{aligned} \tag{9-14}$$

These relations represent recurrence relations which can be used to obtain further terms of the approximation, provided that $\int_0^\infty F^{(k)}(t)\,dt$ exists, as will normally be the case. Thus, for example, to obtain further terms for $I$ in Equation (9-14), we substitute $F''(x)$ in place of $F(x)$ on the left side, then put the new result in on the right side. Doing this to obtain the first three terms plus remainder for both $I$ and $J$ gives

$$\begin{aligned} I &= \frac{F(0)}{k} - \frac{F''(0)}{k^3} + \frac{F^{iv}(0)}{k^5} - \frac{1}{k^6}\int_0^\infty \sin kxF^{vi}(x)\,dx \\ J &= -\frac{F'(0)}{k^2} + \frac{F'''(0)}{k^4} - \frac{F^{v}(0)}{k^6} - \frac{1}{k^6}\int_0^\infty \cos kxF^{vi}(x)\,dx \end{aligned} \tag{9-15}$$

Provided that the integrals $\int_0^\infty \left|F^{(k)}(x)\right| dx$ converge, the error in the above expressions is less than (constant)/$k^6$, so the expansions are indeed asymptotic for large $k$. Obviously, the same process can be continued to determine as many terms as desired. It may also be advantageous, for smaller values of $k$, to attempt increasing the region of validity by using acceleration devices such as Padé approximation or Shanks transformation.

It should be mentioned that trigonometric integrals of the type considered in this section, especially for high frequency and slow decay behavior, are very difficult to handle numerically. One important practical approach is to use the recursive-integration-by-parts procedure given above to convert the original problem to one having an integrand that decays more rapidly.

**Example 9.12 Evaluation of $M = \int_x^{\infty} (\sin t/t)\, dt$ for Large $x$**

This is a form of the Sine integral, which occurs in a number of engineering and scientific problems, especially those involved with wave phenomena. The integral is not initially in our "Fourier integral" form, and we note that it could be handled directly using repeated integration by parts. However, we can put $M$ in the proper form for our recipe, Equation (9-14), by using the translation and scaling transformation $t - x = xu$. When this is used, along with the trigonometric identity $\sin(x + xu) = \sin x \cos xu + \cos x \sin xu$, we obtain

$$M = \sin x \int_0^{\infty} \frac{\cos xu}{1+u} du + \cos x \int_0^{\infty} \frac{\sin xu}{1+u} du \qquad \text{(a)}$$

Our transformation has produced an expression involving *two* Fourier integrals, which can be treated using Equation (9-15). This leads to the following asymptotic series for $M$:

$$\begin{aligned} M = \sin x &\left[ \frac{1}{x^2} - \frac{3!}{x^4} + \frac{5!}{x^6} - \frac{6!}{x^6} \int_0^{\infty} \frac{\cos xu}{(1+u)^7} du \right] + \\ \cos x &\left[ \frac{1}{x} - \frac{2!}{x^3} + \frac{4!}{x^5} - \frac{6!}{x^6} \int_0^{\infty} \frac{\sin xu}{(1+u)^7} du \right] \end{aligned} \qquad \text{(b)}$$

It is easy to see that each of the remainder integrals is bounded in magnitude by 1/6.

## 9.7 COMPUTER PROGRAMS

In this section, we discuss the computer programs that are applicable to various problems in the chapter. For each program we give a brief discussion and a concise listing. Detailed program listings can be obtained from the diskette.

### Trapezoidal or Midpoint Quadrature with Richardson Extrapolation: Subroutine TRMPQUAD

This is one of the few programs that we have left in subroutine form. This routine does the trapezoidal integration in programs RATINTG, LSTSQGR, and RATLSGR. To run this subroutine by itself, the user need only provide a simple calling program. A sample calling program is given ahead of the subroutine comments, which constitute the concise listing of the subroutine.

The calling program shows the FOR-NEXT loops required for either trapezoidal or midpoint integration. The user must deactivate one of these loops (here we have commented out the midpoint loop, so the present program does trapezoidal integration). As seen in the calling program, the user must specify the limits of integration, $a$ (lower) and $b$ (upper); the function values $f(i)$, either as equally spaced data or by a computed function [f(t) = t^5 in the calling program]; and the minimum number of parts into which interval is to be divided, in response to the prompt. In the print statement "int" is the estimate of the integral, "error" is the *estimated* relative error, and "ratio" is the normalized estimated error ratio, which should approach unity when extrapolation is behaving properly. In general, if $0.8 < \text{ratio} < 1.2$, we can be confident that the estimated error is larger than the true unknown error.

### Program Sub TRMPQUAD with Calling Program.

```
!This is a calling program to exercise subroutine TRMPQUAD.tru
!*** denotes user-specified statements
!Here we are integrating the function f(t)=t^5 from 1 to 2 using the
!trapezoidal rule

DIM f(1), t(1)
INPUT PROMPT"min. no. parts interval divided into (m) = ?": m
MAT REDIM f(0 to 8*m), t(0 to 8*m)
LET a = 1                  !***lower limit
LET b = 2                  !***upper limit
LET hs = (b-a)/(4*m)

!FOR i = 1 to (8*m - 1)    !for midpoint rule(commented out here)
!  LET t(i) = a + i*hs/2
!  LET f(i) = t(i)^5       !***user function
!NEXT i

FOR i = 0 to 4*m           !trapezoidal rule
  LET t(i) = a + i*hs
  LET f(i) = t(i)^5        !***user function
NEXT i

!Next line is call for trapezoidal ("t") quadrature
CALL TRMPQUAD("t",f(),m,a,b,integral,error,ratio$)
PRINT"int = ";integral;Tab(25);"error = ";error;Tab(50);"ratio = ";ratio$
END                          !end of calling program

SUB TRMPQUAD(meth$,f(),m,a,b,integral,error,ratio$)
!This is program trmpquad.tru for trapezoidal or midpoint quadrature with
!2 Richardson extrapolations, error estimate and error ratio by O.T.Hanna.
!The program is in the form of a subroutine.
!Method(meth$) is either trapezoidal (t) or midpoint (m).
!Method (t) is generally about 2 times more efficient than method (m);
!however, method m is useful when function f(t) has endpoint singularities.
!Interval is divided into a minimum no. of m parts; either method
!uses (m,2*m,4*m) intervals for step sizes (h,h/2,h/4).
!a and b are limits of integration; program produces values of integral,
!error and ratio.
END SUB
```

### *Program IEX Adapted to Quadrature*

This program is designed primarily for the integration of initial value systems of differential equations, and is discussed in detail in that chapter. However, we note that the integral $I = \int_a^b f(t)\,dt$ can be evaluated by integrating the differential equation $y' = f(x)$ from $x = a$, $y = 0$, to $x = b$ (then $y(b) = I$). If we specify $q = 4$ in Sub Init (for initial conditions) in program IEX, the program carries out exactly the same trapezoidal quadrature plus two Richardson extrapolations that is done in program TRMPQUAD above. Thus, for a given known function to be integrated, it is usually simplest to use program IEX, since no calling program is required and the function need only be specified in Sub Deriv (for derivative).

## PROBLEMS

**9.1.** Use program IEX to evaluate the integral $\int 5du/(1 + u^5)$ between the limits of zero and unity. Monitor the error to achieve six-figure accuracy.

**9.2.** Consider the integral of problem 9.1. Use the change of variable $u^5 = x$ and evaluate the integral numerically in these coordinates. Attempt to achieve a validated six-figure accuracy. Which seems to be more convenient, the variable of this problem or that of problem 9.1?

**9.3.** It is desired to numerically evaluate the integral $\int \ln t\,dt$ between the limits of zero and unity. Evaluate this to six-figure accuracy, using any variable that seems convenient.

**9.4.** The integral $\int y\,dx$ from $x = 0.2$ to 3 is to be determined numerically from the data below. Carry out this calculation:
(a) By spline approximation
(b) By polynomial interpolation
(c) By rational interpolation
(d) By a "good" (Emax $< 0.01$) polynomial least-squares approximation
(e) By a "good" (Emax $< 0.01$) rational least-squares approximation
Estimate accuracy where possible, and compare the results.

| $y$ | 0.720 | 0.563 | 0.667 | 0.985 | 1.64 | 2.77 | 4.00 |
|---|---|---|---|---|---|---|---|
| $x$ | 0.2 | 0.5 | 1.0 | 1.4 | 1.9 | 2.5 | 3.0 |

**9.5.** It is desired to evaluate the following bulk-concentration integral for cylindrical geometry:

$$I = \int_0^{1.5} yr\,dr$$

Here, $y$ is concentration and $r$ is radius. Measured values of $y$ versus $r$, adapted from the thesis of James [1984], are shown below. Since this is a cylindrical geometry, the condition $dy/dr$ $(r = 0) = 0$ should be satisfied.

| $r$ | 0.25 | 0.5 | 0.75 | 1.0 | 1.25 | 1.50 |
|---|---|---|---|---|---|---|
| $y$ | 0.29 | 0.30 | 0.305 | 0.29 | 0.26 | 0.225 |

(a) Derive an apparent upper bound for the integral.
(b) Assuming that the data points are known exactly, use trapezoidal integration with extrapolation to evaluate the integral and provide an error estimate.
(c) Use midpoint integration as in part (b).
(d) Use spline approximation to evaluate the integral. To do this, get a spline approximation to $(ry)$. Estimate the value of $y(0)$. Use both the given data and data supplemented with a point at $x = 0$. Compare the results.
(e) Suppose that the $y$ data are not known exactly, but rather are contaminated with experimental error. Try fitting the $y$ data by *least squares* with both cubic and quartic polynomials, each constrained to satisfy the appropriate centerline condition $dy/dr\ (r = 0) = 0$. Use the "error" values plus graphical representation to decide which of these is best. For the best polynomial, fit the $(xy)$ data to the appropriate model and evaluate the integral.

**9.6.** Test the following integrals for convergence:

$$\text{(a)} \quad \int_0^\infty \frac{\cos x dx}{1+x} \qquad \text{(b)} \quad \int_0^\infty \frac{x^{-1/2}dx}{1+x^{4/5}}$$

**9.7.** Find numerical upper bounds for the integrals of problem 9.6.

**9.8.** Express the following integrals in terms of gamma functions:

$$\text{(a)} \quad \int_0^\infty 2t^{1/2}e^{-3t}dt \qquad \text{(b)} \quad \int_0^1 (1-x)^{3/2}x^4dx \qquad \text{(c)} \quad \int_0^\infty e^{-t^3}dt$$

**9.9.** Consider the integral

$$\int_0^x \frac{du}{(1-u^3)^{1/2}}$$

(a) Determine an analytical approximation for the range $0 \le x \le 1$.
(b) For $x = 1$, evaluate the integral numerically to six figures and compare with the result of (a).

**9.10.** Develop (a) small $x$ and (b) large $x$ approximations for the integral

$$\int_0^x \frac{t^2dt}{1+t^4}$$

**9.11.** By forming (and integrating) a differential equation for $S(x)$, show that the series

$$S(x) = \sum_{n=0}^\infty \frac{(-1)^n x^n}{a+n}$$

can be expressed in terms of the integral

$$S(x) = \frac{1}{x^a}\int_0^x \frac{t^{a-1}dt}{1+t}$$

Expand the integral to verify this relationship. This integral formula offers a very convenient way to sum the above often-occurring series. Compute the value of $S(x = 1)$ to six figures for (a) $a = 9$, (b) $a = 1/5$.

**9.12.** The well-known Error function has many applications in transport theory as well as in probability. Apart from a constant factor that varies with the normalization convention being used, it can be expressed as

$$E(x) = \int_0^x e^{-t^2}dt$$

(a) Determine a small $x$ approximation to $E(x)$ by expanding the exponential and integrating term by term. What is the error of the approximation?

(b) Use gamma functions to calculate $E(\infty)$.

(c) Derive a large $x$ approximation for $E(x)$ in the following way. Write

$$E(x) = \int_0^x e^{-t^2}\,dt = \int_0^\infty e^{-t^2}\,dt - \int_x^\infty e^{-t^2}\,dt$$

Then for the last integral on the right, use the change of variable $t^2 = au + b$ with appropriate choices for $a$ and $b$ to show that

$$\int_x^\infty e^{-t^2}\,dt = \frac{xe^{-x^2}}{2} \int_0^\infty \frac{e^{-x^2 u}\,du}{(1+u)^{1/2}}$$

The integral on the right is now in the form to which Watson's lemma can be applied. What is the error of the approximation?

**9.13.** The integral below occurs in the study of transport processes in laminar flow near the separation point. Following the procedures of problem 9.12, determine an approximation for it in the cases of small $x$ and large $x$.

$$\int_0^x e^{-u^4}\,du$$

**9.14.** Use the Laplace method to derive an asymptotic approximation to the following transport integral in the case of large, positive $A$:

$$J(A) = \int_0^\infty e^{-t^2 + At}\,dt$$

Compare this approximation to the numerical value of $J$ for $A = 5$.

**9.15.** Perform a numerical evaluation of the integral

$$\int_0^\infty \frac{\cos x dx}{1+x}$$

(a) Try this by direct evaluation of the original form, using program IEX and integrating as far as necessary.

(b) Use the integration-by-parts technique as shown in the discussion of Fourier integrals. Directly integrate the residual integral, whose integrand decreases more rapidly than that of the original form.

## REFERENCES

ABRAMOWITZ, M. and STEGUN, I. (ed.) (1965), *Handbook of Mathematical Functions*, Dover, New York

BENDER, C. and ORSZAG, S. (1978), *Advanced Mathematical Methods for Scientists and Engineers*, McGraw-Hill, New York

BROMWICH, T. J. I'A. (1964), *Infinite Series,* 2nd ed., Macmillan, New York

CARNAHAN, B., LUTHER, H. A., and WILKES, J. O. (1969), *Applied Numerical Methods*, Wiley, New York

GOODWIN, E. T. (1961), *Modern Computing Methods*, 2nd ed., Philosophical Library, New York

JAMES, D. E. (1984), *Mass Transfer in a Packed Distillation Column with Significant Liquid Phase Resistance*, M.S. Thesis, Univ. of California-Santa Barbara.

KAHANER, D., MOLER, C., and NASH, S. (1989), *Numerical Methods and Software*, Prentice-Hall, Englewood Cliffs, NJ

MEKSYN, D. (1961), *New Methods in Laminar Boundary Layer Theory*, Pergamon Press, New York

SCHEID, F. (1988), *Numerical Analysis*, 2nd ed., McGraw-Hill, New York

SQUIRE, W. (1970), *Integration for Engineers and Scientists*, Elsevier, New York

# 10

# Ordinary Differential Equations: Analytical Methods

Differential equations of various types probably represent the main "modeling" tools of chemical engineering. This is true because so many chemical and physical processes can be described by equations involving rates of change, or derivatives. For this reason differential equations are very important, and the elements of this subject have been studied in earlier courses. In this chapter we investigate the most important practical procedures for solving differential equations *analytically*. This is always desirable when it can be accomplished with a reasonable effort, since this allows us to see the actual behavior of the solution as a function of the independent variable and any parameters that may be present. Unfortunately, since there is no calculus chain rule for integration, getting an analytical solution is not always possible.

Even if it is necessary to proceed numerically, however, it may be useful to correlate the results in terms of an equation by the data-fitting techniques we have discussed in interpolation and least squares. Of these techniques, the one that we believe to be of the greatest practical value is that of rational least squares. Before starting our discussion, we stop to review some basic concepts and results from ordinary differential equations which should already be familiar to the reader.

## 10.1 REVIEW OF ELEMENTARY METHODS

**Definitions.** An ordinary differential equation is merely an equation that involves one or more derivatives of some unknown function $y$ which depends on a single variable $x$. The variable $y$ is said to be the dependent variable, while $x$ is said to be independent. The distinction between dependent and independent variables is artificial, since their roles can be interchanged, but the distinction has important mathematical consequences. The order of a differential equation is defined as the order of the highest derivative occurring in the equation. The most general first-order differential equation can be written as $F(dy/dx, y, x) = 0$. In most cases it is possible to express the equation explicitly as $dy/dx = f(x, y)$. The fundamental problem associated with solving this general equation is that the derivative of the unknown function $y$ depends on itself. In general, a differential equation problem is not completely prescribed until a number of initial and/or boundary conditions equal to the order of the equation has been specified.

### 10.1.1 Separable Equations

If a first-order differential equation can be written in the form $dy/dx = g(x)/h(y)$, it is said to be separable. In this case the solution is obtained by "separating" the variables according to $h(y)dy = g(x)dx$ and integrating both sides. The constant of integration is determined from the initial condition. A separable solution of the above type may lead to an implicit relation between $y$ and $x$ which could be solved by nonlinear algebraic equation techniques.

### 10.1.2 The First-Order Linear Equation

The concept of linearity is important in the theory of differential equations. A differential equation is said to be linear if it can be written as a linear combination of the dependent variable $y$ and its derivatives. Thus, the most general linear equation of the first order can be written as $dy/dx + p(x)y = q(x)$. Any first-order equation that cannot be written in this form is said to be nonlinear. It is very important to note that linearity depends only on the *dependent* variable. Thus, the equation $dy/dx + (\sin x)y = 1$ is linear, while $dy/dx + (\sin y)x = 1$ is nonlinear. Linearity is of importance because the existence and uniqueness of solutions is well understood for such problems, and because different solutions of linear equations can be superimposed to form additional solutions.

The solution of a first-order linear equation can be effected as follows. The equation is $dy/dx + p(x)y = q(x)$. We seek to find a function $g(x)$ which, when multiplied by the original equation, reduces it to a separable type. Thus, we wish to determine $g(x)$ so that

$$g(x)\left(\frac{dy}{dx} + p(x)y\right) = \frac{d}{dx}[h(x)y] \tag{10-1}$$

Here, $h(x)$ is also an unknown function. If the above expression is expanded by the product rule, we wish to have $g(y' + py) = hy' + h'y$. If this is to be an identity, we must take $g(x) = h(x)$ and $g(x)p(x) = h'(x) = h(x)p(x)$. But this last equation is simply a separable equation for $h(x)$, whose solution is $h(x) = g(x) = \exp(\int_a^x p(t)\,dt)$. Here, for convenience, the lower limit, $t = a$, has been used in the integral. The initial condition for the differential equation is assumed to be specified at $x = a$. With $g(x)$ and $h(x)$ determined in this way, the original equation can be written as

$$\frac{d}{dx}(h(x)y) = h(x)q(x); \quad h(x) = e^{\left(\int_a^x p(t)\,dt\right)} \tag{10-2}$$

But this equation is separable, so if we integrate both sides from $a$ to $x$, assuming $y(a)$ is the known initial condition, we finally obtain

$$y(x)e^{\int_a^x p(t)\,dt} - y(a) = \int_{t=a}^{x} q(t)e^{\int_{z=a}^{t} p(z)\,dz}\,dt \tag{10-3}$$

As an example, consider the equation $y' - 2xy = \cos x$ with $y(0) = 0$. Then $\int_0^x p(t)dt = \int_0^x -2t\,dt = -x^2$. By using Equation (10-3), the solution for $y(x)$ is

$$y(x)e^{-x^2} = \int_0^x e^{-t^2} \cos t\,dt \tag{10-4}$$

In this problem, as often happens, the integral cannot be evaluated in terms of a finite combination of elementary functions. In such cases, either numerical or analytical approximation methods must be used.

### 10.1.3 The Second-Order Linear Constant-Coefficient Equation

The second-order linear, constant-coefficient differential equation occurs frequently and can be solved explicitly. A general form of this equation is

$$\frac{d^2y}{dx^2} + a\frac{dy}{dx} + by = h(x) \tag{10-5}$$

This equation is linear since it involves a linear expression in $y$ and its derivatives. The coefficients of $y$ and its derivatives, namely 1, $a$, and $b$, are assumed to be constant. If these coefficients vary with $x$, the methods discussed here are not appropriate and, in fact, no general, explicit solutions are known. This is in contrast to the first-order linear case, where an explicit solution was obtained even for variable coefficients. The term $h(x)$ is referred to as the non-homogeneous term of the differential equation. It will be recalled from a first course in differential equations that the general solution to a second-order linear constant coefficient differential equation can be written as the sum of $y_c$ plus $y_p$, where $y_c$ is the so-called complementary (or homogeneous) solution corresponding to setting $h(x)$ equal to zero in the equation, and $y_p$ is a particular integral of the complete equation. The complementary solution contains arbitrary constants for satisfying the boundary conditions. It is usually convenient to arrange the solution as $y = y_c + y_p$ in this way, since the methods for obtaining $y_c$ and $y_p$ are different.

The complementary solution of Equation (10-5), $y_c$, is by definition the general solution to the equation $y'' + ay' + by = 0$. Since $a$ and $b$ are constants, it is easy to verify that the function $y = C\exp(\lambda x)$ satisfies the complementary equation provided that $\lambda$ is a root of the characteristic equation, $\lambda^2 + a\lambda + b = 0$. This quadratic equation for $\lambda$ has two roots, so that the above procedure produces two different solution functions which, when added together, give the general complementary solution. The form of the complementary solution depends on the nature of the roots of the characteristic equation. The three cases are indicated below, where $C_1$ and $C_2$ are arbitrary constants to be determined from the boundary conditions.

$$
\begin{aligned}
&\text{(a)} \quad y_c = C_1 e^{\lambda_1 x} + C_2 e^{\lambda_2 x} && (\lambda_1, \lambda_2 \text{ real and distinct})\\
&\text{(b)} \quad y_c = C_1 e^{\lambda_1 x} + C_2 x e^{\lambda_2 x} && (\lambda_1, \lambda_2 \text{ real and equal})\\
&\text{(c)} \quad y_c = e^{\alpha x}(C_1 \cos \beta x + C_2 \sin \beta x) && (\lambda_{1,2} = \alpha \pm \beta i)
\end{aligned}
\tag{10-6}
$$

The general solution of Equation (10-5), for arbitrary $h(x)$, is given by $y(x) = C_1 y_1 + C_2 y_2 + y_p = y_c + y_p$. Here, $y_p(x)$ is *any* solution (without regard to boundary conditions) of the full equation that includes $h(x)$. A particular solution can sometimes be "guessed," and a procedure known as the method of undetermined coefficients involves assuming the form of the solution using free, unspecified parameters and then choosing the parameters so that the assumed form really is a solution. As a simple example of this idea, consider the equation $y'' - 3y = 2$. A particular integral is sought in the form $y_p = B$, where $B$ is an unknown constant. Since $y_p'' = 0$ for the assumed form, substitution into the equation gives $-3B = 2$. Thus, $y_p = B$ is indeed a particular integral only if $B = -2/3$.

A very general but rather awkward procedure for obtaining a particular integral to Equation (10-5) is known as the variation of parameters. Without going into details, it can be shown that a particular integral of Equation (10-5) can be expressed as

$$
y_p(x) = y_1(x) \int_{t=a}^{x} \frac{-h(t) y_2(t)}{W[y_1(t), y_2(t)]} dt + y_2(x) \int_{t=a}^{x} \frac{h(t) y_1(t)}{W[y_1(t), y_2(t)]} dt \tag{10-7}
$$

where $y_1(x)$ and $y_2(x)$ are complementary solutions, and $W[y_1, y_2] \equiv y_1 y_2' - y_2 y_1'$ is known as the Wronskian determinant. For cases (a), (b), and (c) above, we get the appropriate $W$ functions from

$$
\begin{aligned}
&\text{(a)} \quad W(x) = (\lambda_2 - \lambda_1) e^{(\lambda_1 + \lambda_2) x}\\
&\text{(b)} \quad W(x) = e^{2\lambda_1 x}\\
&\text{(c)} \quad W(x) = \beta e^{2\alpha x}
\end{aligned}
\tag{10-8}
$$

As an example, consider the equation $y'' + 3y' + 2y = 1/(1 + x)$. Trying $y_c = e^{\lambda x}$, we find that $(\lambda^2 + 3\lambda + 2) = 0$ and $\lambda_1 = -1$, $\lambda_2 = -2$, so that $y_1 = e^{-x}$ and $y_2 = e^{-2x}$. The Wronskian becomes $W(x) = (\lambda_2 - \lambda_1) e^{(\lambda_2 - \lambda_1) x} = -e^{-3x}$. Substitution into Equation (10-7) then gives $y_p(x)$ as

$$
y_p(x) = e^{-x} \int_0^x \frac{e^t}{1 + t} dt - e^{-2x} \int_0^x \frac{e^{2t}}{1 + t} dt
$$

To get an approximate analytical evaluation of the above integrals, we could do the following. For small $x$, use the series quotient algorithm to get ascending Taylor series for the integrals; we might try to accelerate these series with Padé approximations. For large $x$, we could use recursive integration by parts (integrating the exponential terms) to get asymptotic series for the integrals.

Another procedure that is sometimes useful for linear, constant-coefficient differential equations (especially those having a "forcing function") is the Laplace transform. A good discussion of this topic is given by Spiegel [1965].

## 10.2 SERIES SOLUTION

The method of infinite series is one of the oldest techniques for obtaining a solution to a differential equation for which no explicit formula can be found. The technique

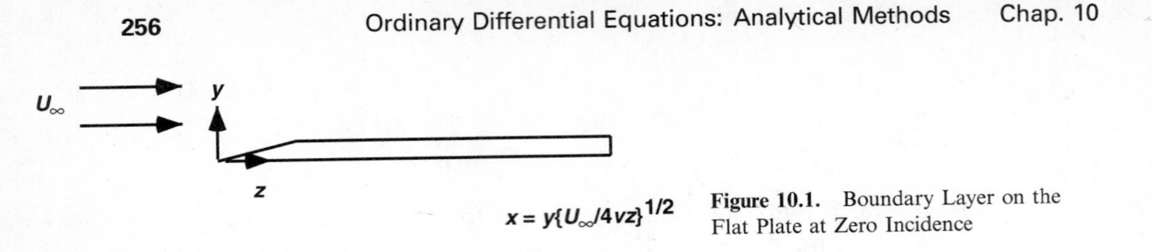


**Figure 10.1.** Boundary Layer on the Flat Plate at Zero Incidence

is a classical one of mathematical physics, and therefore, many solutions of problems in contemporary chemical engineering are expressed in this form. Many special functions, including those of Bessel, Legendre, Laguerre, Hermite, Chebyshev, and so forth, are usually defined as series solutions of differential equations. The series method has been used primarily for linear problems involving variable coefficients, but it can also be employed to advantage in nonlinear problems. Many applications of the series method have been made to partial differential equations through the method of separation of variables.

The series method is nothing more than a procedure for representing a solution of a differential equation as

$$y(x) = \sum_{i=0}^{\infty} a_i x^i = \sum_{i=0}^{\infty} \frac{y^{(i)}(0)}{i!} x^i \tag{10-9}$$

If the desired series expansion point is not zero, it is convenient to translate axes to make it so. A solution does not *necessarily* exist in the above form, but many equations have a solution expressible as a series in powers of $x$, or at least the product of a nonanalytic function (such as $x^{1/3}$, $\ln x$, or $e^{-1/x}$, etc.) times such a series. The differential equation itself can be used to determine the series coefficients directly, and even if we can calculate the full series, it may be useful to compute a few coefficients in this way for checking purposes. If the problem is nonlinear, this direct calculation procedure may be the only way to determine a finite series approximation. In such a case, where only a relatively small number of terms are known, acceleration techniques such as Euler's transformation, Padé approximation, or Shanks transformation may be extremely valuable.

**Example 10.1 The Blasius equation $f''' + ff'' = 0$.**

This famous third-order nonlinear equation describes flow in the laminar boundary layer on a flat plate, as illustrated in Figure 10.1. It is a boundary value problem, with data specified at both ends of the interval, having boundary conditions $f(0) = f'(0) = 0$, and $f'(\infty) = 1$. In this problem $f'$ is the dimensionless axial velocity, $f' = u/U_\infty$, where $U_\infty$ is the velocity in the free stream outside the boundary layer. The independent variable, $x$, is a combination of variables as defined in the figure, where $\nu$ is the kinematic viscosity.

Here we wish to illustrate the direct use of Taylor's theorem to determine several terms of a series solution about $x = 0$. Since we have $f(0) = f'(0) = 0$, the first nonzero coefficient will involve $f''(0)$, and we then get from the differential equation

$$\begin{aligned} f''' &= -ff'' & f'''(0) &= 0 \\ f^{iv} &= (ff''' + f'f'') & f^{iv}(0) &= 0 \\ f^{v} &= (ff^{iv} + 2f'f''' + f''f'') & f^{v}(0) &= -[f''(0)]^2 \end{aligned} \tag{a}$$

If we continue differentiating, the equations become longer and longer, and we are unable to identify a pattern in the coefficients. In fact, continuing to terms through

$f^{viii}$, we find $f^{vi}(0) = f^{vii}(0) = 0$ and $f^{viii}(0) = 11[f''(0)]^3$. This finally gives, for the Taylor expansion through terms to $x^8$,

$$f(x) = \frac{f''(0)}{2!}x^2 - \frac{[f''(0)]^2}{5!}x^5 + \frac{11[f''(0)]^3}{8!}x^8 + \ldots \tag{b}$$

All of the coefficients in the expansion depend on the unknown value of $f''(0)$. This quantity must be found by application of the remaining boundary condition, $f'(\infty) = 1$. Note that the form of Equation (b) cannot accommodate an infinite value for $x$, so the determination of $f''(0)$ must be accomplished in an alternative way. Instead, we determine $f''(0)$ for this problem in the section on successive approximation (Example 10.3). After the value of $f''(0)$ has been determined, the series in Equation (b) is useful for the calculation of the velocity profile and the heat transfer or mass transfer coefficient. Note that had this been an initial value problem, where all the auxiliary data (including $f''(0)$) were known at the expansion point $x = 0$, then the Taylor series would be completely determined by a "brute-force" (one term at a time) calculation such as we have done.

The direct calculation of the Taylor expansion, as we have illustrated above, is normally practical for only a few steps, and it usually does not suggest a pattern for the general coefficient $a_i$. In the case of nonlinear problems, this is usually the only feasible approach. For *linear* differential equations, the situation is different. By working directly with the postulated infinite series, we may often develop a *recurrence* relation for the $a_i$ coefficients. This recurrence relation then allows the calculation of a large number of the $a_i$ in a simple loop on a computer. If the recurrence relation can be solved explicitly to determine $a_i$ in terms of $i$, then the series can be tested for convergence, and a series truncation error bound can be developed. Even if the recurrence relation for the $a_i$ cannot be solved explicitly, we can often obtain convergence and error information through the use of a dominating difference equation.

The series method can be developed in the following way. A solution of the form

$$y(x) = x^s \sum_{i=0}^{\infty} a_i x^i = \sum_{i=0}^{\infty} a_i x^{s+i} \tag{10-10}$$

is postulated. Here, the inclusion of the arbitrary parameter $s$ allows for the case where the solution is a fractional power $x^s$ (which is nonanalytic; it doesn't have a power series expansion about $x = 0$) times a power series. This technique is generally known as the *method of Frobenius*. The assumed solution is substituted into the differential equation, and the coefficients of all powers of $x$ are set equal to zero. Using a change of summation index, an appropriate value of $s$ is chosen, and a recurrence relation is generated for the $a_i$. Whenever possible, this recurrence relation is solved to obtain an explicit expression for the $a_i$. When the $a_i$ are known, the region of convergence of the series can be determined from the ratio test. Finally, the error associated with taking only a finite number of terms in the series can be bounded by using a dominating geometric series. The details of this whole process are illustrated with an example.

**Example 10.2 Chemical Reaction in a Cylindrical Catalyst Pellet**

A simple model for pseudo-homogeneous, isothermal chemical reaction and diffusion in a cylindrical catalyst pellet with irreversible first-order reaction kinetics can be written in the form

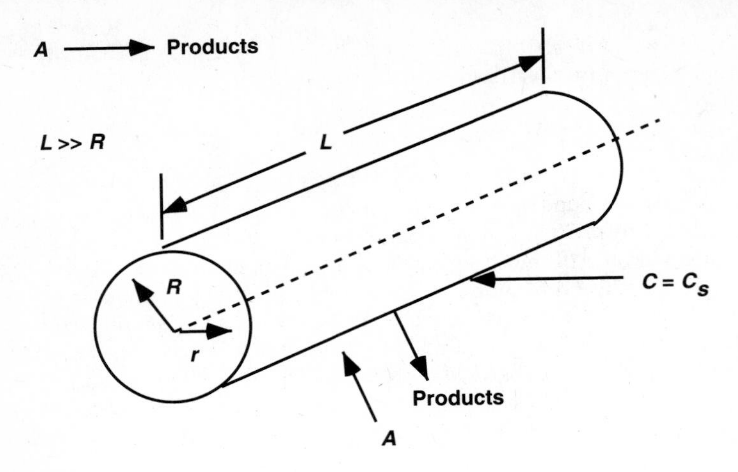


**Figure 10.2.** Cylindrical Catalyst Pellet

$$\frac{D}{r}\frac{d}{dr}\left(r\frac{dC}{dr}\right) = KC = D\left(\frac{d^2C}{dr^2} + \frac{1}{r}\frac{dC}{dr}\right) \tag{a}$$

$$r = 0, \frac{dC}{dr} = 0 \qquad r = R, C = C_s$$

Here, $C$ is the concentration of reactant, $K$ is a rate constant, $D$ is a dispersion coefficient, and $r$ is the radial coordinate. The system considered is illustrated in Figure 10.2.

By introducing the new dimensionless variables $y = C/C_s$ and $x = r/R$, the above equation can be written in the form

$$\frac{1}{x}\frac{d}{dx}\left(x\frac{dy}{dx}\right) = By = \left(\frac{d^2y}{dx^2} + \frac{1}{x}\frac{dy}{dx}\right) \tag{b}$$

$$x = 0, \frac{dy}{dx} = 0 \qquad x = 1, y = 1$$

Here, $B$ is the dimensionless parameter $KR^2/D$. Equation (b) happens to be a particular example of a class of differential equations known as *Bessel functions* Hildebrand [1976]. We wish to develop a power series solution to this problem and use it to calculate $y(0)$ and $y'(1)$ for the cases $B = 1$, 10, and 100. Prior to beginning a *formal* development of the series solution, it is a good idea to "get a feel" for the problem by trying a few terms of the simplest possible Taylor expansion. Thus, we simply write, tentatively, $y = a_0 + a_1x + a_2x^2 + a_3x^3$; then $y' = a_1 + 2a_2x + 3a_3x^2$ and $y'' = 2a_2 + 6a_3x$. Checking the boundary condition at $x = 0$, we see that since $y'(0) = 0$, we must have $a_1 = 0$. Then the constant $a_0$ remains to be determined by the boundary condition at $x = 1$ after the series has been generated. We now substitute the first few terms of the postulated expansion into the differential equation to get

$$y'' + \frac{y'}{x} - By = 0 = (2a_2 + 6a_3x\cdots) + \left(\frac{a_1 + 2a_2x + 3a_3x^2\cdots}{x}\right) - B(a_0 + a_1x\cdots) \tag{c}$$

When we recall that $a_1 = 0$ and collect powers in terms of $x$, Equation (c) gives

$$y'' + \frac{y'}{x} - By = 0 = (2a_2 + 2a_2 - Ba_0) + x(6a_3 + 3a_3 - Ba_1)\cdots \tag{d}$$

By setting the coefficient of each power of $x$ in Equation (d) equal to zero, we get $a_2 = Ba_0/4$ and $a_3 = Ba_1/9 = 0$. It appears from our simple trial calculation that a series expansion of the postulated form should work, so we now develop that formal expansion in hopes of being able to generate a formula for the general $a_i$ coefficient. When we are finished, it will be useful to check back with our simple calculation to see that there is agreement with the $a_2$ and $a_3$ values we have calculated here.

For our systematic analysis, we assume the following form for the series, and then calculate $y'$ and $y''$ accordingly as

$$y(x) = \sum_{i=0}^{\infty} a_i x^i \quad \frac{dy}{dx} = \sum_{i=0}^{\infty} i a_i x^{i-1} \quad \frac{d^2y}{dx^2} = \sum_{i=0}^{\infty} i(i-1) a_i x^{i-2} \tag{e}$$

We then substitute these series expressions into the differential equation, Equation (b), to get

$$\frac{d^2y}{dx^2} + \frac{1}{x}\frac{dy}{dx} - By = 0 = \sum_{i=0}^{\infty} i(i-1)a_i x^{i-2} + \sum_{i=0}^{\infty} i a_i x^{i-2} - B\sum_{i=0}^{\infty} a_i x^i \tag{f}$$

To get a recurrence relation for the $a_i$, we need to combine the three sums in Equation (f) into a single sum. In order to accomplish this, we introduce a change in summation index in the third sum of Equation (f) according to $i = j - 2$, which will then produce like exponents in the three sums. This yields, after switching from $i$ to $j$ in the first two sums, the following:

$$0 = \sum_{j=0}^{\infty} j(j-1)a_j x^{j-2} + \sum_{j=0}^{\infty} j a_j x^{j-2} - B\sum_{j=2}^{\infty} a_{j-2} x^{j-2} \tag{g}$$

Note that the lower limit of summation in the third sum of Equation (g) has now become 2, due to the change in the summation index there. Except for that, we could combine the three sums into a single sum. To get around this problem, we merely take out separately the $j = 0$ and $j = 1$ terms in the first two sums in Equation (g), prior to combining all the sums into one. This gives

$$0 = a_i x^{-1} + \sum_{j=2}^{\infty} x^{j-2}[j^2 a_j - B a_{j-2}] \tag{h}$$

The recurrence relation for the $a_i$ is now obtained by setting the coefficient of each power of $x$ in Equation (h) equal to zero. We have already seen that the boundary condition $y'(0) = 0$ requires that $a_1 = 0$. Setting the term in brackets equal to zero for $i = 2, 3, \ldots$ gives $a_2 = Ba_0/2^2$, $a_3 = Ba_1/3^2 = 0$, $a_4 = Ba_2/4^2 = B^2 a_0/(2^2 4^2)$, and so on. In this way, from an obvious mathematical induction we find that

$$a_k = \frac{B^{k/2} a_0}{2^2 4^2 \cdots k^2} \quad (k \text{ even}) \qquad a_k = 0 \quad (k \text{ odd}) \tag{i}$$

Substitution of the coefficients in Equation (i) back into Equation (e) finally gives the full series for $y(x)$.

$$y(x) = a_0\left[1 + \frac{B}{2^2}x^2 + \frac{B^2}{4^2 2^2}x^4 + \frac{B^3}{6^2 4^2 2^2}x^6 + \frac{B^4}{8^2 6^2 4^2 2^2}x^8 \cdots\right] \tag{j}$$

We still have to determine $a_0$ by the remaining boundary condition, $y(1) = 1$. This gives, for $a_0$, the value

$$a_0 = 1\Big/\left[1 + \frac{B}{2^2} + \frac{B^2}{4^2 2^2} + \frac{B^3}{6^2 4^2 2^2} + \frac{B^4}{8^2 6^2 4^2 2^2} \cdots\right] \tag{k}$$

**TABLE 10.1** VALUES OF SERIES SOLUTION FOR $y'' + (1/x)y' = By$

| $B$ | $y(0)$ | $y'(0)$ |
|---|---|---|
| 1 | 0.790 | 0.446 |
| 10 | 0.179 | 2.60 |
| 100 | 0.000355 | 9.49 |

Now that the power series solution is completely determined, we need to establish its radius of convergence. Obviously, since one of our boundary conditions is at $x = 1$, the series must converge out at least this far to be of utility. Using the ratio test (see Appendix C) in the form $a_k/a_{k-2} = Bx^2/k^2$, taking the limit as $k \to \infty$ with $x$ and $B$ fixed shows that the limit is zero, and therefore the series converges for all $x$ and $B$. It is immediately apparent, however, that although the series converges for any $B$, the convergence will become slower as $B$ gets larger.

We would like to bound the error incurred in truncating the expansion at some point, and this is accomplished as follows. For example, the "tail" $T$ of the series for $y(x)/a_0$ after stopping at the $x^4$ term is expressed as

$$T = \left[\frac{B^3x^6}{6^2 4^2 2^2} + \frac{B^4x^8}{8^2 6^2 4^2 2^2} + \frac{B^5x^{10}}{10^2 8^2 6^2 4^2 2^2}\cdots\right] = \frac{B^3x^6}{6^2 4^2 2^2}\left[1 + \frac{Bx^2}{8^2} + \frac{B^2x^4}{10^2 8^2} + \cdots\right] \tag{l}$$

Inspection of the series in brackets on the right side of Equation (l) shows that it is smaller than the corresponding *geometric* series, which we get by replacing the factor $10^2$ by $8^2$, and so on. Thus, we have

$$T < \frac{B^3x^6}{6^2 4^2 2^2}\left[1 + \frac{Bx^2}{8^2} + \frac{B^2x^4}{8^2 8^2} + \frac{B^3x^6}{8^2 8^2 8^2}\cdots\right] = \frac{B^3x^6}{6^2 4^2 2^2} \cdot \frac{1}{(1 - Bx^2/8^2)} \qquad (Bx^2 < 8^2) \tag{m}$$

Here, the sum of the geometric series $1 + a + a^2 + \cdots = 1/(1-a)$ for $a < 1$, and this is the reason for the restriction $Bx^2 < 8^2$ in Equation (m). To reduce this restriction, we could merely add more terms before neglecting the tail. By repeating this analysis and stopping just before the general term involving $x^k$, we can see that the error would be just this first neglected term multiplied by $1/[1 - Bx^2/(k+2)^2]$, with the restriction that $Bx^2 < (k+2)^2$. The error calculation we have performed here is also directly applicable to the determination of the value of $a_0$; in this case, the above results apply with $x = 1$. Table 10.1 shows the results of some calculations with the series. To get six-figure accuracy for the full range of $B$, we need only about 20 terms of the series.

## 10.3 METHOD OF SUCCESSIVE APPROXIMATION

We mention this method, which is also known as the Picard or iterative method, for the purpose of illustrating the ideas that are of use in asymptotic analysis and the method of weighted residuals. Much more detailed coverage of this topic is given by Collatz [1966]. We consider the simplest case of a single, first-order equation of the general type $dy/dx = f(x, y)$ with initial condition $y(0) = A$. The classical successive approximation approach to solving such a problem is to do a formal integration of the equation (ignoring the fact that $y(x)$ is an unknown function). This yields

$$y(x) - A = \int_{t=0}^{x} f(t, y(t))\, dt \tag{10-11}$$

To proceed, we *assume* that we know the function $y(t)$ in the integral on the right of Equation (10-11), and we call this function our zeroth approximation, $y_0(x)$. The result of performing the integration using $y_0(t)$ is called $y_1(x)$. Then we can systematically generalize this process by defining the following successive approximation scheme:

$$y_{n+1}(x) - A = \int_{t=0}^{x} f(t, y_n(t))\, dt \tag{10-12}$$

By using $y_0(x)$ on the right in Equation (10-12), we get $y_1(x)$; using this on the right, we generate $y_2(x)$; and so on. The traditional idea is to show that this process ultimately converges to the true solution under appropriate circumstances. The usual procedure employs the known initial condition, $y(0) = A$, to define the initial function $y_0(x)$ according to $y_0(x) = A$. This particular approach is a standard one for proving the existence and uniqueness of the solution of a differential equation; but this is *not* the reason we are discussing it here. Instead, we note that although the traditional procedure is generally not practical because of the pedestrian choice of $y_0(x) = A$, more imaginative choices can sometimes cause the method to give useful results in only one or two iterations. This allows the method to be of practical value in getting an approximate solution. We will give an illustration of this a bit later.

Another important aspect of the method of successive approximations is the error analysis associated with it. Let us define the error at the $n$th stage as $e_n(x) = y_n(x) - y(x)$, where $y(x)$ is the true unknown solution. By subtracting Equation (10-11) from (10.12) and using Taylor's theorem to write $f(x, y_n) - f(x, y) = (\partial f/\partial y)_\xi e_n$, where $\xi$ is between $y_n(x)$ and $y(x)$, we get

$$e_{n+1}(x) = \int_0^x \left(\frac{\partial f}{\partial y}\right)_{x,\xi} e_n(t)\, dt \tag{10-13}$$

If we let $E_n = \max |e_n(t)|$, then by carefully applying this definition, taking absolute values and using inequalities in Equation (10-13), we get the following inequality:

$$E_{n+1} < E_n \int_0^x \left|\frac{\partial f}{\partial y}\right|_{x,\xi} dt < E_n \cdot x \cdot \max\left|\frac{\partial f}{\partial y}\right| = K(x) \cdot E_n \tag{10-14}$$

This gives a bound for the maximum error at stage $(n + 1)$ in terms of a maximum bound at stage $n$. We now define the difference between two successive approximations as $d_{n+1}(x) = y_{n+1}(x) - y_n(x) = E_{n+1}(x) - E_n(x)$. By taking $D_{n+1} = \max |d_{n+1}(x)|$ and using the triangle inequality, we have that $E_n \leq E_{n+1} + D_{n+1}$. When this is combined with Equation (10-14) and rearranged, we finally obtain the useful result

$$E_{n+1} \leq \frac{K(x) D_{n+1}}{1 - K(x)} \quad \text{(for } K(x) < 1\text{)} \tag{10-15}$$

Equation (10-15) shows that the maximum error at stage $(n + 1)$ is bounded by a quantity containing $D_{n+1}$, the maximum difference between $y_{n+1}$ and $y_n$. This result is very similar to one derived for successive approximation applied to algebraic equations. Thus, a way to make the error small is to *find a* $y_0(x)$ *that produces a nearly equal* $y_1(x)$. We shall come back to this important idea when we discuss the method of weighted residuals.

**Example 10.3 Successive Approximation for the Blasius Equation**

We reconsider the Blasius equation, which we treated by Taylor approximation in Example 10.1. The equation and boundary conditions are $f''' + ff'' = 0$ with $f(0) = f'(0) = 0$; $f'(\infty) = 1$. *Classical* successive approximation would involve, to get $f''$, a single integration followed by use of the boundary conditions at $x = 0$ to generate a polynomial approximation for $f''$ that depends on the unknown $f''(0)$. This polynomial would be similar to the truncated Taylor expansion we derived earlier. Thus, it would be impossible to approximate the unknown $f''(0)$ because of the infinite boundary condition $f'(\infty) = 1$.

But suppose we proceed by noting that the new variable $p = f''$ reduces the original Blasius equation to $dp/dx = -f \cdot p$. One simple integration of this, assuming $f$ to be known, yields

$$f''(x) = f''(0)e^{-\int_0^x f(u)du} \tag{a}$$

We now integrate Equation (a) with respect to $x$, using $f'(0) = 0$, to get an expression for $f'(x)$.

$$f'(x) = f''(0)\int_{t=0}^{x} e^{-\int_0^t f(u)du}dt \tag{b}$$

Equation (b) gives an expression for $f'(x)$, the velocity profile, if only the value of $f''(0)$ is known. Now we approximate $f(u)$ in the exponent integral by the truncated Taylor series $f(u) = f(0) + f'(0)u + f''(0)u^2/2 = f''(0)u^2/2$, since both $f(0)$ and $f'(0) = 0$. If we replace $f(u)$ in Equation (b) by this approximation, we obtain

$$f'(x) = f''(0)\int_{t=0}^{x} e^{-\int_0^t f''(0)u^2/2du}dt = f''(0)\int_{t=0}^{x} e^{-(f''(0)/6)t^3}dt \tag{c}$$

If we now apply the boundary condition $f'(\infty) = 1$ in Equation (c), we get an expression for $f''(0)$.

$$1 = f''(0)\int_{t=0}^{\infty} e^{-(f''(0)/6)t^3}dt = f''(0)\left(\frac{6}{f''(0)}\right)^{1/3}\frac{\Gamma(1/3)}{3} \tag{d}$$

Here, we have evaluated the infinite integral in Equation (d) by rescaling and using gamma functions. By solving this simple equation for $f''(0)$, we find that $f''(0) = 0.48$ versus the known accurate value of 0.47. For further details on this approach, see Meksyn [1961].

This simple application of successive approximation is seen to give remarkably good results. Knowing the value of $f''(0)$, we can use Equation (c) to get values of the velocity profile $f'(x)$. For moderate values of $x$, we can simply expand the exponential and integrate term by term to get a convergent series. For large $x$, we would use an asymptotic expansion of the integral.

## 10.4 METHOD OF WEIGHTED RESIDUALS

The method of weighted residuals (MWR) is a class of methods for obtaining approximate solutions for problems in differential equations, both ordinary and partial. It is generally directed toward the solution of *boundary value* problems. Here, we will outline the idea of the procedure and give a brief, but typical, example for an

ODE boundary value problem. Extensive discussions of the method are given in Finlayson [1980,1972] and Villadsen and Michelsen [1978].

We considered earlier the approximation of *known* functions and saw, among various possibilities, both the Taylor and distributed error methods (including interpolation and least squares). The MWR can be considered to be a procedure for finding a distributed error approximation to an *unknown* function. This means, for example, that the MWR technique usually generates an approximation of given accuracy that has fewer terms than the Taylor approximation, but its error will not become asymptotically small at some specified point.

To implement the MWR, we proceed as follows. A "trial" solution function is postulated, containing one or more unknown coefficients. The trial function is forced to satisfy the boundary conditions of the problem. The trial function is then substituted into the differential equation to form the "residual," which is the amount by which the trial function fails to satisfy the equation. If the residual should be *identically* zero, the trial function is an exact solution. The idea of MWR is that to get a good approximation, the residual must be made small in some global sense, and the particular MWR methods correspond to different (but similar) ways of doing this. Various MWR methods go by the names of collocation, least squares, subdomain, Galerkin, moments, and orthogonal collocation. We will illustrate here the method that we consider to be most useful in ODE boundary value problems, namely that of orthogonal collocation.

**Example 10.4 A Nonlinear Catalyst Pellet Problem**

We consider the nonlinear boundary value problem $d^2C/dx^2 = BC^2$, with $C(0) = 1$ and $C'(1) = 0$. This problem represents diffusion in a catalyst pellet with a second-order irreversible reaction. The process is described in Cartesian coordinates. We assume the trial function, $C_a$, to be given by the cubic polynomial

$$C_a(x) = a_0 + a_1x + a_2x^2 + a_3x^3 \tag{a}$$

By fitting the boundary conditions, we find that $a_0 = 1$ and $a_1 + 2a_2 + 3a_3 = 0$. We eliminate $a_0$ and $a_1$ from the trial function using these conditions. The altered trial function, which now satisfies the boundary conditions, is

$$C_a(x) = 1 + a_2(x^2 - 2x) + a_3(x^3 - 3x) \tag{b}$$

The next step is to form the residual $R(x, a_2, a_3)$, which is equal to what is left over when we substitute $C_a$ in place of the unknown $C$ in the differential equation. For this problem, we have for the residual

$$R = \frac{d^2C_a}{dx^2} - BC_a^2 = 2a_2 + 6a_3x - B[1 + a_2(x^2 - 2x) + a_3(x^3 - 3x)]^2 \tag{c}$$

As mentioned earlier, the different MWR methods correspond to different ways of choosing the coefficients $a_2$ and $a_3$ by trying to make the residuals small. One natural procedure we could use is to minimize the square of the residual integrated over the region of the problem. That is, we could write

$$\min_{a_2, a_3} \int_0^1 R^2(x, a_2, a_3)dx \tag{d}$$

and use this expression to determine $a_2$ and $a_3$. However, as it stands, implementation of Equation (d) would require an integration followed by minimization. Thus, in this

**TABLE 10.2** MWR ORTHOGONAL COLLOCATION SOLUTION OF $C'' = BC^2$: $B = 1$, $C(0) = 1$, $C'(0) = 0$

| $x$ | 1 parameter | 2 parameters | 3 parameters (converged) |
|---|---|---|---|
| 0.0 | 1.000 | 1.000 | 1.000 |
| 0.25 | 0.869 | 0.864 | 0.865 |
| 0.50 | 0.775 | 0.775 | 0.778 |
| 0.75 | 0.719 | 0.727 | 0.728 |
| 1.0 | 0.700 | 0.712 | 0.712 |

two-parameter problem, by differentiating with respect to $a_2$ and $a_3$ under the integral sign we would get

$$0 = \int_0^1 2R\frac{\partial R}{\partial a_2}dx \qquad 0 = \int_0^1 2R\frac{\partial R}{\partial a_3}dx \tag{e}$$

Solving Equations (e) requires two integrations and the solution of two nonlinear algebraic equations; quite a messy problem. However, we recall from our discussion of Gauss quadrature in Chapter 9 that the Gauss procedure has the optimal property of yielding the best approximation with the fewest quadrature points (within the class of polynomials). Thus, the so-called orthogonal collocation method applies two-point Gauss quadrature to each of the two integrals in Equations (e). This then yields the very nice result that an approximate solution to these equations corresponds to solving the two nonlinear algebraic equations

$$R(x_1, a_2, a_3) = 0 \qquad R(x_2, a_2, a_3) = 0 \tag{f}$$

Thus, for our example, we see that the orthogonal collocation procedure corresponds to the neat formulation in which we *solve the algebraic equations for the residuals set equal to zero at the Gauss quadrature points.* These points happen to be zeroes of orthogonal polynomials, and hence the name "orthogonal collocation." Useful tables for these zeroes are given, for example, in Finlayson [1980]. To get actual numerical results in this problem, we use $B = 1$ and make use of computer program OC, which is discussed at the end of the chapter. For this problem, the Gauss quadrature points are $x_1 = 0.2113$ and $x_2 = 0.7887$. Results of this calculation for the present problem, as well as for the inclusion of additional terms in the trial function, are shown in Table 10.2. We see that, based on comparing successive trial functions, the results indicate that an accuracy of three figures for $C_a(x)$ is achieved by a fourth-degree polynomial trial function, which requires the use of three parameters. The one-, two- (the one we have derived), and three-parameter trial functions, all of which satisfy the boundary conditions, are shown below.

$$\begin{aligned} y_1(x) &= 1 + b(1,1)(x^2 - 2x) \\ y_2(x) &= 1 + b(1,2)(x^2 - 2x) + b(2,2)(x^3 - 3x) \\ y_3(x) &= 1 + b(1,3)(x^2 - 2x) + b(2,3)(x^3 - 3x) + b(3,3)(x^4 - 4x) \end{aligned} \tag{g}$$

The notation in these equations means, for example, that $b(2,3)$ is the coefficient of the second component of the three-parameter trial function. The values of the calculated parameters are

$$\begin{aligned} &b(1,1) = 0.300198 \\ &b(1,2) = 0.435227 \quad b(2,2) = -0.0736724 \\ &b(1,3) = 0.489440 \quad b(2,3) = -0.172600 \quad b(3,3) = 0.0478391 \end{aligned} \tag{h}$$

As indicated in the example problem, a common practice for estimating errors in the MWR method is to compare the results of successive trial functions (each containing one parameter more than the one before) to look for convergence of the approximations. We close our brief discussion of MWR for ODE boundary value problems by noting several points. Possible advantages of MWR include development of a relatively simple approximation to the solution, the rapid convergence characteristics of distributed error approximation, and flexibility in the choice of trial functions. A disadvantage is that nonlinear boundary conditions are awkward to handle. MWR procedures are also employed in the *finite element method* for the numerical solution of partial differential equations, which will be mentioned briefly in Chapter 14.

## 10.5 ASYMPTOTIC METHODS

We briefly consider here an introduction to some of the most important *asymptotic* analytical approximation methods for ordinary differential equations. These methods are often more powerful than those we have discussed previously, but they generally require more analytical skills for their implementation. The interested reader can find more information on these topics in the references by Bender and Orszag [1978], Meksyn [1961], Nayfeh [1981], and Zwillinger [1989].

### 10.5.1 Regular Perturbation

In applications of mathematics, the term "perturbation methods" normally refers to the idea of viewing a particular problem as a "slightly changed" or perturbed form of a well-known problem. Usually, this concept is employed by considering a troublesome term in an equation to be small so that it can be partially ignored. The so-called regular perturbation method, which will be considered here, corresponds to the case where a single, straightforward expansion represents the solution uniformly throughout a given region. The regular perturbation method is usually employed by regarding an equation to have a small parameter, and using Taylor's theorem to obtain an expansion of the solution in terms of this parameter. Thus, the perturbation method aims at obtaining $y(x)$ as an expansion in a parameter for $x$ fixed, in contrast to the usual series method where $y$ is determined by an expansion in $x$ with the parameter fixed. One of the important advantages of the perturbation method is that it is not restricted to linear problems. To indicate in detail the workings of the regular perturbation method, we reconsider the nonlinear boundary value problem considered in the previous example, $C'' = BC^2$ with $C(0) = 1$ and $C'(1) = 0$. This is the problem we have treated above by weighted residuals. We seek a solution to this equation for $B$ near (but not equal to) zero. Small $B$ corresponds to the case of slow kinetics for the reaction in the catalyst pellet. Of course, if $B$ is equal to zero, the equation reduces to $C'' = 0$. Thus, we expect that the first approximation for small $B$ may correspond to this case. To systematically find appropriate correction terms, we write the Taylor expansion for $C(x, B)$ near $B = 0$ as

$$C(x,B) = C(x,0) + \left.\frac{\partial C}{\partial B}\right|_{x,0} \cdot B + \left.\frac{\partial^2 C}{\partial B^2}\right|_{x,0} \cdot \frac{B^2}{2!}\cdots \tag{10-16}$$
$$= C_0(x) + C_1(x)B + C_2(x)B^2\cdots$$

The functions $C_0, C_1$, and so on depend only on $x$, since they represent partial derivatives of $C$ with respect to $B$ evaluated at $B = 0$. Next, we form the derivatives $C'$ and $C''$ from the assumed series according to

$$\begin{aligned} C' &= C_0' + C_1'B + C_2'B^2\cdots \\ C'' &= C_0'' + C_1''B + C_2''B^2\cdots \end{aligned} \tag{10-17}$$

The derivatives in Equations (10-17) are determined by differentiating with respect to $x$, while holding $B$ fixed, since this is the interpretation of the derivatives in the original differential equation. It is also necessary to develop $C^2$ as a series in $B$. This is done in the following way, using the series multiplication algorithm (see Appendix D).

$$C^2(x,B) = (C_0 + C_1B + C_2B^2\cdots)(C_0 + C_1B + C_2B^2\cdots) = C_0^2 + 2C_0C_1B\cdots \tag{10-18}$$

When the series in $B$ for $C''$ and $C^2$ are substituted into the original equation $C'' - BC^2 = 0$, we get

$$(C_0'' + C_1''B + C_2''B^2\cdots) - B(C_0^2 + 2C_0C_1B\cdots) = 0 \tag{10-19}$$

We now rearrange this series in powers of $B$.

$$(C_0'')B^0 + (C_1'' - C_0^2)B + (C_2'' - 2C_0C_1)B^2\cdots = 0 \tag{10-20}$$

The coefficients of the powers of $B$ in Equation (10-20) are now required to be zero. This yields a set of ordinary differential equations for $C_0, C_1, C_2$, and so on, as follows:

$$\begin{aligned} B^0&: \quad C_0'' = 0 \\ B^1&: \quad C_1'' = C_0^2 \\ B^2&: \quad C_2'' = 2C_0C_1 \end{aligned} \tag{10-21}$$

Each of the preceding ordinary differential equations requires two boundary conditions for its solution. These are obtained by substituting the assumed series for $C$ in terms of $B$ into the original boundary conditions. This yields

$$\begin{aligned} x = 0&: \quad C(0,B) = 1 = C_0(0) + C_1(0)B + C_2(0)B^2\cdots \\ x = 1&: \quad C'(1,B) = 0 = C_0'(1) + C_1'(1)B + C_2'(1)B^2\cdots \end{aligned} \tag{10-22}$$

Since this must be true for all $B$ in the range of interest, we equate the powers of $B$ of both sides of Equations (10-22). This shows that we must have $C_i'(1) = 0$ for $i = 0, 1, 2$, and so on, at $x = 1$. The other boundary condition at $x = 0$ yields the requirements $C_0(0) = 1$; $C_i(0) = 0$. Note that the *original* boundary conditions are required to hold for all $B$ in the range considered. This fact leads to the boundary specifications for the mathematical functions $C_0$, $C_1$, $C_2$ as given above. To complete the formal solution, we solve for as many of the $C_0$, $C_1$, and so forth as we wish, using the boundary conditions indicated. The functions $C_0(x)$, $C_1(x)$, and so on are then substituted back into the assumed series $C(x,B) = C_0(x) + C_1(x)B\ \cdots$.

We now pause to determine explicit expressions for the perturbation coefficients $C_0(x)$, $C_1(x)$, and $C_2(x)$. The solution for each of these equations involves integration of a simple polynomial function. When the appropriate boundary conditions are applied as outlined above, substitution of the expressions for $C_0(x)$, $C_1(x)$, and $C_2(x)$ back into the original assumed perturbation expansion gives

$$C(x, B) = 1 + \left(\frac{x^2}{2} - x\right)B + \left(\frac{x^4}{12} - \frac{x^3}{3} + \frac{2x}{3}\right)B^2 + O(B^3) \tag{10-23}$$

It will be observed that the leading term of the perturbation series, $C_0(x)$, corresponds to the solution of the original equation with $B = 0$. This was to be expected. Since the perturbation process leads to an infinite system of differential equations, it might seem that the method involves difficulties which are more severe than those associated with the original equation. However, since the whole purpose is to get a solution valid for small $B$, only a relatively few of the $C_i$(x) functions are normally determined. Note that the original equation is nonlinear, while the perturbation functions $C_i(x)$ each satisfy a linear differential equation. Moreover, the perturbation method leads to an analytical expression of the solution (in terms of *both* variables $x$ and $B$) that is convenient for parametric studies. The MWR "analytical" solution of this problem presented earlier was a function *only* of $x$ (for a fixed $B$). Similarly, a series solution in terms of $x$ would be quite awkward, both because the equation is nonlinear and because of the boundary value nature of the problem. Also, if the above problem was solved by purely numerical techniques, parametric studies would require either correlation of the computer output or numerical interpolation.

An important question that arises in connection with perturbation methods concerns the region of validity of the results. In our example, the parameter $B$ was said to be "small" in some sense. The question is, how accurate is a given perturbation approximation for a given value of $B$? This question can be given a rigorous answer only with great difficulty. The usual procedure in practice is to assume that the error is roughly the magnitude of the last term retained, in analogy to convergent or asymptotic series. Thus, if the terms in the perturbation series decrease rapidly, we expect good accuracy. If not, we should be cautious about the results. Another very useful approach is to perform selected numerical integrations to experimentally determine the region of validity of the perturbation approximation.

In Table 10.3, a comparison is made between the numerical solution and various perturbation solutions of our example problem. We compare the values of $C(1)$ as a function of $B$ for the regular perturbation solution Equation (10-23), the (1,1) Padé version of this from program PADE, an asymptotic rational least-squares fit, and the numerical solution based on program IEXVSSH. The Padé approximation is based on exactly the same information as the perturbation solution. The data of Table 10.3 (except for rational least squares) are shown plotted in Figure 10.3. At very small values of $B$, the agreement of the perturbation solution is essentially perfect. As $B$ gets larger, the deviation between the perturbation solution and numerical values becomes larger. It is very interesting to observe the (1, 1) Padé version of the three-term perturbation expansion, which uses exactly the same information. The table shows that the Padé acceleration gives results that are qualitatively in agreement with the numerical values, and also agree quantitatively over a much wider range of $B$

**TABLE 10.3** REGULAR PERTURBATION SOLUTION OF $C'' = BC^2$

$C(1)_{\text{Pert}} = 1 - B/2 + 5B^2/12 \qquad C(1)_{\text{Padé}} = [1 + B/3]/[1 + 5B/6]$

| B | $C(1)_{\text{Pert}}$ | $C(1)_{\text{Padé}}$ | $C(1)_{\text{RatLS}}$ | $C(1)_{\text{Num}}$ |
|---|---|---|---|---|
| 0.1 | 0.954 | 0.954 | 0.954 | 0.954 |
| 0.2 | 0.917 | 0.914 | 0.914 | 0.914 |
| 0.5 | 0.854 | 0.824 | 0.819 | 0.820 |
| 1.0 | 0.917 | 0.727 | 0.713 | 0.712 |
| 2.0 | 1.67 | 0.625 | 0.582 | 0.581 |
| 3.0 | 3.25 | 0.571 | 0.498 | 0.500 |

than do the perturbation results. The moral here is, whenever dealing with a slowly convergent series, always remember to try acceleration.

It is also interesting to see how rational least squares fares in representing the numerical relationship between $C(1)$ and $B$ shown in Table 10.3. In order to retain the asymptotic character that $C(1) \to 1$ for $B \to 0$, we proceed as follows. Using $y = C(1)$, $x = B$, we define the three-parameter rational least-squares prediction function $z_p$ as $z_p = (y - 1)/x = (p_0 + p_1 x)/(1 + q_1 x)$. Then, using the numerical $y - x$ data in Table 10.3, we employ program RATLSGR to determine the values of $p_0$, $p_1$, and $q_1$. This calculation produces $p_0 = -0.494492$, $p_1 = -2.72426 \times 10^{-2}$, and $q_1 = 0.814875$. The calculated values produced by the rational least-squares fitting formula are shown in Table 10.3. At the given data points, the maximum error is seen to be no larger than 0.002, which is *excellent* agreement compared to the other representations. However, in defense of the perturbation/Padé approximation, the following points must be noted. First, the perturbation calculation produces a result that depends on the *two* variables, $x$ and $B$. Also, the rational least-squares approximation could not be obtained without numerical values, whereas the perturbation solution has no such requirement. The choice of which procedure is "best" depends on the information available and the requirements of the problem.

We give a few final remarks concerning the perturbation method. Sometimes a *large* parameter problem can become a regular perturbation problem through rescaling. If the original problem does not contain a parameter, we can put one in, do a small or large parameter expansion, and at the end replace the parameter by unity. We may be able to *numerically* (rather than analytically) compute many terms of the perturbation series. And finally, if we have enough terms, regular perturbation may be useful if we combine it with an acceleration technique such as Padé approximation.

The simple perturbation method discussed here was shown to lead to useful results for the example presented. In some problems, use of this technique does *not* lead to a useful solution, and modifications of the method are required. Such problems are often referred to as "singular perturbation" problems. A relatively common example of this occurs when the small parameter of an equation is multiplied by the highest derivative. In such a case the natural first approximation, corresponding to placing the small parameter equal to zero, cannot normally satisfy all of the required boundary conditions.

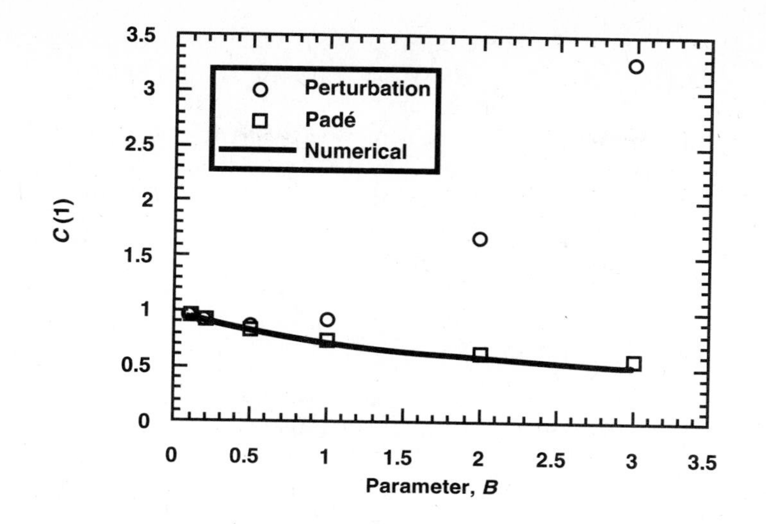


**Figure 10.3.** Comparison of Perturbation and Padé Solutions with Exact Solution

### 10.5.2 Other Asymptotic Procedures

There are several other rather general asymptotic procedures that are useful in various situations, and we will make brief mention of some of these here. First let us consider a rather simple problem that illustrates some important ideas. For concreteness, we take the first-order equation

$$\frac{dy}{dx} + Ky = f(x) \tag{10-24}$$

This equation can be solved explicitly by using the first-order integral formula which was given in the section on review of ODE. The solution is expressed as

$$y(x) = e^{-Kx}\left[y(0) + \int_0^x e^{Kt} f(t)\, dt\right] \tag{10-25}$$

For either large $K$ and fixed $x$ or large $x$ and fixed $K$, through continued integration by parts (or by Watson's lemma in the integral if we translate axes), we can develop the following asymptotic expansion for $y(x)$:

$$y(x) = e^{-Kx}\left[y(0) - \frac{f(0)}{K} + \frac{f'(0)}{K^2}\cdots\right] + \left[\frac{f(x)}{K} - \frac{f'(x)}{K^2}\cdots\right] \tag{10-26}$$

In Equation (10-26), the exponentially small term would normally be dropped in this asymptotic approximation. The interesting thing to be noted in this problem is that we can obtain the asymptotic approximation in a much simpler and more direct manner which does not depend on having an explicit solution to the differential equation. To do this, we proceed as follows. In the equation $dy/dx + Ky = f(x)$, we look for a *dominant balance*. This is the same idea that was used previously in the asymptotic solution of nonlinear algebraic equations. We consider pairs of terms that may be important (dominant) in the equation, and get the solution based on this balance. Then we use this computed approximate solution to see if the neglected terms are indeed small. If so, the balance is consistent, and we hopefully have an asymptotic solution to our equation. When we have obtained such a solution, it is always wise to check it with selected numerical comparisons.

Proceeding with our example problem, we *assume* that the term $dy/dx$ is unimportant compared to $Ky$ or $f(x)$ when either $K$ or $x$ is large. This assumption leads to the approximate solution for $y(x)$ of the form $y(x) = f(x)/K$. To test for consistency, we differentiate this equation to get $y'(x) = f'(x)/K$. If we substitute for $y$ and $y'$ in the original equation, we have

$$\frac{dy}{dx} + Ky - f(x) \cong \frac{1}{K}\frac{df}{dx} + Ky - f(x) = 0 \tag{10-27}$$

In Equation (10-27), we see that the term $(1/K)(df/dx)$ will be small compared to the other term if either $f(x)$ and $df/dx$ are bounded and $K$ is large, or $K$ is fixed and $df/dx$ gets small compared to $f(x)$ for large $x$. To continue, we can simply solve for $y$ on the right-hand side of Equation (10-27) to get

$$y(x) \cong \frac{f(x)}{K} - \frac{f'(x)}{K^2}\cdots \tag{10-28}$$

This can be continued and is seen to be the same as the right-hand side of Equation (10-26) (except for the exponentially small term). What this neat and simple approximate solution does *not* do is allow for satisfaction of any initial conditions. But the idea is important in more complicated problems.

### 10.5.3 The WKB Method

As a brief illustration of this, we make a few remarks about the classical and famous WKB approximation method, whose name comes from three people who made contributions to the procedure (Wentzel, Kramers, and Brillouin). For further detailed information, see Zwillinger [1989], Bender and Orszag [1978], and Nayfeh [1981]. This method is applicable to second-order, variable-coefficient, linear differential equations of the type that arise in the application of separation of variables to partial differential equations. It is valuable for determining approximate solutions for either large $x$ or a large parameter. There are (at least) two ways that the WKB procedure can be developed. We first consider the simplest method, which is analogous to that employed to get Equation (10-28). To show the approach for the case of large $x$, we consider the equation

$$\frac{d^2y}{dx^2} = a^2(x)y \tag{10-29}$$

Note that Equation (10-29) automatically requires a *positive* factor multiplying $y$; the negative factor case will be mentioned later. Any second-order linear homogeneous ODE of the form $y'' + p(x)y' + q(x)y = 0$ can be put into this form (with the possible exception of having a minus sign on the right-hand side of (10-29)) by an appropriate change of variable, which will be discussed below. We now introduce the so-called Ricatti transformation to replace the dependent variable $y$ by $u$ according to $y'/y = u$. This causes Equation (10-29) to become

$$\frac{du}{dx} + u^2 = a^2 \tag{10-30}$$

Next, we use the dominant balance approach on Equation (10-30). The fortunate result is that it often turns out that an appropriate balance corresponds to $u^2 \approx a^2$, which gives the two solutions $u \approx \pm a$. Since the original equation is second-order, we need both solutions. To check that the balance is correct, we substitute (for example) $u \approx a$ back into Equation (10-30) by using $du/dx \approx da/dx$. The requirement is that $da/dx$ be small compared to $a^2$ for large $x$. If this is true, the balance is consistent, and we can use the approximation of $du/dx \approx da/dx$ to improve the approximation according to

$$u(x) = a(x)\left[1 - \frac{1}{a^2}\frac{du}{dx}\right]^{1/2} \cong a\left[1 - \frac{1}{2a^2}\frac{da}{dx}\right] = a - \frac{1}{2a}\frac{da}{dx} \tag{10-31}$$

In Equation (10-31), we have replaced $du/dx$ by $da/dx$ and used Taylor's theorem, since $(1/a^2)(da/dx)$ is small. From the definition of $u$, $y$ is given by

$$y(x) = e^{\int u(t)\,dt} = e^{\int (a-a'/2a)dt} = e^{\int a dt-(1/2)\ln a} = a^{-1/2}e^{\int^x a dt} \tag{10-32}$$

The second solution corresponding to taking the minus sign, $u \approx -a$, gives Equation (10-32) *with a minus sign in the exponential.* The sum of these two solutions, each multiplied by a constant of integration, constitutes the general asymptotic solution. Usually, a more accurate solution can be obtained by continuing the process, which often involves a factor such as in Equation (10-32) multiplied by a descending series in $x$. These series are usually divergent but asymptotic for large $x$.

The WKB asymptotic approach given above can also be pursued successfully by means of appropriate changes of both the independent and dependent variables. Erdelyi [1956] gives an account of this and shows that for the general equation $y'' + [\lambda^2 p(x) + r(x)]y = 0$, for large $\lambda$ we get the leading asymptotic approximations

$$p(x) > 0: \quad y(x) = \frac{1}{p(x)^{1/4}}\left\{C_1 \sin\left(\lambda\int_{x_0}^{x} p(t)^{1/2}dt\right) + C_2 \cos\left(\lambda\int_{x_0}^{x} p(t)^{1/2}dt\right)\right\} \tag{10-33}$$

or

$$p(x) < 0: \quad y(x) = \frac{1}{[-p(x)]^{1/4}}\left\{C_3 \exp\left(-\lambda\int_{x_0}^{x} [-p(x)]^{1/2}dt\right) + C_4 \exp\left(\lambda\int_{x_0}^{x} [-p(x)]^{1/2}dt\right)\right\} \tag{10-34}$$

We may note that Equation (10-34) produces the same result that was derived in Equation (10-32) if we identify $-p(x)$ with $a^2(x)$ and take $\lambda = 1$.

The WKB results quoted above apply only for the case where $p(x) \neq 0$ within the interval of interest. If this condition is not satisfied, significant modifications are required. The results indicated here are especially useful for *Sturm-Liouville*-type problems, which are associated with separation of variables solutions of partial differential equations. *Sturm-Liouville* problems will be discussed in Chapter 13.

## 10.6 COMPUTER PROGRAM

In this section we discuss the computer program OC, the method of weighted residuals program that is applicable to problems in this chapter. We give a brief discussion and a concise listing, which includes (i) a program description, (ii) how to run, and (iii) user-specified subroutines, including a sample problem. Detailed program listings can be obtained from the diskette.

### Weighted Residuals Orthogonal Collocation Program OC

Instructions for running the program are given in the concise listing. The program requires the domain of the independent variable to be from 0 to 1. The problem can be linear or nonlinear, and up to six coefficients in a trial function can be determined. There is an option for Shanks acceleration of the computed approximations. The trial function and its required derivatives, along with the expression for the residual, must be specified in Sub Func. The problem shown in the concise listing is the same one discussed in the chapter, namely, $C'' = BC^2$, with $C(0) = 1$ and $C'(1) = 0$. Notice that the trial function and its derivatives are programmed in a loop, which is very convenient for the user. When the program is run, the user is prompted for the number of parameters to be used (1 to 6), the desired output print increment, and a first estimate of the one-parameter coefficient $b(1, 1)$. For higher approximations, no further initial estimates are necessary. The program produces values of the coefficients $b(i,j)$ for all approximations computed and gives an option for Shanks acceleration of the results if at least three approximations are used. Finally, all approximate solutions are shown, which assists in determining how far the results have "converged."

#### Program OC.

```
!This is program OC.tru, orthogonal collocation for ODE BVP by    O.T.Hanna
!The problem domain must be t=0 to t=1;problem can be linear or nonlinear.
!Program solves for n o.c. coeff.(n = 1 to 6) via full Newton method.
!There is optional Shanks acceleration of computed o.c. approximation.
!n = highest degree of approx.(1 pt.,2pt... up to 6 pt.)
!h = t print increment for solution.
!Trial function, derivatives,residual must be specified in Sub Func.
!User must supply 1st guess of b(1,1) at prompt; only one guess value is
!needed regardless of no. of parameters. If y can be estimated, we can use
!guess = [y-g0]/g(1) at x = 0.5 (k=1).
!Program determines other initial estimates as necessary.
!b(i,k) is parameter vector for ith coeff. in k parameter approx.-in Sub
!Func,x() takes the place of b(i,k).

!************************************************************************
!*To run,specify y(t),y'(t),y"(t),..and residual in Sub Func as shown    *
!*Program prompts for function verification,n,h,guess,Shanks option      *
!************************************************************************
```

```
SUB Func(i,x(),f(),mx,norm_tol,k,t,y)
  !Residual function subroutine;x() is parameter vector,
  !t is indep't variable
  !After use of boun.cond.,trial function and derivatives are as follows.
  !y(t) = g0 + b(1,k)*g(1) + b(2,k)*g(2)...+ b(k,k)g(k)
  !y'(t) = g10 + b(1,k)*g1(1) + b(2,k)*g1(2)...+ b(k,k)*g1(k)
  !y"(t) = g20 + b(1,k)*g2(1) + b(2,k)*g2(2)...+ b(k,k)*g2(k)
  !In these eqns. only various g functions are to be programmed.
  !User specified lines are denoted with ***
  DIM g(1),g1(1),g2(1)
  MAT redim x(k),g(k),g1(k),g2(k)
  LET mx = 20                  !max allowed iterations ***
  LET norm_tol = 1e-7          !***
  LET B = 1                    !***

  !bound.cond. are y(0) = 1, y'(1) = 0
  LET g0 = 1                   !***
  LET g10 = 0 !fns of t        !***
  LET g20 = 0                  !***
  FOR j = 1 to k
    LET g(j) = t^(j+1) - (j+1)*t          !***
    LET g1(j) = (j+1)*t^j - (j+1)         !fns of t ***
    LET g2(j) = (j+1)*j*t^(j-1)           !***
  NEXT j
  LET y = g0 + Dot(x,g)      !y(t)
  LET y1 = g10 + Dot(x,g1)   !y'(t)
  LET y2 = g20 + Dot(x,g2)   !y"(t)
  LET f(i) = y2 - B*y^2                   !Residual(t) ***
END SUB
```

## PROBLEMS

**10.1.** Solve the following differential equations:

(1) $y'' + 2y' = 0$
(2) $y'' - 2y' + y = 0$
(3) $y'' + y = 0$
(4) $y'' - y = 0$
(5) $y'' = 0$
(6) $y'' + y' + 3x = 0$
(7) $y' + 2y = 0$
(8) $y'' + 3y' + y = A$
(9) $y'' - (\sin x)y' = e^x$  (let $z = y'$)

**10.2.** The following differential equation is a form of *Bessel's* equation and occurs in problems involving cylindrical geometry.

$$\frac{1}{x}\frac{d}{dx}\left(x\frac{dy}{dx}\right) + Ky = 0 \quad \text{or} \quad x^2\frac{d^2y}{dx^2} + x\frac{dy}{dx} + Kx^2y = 0$$

(i) Here, $K$ is a parameter. Develop a full series solution to this problem for the particular boundary conditions $dy/dx(0) = 0$. This condition must always be satisfied if the equation is to apply at $x = 0$ (centerline condition).
(ii) Use the ratio test to find the region of convergence of the series.
(iii) How do we bound the error if we stop at some particular term?

**10.3.** The following nonlinear differential equation is encountered in the design of a chemical flow reactor:

$$\frac{dC}{d\phi} = \frac{C}{1 + \beta C}, \; C(0) = 1$$

Here, $\beta$ is a positive parameter.
(i) Develop directly the first three terms of the series expansion for $C(\phi)$.
(ii) Convert this expression to a Padé (1, 1) approximation.

**10.4.** Consider the nonlinear differential equation $dy/dx = (x^2 + y^2)$ subject to the initial condition $y(0) = 1$.
(i) Determine directly the first four terms of the Taylor expansion of the solution. Does the convergence seem rapid?
(ii) Observe that there can be no minimum, and the solution for $y$ necessarily becomes very large. Show from the equation that for $y \to \infty$, $x$ cannot exceed unity, but rather must approach some unknown limiting value $x_{max}$. Look for a simple asymptotic solution of the form $y \approx 1/(x_{max} - x)$ which is valid for large $y$.
(iii) The result of (ii) suggests that it may be useful to introduce the new variable $u = 1/y$. Do this, and use successive approximation to develop solutions for $u(x)$. Use both $u_0(x) = 1$ and $u_0(x) = 0$ as starting approximations (which seems most useful?). Can you say anything about the error of these approximations? Note that a simple way to find the value of $x_{max}$ precisely is to numerically integrate the equation for $x(u)$ from $u = 1$ (where $x = 0$) to $u = 0$ (where $x = x_{max}$).

**10.5.** It is desired to obtain a weighted residuals solution for the cylindrical catalyst pellet problem defined by the equation

$$\frac{d^2y}{dx^2} + \frac{1}{x}\frac{dy}{dx} = Ky \text{ with } \frac{dy}{dx}(0) = 0 \text{ and } y(1) = 1$$

Use a polynomial trial function that satisfies the boundary conditions. Use program OC, and verify the computer solution by solving the quadratic polynomial case by hand. Carry out the computer solution for up to the maximum number of parameters, and estimate the errors in the solution. Do this for (i) $K = 1$ and (ii) $K = 10$.

**10.6.** One model equation for an isothermal chemical flow reactor with dispersion is written in the dimensionless form

$$\epsilon\frac{d^2C}{dx^2} - \frac{dC}{dx} - Bf(C) = 0 \text{ with } \frac{dC}{dx}(1) = 1 \text{ and } \epsilon\frac{dC}{dx}(0) = C(0) - 1$$

Here, $\epsilon$ and $B$ are parameters and $f(C)$ is the general reaction rate function.
(i) It is desired to determine the first two terms of a regular perturbation solution (for small $B$) of this equation.
(ii) Put the solution of part (i) in Padé form.

**10.7.** (i) For the equation of problem 10.3, determine the first two terms of a regular perturbation solution for small $\beta$. (ii) By integrating the differential equation *exactly* in the form $\phi = F(C, \beta)$, develop a two-term asymptotic solution for the case of large $\beta$.

**10.8.** Consider the following two very common forms of the famous Bessel equation, which often apply to transport problems in cylindrical coordinates:

$$\text{(i)}\quad y'' + \frac{y'}{x} - K^2 y = 0$$

$$\text{(ii)}\quad y'' + \frac{y'}{x} + K^2 y = 0$$

Employ the transformation of the dependent variable $y = g(x)z$, and show that if we choose $g(x)$ in order to eliminate the term in $dz/dx$, the equations become

$$\text{(ia)}\quad z'' - z\left[K^2 + \frac{1}{x^2}\right] = 0$$

$$\text{(iia)}\quad z'' + z\left[K^2 + \frac{1}{x^2}\right] = 0$$

In the new form, it appears that for *either* large $K$ (except when $x$ is near zero) or large $x$, the terms in $1/x^2$ may be neglected and the solution expressed in terms of either (a) simple exponential or (b) trigonometric functions for $z(x)$. This example is closely related to the WKB method given in the text.

**10.9.** A powerful practical technique for representing a function (e.g., the solution of an ODE) in analytical form is that of fitting numerical values to a rational function by means of rational least squares. Use program RATLSGR to fit the numerical solution of the equation $y' = -y/(1 + 2 * y)$; $y(0) = 1$, for which data is given below. Consider degrees of the numerator ($k$) and denominator ($m$) corresponding to the cases $(k, m) = (2, 0), (1, 1), (4, 0), (2, 2), (6, 0,), (3, 3)$. Compare the results.

| x | y |
|---|---|
| 0.0 | 1.0 |
| 0.5 | 0.8382 |
| 1.0 | 0.6874 |
| 1.5 | 0.5494 |
| 2.0 | 0.4263 |
| 2.5 | 0.3199 |
| 3.0 | 0.2315 |
| 3.5 | 0.1615 |
| 4.0 | 0.1089 |
| 4.5 | 0.07119 |
| 5.0 | 0.04546 |

## REFERENCES

BENDER, C., and ORSZAG, S. (1978), *Advanced Mathematical Methods for Scientists and Engineers*, McGraw-Hill, New York

COLLATZ, L. (1966), *The Numerical Treatment of Differential Equations*, Springer-Verlag, New York

FINLAYSON, B. (1972), *The Method of Weighted Residuals and Variational Principles*, Academic Press, New York

HILDEBRAND, F. B. (1962), *Advanced Calculus for Applications*, Prentice-Hall, Englewood Cliffs, NJ

JENSON, V. G., and JEFFREYS, G. V. (1977), *Mathematical Methods in Chemical Engineering*, 2nd ed., Academic Press, New York

MEKSYN, D. (1961), *New Methods in Laminar Boundary Layer Theory*, Pergamon Press, New York

MICKLEY, H. S., SHERWOOD, T. K., and REED, C E. (1957), *Applied Mathematics in Chemical Engineering*, McGraw-Hill, New York

NAYFEH, A. H. (1981), *Introduction to Perturbation Techniques,* Wiley, New York

SPIEGEL, M. R. (1965), *Laplace Transforms*, Schaum, New York

VILLADSEN, J., and MICHELSEN, M. L. (1978), *Solution of Differential Equation Models by Polynomial Approximation*, Prentice-Hall, Englewood Cliffs, NJ

ZWILLINGER, D. (1989), *Handbook of Differential Equations*, Academic, New York

# 11

# Numerical Methods for ODE Initial Value Problems

In many problems involving ordinary differential equations, it is either necessary or desirable to effect a numerical solution. The major advantages of numerical methods are their great generality, as well as the fact that they can be conveniently mechanized on a computer. Their disadvantage is that they express the solution as a numerical table, rather than as an analytical function of the independent variable and parameters. However, we can use the tabular information to "fit" the numerical solution to an equation if we wish, for example by rational approximation. An *initial value* problem in ordinary differential equations is one where all of the auxiliary conditions are specified at the same point. This is in contrast to a so-called *boundary value* problem, where the data may be specified at two or more points. For the purpose of numerical solution, the initial value problem is much the easier of the two. In this chapter we restrict our discussion to the initial value problem.

## 11.1 NON-STIFF METHODS

We start with a discussion of several so-called *non-stiff* methods. These are methods which do have a stability limitation, but which work well in many problems. Stiff problems and methods correspond to the case of severe stability requirements, and they will be discussed later. We begin with a discussion of the direct use of Taylor's theorem, and this leads to the forward Euler method followed by Runge-Kutta methods. This is followed by our method of choice for non-stiff problems, the so-called IEX algorithm which is based on the new improved Euler method together with global Richardson extrapolation.

### 11.1.1 Direct Use of Taylor's Theorem

Let us consider the initial value problem $y' = f(x, y)$ with $y(0) = B$. If we temporarily suppose that the solution is somehow known at an arbitrary point $x$, then we can use Taylor's theorem to approximate the solution at a new point $(x + h)$, as follows:

$$y(x+h) = y(x) + y'(x)h + \frac{y''(x)h^2}{2!} + \frac{y'''(x)h^3}{3!} \cdots \tag{11-1}$$

Since $y(x)$ and its derivatives are not explicitly known (indeed, finding $y(x)$ is our objective), we resort to the differential equation itself to compute the derivatives in the Taylor expansion. This is accomplished by differentiating the differential equation and evaluating the result at $x$. By using the chain rule with $y' = f(x,y)$, we arrive at the following expressions for $y''(x)$ and $y'''(x)$:

$$\begin{aligned} y''(x) &= f_x + f_y y' = f_x + f_y f \\ y'''(x) &= (f_{xx} + f_{xy} f) + f_y(f_x + f_y f) + f(f_{yx} + f_{yy} f) \end{aligned} \tag{11-2}$$

Here, subscripts denote partial differentiation with respect to the indicated variable while holding the other variables constant. Substitution into the Taylor expansion, and using the initial condition allows estimation of the solution at $(x + h)$.

One question that arises immediately with regard to application of Taylor's formula concerns the number of terms to be used. We can see that in general, for a single integration step, there must be some kind of trade-off between using a small number of terms with a small $h$, or more terms with a larger $h$. Since many Taylor expansions have only a limited region of convergence, it is normally found practical to retain only a few terms of the Taylor expansion at each step of a numerical solution. Then, after $y(x + h)$ has been estimated, a new Taylor expansion about $(x + h)$ is formed to estimate $y(x + 2h)$, and so on. We can see from the formulas shown for $y''(x)$ and $y'''(x)$ that, in general, it is not easy to directly determine many of the Taylor coefficients. Moreover, the implementation is very problem-dependent, so a great effort is required on the part of the user, even for a single first-order equation.

The order of a Taylor approximation refers to the exponent of $h$ in the overall cumulative (*global*) error of the solution at a point. This global error is defined as $[y_{\text{exact}}(x) - y_{\text{approx}}(x)] = O(h^p)$, where $O(h^p)$ means "proportional to $h^p$ as $h \to 0$." Since a *local* (single-step) error of $O(h^{p+1})$ normally corresponds to a global error of $O(h^p)$, the order of a method is the same as the highest power of $h$ in its Taylor expansion. Thus, a second-order method involves retaining terms in the Taylor expansion up to $h^2$, and so on. In general, methods of higher order give more accuracy for a given number of steps, but each step is more costly. We postpone a detailed discussion of errors and efficiency until after we have introduced the fundamental methods.

To start the solution process using Taylor's formula, we begin at the initial point $x = 0$ where $y$ is known, compute the derivatives, and substitute to get, for example,

$$y(h) = y(0) + hf(0, B) + \frac{h^2}{2!}(f_x + f_y f)_{x=0, y=B} + \cdots \tag{11-3}$$

We then use this estimate of $y(h)$ to compute the derivatives at $x = h$ in order to estimate $y(2h)$. This process is then continued in a recursive manner. As an example, consider the equation $y' = 2 + x \sin y$ with $y(0) = 1$. This is a nonlinear equation for which an analytical solution is not available. Since $y'' = x(\cos y)y' + \sin y = x(\cos y)(2 + x \sin y) + \sin y$, the second-order Taylor approximation is given by

$$y(x+h) \cong y(x) + h(2 + x \sin y(x)) + \frac{h^2}{2!}\{x \cos y(x)[2 + x \sin y(x)] + \sin y(x)\} \tag{11-4}$$

The first step, using the initial condition, is

$$y(h) = 1 + 2h + \frac{h^2}{2!} \sin 1 \tag{11-5}$$

### 11.1.2 The Euler Method

A simple but important method, which can be considered to be a special case of the Taylor approximation, is known as the forward Euler method. This corresponds to retaining terms only through $y'(x)h$ in the Taylor formula, Equation (11-1), and in this sense it represents the simplest nontrivial solution method. Its simplicity makes it ideal for theoretical studies involving error analysis. The basic Euler computation scheme for $y' = f(x, y)$ is

$$\text{forward Euler} \qquad y_E(x+h) = y_E(x) + hf(x, y_E(x)) \tag{11-6}$$

Equation (11-6) defines the forward Euler recurrence relation, which is fully explicit in the sense that $y_E(x+h)$ is calculated directly from a knowledge of $y_E(x)$. It should be noted that the Euler method could also be interpreted as the result of either a forward numerical differentiation approximation to $y'(x)$ or left-hand rectangular integration applied to $y(x+h) - y(x) = \int f(t, y(t)\, dt$ over the step from $x$ to $(x+h)$.

### 11.1.3 The Runge-Kutta Procedure

We can see from the formulas shown previously for $y''(x)$ and $y'''(x)$ that, in general, it is quite difficult to calculate even a few derivatives of a differential equation $y' = f(x, y)$ for use in Taylor's formula. The direct implementation of Taylor's theorem requires an entirely different set of calculations for each differential equation. There is, however, an ingenious way to circumvent this problem, which is attributed to Runge and Kutta. This approach involves the determination of higher derivatives *indirectly*, as a combination of first derivatives evaluated at appropriate points (not necessarily on the solution trajectory). In this sense, it amounts to a kind of numerical differentiation. The Runge-Kutta procedure is extremely useful, for it allows us to effectively compute the Taylor expansion of a solution in the same way, *regardless of the equation.* This means that one set of generalized instructions can be used by a calculator or computer for all problems. Although it allows us to effectively compute the Taylor expansion, no Runge-Kutta formula is *identical* to the corresponding Taylor expansion, but every Runge-Kutta formula has a Taylor expansion that agrees with the true Taylor expansion of the unknown solution, up to and including terms of order $h$. The highest value of $n$ for which this is true is referred to as the order of a Runge-Kutta formula. There are, in principle, infinitely many different Runge-Kutta formulas of a given order. This point will be clarified with an example.

We now consider the development of some second-order Runge-Kutta formulas. For the general equation $y' = f(x,y)$, since $y''(x) = (f_x + f_y f)$ by the chain rule, the second-order Taylor approximation is given by

$$y(x+h) = y(x) + hf(x, y(x)) + \frac{h^2}{2!}(f_x + f_y f)_{x,y(x)} + O(h^3) \tag{11-7}$$

We wish to express $y''(x) = (f_x + f_y f)$ as a linear combination of $f$ values, evaluated at appropriate values of $x$ and $y$. By Taylor's theorem,

$$f(x+h_1, y+k_1) = f(x,y) + f_x h_1 + f_y k_1 + O(h_1^2 + h_1 k_1 + k_1^2) \tag{11-8}$$

By considering Equations (11-7) and (11-8), we see that $(f_x h_1 + f_y k_1)$ can be made proportional to $(f_x + f_y f) \equiv y''$ by taking $k_1 = f h_1$. Then, if we take $h_1 = ah$, we have

$$f(x+ah, y+ahf) = f(x,y) + ahy''(x) + O(h^2) \tag{11-9}$$

Substitution for $y''(x)$ from Equation (11-9) into the Taylor expansion, Equation (11-7), produces

$$\begin{aligned} y(x+h) = y(x) &+ hf(x,y(x))\left(1 - \frac{1}{2a}\right) \\ &+ \frac{h}{2a} f(x+ah, y+ahf(x,y)) + O(h^3) \end{aligned} \tag{11-10}$$

Equation (11-10) agrees, through terms of $O(h^2)$, with the true Taylor expansion, Equation (10-7), of the unknown solution. Thus, for $h$ sufficiently small, the Runge-Kutta approximation of Equation (11-10) gives results arbitrarily close to those of the true three-term Taylor expansion (in practice, this is true provided roundoff error can be neglected). From Equation (11-10), we see that there are, in principle, infinitely many Runge-Kutta formulas of order two, corresponding to different values of $a$. We shall select $a = 1$ (RK2 trapezoidal) for our discussion, but the values $a = 1/2$ (RK2 midpoint) and $a = 2/3$ have also found application. If the selection $a = 1$ is made, the Taylor expansion, Equation (11-10), becomes

$$\text{RK2T} \quad \begin{aligned} y(x+h) = y(x) & \\ &+ \frac{h}{2}\Big\{f[x, y(x)] + f\big[x+h, y+hf(x,y(x))\big]\Big\} + O(h^3) \end{aligned} \tag{11-11}$$

By dropping the error term $O(h^3)$ in Equation (11-11), we get the RK2T recurrence relation. This method is seen to require two derivative evaluations per step, in contrast to only one for forward Euler. It is very convenient to use formula (11-11) recursively within a calculation loop on a computer. The $f(x,y)$ function must be defined for a particular problem in a subprogram.

The Runge-Kutta method used most in the past is a fourth-order approximation (RK4) which requires four derivative evaluations per step, as follows:

$$\text{RK4} \qquad y(x+h) = y(x) = \frac{h}{6}[k_0 + k_1 + k_2 + k_3] \tag{11-12}$$

where

$$k_0 = f(x, y(x))$$
$$k_1 = f\left(x + \frac{h}{2}, y(x) + \frac{k_0}{2}\right)$$
$$k_2 = f\left(x + \frac{h}{2}, y(x) + \frac{k_1}{2}\right)$$
$$k_3 = f(x + h, y(x) + k_2)$$

The derivation of the RK4 formulas is similar to that for RK2, but much longer and more tedious Scheid [1988]. The RK4 method has often been a method of choice in the past because it is self-starting and has a high accuracy. However, as we will discuss and illustrate in detail below, it is, in general, very useful to employ *global Richardson extrapolation* to monitor the overall accuracy of a numerical solution. In this context, it is particularly efficient to use a lower-order method as a building block in constructing these extrapolation approximations, which themselves have both higher accuracy and known error characteristics. For this purpose, we will use the new improved Euler method, which is discussed next.

### 11.1.4 The New Improved Euler Method

A relatively new method Hanna [1988], which combines the best features of the forward Euler and RK2T methods, is called the "new" improved Euler method. This name is to contrast the new method from the several "improved Euler" methods in the literature, which are some form of RK2 requiring two derivative evaluations per step. The new improved Euler procedure is a single-step method that requires only one derivative evaluation per step (as Euler), but yields second-order global accuracy (as RK2). This method is based on the idea of a successive approximation quadrature improvement of the forward Euler solution and it is defined by the following simple recurrence relations.

***Improved Euler Method.***

$z(x_0) = y(x_0)$ Initial Condition

1. Euler step $z(x+h) = y(x) + hf[x, z(x)]$
2. Evaluate derivative $f[x+h, z(x+h)]$
3. Trapezoidal correction $y(x+h)$

$$= y(x) + \frac{h}{2}\{f[x, z(x)] + f[x+h, z(x+h)]\} \tag{11-13}$$

The initial condition for $z(0)$ is just the exact value of $y(0)$, $z(0) = y(0)$. The calculation sequence is just a forward Euler step, starting from the improved value

**TABLE 11.1** NUMERICAL SOLUTION OF $y' = xy + 1;\ y(0) = 0;\ h = 0.2$

| $x$ | $y_{\text{Euler}}$ | $y_{\text{RK2}}$ | $y_{\text{imp. Euler}}$ | $y_{\text{exact}}$ |
|---|---|---|---|---|
| 0.0 | 0.0 | 0.0 | 0.0 | 0.0 |
| 0.2 | 0.200 | 0.204 | 0.204 | 0.203 |
| 0.4 | 0.408 | 0.425 | 0.424 | 0.422 |
| 0.6 | 0.641 | 0.681 | 0.680 | 0.678 |
| 0.8 | 0.918 | 0.999 | 0.997 | 0.995 |
| 1.0 | 1.264 | 1.415 | 1.408 | 1.411 |

$y(x)$, and using a slope $f[x, z(x)]$ evaluated using the Euler value $z(x)$. After the single derivative evaluation $f[x+h, z(x+h)]$ we apply trapezoidal quadrature, where *both slopes are evaluated using the Euler values* z(x), $z(x+h)$. The key point here is that a slope at a particular $x$ is evaluated only once, using for $y(x)$ the Euler value $z(x)$. The slopes are never subsequently corrected after the improved value $y(x)$ has been obtained by trapezoidal quadrature. *By deliberately failing to compute the "corrected" slopes (using* y(x)*), a very substantial efficiency is achieved.* For the "cost" of only one derivative evaluation per step, it is easily seen that the new method achieves a global accuracy of $O(h^2)$, compared to $O(h)$ for forward Euler. Both the Euler and new improved Euler methods cost one derivative evaluation per step. It should be observed that the calculation sequence in Equations (11-13) would be just the RK2 method, if the derivative were updated after the correction of $y$. However, this would then require two derivative evaluations per step. Since the RK2 method has a global accuracy of $O(h^2)$, it would be expected that RK2 and the new method would have comparable accuracy in practice. This can be shown theoretically; it will be illustrated here by a typical numerical comparison.

To illustrate the performance of the Euler, Runge-Kutta-two trapezoidal (RK2), and improved Euler methods, we consider the simple problem $y' = xy + 1$ with $y(0) = 0$. In Table 11.1, we show the results of using these methods to integrate from $x = 0$ to $x = 1$ for the step-size $h = 0.2$. Also shown is the true solution. It is seen that the Euler method provides the least accurate approximation. The RK2 and improved Euler methods are seen to achieve much higher, and comparable, accuracy. Keep in mind that RK2 costs *two* derivative evaluations per step, while Euler and improved Euler cost only one. Thus, on this basis it would be reasonable to compare the Euler method at $h = 0.1$ with RK2 at $h = 0.2$. Even on this basis, RK2 is typically superior as it is in this problem. But the most significant thing to note regards the improved Euler solution. For the same number of derivative evaluations, it is *much* more accurate than Euler, and about four times as accurate as RK2. This behavior is not evident in Table 11.1, since for the given step-size, $h = 0.2$, the RK2 method uses twice as many derivative evaluations as the improved Euler method. When combined with global Richardson extrapolation, the "Improved Euler Extrapolation" (IEX) method will be our method of choice for non-stiff problems.

To gain a clear understanding of how the above numerical methods are implemented, it is useful to write down the details of two successive computational steps. We therefore do this now for the first two steps contained in Table 11.1.

Euler

$$f(0, y(0)) = 0(1) + 1 = 1$$
$$y(0,2) = y(0) + hf(0,0) = 0 + 0.2[1] = 0.2$$
$$f(0.2, y(0,2)) = 0.2(0.2) + 1 = 1.04$$
$$y(0.4) = y(0.2) + hf(0.2, 0.2) = 0.2 + 0.2[1.04] = 0.408$$

RK2

$$f(0, y(0)) = 0(1) + 1 = 1$$
$$z(0.2) = y(0) = hf(0, y(0)) = 0 + 0.2(1) = 0.2$$
$$f(0.2, z(0.2)) = 0.2(0.2) + 1 = 1.04$$
$$y(0.2) = y(0) + \frac{h}{2}\left[f(0, y(0)) + f(0.2, z(0.2))\right]$$
$$= 0 + \frac{0.2}{2}[1 + 1.04] = 0.204$$

$$f(0.2, y(0.2)) = 0.2(0.204) + 1 = 1.0408$$
$$z(0.4) = y(0.2) + hf(0.2, y(0.2))$$
$$= 0.204 + 0.2(1.0408) = 0.41216 \tag{11-14}$$
$$f(0.4, z(0.4)) = 0.4(0.41216) + 1 = 1.1649$$
$$y(0.4) = y(0.2) + \frac{h}{2}\left[f(0.2, y(0.2) + f(0.4, z(0.4))\right]$$
$$\text{"} \quad = 0.204 + \frac{0.2}{2}[1.0408 + 1.1649] = 0.42457$$

Since $z(0) = y(0)$, the first step is identical to RK2; after that there is only one derivative evaluation per step.

Improved Euler

$$f(0.2, z(0.2)) = 1.04 \quad \text{(calculated in first step)}$$
$$z(0.4) = y(0.2) + hf(0.2, z(0.2)) = 0.204 + 0.2(1.04) = 0.412$$
$$f[0.4, z(0.4)] = 0.4(0.412) + 1 = 1.1648$$
$$y(0.4) = y(0.2) + \frac{h}{2}\left[f(0.2, z(0.2)) + f(0.4, z(0.4))\right]$$
$$\text{"} \quad = 0.204 + \frac{0.2}{2}[1.04 + 1.1648] = 0.42448$$

A careful study of the above numerical computations shows the rather subtle difference between RK2 and improved Euler. Notice that for the first step of any integration, since in an improved Euler step we always assume that we know the derivative *before the step begins*, the RK2 and improved Euler values are identical. In the first step, improved Euler requires two derivative evaluations; after the first step, it requires only one derivative evaluation per step.

### 11.1.5 Integration of Systems of Ordinary Differential Equations

We now extend the foregoing ideas to systems of differential equations, which occur frequently in practice. First, it is very important to note that *a single higher-order differential equation is equivalent to a system of first-order equations* of size equal to the order of the single equation. Thus, for example, the second-order equation $y'' + x(y')^2 + y = 0$ can be converted, by the change of dependent variables $y = z_1$, $y' = z_2$ to the system $z_2' + xz_2^2 + z_1 = 0$ and $z_1' = z_2$. The boundary conditions must also be converted to the new variables. Similar transformations apply when there is more than one equation of order higher than first. In most cases, it is desirable to convert the original system to a first-order system for purposes of numerical integration.

All of the foregoing methods can easily be made applicable to systems of equations. To derive numerical procedures for systems analogous to those derived previously for a single equation, we could merely write a corresponding Taylor expansion for each dependent variable. Thus, for example, in the case of the second-order system $y_1' = f_1(x, y_1, y_2)$, $y_2' = f_2(x, y_1, y_2)$, we would write

$$\begin{aligned} y_1(x+h) &= y_1(x) + y_1'(x)h + \frac{y_1''(x)h^2}{2!} + O(h^3) \\ y_2(x+h) &= y_2(x) + y_2'(x)h + \frac{y_2''(x)h^2}{2!} + O(h^3) \end{aligned} \tag{11-15}$$

These expressions could be used for implementation of Taylor's formula directly, for the Euler method, or for the Runge-Kutta-two method. To use Taylor's formula directly, we would merely use the two differential equations to compute the required derivatives at any point where the solution is known, and proceed recursively. For the Euler method, we would retain terms only up to $O(h)$ to get

$$\text{Euler} \quad \begin{aligned} y_1(x+h) &= y_1(x) + hf_1[x, y_1(x), y_2(x)] + O(h^2) \\ y_2(x+h) &= y_2(x) + hf_2[x, y_1(x), y_2(x)] + O(h^2) \end{aligned} \tag{11-16}$$

These formulas are used recursively, starting at the known initial point. To develop the Runge-Kutta-two formulas for our second-order problem, as before, we can use Taylor's theorem to show that

$$\begin{aligned} &f_i(x+h, y_1 + hf_1, y_2 + hf_2) \\ &\quad = f_i(x, y_1, y_2) + h\left(\frac{\partial f_i}{\partial x} + \frac{\partial f_i}{\partial y_1}f_1 + \frac{\partial f_i}{\partial y_2}f_2\right)_{x, y_1, y_2} + O(h^2) \end{aligned} \tag{11-17}$$

Equation (11-17) applies for both $i = 1$ and $i = 2$. Since we have

$$y_i''(x) = \frac{\partial f_i}{\partial x} + \frac{\partial f_i}{\partial y_1}f_1 + \frac{\partial f_i}{\partial y_2}f_2 \tag{11-18}$$

combination of Equations (11-18), (11-17), and (11-15) yields the RK2 formula

$$\begin{aligned} \text{RK2} \quad y_i(x+h) &= y_i(x) \\ &+ \frac{h}{2}[f_i(x, y_1, y_2) + f_i(x+h, y_1 + hf_1, y_2 + hf_2)] + O(h^3) \quad (i = 1, 2) \end{aligned} \tag{11-19}$$

These equations apply for both $i = 1$ and $i = 2$. For comparison with the improved Euler calculations, we now show the form of the RK2 recurrence relations, with their two derivative evaluations per step.

**RK2 Method for a System of Two ODEs**

1. Calc. 1st deriv. $f_i[x, y_1(x), y_2(x)]$
2. Euler step $z_i(x+h) = y_i(x) + hf_i[(x, y_1(x), y_2(x)]$
3. Calc. 2nd deriv. $f_i[x+h, z_1(x+h), z_2(x+h)]$
4. Trap. corr. $y_i(x+h) = y_i(x) + \frac{h}{2}\{f_i[x, y_1(x), y_2(x)] + f_i[x+h, z_1(x+h), z_2(x+h)]\}$ (11-20)

Extension of the two-dimensional results to higher-order systems follows the same pattern. The formulas are used recursively in a calculation loop.

For improved Euler, the recurrence relations are *nearly* the same as for RK2, but involve *only one derivative evaluation per step* as follows.

**Improved Euler Method for a System of 2 ODEs**

1. Euler step $z_i(x+h) = y_i(x) + hf_i[x, z_1(x), z_2(x)]$
2. Calc. deriv. $f_i[x+h, z_1(x+h), z_2(x+h)]$
3. Trap. corr. $y_i(x+h) = y_i(x) + \frac{h}{2}\{f_i[x, z_1(x), z_2(x)] + f_i[x+h, z_1(x+h), z_2(x+h)]\}$ (11-21)

Note that the $z_1(x)$, $z_2(x)$ values in Equations (11-21) are already known at the beginning of the new step (from the previous step) and that the values $z_1(x+h)$, $z_2(x+h)$ are also used in the next step, thus requiring only *one* derivative evaluation per step by improved Euler.

**Example 11.1 A Nonlinear Chemical Reaction System**

The unsteady-state concentration in a constant volume batch reactor for a *nonlinear* chemical reaction of the form A + B ↔ C → D + E might be described by the ODE system

$$\frac{dy_1}{dt} = -k_1 y_1(y_1 - K) + k_2 y_2, \quad y_1(0) = 1$$
$$\frac{dy_2}{dt} = -(k_2 + k_3)y_2 + k_1 y_1(y_1 - K), \quad y_2(0) = 0 \qquad \text{(a)}$$

Here, $y_1$ represents concentration of species $A$ and $y_2$ represents concentration of $C$. The quantities $k_1$, $k_2$, and $k_3$ are rate constants, while $K$ depends on the initial composition of the mixture. For the case where $K = 0$, and $k_1 = k_2 = k_3 = 1$, the errors in the numerical results of integration using the Euler, RK2, and improved Euler methods are shown in Table 11.2. To conserve space, only errors in the $y_1$ value at $t = 0.8$ are shown in the comparisons. The effects of changing step size are also shown.

The results shown in Table 11.2 are typical of similar comparisons for other systems of equations. The following trends are observed: The accuracy of RK2 and improved Euler are comparable, with RK2 generally being slightly more accurate. For the same step size, the accuracy of the Euler method is generally much lower than that of either RK2 or improved Euler. In terms of accuracy, for the same number of derivative evaluations, we must compare the errors of Euler and improved Euler for a given step size versus those of RK2 at twice this value. On this *fixed-cost* basis (in terms

**TABLE 11.2** ERRORS IN $y_1$ (0.8) FOR SOLUTION OF NONLINEAR REACTION SYSTEM

| Step size | Euler | Improved Euler | RK2 |
|---|---|---|---|
| 0.4 | 0.046 | −0.0212 | −0.0403 |
| 0.2 | 0.023 | −0.0101 | −0.0055 |
| 0.1 | 0.011 | −0.0024 | −0.0011 |
| 0.05 | 0.0057 | −0.0004 | −0.0003 |
| 0.025 | 0.0028 | −0.00013 | −0.00006 |
| 0.0125 | 0.0014 | −0.00004 | −0.00002 |
| 0.00625 | 0.0007 | −0.00001 | −0.00001 |

$y_1' = -y_1^2 + y_2;\ y_1(0) = 1;\ y_2' = -2y_2 + y_1^2;\ y_2(0) = 0$

of derivative evaluations), it is seen that the accuracy of improved Euler far exceeds that of Euler and is significantly higher than that of RK2. These results are in accord with theory, since the Euler method has first-order accuracy, while the RK2 and improved Euler methods are of second-order accuracy.

### 11.1.6 Other Non-Stiff Methods

A procedure known as the modified midpoint method has received attention Press *et al.* [1986]. The basic two-step midpoint rule involves only one derivative evaluation per step, although it also requires periodic stabilization. When coupled with either polynomial or rational extrapolation, the method exhibits some attractive features. Traditionally, various predictor-corrector methods have been used in some circumstances, but their practical importance now seems diminished.

## 11.2 LOCAL ERROR ESTIMATION AND VARIABLE STEP-SIZE INTEGRATION

In many difficult ODE problems, it is desirable to vary the step size within the integration process. This often happens when there is a large variation of the solution within the domain of the problem. A particularly important example of this situation occurs when diffusion-like partial differential equations are solved using the ODE "method of lines." Such applications will be discussed in the chapter on the numerical solution of partial differential equations.

To implement a practical variable step-size ODE procedure, it is customary to somehow estimate the *local* solution relative error. This is the estimated error for one step, assuming that we started with the correct values at the beginning of the step. This local estimated relative error is then compared to the user-specified desired local relative error, and a new step size is calculated accordingly. If the estimated error meets the desired error tolerance, we accept the step and continue. If the estimated error does *not* meet this tolerance, we repeat the step using the new estimated step size (which will naturally be smaller than the current step size). Some additional "safety factor" devices often accompany a variable step-size algorithm.

Various ways to estimate local error include so-called imbedding techniques, one-step/two half-step methods, local Richardson extrapolation, and so on.

We now apply these ideas specifically to the new improved Euler method, in order to produce the variable-step algorithm IEXVS, which is discussed at the end of the chapter. The per-step errors of $z_i(x)$ and $y_i(x)$ are $O(h^2)$ and $O(h^3)$, respectively. It is therefore convenient to estimate the local error in terms of the difference $z_i(x) - y_i(x)$. Since this difference is $O(h^2)$ for small $h$, one can adjust the step-size on the basis of the formula

$$h_{\text{new}} = \left(\frac{\text{desired error}}{\text{current error}}\right)^{1/2} \cdot h_{\text{current}} \tag{11-22}$$

Here, the value $|z_i(x) - y_i(x)|$ is used for "current error." This procedure, accompanied by some bookkeeping devices, is used in program IEXVS to adjust the step size. The only extra parameter that the user need specify in the case of error control is the tolerance value (tol), which corresponds to the desired local relative error for all variables. Our experience has shown this error control strategy to be very successful.

## 11.3 ERRORS AND THE USE OF GLOBAL EXTRAPOLATION

The subject of error analysis in the numerical solution of differential equations is difficult, and there are relatively few practical results. This is because the error is a complicated mixture of truncation at each step and the "propagation" or "inherited" error associated with using the wrong initial condition at each step after the first. The simplest approach, also known as "brute force," is to merely reduce the step size of integration until the results are in agreement. This method is sometimes the only one possible. A much more sophisticated and efficient approach to error estimation can be based on the Richardson extrapolation ideas used previously for numerical quadrature. This approach offers the very considerable advantages of both estimation of the global, or overall, error (as opposed to the local, or single-step, error) and simultaneous improvement of the accuracy. To accomplish this, it is very efficient to calculate several independent solution trajectories within a single computer run.

### 11.3.1 Basic Formulas of Global Extrapolation

We have seen application of the technique of Richardson extrapolation in our discussions of numerical differentiation and quadrature (also see Appendix F). This method is also of great value in the numerical solution of differential equations. It is possible to use the extrapolation idea in two fundamentally different ways in solving differential equations. We can use it in a *local* (or active) way by starting over after an extrapolation cycle, using the single best estimate of the solution to repeat the process for the next step, and so on. In this way, this local extrapolation can increase accuracy and estimate *local* errors, which are those incurred within a single extrapolation cycle.

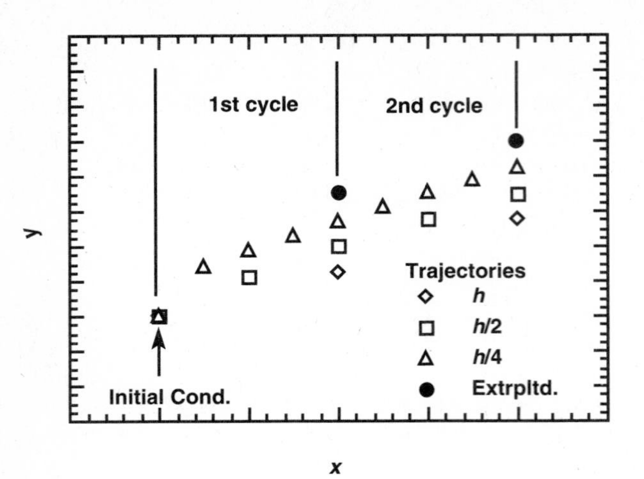


**Figure 11.1.** Calculation Pattern for Global Extrapolation

An alternative procedure is to use *global* (or passive) extrapolation by deliberately *not* updating the solution within a trajectory as we compute. This offers the great advantage of allowing estimation of the global, or overall, error, which is what is really important, in addition to increasing accuracy. Thus, in global extrapolation, the *result* of the extrapolation is the same as if all solution trajectories were already computed before the extrapolation was carried out. But it is much more efficient to compute these global results in parallel by simultaneously calculating several solution trajectories in one run as we go along. In this way the bookkeeping is vastly simplified, we need only calculate the extrapolation values at points where we wish to print the solution, and we can monitor the results (and stop if they are not useful) without needing the entire solution. For these reasons we will adopt the use of *global* extrapolation in our development. The calculation pattern for global extrapolation using three trajectories is illustrated in Figure 11.1. Each trajectory is *independent* of the others; thus, there is no updating of the $y$ values after extrapolation.

Our discussion follows that given by Hanna [1990], which can be consulted for further details and generalizations. The implementation of global extrapolation is essentially the same, regardless of the basic "building-block" integration method used for the independent trajectories. Here, we will be primarily concerned with application of extrapolation to the improved Euler method. However, with minor modifications that we will point out, the global extrapolation procedure can be applied to most single-step methods such as Euler and Runge-Kutta as well as to stiff methods discussed later, such as backward Euler, trapezoidal rule, and implicit improved Euler.

Global extrapolation can be applied whenever the solution $Y$ depends on a known approximation $y_a(h)$ which varies, for example, with the quantity $h$ according to the asymptotic expansion

$$Y = y_a(h) + C_1 h^p + C_2 h^q + O(h^r) \qquad h \rightarrow 0 \quad (p < q < r) \tag{11-23}$$

The values of $p$, $q$, and $r$ are different for various methods. For the improved Euler method, which is second-order, we have $p = 2$, $q = 3$, and $r = 4$. In Equation (11-23)

$C_1$, $C_2$, $C_3$, and so on must be *independent of* $h$, and we assume that $p$, $q$, and $r$ are known. It is known that Equation (11-23) is applicable for integration by many methods, at a given $x$, to every component $Y_i$ of a differential equation system which has appropriate smoothness properties. Equation (11-23) can be viewed at a given $x$ as a Taylor expansion *across* solution trajectories that have different integration step sizes $h$. Here, we will use Equation (11-23) for global extrapolation by calculating, within a given computer run, a number of independent trajectories in parallel.

If we calculate two independent approximate numerical trajectories for the problem $Y' = f(x, Y)$, corresponding to $y_a(h)$ and $y_a(h/2)$, Equation (11-23) is used to estimate the value of $C_1h^p$. Defining $E(h, h/2) \equiv (y_a(h/2) - y_a(h))/(2^p - 1)$ and $y_a(h, h/2) \equiv y_a(h/2) + E(h, h/2)$, we get

$$C_1h^p = E(h, h/2) - \frac{2^q - 1}{2^p - 1}C_2h^q + O(h^r) \tag{11-24}$$

and

$$Y = y_a(h/2) + E(h, h/2) + Dh^q + O(h^r) = y_a(h, h/2) + Dh^q + O(h^r) \tag{11-25}$$

where we have written $(2^p - 2^q)/(2^p - 1) \cdot C_2h^q \equiv Dh^q$. Equation (11-25) shows that the once-extrapolated approximation $y_a(h, h/2)$ is now in error by $O(h^q)$, so that when $h$ is sufficiently small, $y_a(h, h/2)$ will be more accurate than $y_a(h)$. We observe that the form of Equation (11-25) is the same as that of Equation (11-23), with $y_a(h, h/2)$ now playing the role of the base approximation. Therefore, we may now use Equation (11-25) twice, with both $y_a(h, h/2)$ and $y_a(h/2, h/4)$, to estimate the value of $Dh^q$. As before, we now define

$$E(h, h/2, h/4) \equiv [y_a(h/2, h/4) - y_a(h, h/2)]/(2^q - 1)$$

and

$$y_a(h, h/2, h/4) \equiv y_a(h/2, h/4) + E(h, h/2, h/4)$$

so that we can write, using Equation (11-25),

$$Y = y_a(h/2, h/4) + E(h, h/2, h/4) + O(h^r) = y_a(h, h/2, h/4) + O(h^r) \tag{11-26}$$

In this case, we see from Equation (11-26) that the twice-extrapolated approximation $y_a(h, h/2, h/4)$ is in error by $O(h^r)$, so when $h$ is sufficiently small, $y_a(h, h/2, h/4)$ will be more accurate than either $y_a(h, h/2)$ or $y_a(h)$. This process could be continued (in which case it is similar to the "Romberg" method of numerical quadrature), but for reasons given below, it is convenient to terminate the procedure at this point and to make good use of the results of only two extrapolations.

The Richardson extrapolation process described above produces consecutive estimates of the quantities $C_1h^p$, $C_2h^q$, and so on. It is useful to note explicitly that $C_1h^p$ is a realistic estimate of the error in $y_a(h)$ and a conservative estimate of the error in $(y_a(h) + C_1h^p)$. Similarly, $C_2h^q$ is a realistic estimate of the error in $(y_a + C_1h^p)$ and a conservative estimate of the error in $(y_a + C_1h^p + C_2h^q)$. These interpretations are valid *provided that* h *is sufficiently small* (and in the absence of roundoff error). In principle, any number of independent trajectories could be used. With additional global extrapolations, both the order of global accuracy and the integration cost are increased. More extrapolations also result in greater program

**TABLE 11.3** ACCURACY IMPROVEMENT OF $y_1(0.8)$ BY IMPROVED EULER WITH EXTRAPOLATION IN EXAMPLE 11.1

| $h$ | $y(h)$ | $y(h, h/2)$ | $y(h, h/2, h/4)$ |
|---|---|---|---|
| 0.2 | 0.6563 | 0.6460 | 0.646231 |
| 0.1 | 0.6486 | 0.64621 | 0.646233 |
| 0.05 | 0.6468 | 0.646230 | 0.646234 |
| 0.025 | 0.6464 | 0.646234 | 0.646234 |
| 0.0125 | 0.6463 | 0.646234 | |
| 0.00625 | 0.6462 | | |

$y_{\text{exact}}(0.8) = 0.646234$

complexity, more computational effort, and more roundoff error. It is clear that there is a trade-off regarding the optimal number of independent trajectories to be used. For reasons given below, it is the authors' experience that the optimum number of trajectories is generally three. However, it is useful to have the option of running either one, two, or three trajectories, and this is available in the computer programs IEX and IEXVS, which are discussed at the end of the chapter. We next show the typical accuracy improvement effected by extrapolation. We consider again the problem of Example 11.1, and indicate values of $y_1(0.8)$ as calculated by the improved Euler value $y(h)$, improved Euler with one extrapolation $y(h, 2h)$, and improved Euler with two extrapolations.

The results of Table 11.3 show the powerful acceleration effect of global Richardson extrapolation combined with the improved Euler method. The crudest (and least expensive) step-size sequence of $(0.2, 0.1, 0.05)$ is seen to produce far more accurate results than those of the unextrapolated improved Euler at a step size of 0.00625.

### 11.3.2 Determination of Validity—The Error Ratio

Perhaps the most important aspect of the use of global Richardson extrapolation is the determination of the "region of validity" for extrapolation, that is, the region of $h$ where Equation (11-23) is applicable. This is of crucial importance since we must operate in this region in order for the extrapolation process to produce improved accuracy and useful global error estimates. Practically, this translates into the question, *When can we be confident that the true, unknown global error (for any variable, at any* x*) is smaller in magnitude than the estimated global error* $E(h, h/2, h/4)$? This is a meaningful question since applicability of Equation (11-23) with $h$ small enough for the $O(h^r)$ term to be negligible would imply that $|E_{\text{actual}}| < |E(h, h/2, h/4)|$. To answer the above question, we introduce the normalized error ratio, $R$, which is defined as follows for each variable at any given time:

$$R \equiv \frac{E(h, h/2)}{2^p E(h/2, h/4)} \equiv \frac{[y_a(h/2) - y_a(h)]}{2^p [y_a(h/4) - y_a(h/2)]} \tag{11-27}$$

This quantity can be evaluated whenever we have the results $y_a(h)$, $y_a(h/2)$, $y_a(h/4)$ for three independent trajectories. Since Equation (11-24) shows that if extrapolation is valid, $E(h, h/2) \approx C_1 h^p$, we see that when extrapolation is valid, we should have $R \approx 1$. Thus, we would normally expect that if $R$ is near unity, we could be confident that the extrapolation formula is applicable and that the usual interpretations would apply. However, since, in general, $C_2 h^q \neq 0$, the above conjecture begs the question of *how close to unity should* R *be in order to be confident that extrapolation applies?* The answer to this question fortunately turns out to be quite simple and broadly applicable, as explained in detail below. Based on these considerations, both theoretical and experimental, we offer the following result:

> *An appropriate conservative range for the error ratio* R*, where it seems reasonable to expect that the actual unknown global errors will be smaller in magnitude than the estimated errors, is* $0.8 \leq R \leq 1.2$.

We now proceed to a detailed development of the error ratio $R$ and its interpretation.

**11.3.2.1 Theoretical Region for Error Ratio.** We begin by developing certain relationships that will be satisfied by the error ratio $R$, considered generally as a function of the extrapolation exponents $p$ and $q$. First, note the following important *identity*, which involves the error ratio, $R$, and the ratio of the twice- and once-extrapolated errors, $E(h, h/2, h/4)/E(h, h/2)$. This identity results from the previous definitions of the error quantities.

$$R \equiv \frac{E(h, h/2)}{2^p E(h/2, h/4)} \equiv 1 - \frac{2^q - 1}{2^p} \cdot \frac{E(h, h/2, h/4)}{E(h/2, h/4)} \tag{11-28}$$

Since this relationship is an *identity* involving the quantities $y_a(h)$, $y_a(h/2)$, $y_a(h/4)$ calculated in three independent trajectories, it does not depend at all on whether or not we are in the region of validity for extrapolation. Equation (11-28) shows the obvious result that $R \to 1$ for $E(h, h/2, h/4)/E(h/2, h/4) \to 0$, but it also tells us much more than that. Presumably, an outside threshold where extrapolation may have a chance of becoming valid corresponds to the situation where $|E(h, h/2, h/4)/E(h/2, h/4)| < 1$. Equation (11-28) shows that for this to happen, we must have $1 - (2^q - 1)/2^p < R < 1 + (2^q - 1)/2^p$. For the case $p = 2$, $q = 3$ (improved Euler), for example, this translates into $-0.75 < R < 2.75$. Thus, for this case, if $R$ is outside this range, the twice-extrapolated error is larger in magnitude than the once-extrapolated error, and this cannot be characteristic of any "region of validity." Typical values of the error ratio computed from this identity for other $p$, $q$ show that to have $E(h, h/2, h/4)/E(h/2, h/4)| < 1$, we generally find that $-1 < R < 3$ (approximately). It is intuitively appealing to suppose that the outside edge of a conservatively defined extrapolation region might correspond to the situation where $E(h, h/2, h/4)$ is at least one order of magnitude (factor of 10) smaller than $E(h/2, h/4)$. Applied to Equation (11-28), this would produce a theoretical acceptability range for $R$ of $0.825 < R < 1.175$ (for the case of $p = 2$, $q = 3$). Considering minor numerical variations for different $(p, q)$ combinations, we can conclude that an appropriate *"theoretical"* acceptability range for the error ratio $R$ which is general, simple, symmetrical, and conservative would appear to be $0.8 \leq R \leq 1.2$. This applies to all of the methods considered. For $R$ values falling (consistently) in this range, it seems plausible to hope that the actual unknown global errors will be smaller in magnitude than the estimated global errors $E(h, h/2, h/4)$.

**11.3.2.2 Experimental Region for Error Ratio.** It is very interesting to find that the above proposition can be tested experimentally in a quite general way, which yields common results regardless of the problem or step size. By running a given problem at a succession of smaller step sizes using the global extrapolation procedure described, we generate for each variable, at each output $x$, a succession of values of $R$, $E(h, h/2, h/4)$, and $y_a(h, h/2, h/4)$ (the best approximation). By continuing to a sufficiently small step-size, we ultimately obtain a very close approximation to the unknown true value $Y$. This can be used to estimate the actual error, $E_{\text{actual}} = Y - y_a(h, h/2, h/4)$. At these particular conditions, we can then make a plot of $R$ versus $|E_{\text{actual}}/E(h, h/2, h/4)|$ corresponding to various $h$ values. In this way, *such plots for any variable at a particular* x, *for any problem, can be superimposed on the same graph*. To define the region of acceptability that is valid for *all* data on such a plot, we merely *determine the smallest range of* R *for which all values of* $|E_{actual}/E(h, h/2, h/4)|$ *are less than or equal to unity*. This defines empirically the conservative range of $R$ for which the estimated error is larger than the true error.

In a series of computational experiments Hanna [1990], it was found that an acceptable ratio range for the problems considered was $0.5 < R < 1.3$. In most problems, the range was considerably broader. It is interesting to note that the above range is close to, and falls within, the heuristic theoretical ranges defined above. This result reinforces the specification of a general conservative range as $0.8 < R < 1.2$. Of course, the above experiments were limited and cannot be considered conclusive. Nevertheless, it appears that there may be an approximate error ratio "region of acceptability" that is essentially independent of the variable considered, or even of the problem.

For problems of a "stiff" nature (to be discussed below) being integrated by a non-stiff method, we find the following. When we are not too close to the stability limit of the method, we observe normal, predictable behavior. When the integration step size approaches the stability limit, $R$ and its variation show erratic behavior which deviates sharply from the norm. As instability is more closely approached, wild variations of $R$ are observed, and ultimately the $E(h, h/2, h/4)$ values, and finally the $y_a(h, 2h, 4h)$ values, show very erratic and unacceptable behavior. *The error ratio* R *is therefore a very sensitive and useful indicator of instability*. Of course, if a stiff method (such as backward Euler, Trapezoidal Rule, etc.) is being used for integration, the approximate solution generally behaves well, and so do the $R$ values. Preliminary calculations of this type suggest that the $R$ values behave better for backward Euler than for Trapezoidal Rule because of the tendency of the latter to show small oscillations in stiff problems.

The recommendations for implementing a simple, efficient global extrapolation algorithm are as follows. A minimum of three independent trajectories is required to produce information (ratio $R$) that can be used to assess the validity of extrapolation. By extrapolating up to three trajectories, we increase the global accuracy of the approximate solution from $O(h^p)$ up to $O(h^r)$. It therefore seems best to use only (up to) three trajectories within a given computer run. In this way, roundoff problems are minimized, higher costs are avoided, and a desirable simplicity is retained in the algorithm. In terms of cost-accuracy trade-offs, it has been the authors' experience that for commonly required accuracies (e.g., relative errors of about $10^{-4}$ to $10^{-5}$), the most efficient procedure is to use a low-order base integration method, such as the new improved Euler. In the unlikely event that significantly greater accuracies (e.g., better than $10^{-7}$) are required, then it may be more efficient to use a higher-order method, such as RK4, for the individual trajectories.

**Example 11.2 A Nonlinear Catalyst Pellet Problem**

We consider the following second-order nonlinear diffusion equation, which arises in connection with a catalyst pellet problem described in Cartesian coordinates. The equation is $C'' = 100C^2$ with $C(0) = 0.057$ and $C'(0) = 0$. For numerical solution, the second-order equation is converted to a first-order system in the usual way, by using $C = y_1$ and $C' = y_2$. Table 11.5 shows results obtained with the Improved Euler/Extrapolation (IEX) procedure as employed in program IEX, which is discussed at the end of the chapter. The values of $Y$, REL ERROR, and $R$ shown in the table are defined by

$$Y \equiv y(h, h/2, h/4)$$

$$\text{REL ERROR} \equiv \frac{E(h, h/2, h/4)}{y(h, h/2, h/4)} \tag{a}$$

$$R \equiv \frac{E(h, h/2)}{2^p E(h/2, h/4)}$$

The values of $Y$, REL ERROR, and $R$ for a given $h$ are computed in one run of the program, for each variable at each output $x$. We show, in Table 11.4, some typical program IEX output from this nonlinear reaction problem, which has a somewhat steep gradient near $x = 1$.

The way we interpret Table 11.4 above is as follows. We first inspect the error ratios. Since they are all in the range $0.8 \le R \le 1.2$, we are confident that the estimated relative errors are conservative; that is, smaller than the true unknown errors. Next, we look at the estimated relative errors. If they are small enough for our purposes, it would appear that we have a successful solution. Finally, we examine the $y_i$ values to be sure that they seem reasonable. In beginning such a calculation, we generally specify a rather large step size and see what happens. If the ratios or relative errors are not satisfactory, we reduce the step size in an attempt to produce satisfactory results. If a considerable reduction in step size still does not yield acceptable results, we have a very difficult (probably stiff) problem. Further discussion on the interpretation of the error ratio is given below.

The results of an extrapolation validation experiment are shown in Table 11.5. The values of $E_{\text{actual}}/E(h, h/2, h/4)$ can only be computed when there is more than one computer run. Table 11.5 shows behavior that is typical of most problems. For decreasing $h$, the solution becomes more accurate, and the error ratios $R$ move toward the "believable" extrapolation region. Thus, we see that, for example, with $R$ equal to about 0.8, the actual global errors are clearly smaller in magnitude than the estimated

**TABLE 11.4** SAMPLE OUTPUT FROM PROGRAM IEX FOR $h = 0.025$

| $y(h, h/2, h/4)$ | Estimated Relative Error | Error Ratio |
|---|---|---|
| $x = .25$ | | |
| $y(1) = 6.77937e-2$ | $e(1) = 2.39302e^{-6}$ | $r(1) = 0.908201$ |
| $y(2) = -9.17917e-2$ | $e(2) = 6.25471e^{-6}$ | $r(2) = 0.935931$ |
| $x = 0.5$ | | |
| $y(1) = 0.110129$ | $e(1) = 8.12951e^{-6}$ | $r(1) = 0.930943$ |
| $y(2) = -0.276947$ | $e(2) = 1.29424e^{-5}$ | $r(2) = 0.935430$ |
| $x = .75$ | | |
| $y(1) = 0.244458$ | $e(1) = 3.31373e^{-5}$ | $r(1) = 0.914703$ |
| $y(2) = -0.980599$ | $e(2) = 4.97373e^{-5}$ | $r(2) = 0.913540$ |
| $x = 1.$ | | |
| $y(1) = 0.992563$ | $e(1) = 3.22325e^{-4}$ | $r(1) = 0.829564$ |
| $y(2) = -8.07341$ | $e(2) = 4.99081e^{-4}$ | $r(2) = 0.819760$ |

$y_1' = -y_2;\quad y_1(0) = 0.057 \quad y_2' = -100y_1^2;\quad y_2(0) = 0$

**TABLE 11.5** ILLUSTRATION OF VALIDATION EXPERIMENT FOR IEX METHOD AT $x = 1$

| $h$ | $i$ | $Y_i$ $(h, h/2, h/4)$ | Relative Error | $R$ | $\left\lvert\frac{E_{\text{actual}}}{E(h,h/2,h/4)}\right\rvert$ |
|---|---|---|---|---|---|
| 0.2 | (1) | 0.9114 | 2.3 $10^{-2}$ | 0.27 | 3.9 |
| | (2) | −7.045 | 3.2 $10^{-2}$ | 0.23 | 4.5 |
| 0.2/2 | (1) | 0.9813 | 9.3 $10^{-3}$ | 0.44 | 1.2 |
| | (2) | −7.925 | 1.4 $10^{-2}$ | 0.40 | 1.3 |
| $0.2/2^2$ | (1) | 0.9920 | 2.2 $10^{-3}$ | 0.65 | 0.24 |
| | (2) | −8.067 | 3.4 $10^{-3}$ | 0.63 | 0.22 |
| $0.2/2^3$ | (1) | 0.9926 | 3.2 $10^{-4}$ | 0.83 | 0.08 |
| | (2) | −8.073 | 5.0 $10^{-4}$ | 0.82 | 0.11 |
| $0.2/2^4$ | (1) | 0.9925 | 3.8 $10^{-5}$ | 0.93 | $\ll 1$ |
| | (2) | −8.073 | 5.8 $10^{-5}$ | 0.93 | $\ll 1$ |
| $0.2/2^5$ | (1) | 0.9925 | 4.3 $10^{-6}$ | 0.97 | $\ll 1$ |
| | (2) | −8.073 | 6.4 $10^{-6}$ | 0.97 | $\ll 1$ |

$y_1' = -y_2;\quad y_1(0) = 0.057 \quad y_2' = -100y_1^2;\quad y_2(0) = 0$

values $E(h, h/2, h/4)$. We conclude that if we did not know the actual errors in this problem, the ratios $R$ would have guided us reliably to the range of the calculation where the customary interpretations of extrapolation are appropriate. The relationship between $R$ and $|E_{\text{actual}}/E(h, h/2, h/4)|$, based on the values in Table 11.5, is shown in Figure 11.2.

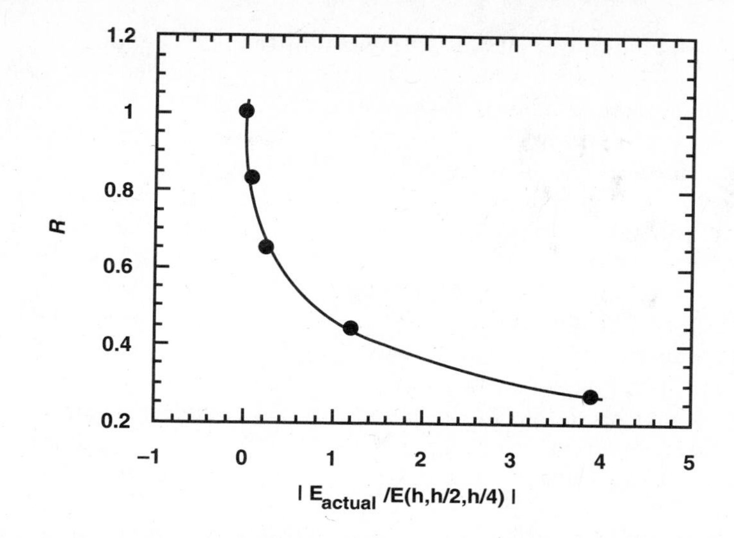


**Figure 11.2.** Results of Validation Experiment for $y(1)$ with the IEX Method Applied to Problem of Example 11.2

### 11.3.3 Interpretation of Extrapolation Results

We now consider further application of the diagnostic tool $R$, which is the key to "believability" of global extrapolation. Of special interest are behaviors that deviate from what is ordinarily expected. In the first place, it normally takes the error ratio several steps to settle down to a value, since at first we have a combination of local and global errors, which differ in their $h$ dependence. Thus, even with step-size reduction, $R$ values for the first few steps often do not approach unity. Also, it is important to note that if the $E(h, 2h)$ and $E(2h, 4h)$ values are *very* small (as they would be in a "near-perfect" solution), then even though the true value of $R$ might be very close to unity, the calculated value could be grossly contaminated due to roundoff error. To easily identify this circumstance, the program simply prints out *** in place of a numerical value for $R$ whenever $|E(h/2, h/4)|$ is smaller than some roundoff threshold value (such as, for example, $10^{-12}$ in double precision). The above situation generally involves the occurrence of a very small value of REL ERROR.

The normal behavior of the error ratio is as follows. For each variable (at a given $h$), $R$ settles down after a few steps to some value, which usually deviates more from unity as $x$ increases (and the global error increases). For a sequence of smaller step sizes, at a given $x$ for a given variable, $R$ approaches unity. This is what normally occurs when truncation error is dominant. If we happen to be in the region where roundoff error is important (this would be very unusual, since this corresponds to a very small step size and high cost), $R$ and its variation behave erratically.

### 11.3.4 Miscellaneous Aspects and Extensions

The case of *nonanalytic behavior* (such as fractional powers, logarithms, etc.) of the differential equation(s) is usually characterized by $R$ values and variations that seem to be approaching a limit, but *not unity*. To illustrate, if the global error in a given problem is varying as $h^{3/2}$, whereas the ratio $R$ is being calculated as if the error varied as $h^2$, then the calculated $R$ value would approach $2^{3/2}/2^2 = 0.707$ as $h \to 0$. Numerical experience shows that the diagnostic features of the error ratio are very important in problems involving nonanalytic behavior, since it is common in these cases for global extrapolation to *underestimate* the actual error. The extrapolation process itself can be a useful tool for the experimental determination of the exponents $p$ and $q$, whenever this might be necessary.

It is also very important to point out that global extrapolation can be applied for *variable step-size* algorithms, as discussed by Dahlquist *et al.* [1974]. It is the authors' experience that this procedure works well when we use local error to control step size and extrapolation to monitor global error. Because of its simple local error properties, IEX is ideally suited as a base integration algorithm for global extrapolation with variable step size. The IEX variable step algorithm is contained in program IEXVS, which is discussed at the end of the chapter.

## 11.4 STABILITY OF NUMERICAL METHODS

We have not said very much thus far about the nature of the error in numerical integration of ordinary differential equations. We now consider this matter in some detail. It is very useful in this regard to fix attention on the forward Euler method, for the analysis in this case is fairly simple, but the results are qualitatively similar to those for other non-stiff methods such as Runge-Kutta and improved Euler. Consider integration of the single equation $dY/dx = f(x, Y)$ using the fixed-step Euler method. Here, $Y(x)$ is the true unknown solution and $y_E(x)$ is the Euler approximation. We write down the equations for one step of the Euler process and one step of the exact Taylor approximation.

$$\begin{aligned} &\text{Euler:} && y_E(x+h) = y_E(x) + hf[x, y_E(x)] \\ &\text{Exact Taylor:} && Y(x+h) = Y(x) + hf[x, Y(x)] + \frac{Y''(\xi_1)h^2}{2!} \end{aligned} \tag{11-29}$$

We let $e(x) = y_E(x) - Y(x)$ so that, by subtraction, we get

$$e(x+h) = e(x) + h\{f[x, y_E(x)] - f[x, Y(x)]\} - \frac{Y''(\xi_1)h^2}{2!} \tag{11-30}$$

But by Taylors theorem, $f[x, y_E(x)] - f[x, Y(x)] = (\partial f/\partial y)_{\xi_2}\, e(x)$, where the partial derivative is evaluated at $x$ and at some unknown $\xi_2$ point between $y_E$ and $Y$. Substituting this into Equation (11-30) gives

$$e(x+h) = \underbrace{e(x)\left[1 + h\frac{\partial f}{\partial Y}\right]}_{\text{"distortion" or propagation of previous error}} - \underbrace{\frac{Y''(\xi_1)h^2}{2!}}_{\text{new truncation error}} \tag{11-31}$$

Equation (11-31) is very revealing. It shows that at any point of the calculation, the new error $e(x+h)$ is composed of the new truncation error *plus* the previous error, which is "distorted" according to the factor $[1 + h(\partial f/\partial Y)]$. This is in contrast to the case of quadrature (where $f(x, Y)$ depends only on $x$, so that $\partial f/\partial Y = 0$), where the overall error is simply the sum of the single-step errors. For differential equation integration, the situation is much more complicated. To illustrate this point, suppose we let $\partial f/\partial Y = K(x)$ and $-Y''h^2/2! = T(x)$. Then we can write $e(x + h) = e(x)[1 + hK(x)] + T(x)$. We now calculate the errors after the first three integration steps, assuming the initial conditions are known exactly.

$$e(h) = e(0)[1 + hK(0)] + T(0) = T(0)$$

$$e(2h) = e(h)[1 + hK(h)] + T(h) = T(0)[1 + hK(h)] + T(h)$$

$$\begin{aligned} e(3h) &= e(2h)[1 + hK(2h)] + T(2h) \\ &= T(0)[1 + hK(h)][1 + hK(2h)] + T(h)[1 + hK(2h)] + T(2h) \end{aligned}$$

The above equations show that as we continue the integration, the error at any point is made up of a complicated sum of "distorted" previous truncation errors plus the newly incurred truncation error. Although we have derived these relationships for the Euler method, the qualitative behavior will be similar for other numerical integration methods. If we assume that $\left|\dfrac{\partial f}{\partial Y} \le K\right|$ and $|Y''| \le T$ in the region of interest, it is not difficult to show that the overall or global error of the Euler method, if we integrate to $x = L$, is bounded as follows:

$$|e(x)| \le \frac{Th}{2K}(e^{Kx} - 1) \tag{11-32}$$

Equation (11-32) shows that the global error of the Euler method is first-order in $h$ (proportional to $h$), even though the per-step (local) error is proportional to $h^2$. It is generally the case for other methods that global errors are one order lower in $h$ than the local errors. Equation (11-32) also shows a *very* major role possibly being played by the factor $K$, which is the upper bound to $|\partial f/\partial Y|$. This will turn out to be true in so-called *stiff* differential equations.

### 11.4.1 Analysis of Prototype Equation $y' = -Ky$, $K$ Positive

The previous discussion shows that for the Euler method, the quantity of crucial importance regarding propagation of error is $h(\partial f/\partial y)$, and this is generally true of all methods. Because of this fact, we can use the simple differential equation $y' = -Ky$ (with $K$ positive) as a standard or prototype equation to study the stability characteristics of other methods. The reason is that any equation $y' = f(x, y)$ can be *linearized* in the vicinity of some point $x_0$, $y_0$ to produce the equation $y' = f(x_0, y_0) + (\partial f/\partial x)_{x0,y0}(x - x_0) + (\partial f/\partial y)_{x0,y0}(y - y_0)$. The error propagation characteristics of this linearized equation depend only on the coefficient of $y$. Thus, if we examine how errors propagate with a method for the simple equation $y' = -Ky$, we are effectively studying the general behavior of the method. The reason for the minus sign in the prototype equation is as follows. The equation $y' = -Ky$ has the solution (for $y(0) = 1$) $y = \text{Exp}(-Kx)$. For $K$ negative, this represents a growing function, and so as long as the error grows less than the solution, the computation is reasonable. However, for $K$ positive the solution decays, so if the error grows, the *relative* error approaches $\infty/0$, which is a computational disaster. Also, in mathematical models

**TABLE 11.6** CHARACTERISTICS OF SOME EXPLICIT INTEGRATION METHODS

| Method | Order of Accuracy | Stability for $y' = -Ky, \|Kh\| < a$ | Derivative Evaluations per Step |
|---|---|---|---|
| Euler | First | $a = 2$ | One |
| Improved Euler | Second | $a = 1$ | One |
| RK2 | Second | $a = 2$ | Two |
| RK4 | Fourth | $a = 2.78$ | Four |
| Euler/RK2hybrid | First | $a = 7.6$ | Two |

in engineering and science, continued growth rarely happens, while continued decay behavior is very common. Thus, the case of $K$ positive is of crucial importance, while that of $K$ negative is not.

We now consider several simple examples of stability analysis for the prototype equation $y' = -Ky$ ($K$ positive). First, we look at the forward Euler method, which for the prototype equation yields

$$\text{Euler} \qquad y(x+h) = y(x) + h[-Ky(x)] = y(x)(1 - Kh) \tag{11-33}$$

We know that the true solution decays, so this would be required in a satisfactory approximate solution. If $Kh$ is between zero and two, the factor $(1 - Kh)$ in Equation (11-33) is less than unity in absolute value, and the solution decays, as required. But *if* Kh *is larger than two*, the factor $(1 - Kh)$ exceeds unity in absolute value, and the approximate solution *grows* in magnitude, rather than decaying, and this is what we call unacceptable *unstable* behavior. Thus, for the forward Euler method, the stability limit is $Kh \le 2$. Thus, the Euler method is said to be *conditionally* stable, in the sense that if we can maintain the step size $h$ so that $Kh \le 2$, the calculation should remain stable. In a similar way, stability analyses can be conducted for other methods using the prototype equation by forming an expression of the type $y(x+h) = y(x)F(Kh)$ and finding the stable range for which $F(Kh)$ does not exceed unity. The so-called non-stiff methods are all (at best) conditionally stable for some range of $Kh$. Table 11.6 shows the stability properties of various methods.

In dealing with problems whose numerical solution is dominated by stability considerations, it is necessary to use methods that are very stable and, if possible, unconditionally stable. Unconditionally stable methods are ones that have no stability restriction on the integration step size. Usually, such methods are *implicit*, meaning that the numerical approximation requires a solution of algebraic equations at every step, in contrast to explicit methods, which do not. Two simple but important examples of such implicit methods are the backward Euler and implicit trapezoidal rule, which we now consider.

The backward Euler method is analogous to the forward Euler in the following way. The forward Euler method is the simplest explicit, conditionally stable method. Backward Euler is the simplest implicit, *unconditionally stable* method. Just as the forward Euler method can be derived from a forward Taylor approximation, backward Euler can be derived from a *backward* Taylor approximation of the form

$$y(x-h) = y(x) - y'(x)h + y''(x)\frac{h^2}{2!} \cdots \tag{11-34}$$

Equation (11-34) is just Equation (11-1) (less the final term) with $h$ replaced by $-h$. If we now delete the $y''(x)$ term in Equation (11-34) and solve for $y(x)$, we get

$$y(x) = y(x-h) + y'(x)h \tag{11-35}$$

In Equation (11-35) we put in $y' = f(x, y(x))$ and replace $x$ everywhere by $(x+h)$ to get the backward Euler (BE) formula

$$y_{BE}(x+h) = y_{BE}(x) + hf[x+h, y_{BE}(x+h)] \tag{11-36}$$

Equation (11-36) is an *implicit* one for $y_{BE}(x+h)$, since this unknown quantity appears on both sides of the equation. The per-step error of backward Euler is the same as forward Euler, except that the sign is reversed. In general, the backward Euler approximation requires solution of nonlinear algebraic equations, a distinctly unattractive feature (which is why we didn't mention it for solving non-stiff equations). However, the saving grace of backward Euler is its strong *unconditional stability*, which we now demonstrate for the prototype equation $y' = -Ky$. Applying the backward Euler formula gives

$$\text{Backward Euler} \quad y(x+h) = y(x) + h[-Ky(x+h)]$$
$$\text{or} \quad y(x+h) = \frac{y(x)}{1+Kh} \tag{11-37}$$

In Equation (11-37) we see that for *any* positive value of $Kh$ the solution is damped, and strongly so in the sense that the approximate solution $\rightarrow 0$ for $Kh \rightarrow \infty$. In terms of stability, the backward Euler method could hardly be improved, but its accuracy, like that of forward-Euler, is only first-order. However, using global extrapolation with three trajectories, the accuracy is increased to third-order. Just as improved Euler represented a considerable advantage over Euler, there is an *implicit improved Euler* method that represents an advantage over backward Euler. We will discuss this method below.

The implicit trapezoidal rule is another unconditionally stable method that is useful. The general algorithm is derived by applying trapezoidal quadrature to the equation $y' = f(x, y)$; it takes the form

$$y_T(x+h) = y_T(x) + \frac{h}{2}\{[f(x, y_T(x)] + f[x+h, y_T(x+h)]\} \tag{11-38}$$

Equation (11-38) is seen to be implicit, since $y_T(x+h)$ appears on both sides of the equation. This method has second-order accuracy, which is an advantage over backward Euler. To examine stability, we apply Equation (11-38) to the prototype equation $y' = -Ky$, and this yields

Implicit Trapezoidal Rule

$$y(x+h) = y(x) + \frac{h}{2}[-Ky(x) - Ky(x+h)] \quad \text{or} \quad y(x+h) = y(x)\left(\frac{1-\dfrac{Kh}{2}}{1+\dfrac{Kh}{2}}\right) \tag{11-39}$$

Equation (11-39) shows that the implicit trapezoidal rule is unconditionally stable, since the factor multiplying $y(x)$ on the right side is less than one in magnitude for any positive $Kh$. However, note that this is just barely so, in the sense that for

$Kh \to \infty$, this factor $\to -1$. Thus, we see that for large $Kh$, the implicit trapezoidal rule damps the solution very little, but causes a sign change (oscillation) at each step. For most very stiff problems in ODE, this is a disadvantage.

### 11.4.2 Experimental Stability Analysis

It is worth mentioning that it is sometimes useful to conduct "experimental" stability analysis using the computer. Of course, this is useful as a check on an analytical analysis, but it can also be valuable before embarking on a standard analysis. This is the case since, for many methods, a stability analysis is not simple. It is often much easier to look for experimental stable behavior *before* putting any time into the analysis, because otherwise a lengthy analysis may just lead to the negative conclusion that the method is not useful.

To carry out an experimental stability analysis, we program the method using the prototype equation $y' = -Ky$, where we can take $K$ to be any convenient positive value (e.g., $K = 100$). Then we do the integrations out to some reasonably large number of steps, looking to see if the solution continues to decay. This is done for a number of step sizes $h$. Then we keep track of the range of $Kh$ for which the solution remains stable in order to define the numerical stability limit.

## 11.5 STIFF EQUATIONS AND METHODS

We have discussed above examples of integration methods which have the property of unconditional stability. These methods are especially important with regard to so-called *stiff* differential equations. There is not unanimous agreement in the literature regarding a technical definition of "stiffness," but a practical definition might be that *a stiff problem is one that is dominated by stability or error propagation considerations.* Thus, in such problems the so-called non-stiff methods (such as forward Euler, Runge-Kutta, improved Euler, etc.), which have limited regions of stability with respect to step size, do not give acceptable performance. Briefly, a non-stiff method applied to a very stiff problem will only give stable results if it creeps along at a miniscule step size, which will take forever to get to the desired $x$. It may be difficult to predict by inspection that a problem will be stiff, but often stiff problems are characterized by having constants (or parameters) of widely different magnitude appearing in the set of equations. For a thorough discussion of stiff equations and methods, the reader is referred to Aiken [1985].

### 11.5.1 Some Useful Implicit Methods

The best-known stiff methods for application to either very difficult or large problems include the Gear backward differentiation (BDF) methods, of which backward Euler is the simplest example, and implicit or semi-implicit Runge-Kutta methods. Here, we will primarily be concerned with the simplest method having very strong unconditional stability. We consider the backward Euler method, along with a variation known as the implicit improved Euler (IIE), from Hanna [1988] and Ashour and Hanna [1990]. As we saw earlier, the backward Euler has excellent stability characteristics but is only first-order accurate. The implicit improved Euler is just backward Euler enhanced with one final step of the trapezoidal rule retaining the backward Euler derivatives, as follows:

Implicit Improved Euler

$$z(x+h) = y(x) + hf[x+h, z(x+h)]$$
$$y(x+h) = y(x) + \frac{h}{2}\{f[x, z(x)] + f[x+h, z(x+h)]\} \tag{11-40}$$

This gives IIE a second-order accuracy and changes stability so that IIE is unconditionally stable and is strongly damped, but less so than backward Euler. Thus, the IIE inherits the best features of backward Euler and trapezoidal rule, since it is strongly stable and has second-order accuracy. Computer program IIEX, which is discussed at the end of the chapter, has options for both backward Euler and implicit improved Euler. Both methods have a variable step-size option, and both can be employed with global extrapolation to estimate error and increase accuracy.

### 11.5.2 The Explicit "Hybrid" Method

Another useful "moderately stiff" method we wish to introduce here is called the Euler/RK2 hybrid method, or just the "hybrid" method Ashour and Hanna [1990]. This is a new explicit method that has a stability limit which is four times that of either the forward Euler or RK2 method. It combines the results of both Euler and RK2 integration in an optimal way, so that *enhanced stability* and first-order accuracy are achieved using two derivative evaluations per step. The algorithm is defined as follows, with $y(x)$ representing the new "hybrid" solution to the equation $y' = f(x, y)$ with $y(x_0) = y_0$, the initial condition:

$$y_{\text{Euler}}(x+h) = y(x) + hf(x, y(x))$$
$$y_{\text{RK2}}(x+h) = y(x) + \frac{h}{2}[f[x, y(x)] + f[x+h, y_{\text{Euler}}(x+h)]] \tag{11-41}$$
$$y(x+h) = \alpha y_{\text{Euler}}(x+h) + (1-\alpha) y_{\text{RK2}}(x+h)$$

Extension of Equations (11-41) to a system of ODEs is made in an obvious way. We see from Equations (11-41) that for $\alpha = 0$, we have RK2, and for $\alpha = 1$, we have Euler. To find the optimal value of $\alpha$ for stability enhancement, we formulate a stability analysis for the above algorithm and optimize with respect to $\alpha$. This yields the optimal value $\alpha = 0.74$ corresponding to $Kh = 7.6$, which is nearly four times larger than the stability limits of both Euler and RK2.

This new "hybrid" algorithm has been found to be particularly well-suited to the numerical solution of partial differential equations, using the method of lines. Our computer program PDEHYBD (in Chapter 14 on the numerical solution of partial differential equations) uses this algorithm, together with options for up to three trajectories with global extrapolation. In addition, an error control option is included which is especially valuable in this type of problem.

**TABLE 11.7** INTEGRATION OF $y' = 50(x^2 - y)$; $y(0) = 0$ BY IMPROVED EULER METHOD (calculated values of $y$ at $x = 2$; $y_{\text{exact}}(2) = 3.9208$)

| Step-size | $y_{\text{IEOnly}}$ | $y_{\text{IEX}}$ $(h, h/2, h/4)$ | Estimated Relative Error | Ratio |
|---|---|---|---|---|
| 0.02 | 3.9212 | 3.9208 | $3.0e^{-16}$ | 1. |
| 0.0222222 | 3.7963 | 3.9180 | $1.8e^{-4}$ | −336. |
| 0.025 | $-4.1232e^{+4}$ | −912.44 | 0.25 | $-8.8e^{+7}$ |
| 0.04 | $-6.35e^{+18}$ | $-2.10e + 13$ | 0.25 | $-7.9e^{+17}$ |

**Example 11.3 Integration of $y' = 50(x^2 - y)$ with $y(0) = 0$**

This is a simple example of a type of stiff problem that illustrates the difficulties of employing a non-stiff method (IEX) compared to using a stiff method (BEX, backward Euler with global extrapolation). The equation has the exact solution $y = x^2 - x/25 + (1/1250)[1 - \exp(-50x)]$. After the exponential transient dies out at about $x = 0.2$, the solution is essentially a simple quadratic, which IEX would integrate exactly. Tables 11.7 and 11.8 show the results of the integrations. We see from Table 11.7 that the IEX program does extremely well as long as $h \le 0.02$. This would be predicted since $\partial f/\partial y = -50$ and the IE stability limit is $Kh = -(\partial f/\partial y)h \le 1$ so that $h$ should be less than $1/50 = 0.02$. Thus, IEX will require many steps to go very far in $x$. We also see that the error ratio and estimated relative error of IEX at $h = 2/90$ (= 0.0222222...) indicate that something is wrong with the calculation, even though the $y$ values are not yet terribly contaminated. Thus, the ratio and estimated error serve here as useful diagnostics which suggest that the calculation is becoming unstable. On the other hand, inspection of Table 11.8 shows that backward Euler and, in particular, BEX are not having the least bit of trouble. In fact, we see the rather remarkable result that if we use *the largest possible step size, $h = 2$*, we get almost five-figure accuracy with good error estimate and ratio. It might seem plausible that if IEX was used *with variable step-size*, perhaps the program could increase the step size after the initial transient had died out. But this is the essential difficulty of instability; *even with variable step-size*, IEX and other non-stiff methods would continue to have the same stability restriction no matter how far the integration is carried out.

In this problem, if the 50 is replaced by $10^6$, IEX must take steps no larger than $1/10^6$, while BEX produces the exact result to six figures with $h = 2$, the largest possible step. This example dramatically shows the difference in behavior that can occur in using non-stiff and stiff methods in a stiff problem. It is *not* typical for the stiff method to produce such good results with a very large step size. In really difficult problems, we have both severe stability and truncation restrictions.

**Example 11.4 A Three-Dimensional, Mildly Stiff Problem**

We consider the following mildly stiff problem, which is discussed by Lapidus and Seinfeld [1971].

$$\begin{aligned} y_1' &= -0.1y_1 - 49.9y_2 \qquad & y_1(0) &= 2 \\ y_2' &= -50y_2 & y_2(0) &= 1 \\ y_3' &= 70y_2 - 120y_3 & y_3(0) &= 2 \end{aligned}$$

**TABLE 11.8** INTEGRATION OF $y' = 50(x^2 - y)$; $y(0) = 0$ BY BACKWARD EULER METHOD (calculated values of $y$ at $x = 2$; $y_{exact}(2) = 3.9208$)

| Step-size | $y_{BEOnly}$ | $y_{BEX}$ $(h, h/2, h/4)$ | Estimated Relative Error | Ratio |
|---|---|---|---|---|
| 2.0 | 3.9604 | 3.9207 | $2.4e^{-5}$ | 0.98 |
| 1.0 | 3.9408 | 3.9208 | $5.1e^{-7}$ | 0.999 |
| 0.4 | 3.9288 | | | |
| 0.2 | 3.9248 | | | |
| 0.1 | 3.9218 | | | |

**TABLE 11.9** COMPARISON OF NON-STIFF AND STIFF METHODS FOR 3-DIMENSIONAL PROBLEM AT $x = 2$

| | IEX Method | | | BEX Method | | |
|---|---|---|---|---|---|---|
| Step-size | $y_1$ | $y_2$ | $y_3$ | $y_1$ | $y_2$ | $y_3$ |
| 0.005 | 0.8187 | $e^{-44}$ | $e^{-44}$ | | | |
| 0.01 | 0.8187 | $e^{-40}$ | $-e^{+18}$ | | | |
| 0.02 | 0.8187 | $e^{-34}$ | $-e^{+45}$ | | | |
| 0.05 | $-1.4e^{+18}$ | $-e^{+18}$ | $e^{+46}$ | | | |
| 0.1 | Unstable | | | 0.8187 | $e^{-18}$ | $e^{-18}$ |
| 0.2 | Unstable | | | 0.8187 | $e^{-11}$ | $e^{-11}$ |
| 1.0 | Unstable | | | 0.8189 | $e^{-4}$ | $e^{-4}$ |

We compare the results of integration using the non-stiff program IEXVS and the BEX (Backward Euler Extrapolation) stiff method in program IIEX. The programming of this problem is shown in the concise listing of program IIEX at the end of the chapter. The integration is carried out to $x = 2$. The results are shown in Table 11.9.

The exact solution components $y_2$ and $y_3$ are virtually zero (about $10^{-44}$) at $x = 2$. We can note the following conclusions from Table 11.9. In this problem, to achieve excellent results with the non-stiff IEX method, we need to have a step size $h$ of about $h = 0.005$. By contrast, the stiff method BEX achieves almost as good results at step size $h = 1.0$, and it is stable for all step sizes. One can appreciate that the computational efficiency of the stiff method would be even much greater in this problem if it was required to integrate to a much larger $x$ value! The efficiency of the stiff method would also be much greater in a more difficult problem.

## 11.6 SPECIAL APPLICATIONS

In this section, we discuss some aspects of the numerical solution of ODE initial value problems that are important but do not fit naturally into the previous material.

### 11.6.1 Numerical Quadrature

In the chapter on evaluation of integrals, we mentioned the use of the ODE initial value program IEX for doing numerical quadrature. This is accomplished in the following way. If our desired integral is $I = \int_a^b f(t)\,dt$, we define $y(x) = \int_a^x f(t)\,dt$ (so that $I = y(b)$) and differentiate $y(x)$ to get the differential equation $y' = f(x)$. The proper initial condition corresponds to $y = 0$ at $x = a$. Solution of this initial value ODE problem out to $x = b$ gives $y(b) = I$. Program IEX is set up to do this quadrature problem with two Richardson extrapolations if we specify $q = 4$ in Sub Init. The $q = 4$ value is used for quadrature rather than the $q = 3$ for ODE because this is the proper extrapolation exponent for *quadrature*. Since we are, in this way, using the Euler-Maclaurin sum formula with two extrapolations, the final result for $y(b)$ is accurate to $O(h^6)$. The estimated error and ratio are interpreted in the usual way.

### 11.6.2 Apparent Singularity at Either End of the Interval of Integration

This problem arises occasionally and is most easily treated by modifying the derivative subroutine to avoid zero denominators. As a typical example, consider the cylindrical catalyst pellet problem $C'' + C'/x = KC$, with one boundary condition being $C'(0) = 0$. This condition is required for boundedness of the solution when the radius is zero ($x = 0$), for otherwise the term $C'(0)/x$ would be infinite. Using the usual new variables $y_1 = C$ and $y_2 = C'$, the original equation becomes the system $y_1' = y_2$ and $y_2' = Ky_1 - y_2/x$. As $x \to 0$, we have $y_2/x \to y_2'(0)$ (by L'Hospital's rule), and therefore, the correct derivative condition at $x = 0$ is $y_2'(0) = Ky_1(0)/2$. The numerical difficulty is that the derivative expression for $y_2'$ blows up at $x = 0$. A simple remedy is to replace the usual single statement `LET d(2) = K*y(1) - y(2)/x` in subroutine Deriv with the True BASIC sequence

```
IF x = 0 THEN
    LET d(2) = K*y(1)/2
ELSE
    LET d(2) = Ky(1) - y(2)/x
END IF
```

### 11.6.3 Functions Defined by Discrete Data

It sometimes happens that some known function in a differential equation is only defined by discrete data (e.g., a forcing function or a variable coefficient). To apply the usual numerical methods, it is convenient to "fit" the discrete function with a formula such as rational interpolation or least squares. This is usually more convenient than introducing some kind of interpolation subroutine into the calculations.

### 11.6.4 Discontinuities

Mathematical models sometimes involve some kind of discontinuity as a function of some variable or parameters. Common examples include problems where some known model function undergoes a "switch" from one type behavior to another (such as a change from laminar to turbulent flow), or where the actual model differential equations undergo a change. The case of a model function switch can usually be accommodated easily with the use of the IF-THEN-ELSE construct. In the case of a model change, we must generally stop right at the point of the change, and use the prior model values as initial conditions for the new model.

### 11.6.5 Differential and Algebraic Equations

Some mathematical models involve coupled systems of algebraic and differential equations. In principle, one could solve the algebraic equations every time their solution is needed as an input to the ODE calculation, but this is usually very awkward and inconvenient. It is generally much more convenient to convert the algebraic equations to differential equations and solve the whole problem as one (larger) set of differential equations. As a simple example, consider the problem $y_1' = f_1(x, y_1, y_2)$ and $f_2(y_1, y_2) = 0$. If we differentiate the algebraic equation for $f_2$ with respect to $x$ and then solve for $y_2'$, we get $y_2' = -[(\partial f_2/\partial y_1)/(\partial f_2/y_2)] \cdot y_1'$. The problem now is *two* differential equations and no algebraic equation. To get the initial condition for $y_2$, we must solve the algebraic equation $f_2$ *once* using the known initial condition for $y_1$.

## 11.7 COMPUTER PROGRAMS

In this section, we discuss the computer programs applicable to various problems in the chapter. For each program, we give a brief discussion and a concise listing, which includes (i) a program description, (ii) how to run, (iii) user-specified subroutines, including a sample problem, and (iv) selected output. Detailed program listings can be obtained from the diskette.

### ODE System Solvers IEX and IEXVS

These two programs differ only in (i) the variable integration step-size option and (ii) the direct error estimation option, which are available in program IEXVS. Therefore, we give only the concise listing for IEXVS, of which IEX is merely a subset (with no error control parameters). Instructions for running the program are given in the concise listing. The parameter $q$ in Sub Init should be set equal to four *only* if *all of the equations are independent of* y (quadrature problem); otherwise, $q = 3$. The program first prompts for a possible derivative check; this should always be used initially in order to ensure that the problem is programmed correctly. Next, the program asks for the number of trajectories (1, 2, or 3) and the initial step size. One trajectory has no extrapolation; two trajectories gives either direct error estimation (ext = 0) comparing trajectories $y(h)$ and $y(h/2)$, or one extrapolation (ext = 1) with these two trajectories. Three trajectories does two single extrapolations and one double extrapolation (the best estimate of $y$) and also calculates the error ratio. We would generally use three trajectories in order to get global error information as well as higher-order accuracy.

The user is asked whether or not to use error control (variable step-size). If the answer is yes, integration begins from $x_0$ and prints at intervals of $h_p$ (which is specified in Sub Init) until $x_1$ is reached. The program varies step size in an attempt to achieve the specified *local* relative desired error tolerance (parameter tol in Sub Init). If error control is not used, fixed-step integration is carried out using the previously specified initial step size, and the user is prompted for the value of the print parameter $pr$. For $pr = 1$ we print every integration cycle, for $pr = 5$, every 5 cycles, and so on. For fixed-step integration we usually coordinate the step size $h$ and print parameter $pr$ so as to get output at the same $x$ values regardless of the step size.

### Program IEXVS.

```
!This is improved Euler/extrapolation/error control(IEXVS.tru) by O.T.Hanna
!n=number of differential equations
!x,y(i) are independent and dependent variables in subroutines & output
!x0,x1 are initial and final values of x
!y1(i,j) is dummy for y(i) in program
!d(i),dold(i,j),dnew(i) are derivatives of y(i)
!h=local step size;hp is print increment;
!pr is print parameter for fixed step
!j1 is number of trajectories;j1=1 for h,j1=2 for h,h/2;j1=3 for h,h/2,h/4
!q=3 for ode; q=4 for quadrature
!**********************************************************************
!* To run,specify q,n,x0,x1,initial y(i),error parameters in Sub Init;   *
!* specify derivatives d(i) in terms of x,y(i) in Sub Deriv;             *
!* program prompts for (a) derivative check? (b) j1,(ext),h,ec,(pr)?     *
!**********************************************************************
!**IMPROVED EULER/EXTRAPOLATION/ERR-CTRL SOLUTION OF PROBLEM BELOW**
!ext is extrapol. parameter(with j=2 only);
!ext=0,j1=2 gives direct error calc.
!error parameter tol is desired local relative error
!parameters eps,sf,are zero denominator and safety factor values
!To print every step,comment out (!DO !LOOP) in Integration section.

SUB Init (q,n,x0,x1,y(),d(),hp,hmaxstep,tol,eps,sf)
    !***denotes user specified quantity
    LET q = 3               !***q=3 for ODE, q=4 for quadrature
    LET n = 2               !***number of equations
    LET x0 = 0              !***initial x value
    LET x1 = 5              !***final x value
    LET y(1) = 0.057        !***initial y values
    LET y(2) = 0.0

    CALL Deriv (x0,y,d)

    !The following is used only with error control(ec=1)
    LET hp = .01                   !***hprint
    LET hmaxstep = hp              !***maximum user-specified step size
```

```
    LET tol = .01                   !***desired local relative errors
    LET eps = .0001                 !zero denominator parameter
    LET sf = 2                      !safety factor
END SUB

SUB Deriv(x,y(),d())
!Cartesian catalyst pellet problem
    LET d(1) = -y(2)                !***equation derivatives
    LET d(2) = -100*y(1)^2
END SUB
```

### *ODE System Solver IIEX for Stiff Differential Equations*

This program offers the option for (i) backward Euler (BE) or (ii) implicit improved Euler (IIE). Both programs use the same information and have options for 1, 2, or 3 trajectories, global extrapolation, and error control (variable step-size). The main differences between the two algorithms are that BE is more stable (both methods are unconditionally stable) but less accurate (BE is first-order, IIE is second-order). For more difficult problems where smaller step sizes are required, IIE is usually more efficient.

Instructions for running are given in the concise listing. This program is operated in almost the same way as program IEXVS discussed above. The main difference is that in program IIEX a Jacobian matrix of partial derivatives $[(\partial f_i/\partial y_j)$ where $y_i' = f_i(x, y_1, \ldots, y_n)$, $i = 1, \ldots, n]$ must be given. This is accomplished by the user specifying these partials analytically in Sub Jacobian. The program output and interpretation are the same as for program IEXVS. One would generally use three trajectories in order to get global error information and higher accuracy (order three for BE, order four for IIE). If error control is used (variable step-size), we generally need to give a conservatively small initial step size. If this does not produce satisfactory results, the error tolerance parameter, tol, should be decreased. The problem shown in the concise listing of program IIEX is a three-dimensional moderately stiff problem taken from Lapidus and Seinfeld [1971].

#### Program IIEX.

```
!This is implicit improved Euler (IIEX.tru) by O.T.Hanna
!Program has option for backward Euler or implicit improved Euler.
!Program is for stiff equations and allows global extrapolation
!and error control.
!n=number of differential equations
!x,y(i) are independent and dependent variables in subroutines & output
!x0,x1 are initial and final values of x
!y1(i,j) is dummy for y(i) in program
!d(i),dold(i,j),dnew(i) are derivatives of y(i)
!h=local step size;hp is print increment;
!pr is print parameter for fixed step
!j1 is number of trajectories;j1=1 for h,j1=2 for h,h/2;j1=3 for h,h/2,h/4
```

```
!*********************************************************************
!* To run,specify n,x0,x1,initial y(i),iteration parameters         *
!* norm_tol,mx and error parameters(optional) in Sub Init,          *
!* specify derivatives d(i) in terms of x,y(i) in Sub Deriv,        *
!* specify Jacobian part.der. matrix of ODE eqns. in Sub Jacobian,  *
!* program prompts for (a) derivative check? (b) j1,h,ec,(pr)?      *
!*********************************************************************

SUB Init (n,x0,x1,y(),d(),hp,hmaxstep,tol,eps,sf,norm_tol,mx)
    !***designate user specified quantity
    LET norm_tol = 1.e-6     !***alg. eqn. sol'n tolerance
    LET mx =8                !***max. no. iterations for alg. eqns.
    LET n = 3                !***number of ODE
    LET x0 = 0               !***init. value of indep't variable
    LET x1 = 2               !***final value of indep't variable
    LET y(1) = 2             !***init. values of dependent variables
    LET y(2) = 1
    LET y(3) = 2
    CALL Deriv (x0,y,d)

    !The following is used only with error control(ec=1)
    LET hp = .2              !***hprint
    LET hmaxstep = hp        !***maximum user-specified step size
    LET tol = .1             !***desired local relative errors
    LET eps = .01            !zero denominator parameters
    LET sf = 2               !safety factors
END SUB

SUB Deriv(x,y(),d())         !***ODE derivatives
!Lapidus-Seinfeld problem
    LET d(1) = -.1*y(1)-49.9*y(2)
    LET d(2) = -50*y(2)
    LET d(3) = 70*y(2) - 120*y(3)
END SUB

SUB Jacobian (x,y(),f(,))    !***df(i)/dy(j) from ODE
    LET f(1,1) = -0.1
    LET f(1,2) = -49.9
    LET f(1,3) = 0.0
    LET f(2,1) = 0.0
    LET f(2,2) = -50
    LET f(2,3) = 0.0
    LET f(3,1) = 0.0
    LET f(3,2) = 70
    LET f(3,3) = -120
END SUB
```

## PROBLEMS

**11.1.** Consider the initial value problem $dy/dx = y$ with $y(0) = 1$, which has a well-known solution. Solve this problem using program IEX by methods of order two, three, and four as follows:
(a) Use one trajectory
(b) Use two trajectories
(c) Use three trajectories
In each case, find the approximate step size $h$ and computer time required to achieve six-figure accuracy for $y$ at $x = 1$.

**11.2.** Determine the numerical solution of the ODE problem $y'' + 4y = 0$ with boundary conditions $y(0) = 0$ and $y'(0) = 2$. Integrate (six-figure accuracy) to $x = 2$, and compare the numerical solution with the analytical solution evaluated with a handheld calculator.

**11.3.** A chemical reaction is being carried out in a batch, isothermal reactor. The rate of reaction is given by $dC/dt$ = rate = $-[C/(1 + C)^{1/2} - 0.3C^2]$ ($\text{hr}^{-1}$). The initial concentration is $C(0) = 0.6$. Calculate the concentration in the reactor at the end of one hour. Also calculate the time required for the concentration to become 0.1.

**11.4.** For the problem defined in Table 11.4, carry out the following calculations:
(a) Do one step of the Euler method by hand.
(b) Do one step of the Improved Euler method by hand.
(c) Compare the results of (a) and (b) with each other and the result of running program IEX (one trajectory) for one step.

**11.5.** A mathematical model for a non-isothermal tubular chemical reactor with heat transfer can be written in dimensionless form as follows. Here, $y$ is dimensionless concentration, and $\theta$ is dimensionless temperature (see Davis [1984]).

$$\frac{dy}{dz} = -D_a \cdot y \cdot \exp\left(\delta\left(1 - \frac{1}{\theta}\right)\right)$$

$$\frac{d\theta}{dz} = \beta \cdot D_a \cdot y \cdot \exp\left(\delta\left(1 - \frac{1}{\theta}\right)\right) - h(\theta - 1)$$

The parameters $D_a$, $\beta$, $\delta$, and $h$ represent dimensionless reaction-rate constant, heat of reaction parameter, Arrhenius constant, and heat transfer coefficient. Set up program IEXVS to solve the set of equations for any values of $D_a$, $\beta$, $\delta$, and $h$. Solve the system from $z = 0$ to $z = 1$ with $y(0) = 1$, $\theta(0) = 1$, and the parameter values $D_a = 3$, $\beta = 10$, $\delta = 0.02$, and $h = 1000$.
(i) Determine by experiment the largest fixed step (one significant figure) that produces acceptable error ratios and estimated errors. What accuracy is achieved in this case?
(ii) Run in variable-step mode (use Tol values of $10^{-6}$ to $10^{-1}$) and observe the results and the corresponding step sizes selected by the program.
(iii) Use the BEX option of ODE stiff program IIEX with step sizes 0.1 and 0.01; compare with the above results.

**11.6.** The Blasius equation of fluid mechanics is a third-order differential equation of the form $d^3f/dx^3 + f \cdot d^2f/dx^2 = 0$. Take as initial conditions $f(0) = df/dx(0) = 0$ and $d^2f/dx^2(0) = 0.5$. To six figures, what is the limiting value of $df/dx$ for "large" $x$? For what value of $x$ is the limit reached (two figures)?

**11.7.** The integral $I(x) = \int \exp(-t^2)dt$ between the limits of 0 and $x$ is, except for a scale factor, known as the Error function, which is important in a number of engineering applications. Although its appearance is simple, this integral cannot be evaluated exactly by the methods of calculus.

(i) Find the value of $I$ for $x = 2$.

(ii) Find the value of $x$ for $I = 0.5$.

**11.8.** A method whose global error exponent $p$ is unknown has been used to generate the data in the table below. Use this data to estimate the apparent value of $p$.

| $x$ | $y(h = 0.2)$ | $y(h = 0.1)$ | $y(h = 0.05)$ |
|---|---|---|---|
| 0 | 1.0 | 1.0 | 1.0 |
| 0.2 | 1.22137 | 1.22140 | 1.22140 |
| 0.4 | 1.49174 | 1.49181 | 1.49182 |
| 0.6 | 1.82196 | 1.82210 | 1.82212 |
| 0.8 | 2.22528 | 2.22551 | 2.22554 |
| 1.0 | 2.71788 | 2.71823 | 2.71827 |

**11.9.** A chemical reaction problem that is quite stiff corresponds to the following set of equations:

$$y_1' = -0.04y_1 + 10^4 y_2 y_3$$
$$y_2' = 0.04y_1 - 10^4 y_2 y_3 - 3 \cdot 10^7 y_2^2$$
$$y_3' = 3 \cdot 10^7 y_2^2$$
$$y_1(0) = 1 \quad y_2(0) = 0 \quad y_3(0) = 0$$

It is required to estimate the solution to four figures at $x = 100$.

(i) Do this by experimenting with stiff program IIEX (use the BEX method). Nominal parameter values include Tol $= 10^{-3}$, hinit $= 10^{-6}$, and hmaxstep $= 1$.

(ii) For comparison, use non-stiff program IEXVS and observe that the allowable step size near the start of integration is comparable to that of program BEX. Then, for contrast, use the solution from (i) at $x = 100$ as initial conditions and see the allowable step size using IEXVS at the end of the integration.

**11.10.** Specify the appropriate derivative subroutine that could be used with program IEX to solve the following differential/algebraic problem:

$$\frac{d^2y}{dx^2} = f_1(x, y, z); \quad y(0) = A, y'(0) = B$$
$$\frac{d^2z}{dx^2} = f_2(x, y, z); \quad z(0) = C, z'(0) = D$$
$$0 = f_3(x, y, w)$$

Also indicate the necessary initial conditions.

**11.11.** Consider the implicit differential equation

$$\frac{dy}{dx} - e^{\left(\frac{dy}{dx}\right)} = x - 2y; \quad y(0) = 1$$

It is desired to obtain a numerical solution to this equation (six-figure accuracy) out to $x = 2$.

**11.12.** For the differential equation $dy/dx = \ln x + xy^2$, with $y(0) = 0$, calculate the value of $y(3)$.

## REFERENCES


AIKEN, R. C. (1985), *Stiff Computation*, Oxford, New York

ASHOUR, S. S., and HANNA, O. T. (1990), *Computers Chem. Engng.*, *14*, p. 267

COLLATZ, L. (1966), *The Numerical Treatment of Differential Equations*, Springer-Verlag, New York

CRANDALL, S. (1956), *Engineering Analysis*, McGraw-Hill, New York

DAVIS, M. E. (1984), *Numerical Methods & Modeling for Chemical Engineers*, Wiley, New York

DAHLQUIST, G., BJORK, A., and ANDERSON, N. (1974), *Numerical Methods*, Prentice-Hall, Englewood Cliffs, NJ

FERZIGER, J. H. (1981), *Numerical Methods for Engineering Application*, Wiley, New York

FINLAYSON, B. (1972), *The Method of Weighted Residuals and Variational Principles*, Academic Press, New York

FINLAYSON, B. (1980), *Nonlinear Analysis in Chemical Engineering*, McGraw-Hill, New York

FRANKS, R. G. E.(1972), *Modeling and Simulation in Chemical Engineering*, Wiley, New York

HANNA, O. T. (1988), *Computers Chem. Engng.*, **12**, p. 1083.

HANNA, O. T. (1990), *Computers Chem. Engng.*, **14**, p. 273.

HENRICI, P. (1964), *Elements of Numerical Analysis*, Wiley, New York

KAHANER, D., MOLER, C., and NASH, S. (1989), *Numerical Methods and Software*, Prentice-Hall, Englewood Cliffs, NJ

LAPIDUS, L., and SEINFELD, J. H. (1971), *Numerical Solution of Ordinary Differential Equations*, Academic Press, Inc., New York

PRESS, W. H., FLANNERY, B. P., TEUKOLSKY, S. A., and VETTERLING, W. T. (1986), *Numerical Recipes*, Cambridge Univ. Press, New York

RATKOWSKY, D. A.(1990), *Handbook of Nonlinear Regression Models*, Marcel Dekker, New York

SCHEID, F. (1988), *Numerical Analysis*, 2nd ed., McGraw-Hill, New York

SCRATON, R. E. (1984), *Basic Numerical Methods*, Edward Arnold, Baltimore

SCRATON, R. E. (1987), *Further Numerical Methods in BASIC*, Edward Arnold, Baltimore

SILEBI, C. A., and SCHIESSER, W. E. (1992), *Dynamic Modeling of Transport Process Systems*, Academic Press, Inc., New York

# 12

# Numerical Methods for ODE Boundary Value Problems

A boundary value problem in ordinary differential equations refers to a situation where the auxiliary conditions are specified at more than one value of the independent variable. Aside from questions of existence and uniqueness, which are more complicated in this case, such problems would pose no special difficulty if the solution of the differential equation could be expressed in a simple, explicit, analytical form; for then we could simply apply the appropriate boundary conditions directly. As we know, however, the solutions of many differential equations cannot be found except by approximation or numerical techniques. Numerical techniques for boundary value problems usually require greater sophistication than those for initial value problems. Note that normally a boundary value problem must be at least second-order, to have two conditions specified at different points. In the discussion that follows, we refer primarily to second-order boundary value problems.

## 12.1 DIRECT USE OF TAYLOR'S THEOREM

Just as in the case of initial value problems, Taylor's theorem may be applied directly to a boundary value problem to obtain an approximate solution. If boundary conditions are specified at two points, then a Taylor expansion about either point, when evaluated at the other point, provides a single algebraic equation for the unknown coefficient in the expansion. The algebraic equation will be linear if the differential equation and boundary conditions are linear; otherwise, it will be nonlinear. This situation remains true no matter how many terms are taken in the Taylor expansion, since all higher derivatives can be expressed in terms of the single unknown by means of the differential equation. One possible limitation of the foregoing procedure is that the Taylor expansion which determines the unknown coefficient may either diverge or converge very slowly. To decrease these hazards, it is sometimes desirable to write the Taylor expansion about the midpoint between the two specified boundary points, rather than about one of these points. This has the effect of producing a Taylor expansion with generally improved convergence properties, but the price paid is that now evaluation at the two specified points results in two algebraic equations for the two unknown Taylor coefficients in the expansion.

These ideas will now be illustrated for the boundary value problem $d^2C/dx^2 = BC^2$, with $C(0) = 1$, $C'(1) = 0$, and $B$ a parameter. We have considered this nonlinear catalyst problem earlier. It is possible to expand the solution by Taylor's theorem about either boundary point, and we illustrate both cases to show the technique involved. First, let us expand about $x = 1$. We then have

$$C(x) \cong C(1) + C'(1)(x-1) + \frac{C''(1)(x-1)^2}{2!} = C(1) + \frac{BC(1)^2(x-1)^2}{2!} \tag{12-1}$$

In Equation (12-1) we have used the boundary condition $C'(1) = 0$, and $C''(1)$ has been replaced by $BC(1)^2$ from the differential equation. The unknown in the Taylor expansion to be determined by using the boundary condition at $x = 0$ is $C(1)$. The boundary condition $C(0) = 1$ is now applied to Equation (12-1), and this gives the following equation for $C(1)$:

$$1 = C(1) + \frac{BC(1)^2}{2} \tag{12-2}$$

Equation (12-2) is a quadratic equation which could be solved explicitly for $C(1)$ as a function of $B$. When this is substituted back into Equation (12-1), we have the three-term Taylor approximation to the solution. Rather than carrying out the solution of the quadratic, however, we merely observe that for $B \to 0$, $C(1) \to 1$ (no reaction); and for $B \to \infty$, $C(1) \to 0$ (complete reaction). Thus, for this simple approximation we obtain what seems to be reasonable qualitative behavior. Of course, we could have taken further terms in the Taylor expansion in an attempt to increase accuracy, or we could have tried acceleration by means of Padé approximation, but we merely wish to illustrate the ideas involved. Developing a simple Taylor expansion in boundary value problems is useful for checking the results of a computer program, or getting approximate starting values for a numerical algorithm such as shooting or finite difference.

We now consider doing the Taylor expansion about $x = 0$. Then we would get

$$C(x) \cong C(0) + C'(0)x + \frac{C''(0)x^2}{2!} = 1 + C'(0)x + \frac{Bx^2}{2!} \tag{12-3}$$

In Equation (12-3) we have used the condition $C(0) = 1$, and we have replaced $C''(0)$ with $B$ from the differential equation. The unknown to be determined in this case is $C'(0)$. We now apply the boundary condition at $x = 1$, namely $C'(1) = 0$, by differentiating Equation (12-3) and evaluating it at $x = 1$. This gives $C'(0) = -B$. When this value of $C'(0)$ is substituted back into Equation (12-3), we have the three-term Taylor approximation about $x = 0$. The evaluation of this at $x = 1$ gives $C(1) = 1 - B/2$. This value of $C(1)$ also goes to unity for $B \to 0$ (no reaction).

However, for larger values of $B$, this expression first goes to zero and then becomes negative, which is physically unrealistic. Once again, for this expansion more terms could be obtained if desired. The technique we have illustrated here works essentially the same way in all boundary value problems.

## 12.2 THE SHOOTING METHOD FOR BOUNDARY VALUE PROBLEMS

An important method for solving boundary value problems through the use of initial value techniques is known as the "shooting" method. In the authors' opinion, this is generally the most efficient numerical technique for such problems. The idea of the method is to use successive approximation to solve for the missing initial condition. This is implemented by initially assuming a value for the missing initial condition, solving the initial value problem to estimate the final boundary condition, and successively refining this process until the required final boundary condition is satisfied. Thus, each initial value trajectory defines one point on a curve of the computed residual of the final boundary condition versus the assumed initial condition. When the residual of this boundary condition is close enough to zero, we have found the desired numerical solution. The refinement of the missing initial condition is usually accomplished by means of linear interpolation, which is the same as solving the nonlinear algebraic equation by the secant method.

A very important advantage of the shooting method is that it can utilize highly efficient initial value procedures. We exploit this possibility here by modifying the initial value solver IEXVS to become the one-parameter automatic boundary value solver IEXVSSH. Here, the SH at the end of the name stands for "shooting." The program takes an initial guess for the unknown initial condition and performs a series of integrations that attempt to adjust the guessed quantity so as to satisfy the boundary condition at the other end. Program IEXVSSH will be explained in detail at the end of the chapter; for the user, the essential way in which it is different from program IEXVS is that now one of variables (the one to be guessed) must be labeled "guess," and a residual for the other-end boundary condition must be defined in the subroutine Sub Residual.

We must be aware that the shooting method may encounter stability difficulties in some problems, especially those of a "stiff" nature which we have encountered in our study of integration of initial value problems. In this connection, it should be borne in mind that "shooting" can normally be tried from either end of the boundary value problem. In some cases, shooting from one end is unstable, while from the other end it is not. We now consider an example of using the shooting method in a boundary value problem by means of program IEXVSSH.

**Example 12.1 Backward Shooting Solution of $C'' = 100\, C^2$; $C(0) = 1$, $C'(1) = 0$**

This is the nonlinear catalyst pellet problem $C'' = BC^2$ that we have considered earlier by means of regular perturbation, and weighted residuals when $B$ was relatively small. The difference is that here we choose the parameter to have the large value $B = 100$, in order to show some of the difficulties that can occur. First, we may note that a natural approach would be to shoot from $x = 0$, by guessing $C'(0)$, and attempting to satisfy the boundary condition residual at $x = 1$, namely $R = C'(1) = 0$. Without showing various disastrous results, we report that even when we start with a correct six-figure value of $C'(0)$, we are generally unable to get convergence to the solution of this problem by integrating forward from $x = 0$ to $x = 1$ using program IEXVSSH. On the other hand, if $B$ is smaller (e.g., $B = 1$), there are no difficulties integrating forward.

We therefore now try integrating *backward* from $x = 1$ to $x = 0$. We could do this directly (*for fixed step integration only*) by simply starting at $x = 1$ using the known initial condition $C'(1) = 0$, guessing the unknown initial value $C(1)$, and integrating backwards to $x = 0$ in an attempt to satisfy the boundary condition residual at $x = 0$, namely $R = C(0) - 1 = 0$. However, program IEXVSSH does not allow negative step sizes for variable-step integration, so we will proceed to modify the problem so that we can shoot backwards using variable-step integration, if desired. First, we convert the original second-order equation to a system of first-order equations in the usual way (as we would do for forward integration). Thus, we let $u_1 = C$ and $u_2 = C'$. When we transform the original equation and the boundary conditions, we get

$$\frac{du_1}{dx} = u_2 \qquad u_1(x = 0) = 1$$
$$\frac{du_2}{dx} = 100u_1^2 \qquad u_2(x = 1) = 0 \tag{a}$$

Now we wish to change the independent variable from $x$ to $t$ so that $x = 1$ corresponds to $t = 0$. Thus, we introduce $t = (1 - x)$ so that $du_i/dx = -du_i/dt$ for both $i = 1$ and 2. When we substitute into Equations (a) and transform the boundary conditions, we obtain

$$\frac{du_1}{dt} = -u_2 \qquad u_1(t = 1) = 1$$
$$\frac{du_2}{dt} = -100u_1^2 \qquad u_2(t = 0) = 0 \tag{b}$$

In this form, Equations (b) are ready for shooting from $t = 0$ $(x = 1)$. Thus, we guess the value of $u_1$ at $t = 0$ and integrate toward $t = 1$ in an attempt to satisfy the boundary condition residual at $t = 1$, which is $R = u_1(t = 1) - 1 = 0$. (In the computer program, the independent variable is always called $x$ and the dependent variables $y(i)$). We now show, in Table 12.1, some results of carrying out this calculation using program IEXVSSH. It must be remembered that each iteration of the shooting method involves an integration of the system of differential equations. Also, the "Number of Trajectories" refers to whether, in a given secant iteration, we are conducting the integration with one trajectory ($j1 = 1$, no extrapolation) or three trajectories ($j1 = 3$, two extrapolations).

The rather extensive results of Table 12.1, as shown in Figure 12.1, are illuminating and suggest the proper way of solving a relatively "difficult" boundary value problem. One salient feature of boundary value problems solved by the shooting method is that there are two independent sources of error: error in solving the algebraic equation to find the "guess" variable, and error in the numerical integration of the differential equations. If we get convergence with a sensible integration (one that is not obviously misbehaving), then this convergence should be the same for a given integration step size regardless of the initial value of the guessed variable $C(1)$. We see this behavior occurring in Table 12.1. It is interesting and significant to note that *convergence is achieved for a much wider range of initial guesses if the integration is very crude.* Thus, even by taking only *one step* ($h = 1$), we get convergence to a value that could then be used as a new starting value. Notice that for

**TABLE 12.1** BACKWARD SHOOTING SOLUTION OF $C'' = 100C^2$ WITH $C(0) = 1,\ C'(1) = 0$

| Initial Guess for C(1) | Number of Integration Trajectories $j1$ | Integration Step Size | Iterations Required | Converged Value of C(1) |
|---|---|---|---|---|
| 1.0 | 1 | 1.0 | 11 | 0.131774 |
| 0.0 | 1 | 1.0 | 10 | 0.131774 |
| 0.1 | 1 | 1.0 | 7 | 0.131774 |
| 1.0 | 1 | 0.2 | 25 | 0.0790796 |
| 0.0 | 1 | 0.2 | 62 | 0.0790796 |
| 0.1 | 1 | 0.2 | 7 | 0.0790796 |
| 1.0 | 3 | 0.1 | Overflow | Overflow |
| 0.0 | 3 | 0.1 | Overflow | Overflow |
| 0.1 | 3 | 0.1 | 19 | 0.0572166 |
| 0.1 | 3 | 0.01 | Overflow | Overflow |
| 0.06 | 3 | 0.01 | 7 | 0.0570842 |
| 0.0570842 | 3 | 0.005 | 3 | 0.0570842 |

Correct value of $C(1)$ is 0.0570842

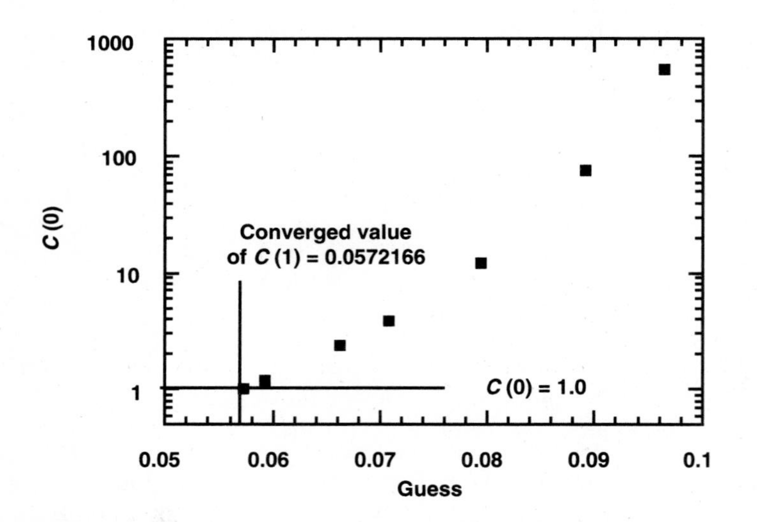


**Figure 12.1.** Convergence of Secant Method for Case of Three Integration Trajectories ($h = 0.1, 0.05, 0.025$)

a starting value of $C(1)$ equal to zero, we get convergence with $h = 1$ or $h = 0.2$, but not with $h = 0.1$, which is a more accurate integration. Similarly, for a starting value of $C(1) = 0.1$ with three trajectories, we get convergence with $h = 0.1$, but overflow with $h = 0.01$. Thus, it appears that the way *not* to proceed is to make a crude estimate of the guessed variable and then immediately do accurate integration. Instead, we should *begin with crude integration* to successively refine our estimate of the guessed value so that we can ultimately do accurate, error-monitored integration using three trajectories. At the end of the calculation we use the error information from the integration to be sure that we have the desired accuracy. It is also a good idea to take the ultimate result and use it as the first guess to be sure that we get convergence to this same value. We have done this in Table 12.1 in the case of three trajectories and $h = 0.005$ to verify that we have six-figure accuracy (with good error ratios).

For the case of higher-order boundary value problems, it may happen that there are more than one missing initial conditions which need to be "guessed." In such a situation, there must generally be one boundary residual corresponding to each guessed initial condition. This then involves the solution of a system of algebraic equations, where each iteration requires the full solution of the system differential equations (as well as possible derivatives of the residual/guess relationship, if a Newton-type method is being used). This is obviously a computationally intensive problem.

In the simplest special case where we have two guessed initial conditions and two boundary residuals, a simple brute-force calculation employing program IEXVSSH may be convenient. This procedure is a kind of inner and outer loop calculation, where the inner loop involves guess variable one (G1) and residual one (R1), and the outer loop involves guess variable two (G2) and residual two (R2). The inner loop is satisfied by program IEXVSSH, while the outer loop is satisfied directly by the user. To conduct the calculation, we select some particular value of G2 and calculate (via program IEXVSSH) the corresponding value of G1 that satisfies $R1 = 0$. In general, this value of G2 will *not* satisfy $R2 = 0$. Then we repeat this process a number of times, keeping track of the numerical values of G2 versus the corresponding residual R2. Finally, we can interpolate the $G2 - R2$ relationship to find the value of G2 that causes R2 to become equal to zero. This is the solution to the original problem. This procedure was used by Davis and Sandall [1991] to interpret mass transfer data for a membrane separator.

## 12.3 THE FINITE DIFFERENCE METHOD FOR BOUNDARY VALUE PROBLEMS

Another numerical method that is quite important for the solution of boundary value problems is known as the finite difference method. Although this approach is perhaps not as powerful as the shooting method for ODE boundary value problems, it is *very* important for the numerical solution of partial differential equations, which will be discussed later. The finite difference method involves approximating the unknown solution at a discrete set of points through the replacement of derivatives by differences. That is, numerical differentiation formulas are used to replace the derivatives in the equation with unknown values of the dependent variable.

To derive the appropriate difference relationships, a numerical differentiation approximation is used to replace derivatives at each of the equidistant points that divide the interval of interest. In addition, the boundary conditions must be replaced by differences where necessary. This process results in a system of algebraic equations for the unknown dependent variables at the so-called node points. If the differential equation and boundary conditions are linear, the algebraic equations are also linear. Otherwise, they are nonlinear. It is possible to take two rather different approaches to the formulation and solution of finite difference problems: a basic approach, which is conceptually very simple, and a more refined approach, which greatly improves computational efficiency. Here, we shall begin with the basic approach and then move on to some refinements. But first, we stop to show the numerical differentiation formulas we will need.

### 12.3.1 Numerical Differentiation Formulas

We briefly discuss here the numerical differentiation formulas that we will need to solve second-order ODE boundary value problems. These include the central, three-point forward and three-point backward formulas for $y'(x)$, as well as the central approximation to $y''(x)$. Other, more complicated (and accurate) formulas can be derived (see Appendix G on numerical differentiation), but they will not be needed in our discussion. The fundamental equation that can be used to derive all standard numerical differentiation approximations is Taylor's formula, in the form

$$y(x+ah) = y(x) + y'(x)ah + \frac{y''(x)(ah)^2}{2!} + \frac{y'''(x)(ah)^3}{3!} + \cdots \qquad (12\text{-}4)$$

By using Equation (12-4) as many times as necessary with different integer $a$ values, we can derive sets of linear algebraic equations having various derivatives at position $x$ as unknowns. Numerical differentiation expresses a derivative at location $x$ in terms of $y$ values at both $x$ and other locations. If all other $x$ values are ahead of location $x$, we have a *forward* approximation. If all other $x$ values are behind location $x$, we have a *backward* approximation. If all other $x$ values are located symetrically ahead of and behind location $x$, we have a *central* approximation. As a simple illustration, if we use Equation (12-4) with $a = 1$ and $a = -1$, and then subtract these equations, we obtain the central approximation to $y'(x)$, which is $[y(x+h) - y(x-h)]/2h$.

It is very convenient to use a *subscript* notation to specify numerical differentiation formulas. Subscript $i$ indicates location $x$, $i+1$ indicates $(x+h)$, and $i-1$ indicates $(x-h)$, and so on. Using the subscript notation, the above central approximation to $y'(x)$ would be written as $(y_{i+1} - y_{i-1})/2h$. Using Equation (12-4) with appropriate $a$ values and introducing subscript notation produces the results shown in Table 12.2. All of the results shown in this table have second-order accuracy, which means that their error is proportional to $h^2$. This error behavior is often indicated by the symbol $O(h^2)$ (read "big oh of $h^2$").

### 12.3.2 The Basic Finite Difference Method

We begin by illustrating the basic finite difference method for the very simple linear problem $d^2C/dx^2 = 2C$ with $dC/dx = 0$ at $x = 0$ and $C = 1$ at $x = 1$. It is first necessary to replace the derivative $d^2C/dx^2$ by a numerical differentiation approximation at an arbitrary interior point of the $x$ interval. This is accomplished

**TABLE 12.2** USEFUL NUMERICAL DIFFERENTIATION FORMULAS IN SUBSCRIPT NOTATION

| | |
|---|---|
| $y'(x)$ central | $\dfrac{y_{i+1} - y_{i-1}}{2h}$ |
| $y'(x)$ three-point forward | $\dfrac{-y_{i+2} + 4y_{i+1} - 3y_i}{2h}$ |
| $y'(x)$ three-point backward | $\dfrac{3y_i - 4y_{i-1} + y_{i-2}}{2h}$ |
| $y''(x)$ central | $\dfrac{y_{i+1} - 2y_i + y_{i-1}}{h^2}$ |

by using the second-order central approximation $C''(x) = [C(x+h) - 2C(x) + C(x-h)]/h^2 + O(h^2)$ (we will avoid using the subscript notation for a little while). Then the differential equation becomes

$$C''(x) - 2C(x) = 0 = \frac{C(x+h) - 2C(x) + C(x-h)}{h^2} + O(h^2) - 2C(x) \tag{12-5}$$

To illustrate the solution procedure, we now choose to divide the interval $(x = 0, 1)$ into four equal parts so that $h = 1/4$. Then we apply Equation (12-4) at all of the *interior* $x$ points of the region, *dropping the unknown error term* $O(h^2)$. This gives three equations for the values $C(0.25)$, $C(0.5)$, and $C(0.75)$. At the boundaries of the region of interest, the boundary conditions are also replaced by difference approximations where necessary. In this example, since $C(1) = 1$ is specified, this condition can be used as it stands. On the other hand, the condition $C'(0) = 0$ must be replaced by a difference approximation. In this example, a very crude but natural way to do this is by means of the simple two point forward approximation for $C'$. This would give $0 = C'(0) = (C(h) - C(0))/h + O(h)$, which yields, after dropping the error term, the difference approximation $C(0) = C(h)$. In the present problem, this very crude approximation would be $C(0) = C(0.25)$. If we now collect the interior and boundary equations together, we have

$$\begin{aligned}
&x = 0.0 && C(0) = C(0.25)\\
&x = 0.25 && \frac{C(0.5) - 2C(0.25) + C(0)}{(0.25)^2} = 2C(0.25)\\
&x = 0.5 && \frac{C(0.75) - 2C(0.5) + C(0.25)}{(0.25)^2} = 2C(0.5)\\
&x = 0.75 && \frac{C(1) - 2C(0.75) + C(0.5)}{(0.25)^2} = 2C(0.75)\\
&x = 1.0 && C(1) = 1
\end{aligned} \tag{12-6}$$

Equations (12-6) represent a system of five linear algebraic equations for the five unknowns $C(0)$, $C(0.25)$, $C(0.5)$, $C(0.75)$, and $C(1)$. These equations can be solved to yield the solution $C(0) = C(0.25) = 0.546$, $C(0.5) = 0.615$, $C(0.75) = 0.760$, and $C(1) = 1$.

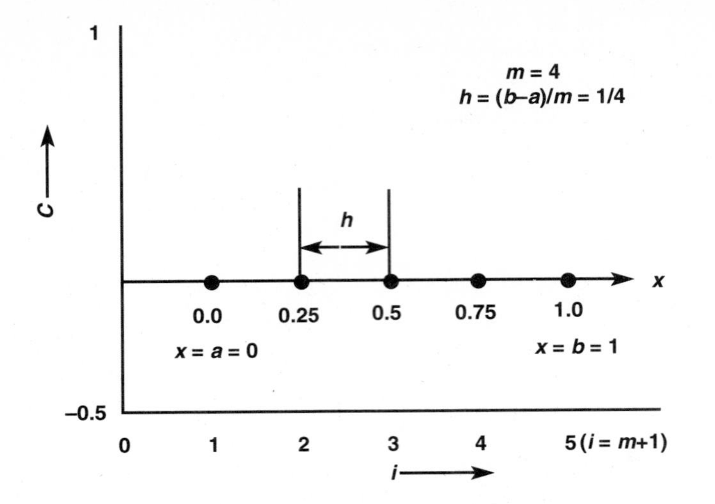


**Figure 12.2.** Use of Subscripts in Finite Difference Problems. Interval $(a, b)$ is divided into $m$ parts.

The foregoing calculation illustrates the general ideas of the discretization of a differential equation by the difference method. However, rather than proceeding as above, in order to prepare for a computer solution, it is now much more convenient to introduce a subscript notation for the $C$ values. Thus, we number the so-called node points (locations where the approximate solution is computed) beginning with $i = 1$ at $x = 0$ and numbering consecutively until $i = (m + 1)$ at the right-hand boundary, where $m$ is the number of parts into which the interval is divided. The step size $h$ is then the length of the interval divided by $m$. For our example, $m = 4$, $h = 1/4$, and we have the correspondences between $x$ and $i$ shown in Figure 12.2.

The great advantage of the subscript notation is that all of the interior equations can be written as a single "generalized" equation, which can then be programmed in a loop on the computer. Thus, for our present example, using the subscript notation, Equations (12-6) can be expressed in the compact form

$$\begin{aligned} &C_1 = C_2 \\ &C_{i+1} - 2C_i + C_{i-1} = h^2 2C_i \qquad (i = 2, 3, 4) \\ &C_5 = 1 \end{aligned} \tag{12-7}$$

Equations (12-7) are in a form very convenient for computer use, not only for the particular case considered here of $m = 4$, $h = 1/4$, but also for *any* value of $m$, so long as $h$ is calculated as (length of interval)/$m$ and so long as the numbering of the subscripts is correct. For our problem (and in general), the subscripts for *interior* nodes should go from 2 to $m$.

A remark must be made here concerning the implementation of *boundary conditions*. In our discussion above, we employed the simple two-point forward approximation to $C'(0)$. This has error of $O(h)$ compared to the errors $O(h^2)$ which occurred in the differential equation approximation, Equation (12-4). Roughly speaking, it is generally true that the accuracy of a finite difference calculation is governed by the "weak link" of the approximation chain; that is, the results are limited by the

**TABLE 12.3** BASIC FINITE DIFFERENCE SOLUTION OF $C'' = 2C$, WITH $C'(0) = 0$, $C(1) = 1$

| Two-Point Forward Approximation for $C'(0)$ | | | | |
|---|---|---|---|---|
| | $h = 1/2$ | $h = 1/4$ | $h = 1/8$ | Exact |
| C(0) | 0.667 | 0.546 | 0.499 | 0.459 |
| C(0.5) | 0.667 | 0.615 | 0.595 | 0.579 |
| **Three-Point Forward Approximation for $C'(0)$** | | | | |
| | $h = 1/2$ | $h = 1/4$ | $h = 1/8$ | Exact |
| C(0) | 0.429 | 0.457 | 0.459 | 0.459 |
| C(0.5) | 0.571 | 0.579 | 0.579 | 0.579 |

lowest-order accuracy approximation occurring in the calculation. For this reason, it is usually advisable to use approximations for both the differential equation and boundary conditions which are of the same order of accuracy in $h$. For the example we are considering, it would be better to use a second-order approximation for the derivative boundary condition. The simplest way of doing this is to use the three-point forward approximation to $C'(0)$, which is of the form (see Table 12.2) $C'(0) = 0 = (-C_3 + 4C_2 - 3C_1)/(2h)$. This yields $C_1 = (4C_2 - C_3)/3$, which, for the improved calculation, would replace the less accurate previous equation, $C_1 = C_2$.

The actual solution of the algebraic Equations (12-7) can be carried out directly, iteratively, or by the tridiagonal algorithm. *The tridiagonal algorithm is by far the most efficient method*, but we save that for discussion when we consider the refined finite difference method. For the basic finite difference method we are now discussing, it is useful to have a general computer program for either linear or nonlinear differential equations. We therefore use the *iterative* procedure to solve the algebraic equations, since this can be used in either case. This is implemented in program FINDIF, which is adapted from our earlier program DELSQ for the solution of nonlinear (or linear) algebraic equations. Program FINDIF will be discussed in detail at the end of this chapter. In finite difference problems where we wish to arrange the algebraic equations in the form $x_i = g_i(x_1, x_2, \ldots, x_n)$, the best systematic procedure is to *solve the equation at node* i *for the* i*th variable*. For the example being considered here in Equations (12-6), we would solve the $i$th equation for $C_i$ to get

$$C_1 = \frac{4C_2 - C_3}{3}$$
$$C_i = \frac{C_{i+1} + C_{i-1}}{2 + 2h^2} \qquad (i = 2, 3, \ldots, m) \tag{12-8}$$
$$C_{m+1} = 1$$

Table 12.3 shows the results of using program FINDIF on the problem considered above, using both the crude and improved boundary condition approximations, along with several values of $h$ in each case.

By observing the results of Table 12.3, as shown plotted in Figure 12.3, we see several trends. As we expect, the results generally become more accurate as $h$ decreases. Although this is a very simple problem, the results seem rather accurate

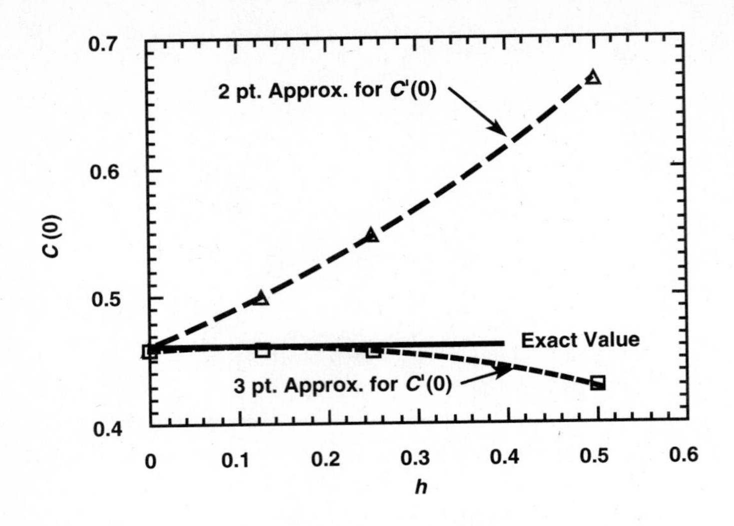


**Figure 12.3.** Comparison of Finite Difference Calculations for $C(0)$

considering the large values of $h$. This is somewhat deceptive; in many problems, the results for larger $h$ seem rather good, but then refinements as $h$ decreases often come with stubborn slowness. Often, Richardson extrapolation is helpful in this context. One conclusion from the table surely stands out: the three-point boundary condition results are *much* better than those for the two-point approximation. This is illustrated in Figure 12.3, where the calculated results for $C(0)$ are shown as a function of $h$ for both the two-point and three-point forward approximations for $C'(0)$.

Let us now consider use of the basic finite difference method for the *nonlinear* problem $C'' = BC^2$ with $C(0) = 1$ and $C'(1) = 0$. Here we discretize essentially as before, using a central difference for $C''$ and a three-point backward approximation for $C'(1)$. This gives, for the interior points, the set of equations

$$C_{i+1} - 2C_i + C_{i-1} = h^2BC_i^2 \quad (i = 2, 3, \ldots, m) \tag{12-9}$$

The main question concerns how to solve each equation for each variable in the iterative solution. As we have said before, it is desirable to solve the $i$th equation for variable $C_i$. But in the case of the present simple nonlinear problem, there are several ways in which this can be done. For example, if $B$ is small, then it may be useful to solve Equation (12-9) for the $C_i$ on the left-hand side. Alternatively, we could solve the equation as a quadratic in $C_i$, but this might be hazardous since, in the course of an iteration, we might generate negative square roots. What seems to be a better procedure is to account, at least partially, for each of the $C_i$ terms in Equation (12-9) without introducing any fractional powers such as a square root. This can be done by rearranging Equation (12-9) as $C_{i+1} - 2C_i + C_{i-1} = (h^2BC_i)C_i$ and solving the *linear* equation obtained by regarding the term in parentheses as a "constant." This would give the iterative scheme

$$C_i = \frac{C_{i+1} + C_{i-1}}{2 + h^2BC_i} \qquad (i = 2, 3, \ldots, m) \tag{12-10}$$

When we add to Equations (12-10) the boundary equations $C_1 = 1$ and the three-point backward approximation $C'(1) = 0 = (3C_m - 4C_{m-1} + C_{m-2})/2h$, we have an appropriate system of equations that can be put into program FINDIF. In this program, the user has the option to use either the ordinary iteration or delta-square methods. It is convenient to use the simple starting values $C_i = 1$ (all $i$). The basic finite difference method described here works reasonably well if $h$ is not too small (otherwise it takes forever!), but it is very computationally inefficient compared to some refinements we shall now discuss.

## 12.4 THE REFINED FINITE DIFFERENCE METHOD: THE TRIDIAGONAL ALGORITHM

The basic finite difference method we have just described fails to take advantage of the special structure of Equation (12-7), which simplifies the solution of these equations. Considering this linear example further, we note that in any one interior equation, only three unknowns appear; for equation $i$, we have only the unknowns $C_{i+1}$, $C_i$, and $C_{i-1}$. Since the boundary condition equations of Equation (12-7) are of suitable form, this system of equations has a *tridiagonal* structure, of the type discussed in the chapter on linear equations. Thus, it is possible to use the *very efficient tridiagonal algorithm* to solve this system of equations. This is particularly important when $h$ is taken small (for higher accuracy), since then there are more equations in the system. If we recall briefly that the standard tridiagonal system is of the form

$$\begin{aligned}
b_1x_1 + c_1x_2 \qquad\qquad\qquad\qquad &= d_1 \\
a_2x_1 + b_2x_2 + c_2x_3 \qquad\qquad &= d_2 \\
a_3x_2 + b_3x_3 + c_3x_4 \qquad &= d_3 \\
\cdots \qquad\qquad\qquad & \\
a_{n-1}x_{n-2} + b_{n-1}x_{n-1} + c_{n-1}x_n &= d_{n-1} \\
a_nx_{n-1} + b_nx_n &= d_n
\end{aligned} \tag{12-11}$$

then, by comparison, we can see that Equations (12-7) are of this form, provided we choose

$$\begin{aligned}
&b_1 = 1, && c_1 = -1, d_1 = 0 \\
&a_1 = 1, b_i = -2 - 2h^2, && c_i = 1, d_i = 0 \\
&a_n = 0, b_n = 1, && d_n = 1
\end{aligned} \tag{12-12}$$

The quantities $a_1$ and $c_n$ do not enter into the calculation, since they must always be zero in the tridiagonal algorithm. The computer program FDTRID (finite difference tridiagonal) uses this algorithm, and also allows for the possibility of application to nonlinear problems through user options for ordinary iteration or delta-square acceleration. This allows a *much* faster calculation, especially when many nodes are being used (small $h$). Program FDTRID will be discussed in detail at the end of the chapter. The tridiagonal structure of finite difference solutions of boundary value problems is very commonly encountered, but sometimes the implementation of the boundary conditions generates first and/or last equations in the system that do not fit the mold of Equations (12-11). This happens, for example, if three-point difference approximations are used for derivatives at the boundaries. Fortunately, the remedy to this problem is simple. *Equations adjacent to the improper boundary equations are used to algebraically eliminate any extra variables in the improper equations before the equations are put into program FDTRID.*

### 12.4.1 Boundary Condition Procedures

Another point relating to refinement of the finite difference method concerns boundary conditions for either linear or nonlinear problems. We have already mentioned, and demonstrated in an example, that it is important to have the discretization errors in the boundary conditions be essentially as small as those of the differential equation. We discussed the three-point forward and backward approximations in this context. An extension of these one-sided approximations might be appropriate if the discretization error in the differential equation was reduced. However, there is another approach that is often used for boundary conditions, which we shall refer to as the *fictitious boundary* method (it also goes under various other names). This method is convenient because *it allows use of central approximations of boundary derivatives.* It might be thought at first that one-sided derivative approximations must necessarily be used at boundaries, since a central approximation would be partly outside the domain of the problem. To get around this difficulty, one can *imagine that the domain of the problem has been extended*, say by one spatial increment $h$. The new imaginary mathematical problem, which is just an extension of the original physical problem, now has a new boundary with an unknown boundary condition. The original boundary condition, which must still be satisfied, now applies at an interior point of the extended problem. Thus, in the new extended problem, we must also have the discrete version of the differential equation satisfied at the original physical boundary point. So this procedure involves having an additional node point (outside the physical boundary) appear in the calculation, along with an additional equation, namely the differential equation evaluated at the original boundary point. Using a central difference for the derivative at the boundary point naturally forms a determinate set of equations. As an example, for the equations $C'' = 2C$, with $C'(0) = 0$ and $C(1) = 1$ considered earlier, we could introduce a new node $i = 0$ at $x = -h$. Then we could replace $C'(0)$ by a central approximation of the form $C'(0) = [C(h) - C(-h)]/2h = 0$, or $C_0 = C_2$. We would also write from the differential equation at $x = 0$ that $[C(h) - 2C(0) + C(-h)]/h^2 = 2C(0)$, or $[C_2 - 2C_1 + C_0]/h^2 = 2C_1$. In the algebraic solution, it is usually convenient to algebraically eliminate the fictitious variable (in this example $C_0$) before starting the computation.

### 12.4.2 The Newton Method for Nonlinear Problems

Another important refinement of the finite difference method, which only applies to *nonlinear* problems, is known as the Newton or quasilinearization method. The result is the same as differencing in the usual way and then using Newton's method to solve the nonlinear algebraic equations, but the formulation is often easier if the following approach is used. The principal use of this method is for difficult problems, where the foregoing combination of the tridiagonal algorithm plus the modified delta-square method does not work well. We can think of the Newton technique as a combination of successive approximation and Taylor's theorem. For concreteness, consider the second-order nonlinear differential equation $y'' = f(x, y, y')$ with linear boundary conditions. The exact equation can be written as

$$\frac{d^2y^{[n+1]}}{dx^2} = f\left(x, y^{[n+1]}, \frac{dy^{[n+1]}}{dx}\right) \tag{12-13}$$

Now we approximate the nonlinearities by writing Taylor's theorem as

$$f\left(x, y^{[n+1]}, \frac{dy^{[n+1]}}{dx}\right) \cong f\left(x, y^{[n]}, \frac{dy^{[n]}}{dx}\right) + \frac{\partial f}{\partial y_n}(y^{[n+1]} - y^{[n]}) + \frac{\partial f}{\partial y_n'}(y'^{[n+1]} - y'^{[n]}) \tag{12-14}$$

Here, we have used Taylor's theorem in the two variables $y$ and $y'$, for any fixed value of $x$. The superscripts and subscript refer to the stage of the approximation. Equations (12-14) and (12-13) are then combined, and the various derivatives are replaced by difference approximations. The variables with $[n+1]$ superscripts are then the current unknowns, while those with superscript $[n]$ correspond to the previous stage and are therefore known. Given reasonable starting values, the iterative solution of these equations converges *quadratically*. Since the linearization causes the discretized system to become linear and tridiagonal, the whole process is computationally very efficient. This kind of calculation is conveniently carried out in program FDTRID. We now consider a nonlinear example to illustrate various refinements.

**Example 12.2 Solution of $C'' = BC^2$ with $C(0) = 1$, $C'(1) = 0$**

The usual discretization of the differential equation is given in Equation (12-9). To put this in the standard tridiagonal form for a *nonlinear* system, we write $C_i^2$ as $C_1^{[n+1]}C_i^{[n]}$ to get

$$C_{i-1}^{[n+1]} - C_i^{[n+1]}(2 + h^2BC_i^{[n]}) + C_{i+1}^{[n+1]} = 0 \qquad (1 = 2, 3, \ldots, m) \tag{a}$$

By regarding $C_i^{[n]}$ in Equation (a) as known, we have a linear tridiagonal form for the unknowns $C_j^{[n+1]}$, $(j = 2, \ldots, m)$. Here the *superscripts* $(n, n+1)$ refer to the iteration number for the nonlinear algebraic equation system, Equations (a). We also use $n$ to represent the total number of algebraic equations, which is always equal to $(m+1)$, where $m$ is the number of parts into which the solution interval is divided. From the boundary condition $C(0) = 1$, we have $C_1 = 1$ for the first equation. For the derivative boundary condition $C'(1) = 0$, we could use either the three-point approximation or the fictitious boundary approach. For the three-point backward approximation we would have $C'(1) = (3C_{m+1} - 4C_m + C_{m-1})/2h = 0$, and this would be the last equation $(m+1)$. However, this last equation contains not only the usual final tridiagonal unknowns $C_m$ and $C_{m+1}$, but also $C_{m-1}$. Therefore, to use the tridiagonal algorithm, we must use $i = m$ in Equation (a) to algebraically eliminate $C_{m-1}$ from equation $(m+1)$. The result of this elimination is shown by the "three-point approx." in Equations (b), which follow.

$$\begin{aligned}
&C_1 = 1 \\
&C_{i-1}^{[n+1]} - C_i^{[n+1]}(2 + h^2 B C_i^{[n]}) + C_{i+1}^{[n+1]} = 0 \qquad (i = 2, 3, \ldots, m) \\
&C_m^{[n+1]}(-2 + h^2 B C_m^{[n]}) + 2C_{m+1}^{[n+1]} = 0 \qquad \text{Three-pt. approximation} \\
&2C_m^{[n+1]} - C_{m+1}^{[n+1]}(2 + h^2 B C_{m+1}^{[n]}) = 0 \qquad \text{Fictitious-pt. approximation}
\end{aligned} \tag{b}$$

Equations (b) show the entire set of required finite difference equations for the problem at hand. For the difference equation corresponding to the derivative boundary condition $C'(1) = 0$ at $x = 1$, we show the equations for both the three-point approximation discussed above and the fictitious boundary approximation to be discussed in the next paragraph.

The alternative approach for the derivative boundary condition at $x = 1$ proceeds as follows. A new fictitious point is introduced, one spatial increment $h$ to the right of $x = 1$, at $x = (1 + h)$. The new node point there is called $C_{m+2}$. The boundary condition is approximated by $C'(1) = 0 = [C(1+h) - C(1-h)]/2h = (C_{m+2} - C_m)/2h$, or $C_{m+2} = C_m$. Then we apply Equation (a) for $i = m + 1$ and replace $C_{m+2}$ in this equation with $C_m$. This then gives the last equation shown in Equations (b) for node $(m + 1)$. This equation automatically contains the proper variables needed for it to be the last equation in the tridiagonal scheme, so no algebraic elimination is required. We now provide starting values (such as $C_i = 1$, for all $i$) and then iterate the system of equations (a), (b), and $C_1^{[n+1]} = 1$ to convergence. This can be accomplished in program FDTRID.

We now show the tridiagonal coefficients $a(i) - d(i)$ that would be required by program FDTRID for solution of this example problem.

$$\begin{aligned}
&b(1) = 1, \quad c(1) = 0, \quad d(1) = 1 \\
&a(i) = 1, \quad b(i) = -2 - h^2 * B1 * y(i), c(i) = 1, \; d(i) = 0 \quad (i = 2, \ldots, n-1) \\
&a(n) = -2 + h^2 * B1 * y(n), \quad b(n) = 2, \qquad d(n) = 0 \quad \text{Three-pt. approximation} \\
&a(n) = 2, \quad b(n) = -2 - h^2 * B1 * y(n), \qquad d(n) = 0 \quad \text{Fictitious-pt. approximation}
\end{aligned} \tag{c}$$

Note that for the tridiagonal system we always have $a(1) = c(n) = 0$, so these values need not be specified. In Equations (c), $n$ represents the last equation of the system, and $y(i)$, $y(n)$ are the $C_i^{[n]}$, $C_n^{[n]}$ values from Equations (b) (remember that Equation ($n$) is the same as Equation ($m + 1$), etc.). Also, we use the symbol $B1$ in place of $B$ in the computer tridiagonal equations, since the symbol $b(i)$ is already used as a coefficient.

As an improvement over the "standard" nonlinear tridiagonal procedure above, the Newton (or quasilinearization) difference form of the equation $C'' = BC^2$ is obtained as follows. We write

$$\begin{aligned}
\frac{d^2 C^{[n+1]}}{dx^2} &= f(C^{[n+1]}) \cong f(C^{[n]}) + \frac{df}{dC_n}(C^{[n+1]} - C^{[n]}) \\
&= B[(C^{[n]})^2 + 2C^{[n]}(C^{[n+1]} - C^{[n]})]
\end{aligned} \tag{d}$$

Now we use the customary differencing of Equation (d) and group the $[n + 1]$ and $[n]$ terms appropriately to obtain the following linear tridiagonal structure:

(Newton)

$$\begin{aligned}
&C_1^{[n+1]} = 1 \\
&C_{i-1}^{[n+1]} - C_i^{[n+1]}(2 + 2h^2 B C_i^{[n]}) + C_{i+1}^{[n+1]} = -h^2 B (C_i^{[n]})^2 \qquad (1 = 2, 3, \ldots, m) \\
&2C_m^{[n+1]} - C_{m+1}^{[n+1]}(2 + 2h^2 B C_{m+1}^{[n]}) = -h^2 B (C_{m+1}^{[n]})^2
\end{aligned} \tag{e}$$

**TABLE 12.4** RESULTS OF DIFFERENCE CALCULATIONS FOR $C'' = BC^2$; $C(0) = 1$, $C'(1) = 0$
FICTITIOUS BOUNDARY METHOD

| | $m = 2$ | $m = 4$ | $m = 8$ | Extrapolated $m = 4, 8$ | Exact |
|---|---|---|---|---|---|
| | | | $B = 1$ | | |
| C(0.5) | 0.782 | 0.779 | 0.778 | 0.778 | 0.778 |
| C(1) | 0.718 | 0.714 | 0.713 | 0.712 | 0.712 |
| | | | $B = 10$ | | |
| C(0.5) | 0.427 | 0.403 | 0.395 | 0.392 | 0.392 |
| C(1) | 0.308 | 0.289 | 0.283 | 0.281 | 0.281 |
| | | | $B = 100$ | | |
| C(0.5) | 0.172 | 0.132 | 0.117 | 0.112 | 0.110 |
| C(1) | 0.0839 | 0.0658 | 0.0597 | 0.0576 | 0.0571 |

The equation for node $(m + 1)$ in Equations (e) corresponds to the use of a fictitious boundary at $x = 1$; this equation was obtained by writing the general equation at node $i$ using $i = (m + 1)$ and replacing $C_{m+2}$ with $C_m$ from the central difference boundary condition at $x = 1$. Equations (e) are in a proper form for iterative tridiagonal solution. We then provide starting values (such as $C_i = 1$, for all $i$) and iterate equations (e) to convergence. This is easily accomplished in program FDTRID.

We now examine some results of these calculations. Table 12.4 gives results based on using the fictitious boundary approach at $x = 1$. In Table 12.4, the column [Extrapolated $m$ = 4,8] refers to using a single Richardson extrapolation with the results for $m = 4$ and $m = 8$. Since second-order approximations are used both in the interior and at the boundaries, the value $p = 2$ is appropriate in the extrapolation relationship $S = A(h) + Ch^p$. The extrapolation values are based on $C$ values involving more significant figures than we show. We observe from the table that the accuracy increases as $h$ gets smaller ($h = 1/m$) and the extrapolation calculation does seem to help. The numerical approximations are *exactly the same* (assuming convergence is obtained) from either the basic finite difference method (program FINDIF), the standard tridiagonal method, or the Newton method (program FDTRID). This is true because the calculated results depend only on the discretizations used in the equation and the boundary conditions. What is different, however, between these schemes is the efficiency of the calculation. The tridiagonal approach is *much* more efficient than the basic iteration procedure (program FINDIF), especially when a large number of node points is used. Table 12.5 shows the difference in efficiency between the standard and Newton formulations of the problem in Example 12.2, both of which use the tridiagonal algorithm of program FDTRID. This table illustrates the advantage of delta-square acceleration in the standard formulation, as well as the usefulness of the Newton approach for nonlinear problems.

The shooting and finite difference methods produce results that are comparable in accuracy when they are employed to the same degree of approximation. But

**TABLE 12.5** NUMBER OF ITERATIONS OF STANDARD AND NEWTON METHODS FOR EXAMPLE 12.2

| $h$ | Standard Finite Difference Ordinary Iteration | Standard Finite Difference Delta-Square | Newton Finite Difference |
|---|---|---|---|
| 1/4 | 51 | 12 | 4 |
| 1/8 | 47 | 14 | 4 |

Results are based on program FDTRID.

in cases where the shooting method is applicable, it is usually the most efficient means for solving boundary value problems. This is because of the built-in global extrapolation and error-monitoring capabilities of the initial value technique (such as in program IEXVSSH), as well as for the ease of using variable step-size if desired. In situations where stability difficulties cause the shooting method to have troubles, the finite-difference method may perform much better. It is sometimes useful to employ these two methods in a complementary way. This can be done by using a crude finite difference calculation to estimate a starting value for the shooting method. Another important reason for the discussion of finite difference methods for ODEs is that it is good preparation for their use in partial differential equations.

It is sometimes useful to employ various higher-order difference approximations in the equation or boundary conditions of a particular boundary value problem. For more information on this and other finite difference topics, see Ferziger [1981], Davis [1984], and Finlayson [1980].

## 12.5 THE METHOD OF WEIGHTED RESIDUALS

The method of weighted residuals is a useful numerical method for the solution of boundary value problems in ODEs. We have already discussed this method in the chapter on analytical solution of ODEs. We discussed it there because this method does allow analytical representation of the solution for a given set of parameters, after the numerical calculations are completed. The method is also useful in the solution of partial differential equations, so we shall discuss it further when we reach that topic.

## 12.6 SPECIAL APPLICATIONS

There are certain types of boundary value problems in ODEs that are somewhat different from those we have previously considered and deserve some discussion.

### 12.6.1 Infinite Boundary Conditions

One type of problem concerns a situation involving an *infinite* boundary condition. We discuss this by means of an example.

**Example 12.3 Integration of the Blasius and Pohlhausen Equations**

These equations occur in the description of laminar boundary layer heat (or mass) transfer from a flat plate. The differential equations and boundary conditions are as follows:

$$f''' + ff'' = 0; \quad f(0) = f'(0) = 0, f'(\infty) = 1$$
$$\theta'' + \sigma f \theta' = 0; \quad \theta(0) = 1, \theta(\infty) = 0 \tag{a}$$

The third-order equation for $f$ is the Blasius equation for the velocity we considered earlier, and the second-order equation for $\theta$ is the Pohlhausen equation for temperature (or concentration). The $\theta$ equation contains the parameter $\sigma$, which represents the Prandtl (heat transfer) or Schmidt (mass transfer) number. The entire system here is fifth-order with two boundary conditions at infinity. However, we see that the equation for $f$ is independent of the one for $\theta$, so it can be solved first. Then $f$ is needed as a known input in the $\theta$ equation. The difficulty with the $f$ equation is the boundary condition at $\infty$. To resolve this, a convenient approach is to use the shooting algorithm (program IEXVSSH) to find the unknown value of $f''(0)$. Thus, we shoot from $x = 0$ toward $x = \infty$ to satisfy the boundary condition residual $R = 0 = f'(\infty) - 1$. To do this, we *experiment* using some finite value to replace $\infty$, doing the integration with a fairly large step size. By trying several values, we find that the results of the integration do not change if we replace $\infty$ by five. Once we essentially know where "infinity" is, we can refine the desired value of $f''(0)$. Following the successful solution of the $f$ equation, we then wish to solve the $\theta$ equation, which contains the function $f$. Since the $f$ function is only known numerically, the easiest way to do this is to solve the $f$ and $\theta$ equations simultaneously, using the previously calculated value of $f''(0)$ as an initial condition for the $f$ equation (and therefore dropping the infinite boundary condition on $f'$). In this simultaneous solution, the value of $\theta'(0)$ is now the unknown to be guessed in program IEXVSSH, and the new boundary condition residual is $R = 0 = \theta(\infty)$. For different values of $\sigma$, the use of the original value of five to replace infinity should be checked. For $\sigma \geq 1$, this works well.

### 12.6.2 Parameter Estimation and Eigenvalue Problems

These problems involve attempting to solve an ODE boundary value problem when a parameter in the equation is unknown. Any of the boundary value techniques we have discussed could be used, but in our opinion the shooting method is by far the most convenient, so we will use it here.

The idea of the simplest (nonstatistical) parameter estimation problem is as follows. Suppose we have the differential equation $y'' = f(x, y, y', p)$ ($p$ is a parameter) and two associated boundary conditions. If the parameter $p$ is known, we have the usual boundary value problem. But sometimes the parameter $p$ is unknown, and there is an additional boundary condition to make the problem determinate. In this case, it is convenient to introduce the unknown parameter (which is an unknown constant) as a new dependent variable that satisfies the ODE $dp/dx = 0$. We then solve this augmented system by the shooting method, using the initial value of $p$ (the unknown constant) as the guessed variable. Since there will now be three boundary conditions, we choose the one that is "by itself" for the boundary condition residual. The final converged value of $p$ is the desired value of the parameter.

Another closely related and important problem is the so-called ODE eigenvalue problem. This problem occurs, in general, as part of the well-known separation of variables method for the solution of certain partial differential equations. The ODE eigenvalue problem is analogous to the algebraic eigenvalue problem discussed in the chapter on linear equations and matrix analysis. It corresponds to the case where an ODE boundary value problem has the trivial solution $y = 0$ (because of homogeneous boundary conditions), but also has nontrivial solutions for particular values of a parameter (called eigenvalues) in the equation. To solve this problem, we take the eigenvalue parameter as a new dependent variable (say, $y_3$) that satisfies the differential equation $dy_3/dx = 0$. Then $y_3$ is the guessed variable in the shooting method. The additional boundary condition, since the problem is homogeneous, is an arbitrary specification of either of the original variables $y(0)$ or $y'(0)$ (whichever one is not specified in the original problem).

The ODE eigenvalue problem generally has many solutions, one corresponding to each (unknown) eigenvalue. The particular one obtained by a shooting trial is governed by the initial guess for the eigenvalue. The solution $y(x)$ corresponding to a given eigenvalue is called an *eigenfunction*. The eigenfunctions generally oscillate more in the region of interest as the eigenvalues increase. That is, the first eigenfunction has no zero, the second has one zero, the third has two zeroes, and so on. This is helpful in verifying that no eigenvalues have been "missed" in a numerical calculation.

**Example 12.4 The Eigenvalue Problem $y'' + \lambda(1 - x^2)y = 0$, $y'(0) = 0$, $y(1) = 0$**

This is the so-called "Graetz" problem for axial laminar transport between parallel planes, with $\lambda$ the eigenvalue Ybarra and Eckert [1985]. We consider finding just the first (or lowest) eigenvalue. To solve this problem using the shooting method (progam IEXVSSH), we define $u_1 = y$, $u_2 = y'$, and $u_3 = \lambda$, with $du_3/dx = d\lambda/dx = 0$. For initial conditions, we use $y(0) = u_1(0) = 1$ (arbitrary), $y'(0) = u_2(0) = 0$, and $u_3(0) =$ guess. At $x = 1$, the boundary condition residual is $R = 0 = y(1) = u_1(1)$. Starting with the first value of guess $= 0$, with one trajectory and $h = 0.1$, we get a new guess of 2.77. Using this as the new guess value, we use three trajectories and $h = 0.02$ to get the six-figure value $\lambda = 2.82776$. As the eigenvalues increase (usually we only need a fairly small number of them), the integration step size must be reduced to maintain accuracy. For example, computing the tenth eigenvalue to six-figure accuracy requires three trajectories and $h = 0.001$.

### 12.6.3 Free Boundary Problems

Some boundary value problems are defined by a boundary condition that is not specified at an explicit location; part of the problem is to find the location of this "boundary." We illustrate how this idea can be usefully implemented in a particular *initial value* problem.

**Example 12.5 Finding the Precise Maximum of Conversion for a Reaction System**

We consider the chemical reaction problem in a constant volume batch reactor for the reaction $A + B \leftrightarrow C \rightarrow$ side products. The objective is to find the maximum yield of $C$. If we write $y_1$ for $C_A$ (concentration of $A$) and $y_2$ for $C_C$, for particular rate constant values the variation of dimensionless concentrations with time can be described by the following differential equations:

$$\frac{dy_1}{dt} = -10y_1^2 + y_2 \qquad y_1(0) = 1$$
$$\frac{dy_2}{dt} = -2y_2 + 10y_1^2 \qquad y_2(0) = 0 \tag{a}$$

It is desired to find *precisely* the maximum value of the desired product $y_2$, as well as the time at which it occurs. Of course, this is just an initial value problem which could be approached by the usual initial value methods (such as program IEX). However, although this would be a straightforward calculation, it is very awkward and tedious since it would require breaking up the integration interval into finer and finer discrete segments, together with using interpolation. This is not because the integration is difficult, but because the information required is not readily determined precisely by an initial value method. Rather, this could really be viewed as a *boundary value* problem where we wish to find the final unknown $x_1$ (time) that corresponds to achieving the maximum of $y_2$, which is characterized by the condition $dy_2/dt = 0$. This calculation is handled very nicely by the shooting method and program IEXVSSH by using the error control option. To proceed, we program the two differential equations and initial conditions in the usual way. The final value of the time, $x1$, is taken as the guessed variable; then the print increment $hp$ in the error control section is defined to be $hp = (x1 - x0)/\text{integer}$. Here integer is any positive integer, which depends on how much output is desired. The boundary condition residual is $dy_2/dt = 0 = -2y_2 + 10y_1^2$ (from the differential equation). This residual condition is programmed by the statement `LET res = -2*y1(2,j2) + 10*y1(1,j2)^2`. A rough preliminary calculation, using initial value program IEX, shows that the maximum for $y_2$ is near $x = 0.25$. We therefore take $x1 = 0.25$ as the initial value for "guess." We then integrate *using error control*. The results using tol = 0.01 (which are the same as those for tol = 0.001) indicate that the maximum value of $y_2$ is equal to 0.552557, and this occurs at time = $x1$ = 0.264217. For tol = 0.01, the global relative errors are estimated to be about $10^{-6}$ for each variable, and the error ratios are about 1.02. Thus, we see that the shooting method, and in particular program IEXVSSH, are useful tools for some types of free boundary problems.

## 12.7 COMPUTER PROGRAMS

In this section we discuss the computer programs applicable to various problems in the chapter. For each program, we give a brief discussion and a concise listing, which includes (i) a program description, (ii) how to run, and (iii) user-specified subroutines, including a sample problem. Detailed program listings can be obtained from the diskette.

### Program IEXVSSH for Shooting in Boundary Value Problems

Instructions for running the program are given in the concise listing. A user should be familiar with program IEXVS for initial value problems before using program IEXVSSH. Program IEXVSSH is very similar to IEXVS, and it includes all the facilities of that program. The difference is that IEXVSSH has been set up for "shooting" through guessing a variable and then iterating, using the secant method, in an attempt to satisfy a specified boundary condition residual. The program prompts for the initial value of the "guess" variable, which can be $y(i)$, or a parameter, or $x1$, and so on. This is followed by prompts for a derivative check, the number of trajectories to be used, the initial step size, and whether or not to use error control (variable-step). If error control is not employed, the user is asked to specify the print cycle parameter, and the integration is carried out with fixed-step using the initial step size. The program iterates until the user-specified error tolerance tolreg (tolerance for relative error in guess) for the secant method is met.

The boundary condition residual is specified in the following way. We write the residual in terms of $y(i)$ values, with $y(i)$ replaced by $y1(i, j2)$. For example, if the boundary condition is $y(1) = 1$, with residual $= y(1) - 1$, we program the statement `LET res = y1(1,j2) - 1.` The reason for this is that, by using this statement, the program is set up to work properly regardless of whether we are using one or two or three trajectories. The output of the program is the same as that for IEXVS with the addition of printing of "guess," relative error in guess, and number of iterations, all after every secant cycle. The problem specified in the concise listing is that from Example 12.1.

### Program IEXVSSH.

```
!This is improved Euler/extrapolation/error control adapted for the
!shooting method in boundary value problems (IEXVSSH.tru) by O.T.Hanna
!Program handles one guessed value, one boundary condition residual.
!n=number of differential equations
!x,y(i) are independent and dependent variables in subroutines & output
!x0,x1 are initial and final values of x
!y1(i,j) is dummy for y(i) in program
!d(i),dold(i,j),dnew(i) are derivatives of y(i)
!h=local step size;hp is print increment;
!p is print parameter for fixed step
!j1 is number of trajectories;j1=1 for h,j1=2 for h,h/2;j1=3 for h,h/2,h/4
!q=3 for ode; q=4 for quadrature
!***********************************************************************
!*To run,specify q,n,x0,x1,initial y(i),"guess" variable,error parameters *
!*in Sub Init; specify residual equation in terms of y(i) in Sub Residual;*
!*specify derivatives d(i) in terms of x,y(i) in Sub Deriv;              *
!*program prompts for guess?,derivative check?,j1?,h?,ec?,(p)?           *
!***********************************************************************
!**IE/EXTRAPOLATION/ERR-CTRL SHOOTING SOLUTION OF BV PROBLEM BELOW**
!error parameter tol is desired local relative integration error
!error parameter tolreg is desired relative error in "guess" variable
!parameters eps,sf,are zero denominator and safety factor values

SUB Init (q,n,x0,x1,y(),d(),hp,hmaxstep,tol,eps,sf,guess,tolreg)
  !***denotes user specified quantity
  LET tolreg = 1.e-7      !***calc.stops when (rel.error guess) < tolreg
  LET q = 3               !***q=3 for ODE, q=4 for quadrature
```

```
  LET n = 2                   !***n=number of eqns
  LET x0 = 0                  !***initial value of x
  LET x1 = 1                  !***final value of x
  !****User must specify one variable as "guess"****--this can be y(i)
  !or a parameter;if a parameter, it should be given value in Sub Init
  LET y(1) = guess            !***specification of guess,y(i) initial
  LET y(2) = 0
  CALL Deriv (x0,y,d)

  !The following is used only with error control(ec=1)
  LET hp = .2                 !***hprint
  LET hmaxstep = hp           !***maximum user-specified step size
  LET tol = .01               !***desired local relative errors
  LET eps = .0001             !zero denominator parameters
  LET sf = 2                  !safety factors
END SUB

SUB Residual(j1,res,y1(,))
  LET j2 = 1 + 3*(j1 - 1) - (1/2)*(j1 - 1)*(j1 - 2)
  !Write b.c. residual in terms of y,with y(i) replaced by y1(i,j2)
  !In this problem, guess = y(1); residual= y(1) - 1 **
  LET res = y1(1,j2) - 1  !***user-specified residual**
END SUB

SUB Deriv(x,y(),d())          !***user specified derivatives
  LET B = 100
  LET d(1) = -y(2)
  LET d(2) = -B*y(1)^2
END SUB
```

### Program FINDIF for the Basic Finite Difference Method

Information and instructions for running the program are given in the concise listing. Program FINDIF is based on program DELSQ, modified specifically for the finite difference algorithm. The program first prompts for the method desired, ordinary iteration or delta-square. The user is then asked for the number of parts, $m$, into which the interval is to be divided. The nodes are numbered from $i = 1$ at $x = x0$ to $i = (m + 1)$ at $x = x1$. The program then gives an option for a function check, followed by a request for printing intermediate results. If many nodes or iterations are involved in the calculation, it is generally more efficient not to print intermediate values. In order to monitor the calculation when intermediate results are not printed, after each iteration the output gives the number of iterations and the current "norm" value. Observation of the norms allows the user to see if there is an apparent convergence. In difficult problems it may be useful to solve first with a very small number of nodes, and then use these results (plus interpolation) to get good starting values for the case of many nodes.

### Program FINDIF.

```
!This is program FINDIF.tru for ODE boundary value probs. by O.T.Hanna
!Program is adapted from delsq.tru for systems of nonlinear alg. eqns.
!x,y(i) are independent,dependent variables;
!difference ODE to get y(i) eqns.
!x0,x1 are initial,final values of x
!Interval (x1-x0) is divided into m parts;prgm prompts for m
!To speed up a "slow" program,avoid printing intermediate results
!**********************************************************************
!*To run define parameters,init. values,node eqns. in Sub Function    *
!*To check Function use mx = 1,method = 0                             *
!*Stop program via (i)Ctrl-Break,(ii)max. no. of                      *
!*iterations(mx),(iii)norm_tol                                        *
!**********************************************************************

!This is ODE BVP finite difference program findif.tru for prob. below
SUB function                        !internal subroutine
!Solution is based on delsq.tru program
  !User specified statements are marked with ***
  !***Problem here is y" = B*y^2; y(0)=1, y'(1)=0
  IF j = 0 then              !for j=0 set parameters,
                              initial y(i) values
    LET mx = 1000            !***maximum number of iterations
    LET norm_tol = 1e-7      !***norm tolerance stop control
    !m is no. of parts interval divided into; n is no. eqns.
    LET n = m + 1            !since i=1 at x=x0, we always have n=m+1
    LET x0 = 0               !***initial x value
    LET x1 = 1               !***final x value
    LET h = (x1-x0)/m
    LET B = 1                !***problem parameter
    FOR i = 1 to (m+1)       !***initial values of y(i)
        LET y(i) = 1
    NEXT i
  END IF

  !now specify node eqns.;be sure to include proper boundary eqns.
  !For convenience we give 3 pt approximations to y'f and y'b
  !y'f(i) = [4*y(i+1)-y(i+2)-3*y(i)] / (2*h)  !forward approx.
  !y'b(i) = [-4*y(i-1)+y(i-2)+3*y(i)] / (2*h) !backward approx.

  !***User specified difference approximations
  !***Problem here is y" = B*y^2; y(0)=1, y'(1)=0
  LET y(1) = 1
  FOR i = 2 to m
    LET y(i) = (y(i+1) + y(i-1))/(2 + h^2*B*y(i))
  NEXT i
  LET y(m+1) = (4*y(m) - y(m-1))/3     !3 pt backward diff. approx. *****
END SUB
END
```

## Program FDTRID for Tridiagonal Solution of Linear or Nonlinear Finite Difference Problems

This program is adapted from programs DELSQ, FINDIF, and TRID. Information and instructions on running are contained in the concise listing below. Specification of the tridiagonal matrix elements in Sub Function is generally straightforward, but care must be taken with the boundary conditions. If we use a fictitious boundary, the end condition will contain the proper variables. But if we use a three-point approximation, we will need to eliminate one of the variables, by using the adjacent equation, to get the proper tridiagonal form.

The program first prompts for the case, which is either (1) tridiagonal-only for either a linear problem or a function check, (2) ordinary iteration for a nonlinear problem, or (3) delta-square for a nonlinear problem. Then the user is asked for the number of parts into which the interval is to be divided, and whether intermediate results are to be printed. If printing is suppressed, the output continuously gives the number of iterations and the norm value. In this way, progress of the calculation can be monitored.

### Program FDTRID.

```
!This is program FDTRID.tru for ODE boundary value probs.
!by O.T.Hanna. Program is adapted from DELSQ, FINDIF
!and TRID and uses tridiagonal algorithm with options for
!ordinary iteration or delta-square. x,y(i) are independent,
!dependent variables; difference ODE to get y(i) eqns.
!x0,x1 are initial,final values of x.
!Interval (x1-x0) is divided into m parts; program prompts for m.
!Finite difference spatial increment is h = (x1-x0)/m.
!Program prompts for print parameter pr; m/pr must be an integer.
!Program prints (i) every h if pr = 1,(ii) every 2h if pr = 2, etc.
!To speed up a "slow" program,avoid printing intermediate results.

!***************************************************************
!*To run define parameters and initial estimates in Sub Init at *
!*end of program. Then define problem tridiagonal matrix        *
!*coefficients a(i),b(i),c(i),d(i) in Sub Function; we          *
!*always must have a(1)=c(n)=0. Program prompts for case        *
!*(1=Tridiag. only, 2=Ord.Iteration, 3=DeltaSquare), m and pr.  *
!*To check Function use case=1. Stop program via                *
!*(i)Ctrl-Break,(ii)max.no.of iterations(mx),(iii)norm_tol.     *
*                                                               *
!***************************************************************

SUB Init(n,mx,norm_tol,x0,x1,y() )
  !***denotes user-specified statement
  LET mx = 500                    !***maximum number of iterations
  LET norm_tol = 1e-7             !***norm tolerance stop control
  LET x0 = 0                      !***initial x value
  LET x1 = 1                      !***final x value
```

```
  FOR i = 1 to n                              !***init.values of y(i) in nonlin.probs.
    LET y(i) = 1
  NEXT i
END SUB

SUB Function(n,h,y(),a(),b(),c(),d() )
  !***denotes user-specified statement
  !Specify coefficients a(i),b(i),c(i),d(i) in a(i)*y(i-1) +
  !b(i)*y(i) + c(i)*y(i+1) = d(i); be sure to include proper
  !boundary eqns. a(1) and c(n) need not be specified since
  !they(=0) are not used.

  !***Problem here is y" = B1*y^2; y(0)=1, y'(1)=0

  LET B1 = 1                                  !***problem parameter

  LET b(1) = 1                                !***
  LET c(1) = 0                                !***
  LET d(1) = 1                                !***

  FOR i = 2 to (n-1)
    LET a(i) = 1                              !***
    LET b(i) = -2 - h*h*B1*y(i)               !***
    LET c(i) = 1                              !***
    LET d(i) = 0
  NEXT i

  LET a(n) = 2                                !***
  LET b(n) = -2 - h*h*B1*y(n)                 !***
  LET d(n) = 0                                !***

END SUB
```

## PROBLEMS

**12.1.** Consider the dimensionless nonlinear catalyst pellet problem $d^2C/dx^2 = KC^2$ with boundary conditions $C(0) = 1$ and $dC/dx(1) = 0$. In this problem, $K = 20$. We wish to determine accurate (to four figures) values of $dC/dx(0)$ and $C(1)$. Do this using the shooting method and program IEXVSSH.

**12.2.** In problem 12.1, use the finite difference method and program FDTRID to obtain a solution. Use a three-point difference approximation for $dC/dx$ at $x = 0$. Divide the interval into two, four, eight, ... parts until convergence to four figures has been achieved for $C(1)$. What is the converged value of $C(1)$?

Use the various difference approximations to $C(1)$ to estimate the value of $p$ in the Richardson extrapolation formula $C = A(h) + Dh^p$. Taking $p = 2$, use Richardson extrapolation to estimate values of $C(1)$ together with error estimates of the approximations.

**12.3.** The Blasius equation for laminar flow over a flat plate is given by the third-order nonlinear boundary value problem $d^3f/dx^3 + fd^2f/dx^2 = 0$ subject to the conditions $f(0) = df/dx(0) = 0$ and $df/dx(\infty) = 1$. The quantity of principal interest is $d^2f/dx^2(0)$. By experimenting to find a practical value for $x$ = "$\infty$," determine this quantity to six figures. Use program IEXVSSH.

**12.4.** The Pohlhausen equation for laminar heat (or mass) transfer from a flat plate is $d^2\theta/dx_2 + (\sigma f)d\theta/dx = 0$ with the boundary conditions $\theta(0) = 1$ and $\theta(\infty) = 0$. Here, $f(x)$ is the Blasius function (see problem 12.3) and $\sigma$ the Prandtl (heat) or Schmidt (mass) number. It is desired to determine $d\theta/dx(0)$ for $\sigma = 0.7$, 50, 100, and $10^5$. As a check, note that for $\sigma = 1$, $d\theta/dx(0) = -d^2f/dx^2(0)$.

**12.5.** A dimensionless model for dispersion in a chemical reactor can be expressed as $\epsilon C'' - C' - KF(C) = 0$ with the boundary conditions $\epsilon C'(0) = C(0) - 1$ and $C'(1) = 0$. The problem to be solved is for the case where $F(C) = C/(1 + C)$, $\epsilon = 0.1$, and $K = 2$. Using the finite difference method with fictitious boundaries for the boundary conditions, calculate the reactor outlet concentration $C(1)$ for $h = 1/2, 1/10, 1/20, 1/40, 1/100$, and $1/1000$.

Use the $h = 1/10$, $1/20$, and $1/40$ results to test whether it is *reasonable* to take $p = 2$ in the usual Richardson extrapolation formula $C(1) = A(h) + Dh^p$, where $A(h)$ is the approximation for $h$, and so on. Using $p = 2$ in the above formula, estimate Richardson extrapolation values/errors for the cases $h = (1/20, 1/10)$ and $(1/40, 1/20)$.

**12.6.** Solve the differential equation problem from 12.5 for $\epsilon = 0.1$ and $K = 100$ using the shooting method. Determine the value of $C(1)$ to four figures. Note that in this problem, backward integration may be useful.

**12.7.** Determine the first three eigenvalues $\lambda$ of the problem $y'' + \lambda(1 - x^2)y = 0$ subject to the boundary conditions $y'(0) = 0 = y(1)$.

**12.8.** Obtain the first two eigenvalues $\lambda$ of the Bessel equation problem $y'' + (y'/x) + \lambda y = 0$ with $y'(0) = y(3.5) = 0$.

**12.9.** Find the first zero of the boundary value problem $y'' + xy = 0$ with $y(0) = 1$ and $y'(0) = 1$ using the shooting method as discussed in the text.

## REFERENCES

COLLATZ, L. (1966), *The Numerical Treatment of Differential Equations*, Springer-Verlag, New York

DAVIS, M. E. (1984), *Numerical Methods and Modeling for Chemical Engineers*, Wiley, New York

DAVIS, R. A. and SANDALL, O. C. (1991), *Chem. Engng. Educ.*, **25**, p. 10.

FERZIGER, J. H. (1981), *Numerical Methods for Engineering Application*, Wiley, New York

FINLAYSON, B. (1980), *Nonlinear Analysis in Chemical Engineering*, McGraw-Hill, New York

HANNA, O. T. (1988), *Computers Chem. Engng.*, **12**, p. 1083

HANNA, O. T. (1990), *Computers Chem. Engng.*, **14**, p. 273

KUBICEK, M. and HLAVACEK, V. (1983) *Numerical Solution of Nonlinear Boundary Value Problems with Applications*, Prentice-Hall, Englewood Cliffs, NJ

PRESS, W. H., FLANNERY, B. P., TEUKOLSKY, S. A., and VETTERLING, W. T. (1986), *Numerical Recipes*, Cambridge Univ. Press, New York

RIGGS, J. B. (1988), *An Introduction to Numerical Methods for Chemical Engineers*, Texas Tech Univ. Press, Lubbock, TX

SCHEID, F. (1988), *Numerical Analysis*, 2nd ed., McGraw-Hill, New York

SCRATON, R. E. (1984), *Basic Numerical Methods*, Edward Arnold, Baltimore

YBARRA, R. M. and ECKERT, R. E. (1985), *AIChE J*, **31**, p. 1755.

# 13

# Analytical Methods for Partial Differential Equations

We have seen that ordinary differential equations (ODEs) play a natural and nearly universal role in the description of a wide variety of chemical engineering problems. These ordinary differential equation models usually, of necessity, represent vast oversimplifications of what we believe to be very complicated phenomena. Fortunately, it turns out that in many instances such simplified mathematical models achieve valuable results. In many problems, however, it is either necessary or very desirable to use more realistic models which allow for more than one independent variable. For example, we must often describe the behavior of some system as a function of more than one space variable, as well as time. Whenever a dependent variable depends on more than one independent variable, the derivatives in a differential equation are partial derivatives, and we call the equation a partial differential equation (PDE).

Linear partial differential equations of second order are frequently classified as being elliptic, hyperbolic, or parabolic. If we consider the general form of a second-order linear PDE to be

$$\sum_{i=1}^{n} A_i \frac{\partial^2 y}{\partial x_i^2} + \sum_{i=1}^{n} B_i \frac{\partial y}{\partial x_i} + Cy + D = 0 \tag{13-1}$$

where the coefficients are taken to be continuous functions of the $x_i$'s and the $A_i$ coefficients are 1, $-1$, or 0. The classification of the PDE depends upon the magnitude of the coefficients. In particular:

1 The PDE is elliptic if all the $A_i$ are nonzero and have the same sign.

2 The PDE is hyperbolic if all the $A_i$ are nonzero and have, with one exception, the same sign.

3 The PDE is parabolic if one $A_i$ is zero (e.g., $A_s$) and the remaining $A_i$ are nonzero and of the same sign, and $B_s$ is nonzero.

Since, in general, the coefficients $A_i$, $B_i$, $C$, and $D$ are variable, the classification of the PDE may change from point to point.

Partial differential equation problems are generally much more difficult than those for ordinary differential equations. This is because of the higher dimensionality of these problems, together with the difficulty of satisfying more complicated boundary conditions. Until the advent of the digital computer, it was natural that a number of analytical techniques were developed to deal with these problems. Two of the common classical techniques, which still remain important in certain situations, include separation of variables and integral transform methods. These methods can be thought of as complementary to numerical methods, since they are useful in showing qualitative behavior and correlating results, as well as checking numerical solutions. The disadvantage of these analytical methods is that they require fairly sophisticated mathematical skills for their use in all but "easy" problems, and they are generally applicable only to *linear* problems, and indeed not even to all of these. The method of integral transforms is beyond the scope of our discussion. Some useful references on this subject include Hildebrand [1976], Wylie [1966], and Churchill [1963].

Other, somewhat more recent, analytical methods for partial differential equations include similarity transformations, the method of weighted residuals, and asymptotic methods. These newer methods are not necessarily restricted to linear problems, but are still limited in the problems to which they are applicable. References concerned with these methods include Finlayson [1972, 1980], Villadsen and Michelsen [1978], and Meksyn [1961].

The numerical methods, which are employed on a digital computer, have the great advantages of relative simplicity and great generality. For this reason, they have largely supplanted the analytical methods in the solution of various real-world problems in PDEs. In this chapter, we will consider an introduction to some of the "easy" problems in PDEs and their solution by various analytical methods. In the next chapter we will take up a consideration of numerical methods for PDEs.

## 13.1 SEPARATION OF VARIABLES

One of the oldest and most famous methods for the solution of certain partial differential equations is known as the "separation of variables," and this applies to a number of *linear* PDEs. Linearity means that the dependent variable and its derivatives must appear in a linear algebraic form. The separation of variables method is applicable to a number of PDEs from classical mathematical physics that are now important in engineering, such as the heat or diffusion equation, Laplace's equation, and the wave equation. The method involves seeking a solution of the PDE as a product of functions, each of which depend on only one of the independent variables. In many problems, such a separation of the dependence of the solution on the independent variables is possible, and then a rather sophisticated superposition process allows a solution to be built up for various boundary conditions. The real essence of the method is this superposition process, so before presenting some examples, we stop to discuss this matter.

### 13.1.1 Approximation Using Orthogonal Functions

We have, in previous chapters, approximated known functions mainly by either Taylor's theorem, interpolation methods, or least-squares approximation. We now reconsider least-squares approximation where we wish to express a *continuous* function, $F(t)$, in terms of a series of other functions. Thus, we wish to use continuous least-squares approximation in the form

$$\min_{a_i} \int_A^B w(t)[F(t) - (a_1\phi_1(t) + a_2\phi_2(t) \cdots + a_m\phi_m(t))]^2\, dt \tag{13-2}$$

Here, we wish to minimize the expression in Equation (13-2) with respect to all of the coefficients $a_i$ $(i = 1, \ldots, m)$. The weight function $w(t)$ is positive. We can find conditions that must be satisfied by the $a_i$ by differentiating Equation (13-2) and setting the derivatives equal to zero. Thus, by differentiating with respect to the general coefficient $a_k$ we get

$$0 = \int_A^B w(t)[F(t) - (a_1\phi_1(t) + a_2\phi_2(t) \cdots + a_m\phi_m(t)]\phi_k(t)\, dt \quad (i = 1, 2 \cdots m) \tag{13-3}$$

We dealt with equations similar to (13-3) in the chapter on least squares, where the term "normal" equation was used. But now we consider the special but very important circumstance where it happens that the $\phi_i(t)$ functions are *orthogonal* to each other with respect to the weight function $w(t)$. The mathematical definition of orthogonality, in words, is that the integral of the product $w(t)\phi_i(t)\phi_k(t)$ over the interval from $A$ to $B$ is equal to zero whenever $i$ is *not* equal to $k$. Of course, if $i = k$, then the integrand of this product is equal to $w(t)\phi_i^2(t)$, which is positive since $w(t)$ is taken to be positive, and therefore the integral is necessarily positive. In equation form, orthogonality is expressed as

$$\text{Orthogonality:} \qquad 0 = \int_A^B w(t)\phi_i(t)\phi_k(t)\, dt \quad (i \neq k) \tag{13-4}$$

If we are approximating $F(t)$ in Equation (13-2) by a set of functions $\phi_i$ which happen to be orthogonal with respect to $w(t)$, then Equation (13-3) shows that the coefficients $a_i$ are determined directly by the simple formula

$$a_k = \frac{\displaystyle\int_A^B w(t)F(t)\phi_k(t)\, dt}{\displaystyle\int_A^B w(t)\phi_k^2(t)\, dt} \tag{13-5}$$

As the reader can verify, because of orthogonality all of the other terms in Equation (13-3) are equal to zero. The optimal coefficients $a_k$ in Equation (13-5) are often

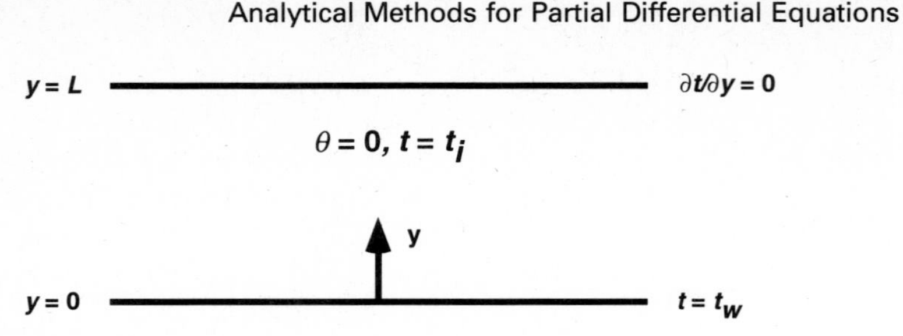


**Figure 13.1.** Transient Heat Conduction in a Slab

called the *Fourier coefficients* of the orthogonal function approximation. It is often convenient, especially in theoretical discussions, to suppose that the $\phi_i$ have been normalized by taking the denominator of Equation (13-5) as unity. We see that the *orthogonality* of the functions in the continuous least-squares approximation produces a vast simplification in the calculation. But what is even more important is that under broad circumstances, *an infinite series of orthogonal functions can represent an arbitrary function* (if the function is reasonably well-behaved). While a general, detailed proof of this is quite involved, we can see that it is quite plausible from the following observations. If we *normalize* the $\phi_i$ by dividing each $\phi_i$ by the denominator integral of Equation (13-5) and put the normalized Fourier coefficients of Equation (13-5) back into Equation (13-2), then, by using orthogonality and the fact that the integral is positive, we see that the following *Bessels inequality* must be true:

$$\sum_{i=1}^{m} a_i^2 \leq \int_A^B F^2(t)\,dt \qquad (13\text{-}6)$$

Equation (13-6) shows that the series $\sum a_i^2$ converges, so that $a_i \rightarrow 0$ as $i \rightarrow \infty$. Also, orthogonal functions are generally *oscillatory* (otherwise, how would their integrals in Equation (13-4) vanish?) and therefore bounded. Thus, it is very *plausible* that a series of the form $\sum a_i \phi_i(t)$, where the $\phi_i$ form a normalized orthogonal function ("orthonormal") system and the $a_i$ are Fourier coefficients, would converge under many circumstances; and indeed, this turns out to be the case. The most famous example of this kind of series is the *Fourier Series* of trigonometric functions which are orthogonal on the interval $(-\pi, \pi)$. However, there are many other orthogonal function systems generated in the course of separation-of-variables solutions of PDEs Hildebrand [1976].

As we move now to some illustrative examples, we will assume that the orthogonal function expansions we encounter do converge. This is true for the examples considered.

**Example 13.1 Transient Heat Conduction in a Slab: The Suddenly Heated Wall**

We consider the problem of transient heat conduction in a "one-dimensional" slab of thickness $L$, where transport in the $x$ and $z$ directions can be neglected (see Figure 13.1). The slab is initially at a uniform temperature of $t_i$ when the face at $y = 0$ is suddenly changed to (and maintained at) $t_w$. The face at $y = L$ is insulated so that $\partial t/\partial y = 0$. We wish to determine the temperature in the slab as a function of time.

The PDE to be satisfied is the following one-dimensional heat conduction equation:

$$\frac{\partial t}{\partial \theta} = \alpha \frac{\partial^2 t}{\partial y^2} \tag{a}$$

where $t$ is temperature, $\theta$ is time, and $\alpha$ is the thermal diffusivity. We first note that the steady-state solution of this problem exists and is equal to $t = t_w$, which corresponds to the solution of $0 = \alpha \partial^2 t/\partial y^2$ subject to the given boundary conditions of $t = t_w$ at $y = 0$ and $\partial t/\partial y = 0$ at $y = L$. A steady-state solution *by definition* is the time-independent solution of the equation corresponding to the specified boundary conditions. Thus, a steady-state solution does not *always* exist (for example, if the temperature at one surface varies with time), but it is useful to look for one at the start of a calculation. An important reason for this is that *it is useful to proceed by taking the unknown temperature to be equal to a transient component plus the steady-state solution.* This is possible since the PDE is linear, and therefore solutions can be added together to form new solutions (principle of superposition).

In this problem it is convenient to start by defining a new dimensionless temperature variable as $T = (t - t_w)/(t_i - t_w)$, which will reduce to unity at time zero and to zero at $y = 0$ and $\theta > 0$. We also define a new dimensionless time variable as $\phi = \alpha\theta/L^2$ and a new dimensionless distance coordinate as $Y = y/L$. Using these definitions causes the original problem to become

$$\frac{\partial T}{\partial \phi} = \frac{\partial^2 T}{\partial Y^2}; \; T(\phi, 0) = 0, \; \frac{\partial T}{\partial Y}(\phi, 1) = 0, \; T(0, Y) = 1 \tag{b}$$

A considerable advantage of using the new dimensionless coordinates is that the resulting mathematical problem is *independent of* $t_w$, $t_i$, *and* $\alpha$, which would be especially valuable for a numerical solution. At this point, it is important to realize that we have made the boundary conditions at $Y = 0$ and $Y = 1$ *homogeneous*. A general linear homogeneous boundary condition is defined to be of the form $c_1 T + c_2 \partial T/\partial Y = 0$. This setting up of homogeneous boundary conditions is an essential ingredient of the separation of variables method. We now try separation of variables in the form

$$T(\phi, Y) = \Phi(\phi) \cdot \Psi(Y) \tag{c}$$

When we perform the differentiations, substitute into the PDE $\partial T/\partial \phi = \partial^2 T/\partial Y^2$, and separate variables we get the equations

$$\frac{1}{\Phi}\frac{d\Phi}{d\phi} = \frac{1}{\Psi}\frac{d^2\Psi}{dY^2} = \text{constant} = -\lambda^2 \tag{d}$$

Since the left-hand side of Equations (d) is a function only of $\phi$, differentiation of this with respect to $Y$ gives zero; this is similar to the derivative with respect to $\phi$ of $(d^2\Psi/dY^2)/\Psi$. Thus, each side of Equation (d) must be equal to some constant. We have written the constant as $-\lambda^2$, which implies that it must be *negative* (or zero). It is apparent on physical grounds in this problem that if the constant was positive, the time dependence of the solution would be exponentially increasing, and therefore the solution would be unbounded in time, an unacceptable behavior. However, a more subtle and more important argument, which applies in cases where physical reasoning cannot be used, is that by choosing the *negative* constant, the character of the $\Psi(Y)$ solution will be oscillatory and *orthogonal functions will be generated,* which is crucial to the success of the method. This permits the addition of many different solutions, each of which satisfies the same homogeneous boundary conditions, to form a *generalized Fourier expansion*, which in turn allows satisfaction of the initial condition. We should also mention that it is sometimes necessary, depending on the boundary conditions of the problem, to add in the solution corresponding to $\lambda = 0$; in this example, that solution is just $K_1 + K_2 Y$ and is not needed. In our example we can easily integrate Equation (d) to find $\Phi(\phi) = \text{Exp}(-\lambda^2\phi)$ and $\Psi(Y) = C_1 \cos \lambda Y + C_2 \sin \lambda Y$, so the overall solution is given by

$$T = \Phi(\phi)\Psi(Y) = e^{-\lambda^2\phi}[C_1 \cos \lambda Y + C_2 \sin \lambda Y] \tag{e}$$

The constants $\lambda$, $C_1$, and $C_2$ are at this point unknown. We now apply the homogeneous boundary conditions at $Y = 0$ and $Y = 1$. First, we use $T = 0$ at $Y = 0$. Since the exponential can never be zero, we must have $0 = C_1 \cos(0) + C_2 \sin(0) = C_1(1) + C_2(0)$. Thus, $C_1$ must be zero. Moving now to the condition $\partial T/\partial Y = 0$ at $Y = 1$, we get $0 = C_2 \lambda \cos \lambda$. This condition could be satisfied by taking $C_2 = 0$, but since $C_1$ is already zero, this would yield the undesired trivial solution $T = 0$. The only other alternative is to choose $\lambda$ so that $\cos \lambda = 0$. Since $\cos x = 0$ for the infinitely many values $x = \pi/2, 3\pi/2, 5\pi/2$, and so on, we have many different possible values of $\lambda$. Denoting these by $\lambda_n = (\pi/2)(2n - 1)$ (with $n = 1, 2, \ldots$) and reverting to the original variables, we obtain the solution in the form

$$T = C_n e^{-\lambda_n^2 \phi} \sin \lambda_n Y = C_n e^{-\left(\frac{\pi(2n-1)}{2L}\right)^2 \alpha\theta} \sin\left(\frac{\pi(2n-1)y}{2L}\right) \tag{f}$$

Equation (f) satisfies the original PDE as well as the boundary conditions at $y = 0$ and $L$ for any positive integer value of $n$ and any value of $C_n$. But it does *not* satisfy the initial condition $T = 1$ for $0 < y < L$ at $\phi = 0$ (time zero).

This is where the orthogonal function expansion comes in. Since the original PDE is *linear*, we may verify that the *sum* of any number of solutions is also a new solution. Thus, by adding together solutions of Equation (f) for $n = 1, 2, 3, \ldots$, we form new solutions. Since $n$ can be arbitrarily large, we form the infinite series solution

$$T = \sum_{n=1}^{\infty} C_n e^{-\left(\frac{\pi(2n-1)}{2L}\right)^2 \alpha\theta} \sin\left(\frac{\pi(2n-1)y}{2L}\right) \tag{g}$$

To satisfy some arbitrary initial condition $T = F(y)$ at $\theta = 0$ ($F(y) = 1$ in our example), we must have that

$$F(y) = \sum_{n=1}^{\infty} C_n \sin\left(\frac{\pi(2n-1)y}{2L}\right) \tag{h}$$

Now, it happens (not by chance!) that the functions $\sin[\pi(2n-1)y/2L]$ $(n = 1, 2, \ldots)$ are *orthogonal* to each other with respect to the weight function $w(y) = 1$, so the $C_n$ are just the Fourier coefficients of this expansion. They are obtained by multiplying Equation (h) by $\sin[\pi(2k-1)y/2L]$ and integrating both sides from $y = 0$ to $y = L$. This gives, for the coefficients $C_n$,

$$C_n = \frac{\displaystyle\int_0^L F(y) \sin\left(\frac{\pi(2n-1)y}{2L}\right) dy}{\displaystyle\int_0^L \sin^2\left(\frac{\pi(2n-1)y}{2L}\right) dy} = \frac{\displaystyle\int_0^L 1 \cdot \sin\left(\frac{\pi(2n-1)y}{2L}\right) dy}{\displaystyle\int_0^L \sin^2\left(\frac{\pi(2n-1)y}{2L}\right) dy} = \frac{4}{\pi(2n-1)} \tag{i}$$

Substituting back into Equation (g), we can finally express the solution as

$$\begin{aligned} T &= \frac{4}{\pi} \sum_{k=1}^{\infty} \frac{e^{-\left(\frac{\pi(2k-1)}{2L}\right)^2 \alpha\theta}}{2k-1} \sin\left(\frac{\pi(2k-1)}{2L}\right) \\ &= \frac{4}{\pi}\left\{e^{-\left(\frac{\pi}{L}\right)^2 \alpha\theta} \sin\left(\frac{\pi y}{2L}\right) + \frac{1}{3} e^{-9\left(\frac{\pi}{L}\right)^2 \alpha\theta} \sin\left(\frac{3\pi y}{2L}\right) \cdots\right\} \end{aligned} \tag{j}$$

To bound the error in the series computation of Equation (j), we would proceed as follows. Suppose we stop just *before* the term corresponding to $k = K$. Then since the sine function is no greater than one in magnitude, and replacing the factor $1/(2k-1)$ by its smallest value $1/(2K-1)$, we have that

$$|Error| \leq \frac{4}{\pi(2K-1)} \sum_{k=K}^{\infty} e^{-(2k-1)^2 w} \tag{k}$$

Here, for convenience, we have replaced $(\pi/2L)^2\alpha\theta$ with $w$. We now observe that the general exponent factor $[2(K+p)-1]^2 = (2K-1)^2 + 2(2K-1)2p + (2p)^2 > (2K-1)^2 + 4(2K-1)p$, which is a useful relationship for $p = 1, 2, \ldots$. Using this, the series in Equation (j) is bounded as

$$\begin{aligned} |Error| \frac{\pi(2K-1)}{4} &\leq \sum_{k=K}^{\infty} e^{-(2k-1)^2 w} = e^{-[2K-1]^2 w} + e^{-[2(K+1)-1]^2 w} + e^{-[2(K+2)-1]^2 w \ldots} \\ &\leq e^{-(2K-1)^2 w}[1 + e^{-4(2K-1)w} + (e^{-4(2K-1)w})^2 \cdots] = \frac{e^{-(2K-1)^2 w}}{1 - e^{-4(2K-1)w}} \end{aligned} \tag{l}$$

The dominating series produced here for the error is geometric and was summed directly. Equation (l) shows that the series in Equation (j) converges *very* rapidly except at very small times.

In the previous example problem, it happened that the so-called characteristic functions $Y_i$ (also called eigenfunctions) have a simple form, and their eigenvalues and orthogonality could be determined directly (e.g., by using integral tables). Usually, this is *not* the case, and a more general verification of orthogonality is required. We outline briefly how this is accomplished in the present problem; very similar arguments are used in much more general cases (see Hildebrand [1976]). The general procedure is based on a consideration of the basic differential equation, which for our example is $Y'' + \lambda Y = 0$. We write this equation for two distinct values of $\lambda$, say $\lambda_n$ and $\lambda_p$. The $\lambda$ values are called characteristic values or eigenvalues. These two equations both satisfy the same homogeneous boundary conditions, $Y_i(0) = Y_i(L) = 0$ (for $i = n, p$). The equations are, specifically,

$$\begin{aligned} Y_n'' + \lambda_n Y_n &= 0 \qquad Y_n(0) = Y_n(L) = 0 \\ Y_p'' + \lambda_p Y_p &= 0 \qquad Y_p(0) = Y_p(L) = 0 \end{aligned} \tag{13-7}$$

If we now multiply the first equation in (13-7) by $Y_p$ and the second equation by $Y_n$, subtract, and integrate over the interval $(0, L)$ (employing integration by parts), we finally get, after simplification,

$$\left[ Y_p Y_n' - Y_n Y_p' \right]_0^L = (\lambda_n - \lambda_p) \int_0^L Y_p Y_n \, dy \tag{13-8}$$

Applying our boundary conditions for both $Y_p$ and $Y_n$ shows the orthogonality of any two functions of the set. Indeed, Equation (13-8) shows that orthogonality would hold as long as all functions of the set satisfied the same homogeneous boundary conditions of the type $c_1 Y + c_2 Y' = 0$ at $y = 0, L$.

The general process of solving linear partial differential equations by means of the separation of variables technique goes under the heading of "Sturm-Liouville" systems. This involves a general consideration of eigenvalues, eigenfunctions, orthogonality, and expansion of arbitrary functions in terms of eigenfunctions. A broad

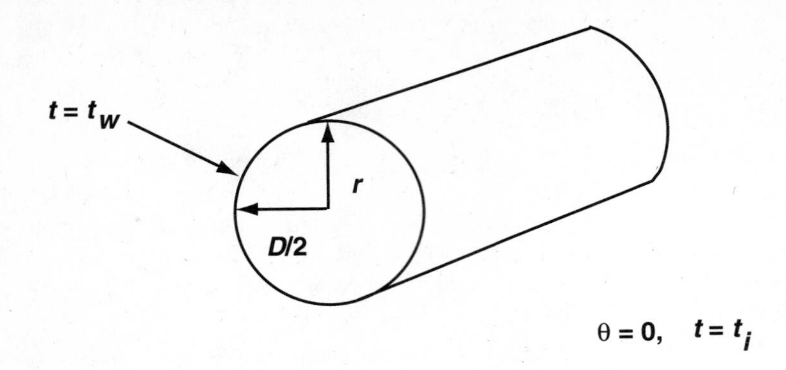


**Figure 13.2.** Transient Heat Conduction in a Circular Cylinder

discussion of this topic is beyond our scope; a good detailed discussion is contained in Hildebrand [1976]. We do consider one further example, involving cylindrical geometry.

**Example 13.2 Transient Heat Conduction in a Circular Cylinder**

Consider transient heat conduction in a long circular cylinder as shown in Figure 13.2. The cylinder is initially at temperature $t = t_i$. At some time the external surface of the cylinder is suddenly changed to, and maintained at, $t = t_w$. It is desired to find a formula for the temperature in the cylinder as a function of time.

We again employ the method of separation of variables. For unsteady radial temperature variations in the cylinder (assuming angular symmetry), the heat conduction equation in cylindrical coordinates is

$$\frac{\partial t}{\partial \theta} = \frac{\alpha}{r}\frac{\partial}{\partial r}\left(r\frac{\partial t}{\partial r}\right) = \alpha\left(\frac{\partial^2 t}{\partial r^2} + \frac{1}{r}\frac{\partial t}{\partial r}\right) \tag{a}$$

Here, $\theta$ is time, $r$ is the radial coordinate, and $\alpha$ is the thermal diffusivity. To ensure homogeneous boundary conditions, we first wish to compute the steady-state solution. Inspection of Equation (a) shows that for $r \to 0$, we must have $\partial t/\partial r = 0$ in order for terms in the equation to be bounded. Thus, our steady-state boundary conditions are $t = t_w$ at $r = D/2$ (where $D$ is the cylinder diameter), and $\partial t/\partial r = 0$ at $r = 0$. This gives the steady-state solution as $t = t_w$. We therefore take as new dimensionless variables

$$\phi = \frac{\alpha\theta}{(D/2)^2}; \quad R = \frac{r}{(D/2)}; \quad T = \frac{t - t_w}{t_i - t_w} \tag{b}$$

When these new variables are substituted into Equation (a), the PDE becomes

$$\frac{\partial T}{\partial \phi} = \frac{1}{R}\frac{\partial}{\partial R}\left(R\frac{\partial T}{\partial R}\right) = \frac{\partial^2 T}{\partial R^2} + \frac{1}{R}\frac{\partial T}{\partial R} \tag{c}$$

with boundary conditions

$$T(\phi, R = 1) = 0; \quad \frac{\partial T}{\partial R}(\phi, R = 0) = 0; \quad T(\phi = 0, R) = 1 \tag{d}$$

We now attempt separation of variables in the form $T = \Phi(\phi)\ \Psi(R)$. Substitution of this expression into Equation (c) and separating variables produces

$$\frac{1}{\Phi}\frac{d\Phi}{d\phi} = \frac{1}{R\psi}\frac{d}{dR}\left(R\frac{d\Psi}{dR}\right) = \text{constant} = -\lambda^2 \tag{e}$$

The solution for $\Phi$ is simply $\Phi(\phi) = C_1 \exp(-\lambda\phi)$. After multiplication by $R^2$, the equation for $\Psi$ in Equation (e) becomes

$$R^2\frac{d^2\Psi}{dR^2} + R\frac{d\Psi}{dR} + \lambda^2 R^2 \Psi = 0 \tag{f}$$

Equation (f) is a form of Bessel's equation, and to satisfy the boundary condition $dT/dR = 0$ at $R = 0$ the solution is $\Psi(R) = C_2 J_0(\lambda R)$, where $J_0$ is a Bessel function of the first kind of order zero. To satisfy the condition $T = 0$ at $r = D/2$ (or $R = 1$), we must have

$$0 = J_0(\lambda_n) \tag{g}$$

Equation (g) has been written in terms of $\lambda_n$ since there is not just one, but infinitely many different $\lambda_n$ values that satisfy this equation, since $J_0(x)$ is an oscillatory function. Thus, for any one value of $\lambda_n$, the solution for $T$ can be written as $T(\phi, R) = C_n \exp(-\lambda_n^2 \phi) J_0(\lambda_n R)$. Since the differential equation and boundary conditions of the problem satisfy the requirements of orthogonality and superposition, it can be shown Hildebrand [1976] that the solution for $T$ can be written in terms of the infinite series

$$T(\phi, R) = \sum_{n=1}^{\infty} C_n e^{-\lambda_n^2 \phi} J_0(\lambda_n R) \tag{h}$$

Since the $J_0(\lambda_n R)$ are orthogonal on the interval from $r = 0$ to $r = D/2$ with respect to the weight function $R$, we determine the coefficients in Equation (h) by using orthogonality to satisfy the initial condition $T = 1$ at $\phi = 0$ for all $R$. At $\phi = 0$, Equation (h) takes the form

$$1 = \sum_{n=1}^{\infty} C_n J_0(\lambda_n R) \tag{i}$$

We then multiply both sides of Equation (i) by $J_0(\lambda_k R)$ and the weight function $R$, and then integrate with respect to $R$ from 0 to 1. All terms on the right become zero due to orthogonality except the term corresponding to $n = k$. The $C_k$ are therefore given by

$$C_k = \frac{\int_0^1 R J_0(\lambda_k R) \cdot 1 dR}{\int_0^1 R J_0^2(\lambda_k R) dR} \tag{j}$$

When the $C_k$ from Equation (j) are substituted into Equation (h), we have the complete formal solution to our original problem.

### 13.1.2 Numerical Computation in Sturm-Liouville Systems

In problems where the particular Sturm-Liouville system under consideration is well known and has its eigenvalues and eigenfunctions extensively tabulated (e.g., Bessel functions, etc.), it is convenient to use this information in numerical computations. However, for problems in which the desired information is not well known, it is very convenient to do the calculations numerically; for example, with computer program IEXVSSH, discussed in Chapter 11. This program allows, within a single successful run, the calculation of a characteristic value and corresponding function, along with the integrals needed for the corresponding Fourier coefficient. This is accomplished as discussed previously for ODE eigenvalues in the chapter on numerical solution of ODE boundary value problems. Thus, we take the eigenvalue parameter as a third differential equation having derivative [$d(3)$] equal to zero, and $y(3)$ is the "guess" value. We can, at the same time, calculate the necessary integrals for the Fourier coefficients in the following way. We need $a_i = \int w(x)F(x)\phi_i(x)\,dx / \int w(x)\phi_i^2(x)\,dx$. In program IEXVSSH (where $y(1)$ is $\phi_i(x)$), we therefore use two additional differential equations, one for each integral: (i) `d(4) = w(x)*F(x)*y(1), y(4)initial = 0`; (ii) `d(5) = w(x)*y(1)^2, y(5)initial = 0`. At the end of the iteration process, $a_i = [y(4)/y(5)]_{x=L}$. As before, the eigenvalue is the final value of "guess," the eigenfunction is $\phi_i(x) \equiv y(1)$, and $\phi_i'(x) \equiv y(2)$. An example of this type of calculation was shown in the chapter on numerical solution of ODE boundary value problems.

## 13.2 SIMILARITY SOLUTIONS

We have seen how the separation of variables technique reduces the solution of certain linear partial differential equations to that of ordinary differential equations. Another technique that accomplishes this in a different way is sometimes applicable to partial differential equations of either linear or nonlinear type. The idea is to seek a solution that has the two independent variables combined into a single *similarity* variable. The technique usually requires that two boundary conditions collapse into a single one when the similarity transformation is employed. Because of the restrictions of the transformation in both the equation and the boundary conditions, the method is of limited applicability, but when it works it is very powerful. We illustrate the technique with an example closely related to the one considered using separation of variables (Example 13.1). The similarity method is also known as "combination of variables."

**Example 13.3 Transient Heat Conduction into a Semi-Infinite Slab; the Suddenly Heated Wall**

We consider the problem of heat conduction into a solid of semi-infinite extent which goes from $y = 0$ at its lower boundary to $y = \infty$ (see Figure 13.3). It can be assumed that there is no significant transport in the $x$ and $z$ directions. The material is initially at $t = t_i$, and at some time the lower boundary is changed to $t_w$ and held at this temperature. We wish to calculate the evolution of the temperature in the solid. What we might expect to find is that just after the imposition of the new temperature on the lower boundary, the influence of the change is gradually felt farther away from this boundary. Hence, there may be a growing region or "boundary layer" where this influence is significant, and above which the temperature remains unchanged and equal to the original value $t_i$.

The partial differential equation describing this process is the one-dimensional transient heat conduction equation,

$$\frac{\partial t}{\partial \theta} = \alpha \frac{\partial^2 t}{\partial y^2} \tag{a}$$

It is convenient to first nondimensionalize the temperature by introduction of the new variable $T$,

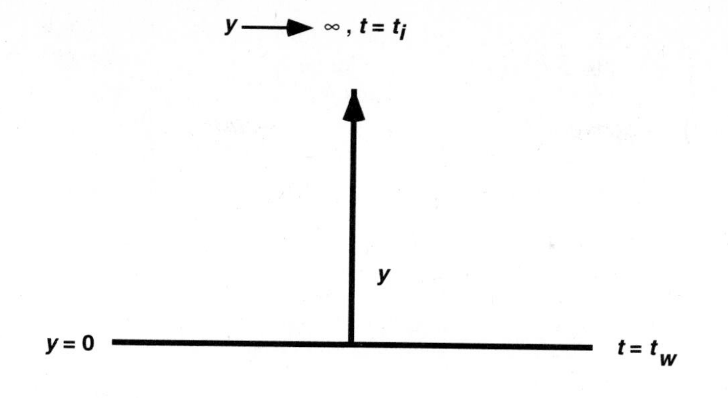

**Figure 13.3.** Transient Heat Conduction in a Semi-Infinite Slab

$$T = \frac{(t - t_i)}{(t_w - t_i)}. \tag{b}$$

Since $t_i$ and $t_w$ are both constant, we have

$$\frac{\partial T}{\partial \theta} = \alpha \frac{\partial^2 T}{\partial y^2} \tag{c}$$

with the new boundary conditions $T = 0$ at $y = \infty$ for all time ($\theta$), $T = 0$ for $y > 0$ at $\theta = 0$, and $T = 1$ at $y = 0$ for $\theta > 0$. We now seek a similarity solution of the type $T = f(\eta)$, where $\eta = y/m(\theta)$ and $m(\theta)$ is a yet-unspecified function of $\theta$. Thus, $\eta$ is a *combination* variable, which combines the effects of $y$ and $\theta$. We *assume* that $T$ can be expressed in the above form; we therefore have to see if this assumption can indeed satisfy the PDE and all of the auxiliary conditions. First, the derivatives necessary for substitution into the PDE must be calculated. These are

$$\frac{\partial T}{\partial \theta} = \frac{df}{d\eta}\left(\frac{-y}{m^2}\right)\frac{dm}{d\theta} = -\frac{\eta}{m}\frac{dm}{d\theta}f'(\eta); \quad \frac{\partial T}{\partial y} = \frac{f'(\eta)}{m}; \quad \frac{\partial^2 T}{\partial y^2} = \frac{f''(\eta)}{m^2} \tag{d}$$

The derivatives in Equations (d) are obtained by using the chain rule for differentiation together with the assumed form for $T$. Derivatives with respect to $\eta$ are denoted with a prime. If we now substitute these relations into our PDE, we get

$$-\frac{\eta f'(\eta)}{m}\frac{dm}{d\theta} = \alpha \frac{f''(\eta)}{m^2} \tag{e}$$

Equation (e) can be conveniently re-arranged to read $-(m\, dm/d\theta)\eta f'(\eta) = \alpha f''(\eta)$. Thus, if we take $(m\ dm/d\theta$ = constant = $2\alpha$, Equation (e) becomes

$$2\eta f'(\eta) = f''(\eta) \tag{f}$$

This choice for the function $m(\theta)$ (apart from the constant) is the only one that allows the possibility of a solution to the original PDE, since any other would result in the $f$ function depending on $\theta$ in addition to $\eta$, which would contradict our original assumption. The choice of $2\alpha$ as the constant is convenient since this choice removes the parameter $\alpha$ and gives a simple form for the differential equation. Since $(m\ dm/d\theta) = 2\alpha$, integration gives

$$m^2 = 4\alpha\theta + C_1 \tag{g}$$

where $C_1$ is the constant of integration.

What we can say so far is that our choice of similarity variables $\eta$ and $m(\theta)$ appears to be successful in reducing the original PDE to a pair of ODEs. But success of the transformation also requires that the *boundary conditions* on $f(\eta)$ must be independent of $\theta$. After choosing $m(\theta)$, the equation for $f(\eta)$ is $f''(\eta) = -2\eta f'(\eta)$. By using $p = f'$, we have $dp/d\eta = -2\eta p$, which integrates to $df/d\eta = C_2 \operatorname{Exp}(-\eta^2)$; therefore, $df/d\eta(0) = C_2$. Integrating again, using the condition $T(\eta = 0) = f(0) = 1$, gives

$$f(\eta) - 1 = C_2 \int_0^\eta e^{-\tau^2} d\tau \tag{h}$$

The previous boundary condition $f(0) = 1$ corresponds to $t = t_w$ $(T = 1)$ at $y = 0$, since $\eta = 0$ when $y = 0$. We note the *crucial* fact that if we choose $C_1 = 0$ in the equation for $m(\theta)$, then $m(\theta) = (\alpha\theta)^{1/2}$; and the two conditions $t = t_i$ for $y = \infty$ and $t = t_i$ at $\theta = 0$ *both* take the form that $t = t_i$ $(T = 0)$ at $\eta = \infty$. Thus, the mathematical boundary condition $f(\infty) = 0$ satisfies these *two* physical conditions; use of $f(\infty) = 0$ in Equation (h) gives

$$C_2 = \frac{-1}{\displaystyle\int_0^\infty e^{-\tau^2} d\tau} = \frac{-1}{\dfrac{\Gamma(1/2)}{2}} = -\frac{2}{\sqrt{\pi}} \tag{i}$$

The value of the integral in Equation (i) was obtained by changing variables and using the gamma function $[\Gamma(1/2) = \pi^{1/2}]$. When we substitute the value of $C_2$ in Equation (h) and rearrange the integral, we finally get

$$T = f(\eta) = \frac{2}{\sqrt{\pi}} \int_\eta^\infty e^{-\tau^2} d\tau \equiv \operatorname{erfc}(\eta) \tag{j}$$

The final integral for $T = f(\eta)$ in Equation (j) is called the *complementary error function*, $\operatorname{erfc}(\eta)$, and it is equal to $1 - \operatorname{erf}(\eta)$, where erf is the error function. For small $\eta$ (large times), the integral is easy to evaluate in terms of the Taylor expansion of $\operatorname{erf}(\eta)$. For large $\eta$ (small times), we can use Watson's lemma to evaluate erfc (see discussion of asymptotic evaluation of integrals). The error functions erf and erfc are tabulated in several places (cf. Abramowitz and Stegun [1964]). The energy flux at the surface due to the imposed change in surface temperature is

$$q_w(\theta) = -k\left(\frac{\partial t}{\partial y}\right)_0 = -k(t_w - t_i)f'(0)/m(\theta) = 0.565k(t_w - t_i)/(\alpha\theta)^{1/2} \tag{k}$$

Thus, the surface energy flux is very large at small times (when the thermal boundary layer is thin), but it gradually diminishes toward zero at large times. The thermal boundary layer thickness $\delta$ corresponds to the value of $y$ where $\operatorname{erfc}(\eta)$ is nearly equal to zero. This definition is necessarily somewhat subjective, but if we choose "zero" to be approximately 0.01, we have that $\eta \approx 2.6 = \delta/(\alpha\theta)^{1/2}$, so that $\delta \approx 2.6(\alpha\theta)^{1/2}$. Thus, the thermal boundary layer is zero at $\theta = 0$ and grows proportional to $\theta^{1/2}$.

It is useful to note that the solution given here for a sudden temperature change at the boundary of a *semi-infinite* slab should also apply, at small times, to the case of a finite slab of thickness $L$ whose upper face is insulated and which undergoes the same temperature change at the lower boundary. The temperature field in the two cases should be virtually the same until the time when the thermal boundary layer has reached the upper surface of the finite slab. This corresponds to the time when $\delta = L$, so from the previous paragraph this is $\theta \approx (L/2.6)^2/\alpha$. Following this time the two solutions will, of course, be different. Considerations of this type apply in many problems and are especially useful since the separation of variables solutions

**TABLE 13.1** COMPARISON OF EXACT AND SIMILARITY SOLUTIONS FOR THE SUDDENLY HEATED WALL AT SHORT TIMES

| $Y = y/L$ | $\eta = y/2\sqrt{\alpha\theta}$ | $T = \text{Erf}(\eta)$ | $T_{\text{exact}}$ |
|---|---|---|---|
| 1.00 | 2.041 | 0.9961 | 0.9922 |
| 0.75 | 1.531 | 0.9696 | 0.9693 |
| 0.50 | 1.021 | 0.8512 | 0.8511 |
| 0.25 | 0.5303 | 0.5295 | 0.5295 |
| 0.00 | 0.0 | 0.0 | 0.0 |

$\phi = 0.06$

are relatively difficult to calculate at small times (e.g., we must take many terms of the series).

### 13.2.1 Small Time Comparison

It is of interest to compare numerical calculations at small times for the exact and similarity solutions of the suddenly heated wall problem considered above in Examples 13.1 and 13.3. We choose to make the comparison at $\phi = 0.06$, which is estimated to be approximately the time when the thermal boundary layer just reaches the upper surface of the wall. Thus, we would expect that the similarity solution for an infinitely thick wall should be a good approximation to the exact separation of variables solution for $\phi = 0.06$ and smaller times. The comparison is shown in Table 13.1. The exact values are from Equation (j) in Example 13.1, while the similarity values are from the error function, erf, of Example 13.3.

The results in Table 13.1 confirm our expectations regarding the agreement of the similarity solution with the exact solution: the *maximum* error at ($Y = 1$) is less than 0.5%.

## 13.3 THE INTEGRAL METHOD

Another analytical method that can sometimes be useful in the solution of PDEs is the so-called integral method. This method is now considered to be a rather crude type of weighted residual method which has various limitations. The general MWR (method of weighted residuals) applied to PDEs is generally considered to be a *numerical* method, and we shall accordingly consider MWR in Chapter 14 on numerical solution of PDEs.

The integral method was popular before the coming of the digital computer. Its advantages are simplicity and generality; its disadvantages are dependence on user "insight," questionable reliability, and difficulty in estimating errors. In spite of these limitations, we feel that the method is often useful in illustrating how the solution depends on variables and parameters. One type of problem where the integral method has been useful is in "boundary layers," for instance, in the problem of Example 13.3. We therefore illustrate the integral method on that problem.

The problem from Example 13.3 is

$$\frac{\partial T}{\partial \theta} = \alpha \frac{\partial^2 T}{\partial y^2} \tag{13-9}$$

with

$$T = 0 \text{ at } y = \infty \text{ and } \theta = 0 \tag{13-10}$$

To use the integral method, we "model" the problem by integrating the PDE with respect to $y$ from $y = 0$ to $y = \delta$ (thermal boundary layer thickness $\delta$, defined as the point where $T = 0$, $\partial T/\partial y = 0$). This gives

$$\int_0^\delta \frac{\partial T}{\partial \theta} dy = \frac{d}{d\theta}\int_0^\infty T dy - T(\delta)\frac{d\delta}{d\theta} = \alpha \int_0^\delta \frac{\partial^2 T}{\partial y^2} dy = -\alpha \frac{\partial T}{\partial Y}(0) \tag{13-11}$$

In Equation (13-11) we have used the Leibnitz rule for differentiating an integral when both the integrand and the upper limit depend on the variable, $\frac{d}{dx}\int_0^{B(x)} f(x,t)\,dt = \int_0^B \frac{\partial f(x,t)}{\partial x} dt + f(x,B)\frac{dB}{dx}$. In Equation (13-11), $T(\delta) = 0$ so that we simply have

$$\frac{d}{d\theta}\int_0^\delta T dy = -\alpha \frac{\partial T}{\partial y}(0) \tag{13-12}$$

The next step in the integral method is to assume a form for $T$ in terms of $y$ and $\delta$. This is accomplished by requiring that the assumed $T$ function satisfy as many "reasonable" boundary conditions as possible. For instance, suppose we take $T = C_1 + C_2 y$, the simplest possible function we could choose. Then we would require that $T = 1$ at $y = 0$ and $T = 0$ at $y = \delta$ so that $T = 1 - (y/\delta)$ is the simplest profile for $T$. This profile yields $\partial T/\partial y = -1/\delta$, and Equation (13.12) then becomes $(1/2)(d\delta/d\theta) = \alpha/\delta$. Integration of this ODE with $\delta = 0$ at $\theta = 0$ gives $\delta = 2(\alpha\theta)^{1/2}$.

It is interesting to observe several results of this very simple (and crude) calculation. The boundary layer thickness given by the integral method is the same function of $\alpha$ and $\theta$ as is found in the exact solution. If we calculate the surface heat flux from the integral method solution, we get $q_w = (\partial T/\partial y)_0 = -0.5k(t_w - t_\infty)/(\alpha\theta)^{1/2}$. The exact solution gives the same result except for a factor of 0.565 instead of 0.5. It is remarkable that such a simple calculation seems to produce rather good results.

Before we become too impressed with the integral method, some comments are in order. First, we note that *any* temperature function $T$ that depends only on $y/\delta$ will produce the correct functional dependence for $\delta(\theta)$. This is because the exact solution of this problem corresponds to the combination of variables $y/(\theta)^{1/2}$. If we wanted to get a better integral solution, we might try to incorporate into the assumed solution other conditions the true solution satisfies, such as $\partial T/\partial y = 0$ at $y = \delta$, and $\partial^2 T/\partial y^2 = 0$ at $y = 0$ (this last condition comes from evaluating the PDE at $y = 0$). The integral method involves trying to make the approximation better and better at the boundaries, rather than in the interior region. This is necessary to keep the calculations simple, but is naturally limited in the accuracy that can be achieved. Also, different kinds of functions other than polynomials might be tried. It is clear that there are many possibilities, but there is no *a priori* way to discriminate between them.

One final word in favor of the integral method is that it is essentially just as simple to implement in problems where the similarity transformation does *not* work.

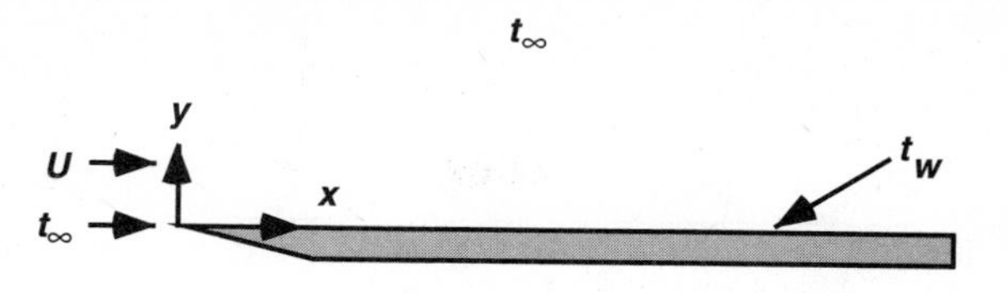


**Figure 13.4.** Heat Transfer in the Laminar Boundary Layer on a Flat Plate

In terms of estimating accuracy, one possibility is to express the true solution as the sum of the integral solution plus an *error* term. Solving the equation for the error *numerically* then gives an indication of the accuracy in the integral solution.

## 13.4 ASYMPTOTIC METHODS

Various kinds of asymptotic methods, both direct and indirect, can be useful in the solution of PDEs. Since the solution of a PDE is often expressed in terms of a series, an integral, or the solution of an ODE, asymptotic procedures for these types of problems apply indirectly to PDEs.

**Example 13.4 Laminar Boundary Layer Heat (or Mass) Transfer at Large Prandtl Number on a Flat Plate**

Steady-state laminar boundary layer transport for flow over a flat plate is described by the partial differential equations Bennett and Myers [1982], Meksyn [1961]

$$\frac{\partial u}{\partial x} + \frac{\partial v}{\partial y} = 0;\ u\frac{\partial u}{\partial x} + v\frac{\partial u}{\partial y} = \xi\frac{\partial^2 u}{\partial y^2};\ u\frac{\partial \theta}{\partial x} + v\frac{\partial \theta}{\partial y} = \alpha\frac{\partial^2 \theta}{\partial y^2} \tag{a}$$

The problem considered here is illustrated in Figure 13.4. The first of Equations (a) is the continuity (or overall mass) equation, the second is the axial momentum equation, and the third is the energy equation. The quantities $u$ and $v$ are the axial and transverse velocity components, and $\xi$ and $\alpha$ are the momentum and thermal diffusivities. Transport properties are assumed to be constant, and $\theta$ is a dimensionless temperature variable, defined by $\theta = (t - t_\infty)/(t_w - t_\infty)$.

**Solution.** For convenience we define the stream function, $\psi$, by the equations

$$u = \partial\psi/\partial y \tag{b}$$

and

$$v = -\partial\psi/\partial x \tag{c}$$

Thus, the continuity equation is automatically satisfied. Next, we introduce the following similarity transformation:

$$\psi = (2\xi U x)^{1/2} f(\eta),\ \eta = \left(\frac{U}{2\xi x}\right)^{1/2} y,\ \theta = \theta(\eta) \tag{d}$$

It can be verified that the transformation of Equations (d) causes Equations (a) to assume the forms

$$\text{momentum:}\quad \frac{d^3 f}{d\eta^3} + f\frac{d^2 f}{d\eta^2} = 0;\ f(0) = f'(0) = 0,\ f'(\infty) = 0$$

$$\text{heat transfer:}\quad \frac{d^2\theta}{d\eta^2} + \Pr f\frac{d\theta}{d\eta} = 0;\ \theta(0) = 1,\ \theta(\infty) = 0 \tag{e}$$

**TABLE 13.2** COMPARISON OF EXACT AND APPROXIMATE VALUES OF $-\theta'(0)$ AS A FUNCTION OF PR

| Pr | $-\theta'(0)_{\text{exact}}$ | $-\theta'(0)_{\text{approx.}}$ Equation(j) |
|---|---|---|
| 0.7 | 0.414 | 0.413 |
| 1 | 0.470 | 0.469 |
| 7 | 0.914 | 0.912 |
| 10 | 1.03 | 1.03 |
| 15 | 1.18 | 1.18 |

where $u/U = f'(\eta)$ and $\text{Pr} = \xi/\alpha$ is the Prandtl number. Our main objective is to calculate $d\theta/d\eta(0)$ as a function of Pr. To do this, we note that the $f$ function is independent of $\theta$ and Pr. The $\theta$ equation in Equation (e) is just first order in $d\theta/\delta\eta$ and linear. With $p = d\theta/d\eta$ we have $dp/d\eta = -\text{Pr} f(\eta)p$, and the solution for $\theta$ can be written as

$$\theta - 1 = \theta'(0) \int_0^{\eta} e^{-\text{Pr}\, F(x)} dx \text{ where } F(x) = \int_0^x f(t)\, dt \tag{f}$$

Applying the boundary condition $\theta(\infty) = 0$ to Equation (f) yields

$$\frac{-1}{\theta'(0)} = \int_0^{\infty} e^{-\text{Pr}\, F(x)} dx \tag{g}$$

Thus, to get $\theta'(0)$, we need to evaluate the integral in Equation (g). The treatment given here follows Meksyn [1961]. The function $f(\eta)$ is known to increase monotonically from zero at $\eta = 0$, so Watson's lemma can be used. To do this we write Equation (g) as

$$\frac{-1}{\theta'(0)} = \int_0^{\infty} e^{-\text{Pr}\, w} \frac{dx}{dw} dw \tag{h}$$

From the Taylor series expansion of $f(\eta)$, using the known value from Meksyn [1961] of $f''(0) = a = 0.4696$, we integrate term by term to get $F(\eta)$ as

$$w = F(\eta) = \frac{a\eta^3}{3!} - \frac{a^2\eta^6}{6!} + \frac{11a^3\eta^9}{9!} - \frac{375a^4\eta^{12}}{12!} \tag{i}$$

We now wish to revert the series in Equation (i), and this is accomplished by letting $w = F(\eta) = s^3$. Then we take cube roots so that $s$ is a power series whose first term is proportional to $\eta$. The reversion algorithm (see the appendix on series and program SERIES) is used to express $\eta$ in terms of $s$, following which $s$ is replaced by $w^{1/3}$. These manipulations finally yield, for the first three terms of the asymptotic series,

$$\begin{aligned} -\frac{3}{\theta'(0)}\left(\frac{a}{6}\right)^{1/3} &= \int_0^{\infty} e^{-\text{Pr}\, w}\left[w^{-2/3} + \frac{w^{1/3}}{15} - \frac{w^{4/3}}{180}\right] dw \\ &= \Gamma(1/3)\,\text{Pr}^{-1/3}\left[1 + \frac{1}{45\,\text{Pr}} - \frac{1}{405\,\text{Pr}^2}\right] \end{aligned} \tag{j}$$

The heat flux at the wall is generally the quantity of interest and is given in terms of $\theta'(0)$ by

$$q_w = -k\left(\frac{U}{2\xi x}\right)^{1/2} (t_w - t_\infty)\theta'(0) \tag{k}$$

The relationship in Equation (j) is *remarkably* accurate, not only for $\text{Pr} \to \infty$ (liquids), but even for Pr as low as $\approx 0.7$ (gases). A number of other problems of this type can be treated in the same way Meksyn [1961]. A comparison of approximate values of $-\theta'(0)$ as given by Equation (j) versus exact numerical values due to Merk [1959] is shown in Table 13.2.

## 13.5 FITTING A NUMERICAL FUNCTION USING RATIONAL LEAST SQUARES

As we have mentioned on several occasions, a very important practical technique for obtaining a useful analytical representation of a function is to *fit numerical values of the function using rational least squares.* This procedure often works very well (especially if any other known information can be worked into the process), and it has the advantage of producing a "parsimonious" (few parameters) result along with a useful error estimate. To accomplish this task we use program RATLSGR.

**Example 13.5: Fitting the Blasius Solution by Rational Least Squares**

As we have seen before, the Blasius nonlinear equation does not have a known analytical solution. We use the following numerical values for the dimensionless velocity $u/U$ as given by Knudsen and Katz [1958].

| $\eta$ | $u/U$ |
|---|---|
| 0.0 | 0.0 |
| 0.2 | 0.1328 |
| 0.4 | 0.2647 |
| 0.6 | 0.3938 |
| 0.8 | 0.5168 |
| 1.0 | 0.6298 |
| 1.2 | 0.7290 |
| 1.4 | 0.8115 |
| 1.6 | 0.8761 |
| 1.8 | 0.9233 |
| 2.0 | 0.9555 |
| 2.2 | 0.9759 |
| 2.4 | 0.9878 |
| 2.6 | 0.9943 |
| 2.8 | 0.9975 |
| 3.0 | 0.9990 |
| 3.2 | 0.9996 |
| 3.4 | 0.9999 |
| 3.6 | 1.0000 |
| 3.8 | 1.0000 |

It is known (but not obvious) that the power series solution for $u/U$ progresses as $u/U = c_1\eta + c_4\eta^4 + c_7\eta^7 + \ldots$ where each succeeding term in the series increases the power of $\eta$ by three. Thus, it is natural to use, as our fitting formula, the expression

$$\left(\frac{u/U}{\eta}\right) = \frac{p_0 + p_1\eta^3 + \cdots}{1 + q_1\eta^3 + \cdots} \tag{a}$$

This expression will automatically yield $u = 0$ at $\eta = 0$ as desired. In program RATLSGR, it is convenient to employ the variable transformation option to use the $y$-$x$ variables $y = (u/U)/\eta$ and $x = \eta^3$. Note that in order to use a data point at $\eta = 0$, since $y$ is indeterminate (0/0) we must use the constant value $d(u/U)/d\eta$ (= 0.6641) for $y(0)$. (If this value was unknown, we could just ignore the data point at $\eta = 0$ or compute it by numerical differentiation.) After inputting the data, we experiment with program RATLSGR by using various $k$ and $m$ (with $k$ close to $m$) and obtain the following results:

| Number of Parameters | $k$ | $m$ | $emax$ | graph |
|---|---|---|---|---|
| 3 | 2 | 0 | $3.4e-2$ | not good |
| 3 | 1 | 1 | $1.6e-3$ | very good |
| 4 | 3 | 0 | $1.3e-2$ | fair |
| 4 | 2 | 1 | $4.6e-4$ | "perfect" |
| 4 | 1 | 2 | $4.7e-4$ | "perfect" |

From the list above, we see that the two polynomial approximations $[(k, m) = 2, 0;\ 3,0]$ are *much* inferior to the rational approximations. The four-parameter approximation with $(k = 2, m = 1)$ is slightly better than $(k = 1, m = 2)$. The parameters in this case for Equation (a) are

$$p_0 = 0.664449$$

$$p_1 = 1.38947e-2$$

$$p_2 = -1.36868e-5$$

$$q_1 = 7.75788e-2$$

## PROBLEMS

**13.1.** Consider one-dimensional transient heat conduction in a rectangular slab geometry. The dimensionless PDE is $\partial T/\partial\phi = \partial^2 T/\partial y^2$, and it is to be solved for $T(\phi, y)$ subject to the boundary conditions $T(\phi, 0) = T(\phi, 1) = 0$ and the initial condition $T(0, y) = f(y)$. Carry out the solution using separation of variables.

**13.2.** We have the following Sturm-Liouville problem (cylindrical coordinates):

$$\frac{d^2y}{dx^2} + \frac{1}{x}\frac{dy}{dx} + \lambda^2 y = 0$$

with boundary conditions $dy/dx(0) = 0$ and $y(1) = 0$. Use the shooting method with program IEXVSSH to determine (to five figures) (i) the first (smallest) eigenvalue, (ii) the second eigenvalue, and (iii) the third eigenvalue.

Hints: Let $y_1 = y$, $y_2 = dy/dx$, $y_3 = \lambda^2 =$ "guessed" variable. Shoot from $x = 0$ with the (arbitrary) initial value $y(0) = 1$. Use crude integration to get rough estimates; then refine these with accurate integration.

**13.3.** For developed-flow heat or mass transfer, a similarity solution known by the name of Leveque often applies. This problem involves the derivation of the Leveque similarity solution. One application of this solution, which was mentioned in the chapter on modeling, concerns solid dissolution into a falling liquid film. The partial differential equation is

$$By\frac{\partial T}{\partial x} = \alpha\frac{\partial^2 T}{\partial y^2}$$

Here, $B$ is a constant related to the surface shear stress, $T$ is the temperature (or composition) dependent variable, $\alpha$ is a constant diffusivity, $x$ is the axial distance, and $y$ is the transverse distance.

Determine the similarity solution which satisfies the boundary conditions $T(x = 0, y > 0) = T_0$, $T(x, y = 0) = T_w$, $T(x, \infty) = T_0$. It is convenient to define $f = (T - T_0)/(T_w - T_0)$ and assume that $f = f(h)$, where $h$ is the similarity variable which combines the effects of $x$ and $y$ according to $h = y/m(x)$.

**13.4.** This problem involves heat conduction in a boundary-layer situation that does *not* admit a similarity solution. Consider radial heat conduction in the annular region *external* to a cylinder of radius $R$. For this situation the heat equation is $\partial T/\partial\theta = (\alpha/r)\partial/\partial r[r\partial T/\partial r]$, where $\alpha$ is the thermal diffusivity. The annular region is initially at a uniform temperature $T_0$; at some later time, the temperature $T_w$ (constant) is imposed at the boundary $r = R$. Use the integral method to derive an expression for the temperature distribution and the thermal boundary layer thickness $\delta$. To keep the calculations simple, use the assumed temperature profile $(T - T_0)/(T_w - T_0) = 1 - y/\delta$.

**13.5.** The similarity solution of a transport problem gives the surface gradient in the form

$$T'(0) = \int_0^\infty e^{-\sigma(y^2+y^3)}dy$$

where $\sigma$ is a parameter.

(i) Use series reversion and the standard Watson's lemma recipe to evaluate $T'(0)$ asymptotically for large $\sigma$.

(ii) Use suitable rescalings to determine asymptotic approximations for both large and small $\sigma$, together with error bounds. Which approach seems best in this problem?

## REFERENCES


ABRAMOWITZ, M., and STEGUN, I. (ed.) (1965), *Handbook of Mathematical Functions*, Dover, New York

AMES, W. F. (1965, 1972), *Nonlinear Partial Differential Equations in Engineering*, I and II, Academic Press, New York

BENNETT, C. O. and MYERS, J. E. (1982), *Momentum, Heat and Mass Transfer*, 3rd ed., McGraw-Hill, New York

CARRIER, G. F., and PEARSON, C. E. (1976), *Partial Differential Equations*, Academic Press, New York

CHURCHILL, R. V. (1963), *Fourier Series and Boundary Value Problems*, 2nd ed., McGraw-Hill, New York

DRESNER, L. (1983), *Similarity Solutions of Nonlinear Partial Differential Equations*, Pitman, Boston

FINLAYSON, B. (1972), *The Method of Weighted Residuals and Variational Principles*, Academic Press, New York

FINLAYSON, B. (1980), *Nonlinear Analysis in Chemical Engineering*, McGraw-Hill, New York

HILDEBRAND, F. B. (1976), *Advanced Calculus for Applications*, 2nd ed., Prentice-Hall, Englewood Cliffs, NJ

HIMMELBLAU, D. M., and BISCHOFF, K. B. (1968), *Process Analysis and Simulation: Deterministic Systems*, Wiley, New York

JENSON, V. G., and JEFFREYS, G. V. (1977), *Mathematical Methods in Chemical Engineering*, 2nd ed., Academic Press, New York

KNUDSEN, J. G. and KATZ, D. L. (1958) *Fluid Dynamics and Heat Transfer*, McGraw-Hill, New York

MEKSYN, D. (1961), *New Methods in Laminar Boundary Layer Theory*, Pergamon Press, New York

MERK, H. J. (1959), *J. Fluid Mechanics*, **5**, p. 460.

MICKLEY, H. S., SHERWOOD, T. K., and REED, C E. (1957), *Applied Mathematics in Chemical Engineering*, McGraw-Hill, New York

SPIEGEL, M. (1965), *Laplace Transforms*, Schaum, New York

VILLADSEN, J., and MICHELSEN, M. L. (1978), *Solution of Differential Equation Models by Polynomial Approximation*, Prentice-Hall, Englewood Cliffs, NJ

WYLIE, C. R. (1966), *Advanced Engineering Mathematics*, 3rd ed., McGraw-Hill, New York

ZWILLINGER, D. (1989), *Handbook of Differential Equations*, Academic Press, Inc., New York

# 14

# Numerical Methods for Partial Differential Equations

We considered in the previous chapter an introduction to some analytical methods for solving partial differential equations (PDEs). These methods are valuable for certain relatively simple problems. But often, contemporary models of interest involve complications that are either very difficult or (more often) impossible to solve with these traditional methods. The advent of the digital computer has allowed consideration of much more complicated PDE problems by means of numerical methods specifically designed for these machines. The numerical methods have a number of very desirable features, such as considerable simplicity and generality, and applicability to very difficult and simple problems alike. Of course, their principal limitation is that they do not yield the functional dependence of the solution on the coordinates and parameters. In addition, there are important questions that must be addressed concerning the accuracy of the numerical solution.

In this chapter we shall consider the development of numerical methods for PDEs. In chemical engineering there is one particular general type of PDE that occurs quite regularly; this is the general *convective-diffusion equation*, which in simplest, one-dimensional, linear form can be written as

$$\frac{\partial c}{\partial \theta} + u\frac{\partial c}{\partial y} = D_{AB}\frac{\partial^2 c}{\partial y^2} - Kc \tag{14-1}$$

where $c$ is concentration, $u$ is velocity, $\theta$ is time, $y$ is a spatial coordinate, $D_{AB}$ is a species diffusion coefficient, and $K$ is a reaction rate constant. It is surprising how general this equation may become if we allow additional space dimensions, other dependent variables (temperature, velocity), other (non-Cartesian) coordinate systems, and various nonlinearities.

We will concentrate here on the finite difference method and (especially) the so-called *method of lines*, which the authors believe is a particularly broad and useful tool for these problems. Several different specific cases of the above equation will be considered. We first consider an introduction to the finite difference method for the transient heat (or diffusion) equation by means of an explicit, and then an implicit, method. This is followed by a consideration of stability analysis. The finite difference method for the Laplace equation is then studied. Following this introduction, we will concentrate mainly on the method of lines for marching problems.

## 14.1 THE FINITE DIFFERENCE METHOD

The finite difference method for partial differential equations proceeds along the same lines as in the case of ordinary differential equations. That is, in the equation(s) of interest, all derivatives are replaced by finite difference (numerical differentiation) approximations (see Appendix G on numerical differentiation for various useful formulas). In addition, wherever derivatives appear in the boundary conditions, they too are replaced by finite difference approximations. Then, finally, the difference approximations are solved by algebraic methods. Of course, we have more independent variables in the case of PDEs than in ODEs, so this poses the obvious computational difficulty of larger systems of equations. In addition, another serious but more subtle potential difficulty lurks in the background, namely the phenomenon of *instability*. This problem, which is that of serious error propagation, is associated with so-called *marching* problems, where at least one independent variable is open-ended, or without a prescribed limit. In time-dependent problems this is the time variable, but the phenomenon can also occur in steady-state *evolutionary* problems such as diffusion or wave propagation. We will therefore devote consideration to the stability analysis which is important in such problems. For purposes of convenient terminology we will distinguish between "marching" problems, where stability may be important, and "steady-state" problems, where stability is not a serious issue.

### 14.1.1 Remarks on Dimensional vs. Nondimensional Calculations

Depending on the particular problem to be solved, for purposes of numerical solution it is often desirable to nondimensionalize (at least partially) the equation and boundary conditions. This is because it is generally very desirable to reduce the number of parameters in the problem and thereby to reduce the required computation. This is usually accomplished by defining various new independent and dependent variables by means of simple linear translation/scaling transformations. This will be illustrated in certain example problems, particularly where comparisons between solution methods are to be made. However, it is important to realize that, in contrast to a technique like separation of variables, when using numerical methods it is not *necessary* to set up homogeneous boundary conditions or dimensionless variables.

## 14.2 THE FINITE DIFFERENCE METHOD FOR MARCHING PROBLEMS

As mentioned above, in the case of *marching* problems the phenomenon of instability is a serious consideration, and we must try to avoid it. We will be concerned with the transient heat conduction (or diffusion) equation, and there are a number of ways that difference approximations can be constructed. These will generally differ in terms of their accuracy and stability behavior. Some procedures are known as *explicit* methods, where each unknown can be updated to the new time by a simple, independent calculation. These explicit methods have the advantage of computational ease, but they normally involve some kind of stability restriction. Other procedures known as *implicit* methods require that all unknowns be updated simultaneously through the solution of coupled algebraic equations. Although not all implicit methods have good stability properties, many of them have no stability restriction at all; that is, the good implicit methods are *unconditionally stable*. Thus, the good implicit methods require considerably more computation, but are very important when stability is of primary concern. The tradeoffs between explicit and implicit methods depend on various circumstances, such as the particular problem, accuracy requirements, user experience, computing facilities, and so on. We will discuss both types of methods and try to illustrate the tradeoffs.

### 14.2.1 The Simple Explicit Method for Transient Heat Conduction

We consider the simplest version of the transient heat conduction equation in the following dimensional form:

$$\frac{\partial t}{\partial \theta} = \alpha \frac{\partial^2 t}{\partial y^2} \tag{14-2}$$

Here, $t$ is temperature, $\theta$ is time, $y$ is the vertical space dimension, and $\alpha$ is the thermal diffusivity of the material. We must keep in mind that $t$ and its derivatives each depend on both $\theta$ and $y$. This then requires that we set up a two-dimensional gridwork of "node" points for both $\theta$ and $y$. If we select the $\theta$ increment to be $\Delta\theta$ and the $y$ increment to be $h_y$, then the simplest difference expressions for $\partial t/\partial\theta$ and $\partial^2 t/\partial y^2$ are

$$\frac{\partial t}{\partial \theta} = \frac{t(\theta + \Delta\theta, y) - t(\theta, y)}{\Delta\theta} + O(\Delta\theta) \tag{14-3}$$

and

$$\frac{\partial^2 t}{\partial y^2} = \frac{t(\theta, y + h_y) - 2t(\theta, y) + t(\theta, y - h_y)}{h_y^2} + O(h_y^2) \tag{14-4}$$

These derivative approximations are merely the usual forward and central approximations, respectively, for $\partial t/\partial\theta$ and $\partial^2 t/\partial y^2$. When these expressions are substituted into $\partial t/\partial\theta = \alpha\partial^2 t/\partial y^2$ and the error terms neglected, we get, after a slight rearrangement,

$$t(\theta + \Delta\theta, y) = rt(\theta, y + h_y) + (1 - 2r)t(\theta, y) + rt(\theta, y - h_y) \tag{14-5}$$

where $r = \alpha\Delta\theta/h_y^2$. The difference solution based on these particular approximations is known as the *simple explicit method* for the heat conduction equation. The quantity $r$ involves both increment sizes $\Delta\theta$ and $h_y$, so it would *appear* that it could be prescribed arbitrarily, so long as $\Delta\theta$ and $h_y$ are both sufficiently small. The choice $r = 1/2$ in Equation (14-5) is suggested, since this simplifies the equation to

$$t(\theta + \Delta\theta, y) = \frac{t(\theta, y + h_y) + t(\theta, y - h_y)}{2} \qquad \left(r = \tfrac{1}{2}\right) \tag{14-6}$$

Equation (14-6) shows that the temperature at any point $y$ at the new time $(\theta + \Delta\theta)$ is calculated to be just the arithmetic average of the temperatures at the two neighboring points at the old time. Thus, if initial values of the temperatures at the node points along with values of the temperatures on the boundary are prescribed, Equation (14-6) can be used to compute the evolution in time of these node temperatures.

**Example 14.1 Solution of the Suddenly Heated Wall Problem by the Explicit Method**

We consider the problem of one-dimensional, transient heat conduction in a wall, which is described by the partial differential equation $\partial t/\partial\theta = \alpha\partial^2 t/\partial y^2$. The wall has thickness $L$ in the $y$ direction and is very long in the $x$ and $z$ directions, so the gradients in the $x$ and $z$ directions can be neglected. The wall is initially at a uniform temperature of $t_i$. At time zero, the lower surface is suddenly changed to and kept at temperature $t_w$, while the upper surface is insulated, so that $\partial t/\partial y = 0$ at that location. We wish to see how the sudden heating at the lower surface penetrates into the wall as a function of time.

To obtain results of some generality, and to illustrate the technique of nondimensionalization, we introduce new variables in the following way. We let $T = (t - t_w)/(t_i - t_w)$, $\phi = \alpha\theta/L^2$ and $Y = y/L$. Then $Y$ goes from 0 to 1, and initially $T = 1$ everywhere, while it suddenly changes to $T = 0$ at the lower surface at time zero, and $\partial T/\partial Y = 0$ at $Y = 1$. We need to compute the derivatives in Equation (14-2) in terms of the new variables. We find that

$$\begin{aligned} \frac{\partial t}{\partial\theta} &= (t_i - t_w)\frac{\partial T}{\partial\theta} = \frac{(t_i - t_w)\alpha}{L^2}\frac{\partial T}{\partial\phi} \\ \frac{\partial^2 t}{\partial y^2} &= (t_i - t_w)\frac{\partial^2 T}{\partial y^2} = \frac{(t_i - t_w)}{L^2}\frac{\partial^2 T}{\partial Y^2} \end{aligned} \tag{a}$$

When the various derivative relationships are substituted into Equation (14-2), we get

$$\frac{\partial T}{\partial\phi} = \frac{\partial^2 T}{\partial Y^2} \qquad T(\phi, Y = 0) = 0; \qquad \frac{\partial T}{\partial Y}(\phi, Y = 1) = 0; \qquad T(\phi = 0, Y) = 1 \tag{b}$$

The situation is depicted in Figure 14.1, where three equally spaced interior node points are shown (nodes 2, 3, and 4), corresponding to $h_y = 1/4$. A considerable advantage of using the new coordinates is that the resulting mathematical problem is independent of $t_w$, $t_i$, and $\alpha$. Thus, the numerical results will be useful for any values of these parameters; if we had not done this, we would need a new set of numerical results for each desired combination of these parameters.

To apply the finite-difference formula (14-6) at the interior points, we need to set up a gridwork of node locations in the wall, starting at the lower surface and extending upward in equal increments to the upper surface (see Figure 14.1). For this calculation,

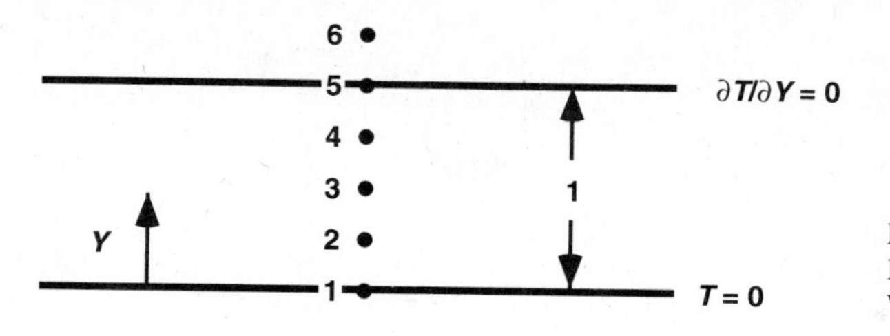


**Figure 14.1.** Node Locations for Finite Difference Solution of Suddenly Heated Wall Problem

we divide the wall into four equal increments. Initially, all temperatures correspond to $T = 1$. We then say, a bit loosely, that $T(Y = 0, \phi = 0) = 0$. The condition just stated corresponds to an infinitely fast "jump" in temperature at $Y = 0$ from $T = 1$ just before $\phi = 0$, to $T = 0$ just after $\phi = 0$. Such an idealization is very convenient, but it causes some computational difficulties. It is certainly necessary to assign some value (other than 1) to $T$ at $Y = 0, \phi = 0$. A natural choice seems to be to set $T(Y = 0, \phi = 0) = 0$. Implementing this choice in Equation (14-6) results in the undesirable behavior that temperatures at a given node point change only every *two* time steps. It has been found in discontinuous problems of this type that is it useful to prescribe the *average* of the initial and new boundary values as the temperature at the boundary at time zero. We therefore use the condition $T = 0.5$ at $Y = 0$ when $\phi = 0$.

Because the boundary condition at the upper surface $Y = 1$ involves a derivative, we must use an appropriate numerical differentiation approximation. Since the spatial discretization error is $O(h_y^2)$ in the equation, we also want an error no larger than this at the boundary. As we found in our treatment of ODE boundary value problems using finite differences, there are two natural ways to do this. We could use a three-point backward approximation, or we could introduce a fictitious node $i = 6$ at $Y = 1.25$. The latter choice is convenient for our purpose. Thus, for this calculation, $t_5$ at the new time is determined as the average of $t_4$ and $t_6$ at the old time. One final point: since we are using the simple explicit method with $r = 1/2 (r = \Delta\phi/h_y^2)$ and we have chosen $h_y = 1/4$, our dimensionless time step will be $\Delta\phi = rh_y^2 = (1/2)(1/4)^2 = 1/32$.

We are now ready to perform the computations shown in Table 14.1. At time zero, all values of $T$ are equal to one except at $Y = 0$, where (as mentioned above) we use the average of the old and new temperatures (0.5) at the "discontinuity." Then, at the end of the first time step, we calculate each interior temperature (including node $i = 5$ at $Y = 1$, which is an "interior" point in the fictitious boundary formulation) by taking the arithmetic average of its adjacent temperatures at the previous time. The fictitious node temperature ($i = 6$) is always equal to $T_4$ at any instant of time, in accordance with the central difference approximation to the boundary condition $\partial T/\partial Y = 0$ at $Y = 1$. Thus, we always determine $T_6$ last in any time step.

Detailed results of the calculations are shown in Table 14.1. We have carried out the solution to seven time steps for purposes of comparison. By using quadratic interpolation with the results for time steps 5, 6, and 7, we have estimated the approximate solution at $\phi = 0.2$ (6.4 time steps). This is compared in the table to the exact solution. Considering the large spatial increment (1/4), the results appear to be surprisingly good.

The finite-difference solution shows that the effect of heating at the boundary propagates upward into the wall. If the wall were of infinite extent, this propagation behavior would continue forever. The effect of a finite thickness is reflected in the prescribed boundary condition at the upper surface, namely $\partial T/\partial Y = 0$. Even for

**TABLE 14.1** HEAT CONDUCTION INTO A SUDDENLY HEATED WALL BY THE SIMPLE EXPLICIT METHOD $T(\phi + \Delta\phi, Y) = (1/2)[T(\phi, Y + h_Y) + T(\phi, Y - h_Y)]$ EFFECT OF "AVERAGING" INITIAL TEMPERATURE $\Delta\phi = 1/32$

| Time | Time Step # | $Y = 0$ $i = 1$ | $Y = 0.25$ $i = 2$ | $Y = 0.5$ $i = 3$ | $Y = 0.75$ $i = 4$ | $Y = 1.0$ $i = 5$ | $Y = 1.25$ $i = 6*$ |
|---|---|---|---|---|---|---|---|
| 0 | 0 | 0.5 | 1.0000 | 1.0000 | 1.0000 | 1.0000 | 1.0000 |
| 1/32 | 1 | 0.0 | 0.7500 | 1.0000 | 1.0000 | 1.0000 | 1.0000 |
| 2/32 | 2 | 0.0 | 0.5000 | 0.8750 | 1.0000 | 1.0000 | 1.0000 |
| 3/32 | 3 | 0.0 | 0.4375 | 0.7500 | 0.9375 | 1.0000 | 0.9375 |
| 4/32 | 4 | 0.0 | 0.375 | 0.6875 | 0.8750 | 0.9375 | 0.8750 |
| 5/32 | 5 | 0.0 | 0.3438 | 0.6250 | 0.8125 | 0.8750 | 0.8125 |
| 6/32 | 6 | 0.0 | 0.3125 | 0.5781 | 0.7500 | 0.8125 | 0.7500 |
| 7/32 | 7 | 0.0 | 0.2891 | 0.5313 | 0.6953 | 0.7500 | 0.6953 |
| $0.2_{inter}$ | $6.4_{inter}$ | 0.0 | 0.3022 | 0.5594 | 0.7272 | 0.7875 | |
| $0.2_{exact}$ | $6.4_{exact}$ | 0.0 | 0.3021 | 0.5532 | 0.7162 | 0.7723 | |

*Fictitious point

a finite wall, however, the infinite wall solution is appropriate *at small times*, before the upper wall has felt the influence of the temperature change at the lower surface. This will be demonstrated in a later section where we utilize the method of lines for this problem.

### 14.2.2 Elementary Stability Analysis of the Simple Explicit Method

Although it is not at all obvious from the previous discussion, it turns out, surprisingly, that success with the simple explicit method outlined above depends in a most important way on the *value of* r selected (regardless of how small the individual $\Delta\phi$ and $h_Y$ may be!). For the problem under consideration, it can be shown theoretically that the simple explicit finite-difference method is useful only for $r \leq 1/2$. For any $r > 1/2$, the unstable character of the method ultimately causes huge magnifications of any errors that occur. Table 14.2 demonstrates the result of using Equation (14-5) with $r = 1$ to solve the previous problem. The calculation clearly produces disastrous growing oscillations, since we know that the true solution should be between zero and one.

The fact that such computational disasters can occur in seemingly harmless problems illustrates that the finite difference method requires caution and careful application. This is especially true for initial/boundary value problems of the "diffusion" or "wave" type. To get some idea of the stability properties of a particular finite-difference scheme (and there are many of these), it is very useful to do an "experimental" stability analysis. This is accomplished in a linear homogeneous problem (which therefore has a trivial zero solution) by using the finite-difference scheme under consideration to see how it propagates an error that is introduced for one time

**TABLE 14.2** HEAT CONDUCTION INTO A SUDDENLY HEATED WALL *UNSTABLE* RESULTS FOR SIMPLE EXPLICIT METHOD WITH
$r = 1, T(\phi + \Delta\phi, Y) = T(\phi, Y + h_Y) - T(\phi, Y) + T(\phi, Y - h_Y) \quad \Delta\phi = 1/16$

| Time Step # | $Y = 0$ $i = 1$ | $Y = 0.25$ $i = 2$ | $Y = 0.5$ $i = 3$ | $Y = 0.75$ $i = 4$ | $Y = 1.0$ $i = 5$ | $Y = 1.25$ $i = 6*$ |
|---|---|---|---|---|---|---|
| 0 | 0.5 | 1 | 1 | 1 | 1 | 1 |
| 1 | 0 | 0.5 | 1 | 1 | 1 | 1 |
| 2 | 0 | 0.5 | 0.5 | 1 | 1 | 1 |
| 3 | 0 | 0 | 1 | 0.5 | 1 | 0.5 |
| 4 | 0 | 1 | -0.5 | 1.5 | 0 | 1.5 |
| 5 | 0 | -1.5 | 3 | -2 | 3 | -2 |
| 6 | 0 | 4.5 | -6.5 | 8 | -7 | 8 |

*Fictitious point

**TABLE 14.3** HEAT CONDUCTION INTO A SUDDENLY HEATED WALL SIMPLE EXPLICIT METHOD-EXPERIMENTAL STABILITY:
$T(\phi + \Delta\phi, Y) = rT(\phi, Y + h_Y) + (1 - 2r)T(\phi, Y) + T(\phi, Y - h_Y)$

| Time Step # | $Y = 0$ $i = 1$ | $Y = 0.25$ $i = 2$ | $Y = 0.5$ $i = 3$ | $Y = 0.75$ $i = 4$ | $Y = 1.0$ $i = 5$ | $Y = 1.25$ $i = 6*$ |
|---|---|---|---|---|---|---|
| | | | $r = 1$ Unstable | | | |
| 0 | 0 | $\epsilon$ | 0 | 0 | 0 | 0 |
| 1 | 0 | $-\epsilon$ | $\epsilon$ | 0 | 0 | 0 |
| 2 | 0 | $2\epsilon$ | $-2\epsilon$ | $\epsilon$ | 0 | $\epsilon$ |
| 3 | 0 | $-4\epsilon$ | $5\epsilon$ | $-3\epsilon$ | $2\epsilon$ | $-3\epsilon$ |
| 4 | 0 | $9\epsilon$ | $-12\epsilon$ | $10\epsilon$ | $-8\epsilon$ | $10\epsilon$ |
| 5 | 0 | $-21\epsilon$ | $31\epsilon$ | $-30\epsilon$ | $28\epsilon$ | $-30\epsilon$ |
| 6 | 0 | $52\epsilon$ | $-82\epsilon$ | $89\epsilon$ | $-88\epsilon$ | $89\epsilon$ |
| | | | $r = 1/2$ Stable | | | |
| 0 | 0 | $\epsilon$ | 0 | 0 | 0 | 0 |
| 1 | 0 | 0 | $\epsilon/2$ | 0 | 0 | 0 |
| 2 | 0 | $\epsilon/4$ | 0 | $\epsilon/4$ | 0 | $\epsilon/4$ |
| 3 | 0 | 0 | $\epsilon/4$ | 0 | $\epsilon/4$ | 0 |
| 4 | 0 | $\epsilon/8$ | 0 | $\epsilon/4$ | 0 | $\epsilon/4$ |
| 5 | 0 | 0 | $3\epsilon/16$ | 0 | $\epsilon/4$ | 0 |
| 6 | 0 | $3\epsilon/32$ | 0 | $7\epsilon/32$ | 0 | $7\epsilon/32$ |

*Fictitious point

only, at some particular location. For example, the results of such a calculation are shown in Table 14.3 for Equation (14-5) with $r = 1/2$ and $r = 1$. An initial error of magnitude $\epsilon$ is introduced at $Y = 0.25$. The simple explicit method is seen to damp an error for $r = 1/2$ but magnify it rapidly for $r = 1$.

### 14.2.3 An Implicit Method for Transient Heat Conduction

We have seen that the simple explicit method for the heat equation is subject to a stability limitation, but the example we showed did not indicate how severe such a limitation can be. In cases where it is necessary to use a *small* spatial increment, or in multidimensional problems, or in the case of a rapid homogeneous chemical reaction, the stability limitations of common explicit methods may preclude their use. In principle, the explicit methods could ultimately achieve desired results, but the marching increment may be required to be so small (for stability) that it might take practically "forever" to complete the task. In addition, when many steps are required, accumulated roundoff error may become a problem.

To overcome stability problems of the above type, it frequently turns out to be useful to employ a so-called implicit finite-difference method. This kind of procedure requires that all node temperatures at a new time be computed simultaneously; this in turn involves solving a set of algebraic equations for the node temperatures at every time step. Obviously, this process requires much more computing effort for each time step; but it is often, nevertheless, a very practical method in cases of severe stability limitations. The utility of the implicit methods is greatly enhanced when it is possible to reduce the algebraic system to *tridiagonal* form. Our consideration of implicit methods will focus primarily on the procedure whereby the marching (time) variable is expressed by a backward numerical differentiation approximation. The procedure, as applied to PDE, is known by various names, such as the fully implicit method or the backward Euler method (especially in the method of lines, to be discussed later). We illustrate the method with an example. We will also briefly mention another popular implicit method known as Crank-Nicholson or the implicit trapezoidal rule.

**Example 14.2 Solution of the Suddenly Heated Wall Problem by an Implicit Method**

We now consider solution of the same suddenly heated wall problem of Example 14.1 by the implicit backward Euler method. We saw in our discussion of stiff ODE methods that backward Euler is very stable for problems which involve rapid decay, and we shall see in our discussion of stability of PDE methods for the diffusion equation that it is unconditionally stable for these kind of problems. For our problem $\partial T/\partial \phi = \partial^2 T/\partial Y^2$, we implement the backward Euler implicit method by using a first-order backward approximation for $\partial T/\partial \phi$, along with the standard central approximation for $\partial^2 T/\partial Y^2$, evaluated at the *new* time. This gives the backward Euler equations as

Backward Euler

$$\frac{T_i^{n+1} - T_i^n}{\Delta\phi} = \left(\frac{1}{h_y^2}\right)[T_{i+1}^{n+1} - 2T_i^{n+1} + T_{i-1}^{n+1}] \qquad (i = 2, \ldots, m) \tag{a}$$

with boundary conditions

$$T_1^{n+1} = 0, \qquad \frac{T_{m+1}^{n+1} - T_{m+1}^n}{\Delta\phi} = \left(\frac{1}{h_y^2}\right)[2T_m^{n+1} - 2T_{m+1}^{n+1}] \tag{b}$$

Here, $\Delta\phi$ is the time increment in $\phi$, and $h_y$ is the spatial increment ($h_y = 1/m$). The interval is divided into $m$ parts. The superscripts indicate time and the subscripts

**TABLE 14.4** IMPLICIT SOLUTION OF SUDDENLY HEATED WALL (FICTITIOUS POINT BOUNDARY CONDITION): $\phi = 0.2$

| $y$ | $T(m = 4)$ | $T(m = 8)$ | $T(m = 16)$ | $T$exact |
|---|---|---|---|---|
| | | $\Delta\phi = 0.001$ | | |
| 1.00 | 0.765157 | 0.770741 | 0.772091 | 0.772312 |
| 0.75 | 0.710924 | 0.715190 | 0.716226 | 0.716227 |
| 0.50 | 0.551586 | 0.553152 | 0.553551 | 0.553176 |
| 0.25 | 0.302603 | 0.302491 | 0.302492 | 0.302084 |
| 0.00 | 0.000000 | 0.000000 | 0.000000 | 0.000000 |
| | | $\Delta\phi = 0.0005$ | | |
| 1.00 | 0.765067 | 0.770637 | 0.771981 | 0.772312 |
| 0.75 | 0.710768 | 0.715024 | 0.716057 | 0.716227 |
| 0.50 | 0.551326 | 0.552894 | 0.553296 | 0.553176 |
| 0.25 | 0.302382 | 0.302281 | 0.302285 | 0.302084 |
| 0.00 | 0.000000 | 0.000000 | 0.000000 | 0.000000 |

spatial locations. The boundary conditions in Equations (b) correspond to $T = 0$ at $Y = 0$, and $\partial T/\partial Y = 0$ at $Y = 1$. Here we have implemented the $Y = 1$ boundary condition by using the fictitious point method, which involves using Equation (a) along with $T_{m+2}^{n+1} = T_m^{n+1}$ in accordance with the central difference approximation to $\partial T/\partial Y$ at $Y = 1$. Hence, we have the indicated difference equation for node $(m + 1)$ at $Y = 1$. The subscripts refer to the node point and the superscripts to the time. The superscript $n$ is the current time when the solution is known, and $n + 1$ is the new time where we wish to calculate the solution. Although Equations (a) are implicit, and therefore require simultaneous algebraic solution for all of the unknowns, they possess the *tridiagonal* structure, which is of great practical importance. The general equation in a tridiagonal system of order $m + 1$ (see Chapter 2) is of the form $a_i x_{i-1} + b_i x_i + c_i x_{i+1} = d_i$ with $i = 1, \ldots, m + 1$, and $a_1 = c_{m+1} = 0$. To put Equations (a) and (b) into this form, we rearrange them as follows:

$$
\begin{aligned}
&T_1^{n+1} = 0 \\
&rT_{i-1}^{n+1} - (2r + 1)T_i^{n+1} + rT_{i+1}^{n+1} = -T_i^n, \quad (i = 2, \ldots, m) \qquad \text{(c)} \\
&2rT_m^{n+1} - (2r + 1)T_{m+1}^{n+1} = -T_{m+i}^n
\end{aligned}
$$

Here, $r = \Delta\phi/h_y^2$. Thus by observing Equations (c), we can pick out the appropriate tridiagonal coefficients. These are shown in the concise listing of program PDEFD (partial differential equations finite difference) at the end of the chapter, provided that we take $k = 0$. Because tridiagonal equations can be solved so rapidly, it is quite practical to use the algorithm of Equation (a), since it does not have a stability requirement. The explicit methods are limited by stability so that if $h_y$ is small enough, the required time step is so small that the calculation may not be feasible.

We now show the results of using the backward Euler method for the suddenly heated wall problem at time $\phi = 0.2$. These were obtained by using program PDEFD, which is discussed in detail at the end of the chapter.

The results in Table 14.4 are interesting and revealing. For many practical purposes they are quite acceptable. When solving PDE problems, it is common for three-figure accuracy to be satisfactory. However, since we wish to perform some Richardson

**TABLE 14.5** TIME EXTRAPOLATION OF RESULTS OF TABLE 14.4 FOR $\Delta\phi = 0.001, 0.0005$, $T_{\text{EXT}} = T(0.0005) + [T(0.0005) - T(0.001)]$

| $y$ | $T_{\text{ext}}(m = 4)$ | $T_{\text{ext}}(m = 8)$ | $T_{\text{ext}}(m = 16)$ | $T_{\text{exact}}$ |
|---|---|---|---|---|
| 1.0 | 0.764977 | 0.770533 | 0.771871 | 0.772312 |
| 0.75 | 0.710612 | 0.714858 | 0.715888 | 0.716227 |
| 0.5 | 0.551066 | 0.552636 | 0.553041 | 0.553176 |
| 0.25 | 0.302161 | 0.302071 | 0.302078 | 0.302084 |
| 0.0 | 0.0 | 0.0 | 0.0 | 0.0 |

extrapolations, we give all six figures from our calculations. We have shown calculations for two different time increments and three different node spacings. By studying the table, going either down or sideways, and comparing with the exact separation of variables solution, we see that the results contain *both* significant time and space errors. This is in contrast to results we shall give later for this problem using the method of lines and program PDEHYBD (explicit method), where extrapolation, local error control, and global error monitoring in *time* allowed removal of time error (time errors are all $\leq 10^{-6}$). The time errors here are quite significant because the backward Euler method is only first-order accurate in time. Although we have not shown it here, the backward Euler method produces good results for both small and large times.

It is useful to use Richardson extrapolation in time (for fixed $m$) to see if this process significantly reduces time errors. For any fixed $m$, the first-order extrapolation formula is $S = A(\Delta t/2) + [A(\Delta t/2) - A(\Delta t)]$. As usual in our extrapolation formula, we deliberately choose to show the error estimate explicitly, and hence do not algebraically simplify. When we apply the extrapolation formula to the results of Table 14.4, we obtain the values shown in Table 14.5.

The estimated *time* errors in Table 14.5 are all about $2 \times 10^{-4}$ and are conservative. It is very interesting to note that the time-extrapolated values here are extremely close to those of the method of lines program PDEHYBD (given later) for the same problem (see Table 14.7). The method of lines results were obtained so that they were virtually devoid of time errors. Since the results of Tables 14.5 and 14.7 are so close, if they are extrapolated in *space* they will produce virtually the same results as the spatial extrapolation of the method of lines results in Table 14.7, which are quite accurate. The essential point is that for the problem considered, extrapolation in both time and space is very useful whether we are using explicit or implicit methods.

### 14.2.4 The Implicit Trapezoidal Rule (Crank-Nicholson)

As we have seen, the backward Euler method is not very accurate in its time integration (first-order), although it can be improved considerably by one extrapolation (to second-order). Therefore, in many applications the implicit trapezoidal rule (also known as the Crank-Nicholson method) is used instead. The implicit trapezoidal rule is second-order accurate in time and is unconditionally stable for the diffusion equation, but barely so. If the time step size is too large, some undesirable oscillations may be present.

The implicit trapezoidal rule for Equation (14-8) can be derived in the following way. We consider the following problem, which is repeated from Example 14.1:

$$\frac{\partial T}{\partial \phi} = \frac{\partial^2 T}{\partial Y^2} \qquad T(\phi, Y = 0) = 0; \qquad \frac{\partial T}{\partial Y}(\phi, Y = 1) = 0; \qquad T(\phi = 0, Y) = 1$$

If we difference the spatial derivative $T_{YY}$, while keeping the time derivative continuous, we get the so-called method of lines equations for this problem. These equations are simply $dT_i/d\phi = [T_{i+1} - 2T_i + T_{i-1}]/h_Y^2 = F_i/h_Y^2$ at node $i$. The implicit trapezoidal rule algorithm is then just $T_i(\phi + \Delta\phi) = T_i(\phi) + (\Delta\phi/2h_Y^2)[F_i(\phi + \Delta\phi) + F_i(\phi)]$ (trapezoidal quadrature applied to $dT_i/d\phi = F_i$). When we put in the detailed expressions for $F_i$, this becomes

Implicit Trapezoidal Rule

$$T_i^{n+1} - T_i^n = \frac{r}{2}\left\{[T_{i+1}^{n+1} - 2T_i^{n+1} + T_{i-1}^{n+1}] + [T_{i+1}^n - 2T_i^n + T_{i-1}^n]\right\} \quad (i = 2, \ldots, m) \tag{14-7}$$

with boundary conditions

$$\begin{aligned} T_1^{n+1} &= 0 \\ T_{m+1}^{n+1} - T_{m+1}^n &= \frac{r}{2}\left\{[2T_m^{n+1} - 2T_{m+1}^{n+1}] + [2T_m^n - 2T_{m+1}^n]\right\} \end{aligned} \tag{14-8}$$

Here, $r = \Delta\phi/h_Y^2$ as before, and the boundary equation at node $(m + 1)$ corresponds to using a fictitious point with the condition $\partial T/\partial Y = 0$. These equations also have a tridiagonal structure and can therefore be used directly in program PDEFD. They will yield results very similar to those of backward Euler with one time extrapolation, since both methods are second-order in time and space. The advantage of the trapezoidal rule over backward Euler is one order of accuracy in time. The advantage of backward Euler is that under severe stability requirements (such as chemical reactions with large rate constants), its behavior is generally better.

Other simple but useful implicit alternatives include an algorithm due to Stone and Brian [1963] and the implicit improved Euler method of Ashour and Hanna [1990], which has some of the best features of both backward Euler and trapezoidal rule.

It is appropriate to conclude with some discussion of the relative advantages of explicit versus implicit (unconditionally stable) methods. The *only* reason that implicit methods are ever used is because explicit methods generally have substantial stability limitations. The implicit methods can take relatively large time steps, so *if they have the efficient tridiagonal structure*, they can utilize many node points to achieve high spatial accuracy in a reasonable time. On the other hand, explicit methods have the great advantages of flexibility and ease of problem formulation. Explicit solvers such as PDEHYBD have built-in extrapolation and error control for the time (continuous) variable. They can easily be set up for problems involving (i) two PDEs, (ii) PDEs with two or three space variables, or (iii) a combination of PDEs, ODEs, and nonlinear algebraic equations.

## 14.3 STABILITY ANALYSIS FOR MARCHING PROBLEMS

This topic, as in the case of ordinary differential equations, refers to the error propagation characteristics of a method. The situation is in general considerably more complicated for PDEs than for ODEs. We can only give an introduction here, but for further information we recommend Carnahan *et al.* [1969], Ferziger [1981], Anderson *et al.* [1984] and Roache [1976]. The problem of stability manifests itself primarily in transient or evolutionary (marching) problems of the diffusion or wave type. For steady-state problems that have no "open-ended" variables, stability difficulties usually do not occur. Stability requirements generally become more stringent for problems (i) with increasing number of space dimensions, (ii) involving chemical reactions, (iii) with convection where the velocity becomes small, and (iv) where it is required to take the spatial discretization interval(s) very small. Stability generally depends on both the problem and the method of solution.

There are several approaches that can be taken to stability analysis. We have already discussed the idea of *experimental* stability analysis. Early in the chapter we illustrated this by both direct numerical experiments and through an analytical approach (in terms of $\epsilon$'s). This experimental method is often useful when employing the method of lines as a solution technique. Another approach is based on developing a simple *sufficient* condition for stability based on inequalities and a dominating difference equation; this is relatively easy and often (but not always) produces useful results. The most common stability analysis is the so-called von Neumann method, in which one considers a difference equation having a typical Fourier solution component of the form $T(\theta, y) = \psi(\theta)e^{i\beta y}$. Here $i$ is the imaginary unit $(-1)^{1/2}$, and $e^{i\beta y} = \cos \beta y + i \sin \beta y$. In the von Neumann method, we substitute the assumed solution component into the difference equation and compute the "amplification factor" $|\psi(\theta + \Delta\theta)/\psi(\theta)|$, which must be $\leq 1$ (for any $\beta$) in order to have stability. Finally, there is a more complicated technique, known as the matrix method, which also takes into account the boundary conditions; this method is discussed in Ames [1977]. We consider now some examples of the sufficient condition and von Neumann methods of stability analysis.

**Example 14.3 An Illustration of Stability Calculations**

**(i) The Equation $\partial c/\partial\theta = D_{AB}\partial^2 c/\partial y^2 - Kc$.** We consider first the important one-dimensional diffusion equation with linear chemical reaction solved by the simple explicit method. The simple explicit method corresponds to using the method of lines and integrating with the forward Euler procedure. We shall first illustrate the sufficient condition analysis for this situation. Using a subscript notation for the spatial nodes (and employing dimensional variables), the difference equations are

$$\frac{c_i(\theta + \Delta\theta) - c_i(\theta)}{\Delta\theta} = \frac{D_{AB}}{h_y^2}[c_{i+1}(\theta) - 2c_i(\theta) + c_{i-1}(\theta)] - Kc_i(\theta) \tag{a}$$

We now rearrange Equation (a) to put $c_i(\theta + \Delta\theta)$ on the left and group the $c_j(\theta)$ terms on the right; this gives

$$c_i(\theta + \Delta\theta) = \lambda c_{i+1}(\theta) + c_i(\theta)(1 - 2\lambda - K\Delta\theta) + \lambda c_{i-1}(\theta) \tag{b}$$

where $\lambda = (D_{AB}\Delta\theta/h_y^2)$. The idea now is to let $M(\theta)$ represent the maximum magnitude of all node values of $c_i$ at time $\theta$. Then by taking absolute values in Equation (b), *provided that* $(1 - 2\lambda - K\Delta\theta) \geq 0$, we get that $M(\theta + \Delta\theta) \leq M(\theta)[\lambda + (1 - 2\lambda - K\Delta\theta) + \lambda] \leq M(\theta)(1 - K\Delta\theta) \leq 1$. Thus, this method is shown to be stable provided that $(1 - 2\lambda - K\Delta\theta) \geq 0$; this inequality can be arranged to the convenient form

$$\frac{D_{AB}\Delta\theta}{h_y^2} = \lambda \leq \frac{1/2}{1 + \dfrac{Kh_y^2}{2D_{AB}}} \tag{c}$$

Equation (c) shows several interesting things. First, for $K = 0$ (no reaction), we see that this result is the same that we found experimentally at the beginning of the chapter. For smaller $h_y$ (more node points), the time step is rapidly required to decrease. For a given $h_y$, the larger the reaction constant $K$, the smaller the time step must be. We must keep in mind that the *sufficient condition* approach we have just demonstrated is based on inequalities; it therefore may yield correct results that are more conservative than necessary. For example, if it is applied to the implicit trapezoidal rule integration of this problem, it gives a *conditional* stability result; the method is actually unconditionally stable. But it also happens that for the problem we have just considered, if $K = 0$, the sufficient condition approach and the more precise von Neumann method give exactly the same result.

Before leaving the sufficient condition method, let us consider its application to the present PDE when the backward Euler method is used for time integration. As we know from our study of ODE methods, for backward Euler the right side of Equation (a) becomes evaluated at the *new* time $(\theta + \Delta\theta)$ rather than the old time $\theta$. In this case, Equation (b) becomes

$$c_i(\theta + \Delta\theta)[1 + 2\lambda + K\Delta\theta] = c_i(\theta) + \lambda[c_{i+1}(\theta + \Delta\theta) + c_{i-1}(\theta + \Delta\theta)] \tag{d}$$

Again taking $M(\theta)$ to be the maximum magnitude of the node values at time $\theta$, if we take absolute values in Equation (d) and use the triangle inequality, we get $M(\theta+\Delta\theta)(1 + 2\lambda + K\Delta\theta) \leq M(\theta) + 2\lambda M(\theta + \Delta\theta)$. Rearrangement of this shows that $M(\theta+\Delta\theta) \leq M(\theta)$ for *any* values of $\Delta\theta$ and $K$, and thus the method is *unconditionally* stable. This is the same result as given by the von Neumann method for this problem. The principal advantage of the sufficient condition method is its simplicity and ease of application. Its disadvantage is that it may yield conclusions that are far more conservative than reality.

We now look at the application of the von Neumann method to the same PDE as above, namely $\partial c/\partial\theta = D_{AB}\partial^2 c/\partial y^2 - Kc$. Consider use of the forward Euler method for time integration, which is the same as the simple explicit method. The difference equation is again Equation (a). We take $c(\theta, y) = \psi(\theta)e^{i\beta y}$. Substitution of this form into Equation (a), together with cancellation of $e^{i\beta y}$ and rearrangement, gives

$$\frac{\psi(\theta + \Delta\theta)}{\psi(\theta)} = (1 - K\,\theta) + \frac{D_{AB}\Delta\theta}{h_y^2}(e^{i\beta\Delta y} + e^{-i\beta\Delta y} - 2) \tag{e}$$

But since $e^{i\phi} = \cos\phi + i\sin\phi$, it is easy to show that $e^{i\beta\Delta y} + e^{-i\beta\Delta y} = 2\cos\beta\Delta y$. Thus, Equation (e) can be written as

$$\frac{\psi(\theta + \Delta\theta)}{\psi(\theta)} = (1 - K\Delta\theta) + \frac{2D_{AB}\Delta\theta}{h_y^2}(\cos\beta\Delta y - 1) \tag{f}$$

Since we are interested in the restrictions on $\Delta\theta$ required to guarantee that $|\psi(\theta + \Delta\theta)/\psi(\theta)| \leq 1$, we examine Equation (f) to see the worst case as a function of $\beta$ (since in the full Fourier solution, many values of $\beta$ would be present). Since $\cos\beta\Delta y$ is between $-1$ and $+1$ for any $\beta$, this worst case corresponds to $\cos\beta\Delta y = -1$, where $\psi(\theta + \Delta\theta)/\psi(\theta) = -1$. Thus, to guarantee stability we must have $[(1 - K\Delta\theta - 4(D_{AB}\Delta\theta/h_y^2)] \geq -1$. Solving this inequality by putting $(D_{AB}\Delta\theta/h_y^2)$ on the left yields the stability criterion

$$\frac{D_{AB}\Delta\theta}{h_y^2} \leq \frac{1/2}{1 + \dfrac{Kh_y^2}{4D_{AB}}} \tag{g}$$

The Fourier result for this problem, Equation (g), gives exactly the same requirement as the sufficient condition method, Equation (c), *provided that* K = 0 (no reaction). However, if $K > 0$, then the requirement of the Fourier method is slightly less stringent. This is in accordance with the idea that the Fourier method will generally give equivalent or better results than the sufficient condition method.

**(ii) The Equation $\partial c/\partial \theta + u\partial c/\partial y = -Kc$.** This equation, which is also clearly a special case of the convection-diffusion equation, is an important prototype for many problems. The quantities $u$ (velocity) and $K$ (first-order reaction rate constant) are assumed here to be constant and positive. The form for $K = 0$ (no reaction) is often called the convection or advection equation; it is also sometimes called the wave equation, since for $K = 0$ it can be transformed to the classical wave equation, $\partial^2 c/\partial \theta^2 = u^2 \partial^2 c/\partial y^2$. (This is shown by differentiating first with respect to $\theta$ and then with respect to $y$ and eliminating the cross-derivative $\partial^2 c/\partial y \partial \theta$.) The equation generally exhibits a wave-type behavior.

At first glance, it would appear that a straightforward approach to this problem would involve replacing $\partial c/\partial y$ with a central difference approximation, and $\partial c/\partial \theta$ with a forward approximation. However, a von Neumann stability analysis shows that *this procedure is unconditionally unstable.* This example illustrates what *can* happen in the development of PDE numerical solution methods; an apparently reasonable procedure sometimes is completely unstable. One point that can be made regarding differencing and stability is that *whenever a difference approximation is of higher order than the derivative it is replacing*, there is the *possibility* of unstable behavior, since extraneous solutions could thereby be introduced. In the case we have just considered, we have replaced a first derivative by a second-order difference approximation, so we must be cautious. However, we may now note that even if we use a first-order *forward* difference to approximate $\partial c/\partial y$, *the result is still unconditionally unstable.* This is even more surprising than the first result! Fortunately, if we replace $\partial c/\partial y$ with a first-order backward approximation, we do achieve conditional stability, and we illustrate the stability calculation for this successful case.

We now give the difference expression for our convection equation using a first-order backward approximation for $\partial c/\partial y$ and a first-order forward approximation for $\partial c/\partial \theta$. This produces

$$\frac{c_i(\theta + \Delta\theta) - c_i(\theta)}{\Delta\theta} + u\frac{c_i(\theta) - c_{i-1}(\theta)}{h_y} = -Kc_i(\theta) \tag{h}$$

We now introduce the Fourier representation $c(\theta, y) = \psi(\theta)e^{i\beta y}$. When this is substituted into Equation (h), the factor $e^{i\beta y}$ cancelled, and the equation rearranged, we get

$$\frac{\psi(\theta + \Delta\theta)}{\psi(\theta)} = (1 - K\Delta\theta) - \frac{u\Delta\theta}{h_y}(1 - e^{-i\beta\Delta y}) \tag{i}$$

From Equation (i), the factor $w = [1 - e^{-i\beta\Delta y}]$ varies between 0 and 2 (since $e^{-i\beta\Delta y}$ varies between $-1$ and $+1$), and we see that the worst case corresponds to $w = 2$ and we must require for stability that $[1 - K\Delta\theta - 2(u\Delta\theta/h_y)] \geq -1$ or

$$\frac{u\Delta\theta}{h_y} \leq \frac{1}{1 + \dfrac{Kh_y}{2u}} \tag{j}$$

Equation (j) shows the stability restriction on $\Delta\theta$ for this problem. When $K = 0$ (no reaction), the restriction increases as either $h_y$ becomes smaller or $u$ becomes larger. When $K > 0$, the restriction also increases. It is interesting to note that in this particular problem, the sufficient condition method produces the same result as Equation (j) for $K = 0$, and a slightly greater restriction when $K > 0$.

Finally, it is important to realize that the successful (backward) spatial differencing in this problem is an example of what is known as *upwind differencing* for convection terms (of the type $u\partial c/\partial y$) in an equation. In the simple problem considered here, we get stability for first-order forward time integration provided that we use a *backward* approximation to $\partial c/\partial y$ if $u$ is positive or a forward approximation if $u$ is negative. In other words, we difference the convection term to go upwind, or against the flow. This idea is important for convection terms in more complicated problems.

## 14.4 THE FINITE DIFFERENCE METHOD FOR STEADY-STATE PROBLEMS

The Laplace equation represents a prototype for steady-state diffusion processes. For instance, the two-dimensional Laplace equation, $\partial^2 t/\partial x^2 + \partial^2 t/\partial y^2 = 0$, represents a description of two-dimensional steady-state heat conduction in a solid, where $t$ is temperature and $x$, $y$ are space coordinates. Equations very similar to this arise in many problems of fluid mechanics, heat transfer, and mass transfer.

In a manner very similar to that described earlier for the heat equation, we can develop finite-difference approximations for the Laplace equation. This again involves the use of numerical differentiation formulas. If the increments $h_x$ in the $x$ direction and $h_y$ in the $y$ direction are chosen, then Equation (14-4) is used to approximate $\partial^2 t/\partial x^2$ and $\partial^2 t/\partial y^2$ by central differences. When this is done with the error terms dropped, and if the results are substituted into the Laplace equation, we get

$$0 = \frac{t(x+h_x, y) - 2t(x,y) + t(x-h_x, y)}{h_x^2} + \frac{t(x, y+h_y) - 2t(x,y) + t(x, y-h_y)}{h_y^2} \tag{14-9}$$

If the grid chosen is *square*, i.e. if $h_x = h_y$ (see Figure 14.2), Equation (14-9) becomes independent of $h_x$ and $h_y$, and we can solve for $t(x, y)$ as

$$t(x,y) = \frac{t(x+h_x, y) + t(x, y+h_y) + t(x-h_x, y) + t(x, y-h_y)}{4} \tag{14-10}$$

Equation (14-10) states that the temperature at any grid point is the arithmetic average of the temperatures of the four surrounding grid points. To calculate the temperature distribution, Equation (14-10) is applied at each interior node point. Either the known boundary temperatures of the gridwork are specified, or numerical differentiation is applied to the boundary conditions. This process determines a set of linear algebraic equations for the unknown temperatures.

**Example 14.4 Steady Conduction with Specified Temperatures and Equal Spacing**

As an illustration of the use of formula (14-10), we consider two-dimensional steady-state heat conduction in the square geometry shown in Figure 14.2. The temperature of the upper side is 100, while those of the other three sides are 0. A square gridwork is

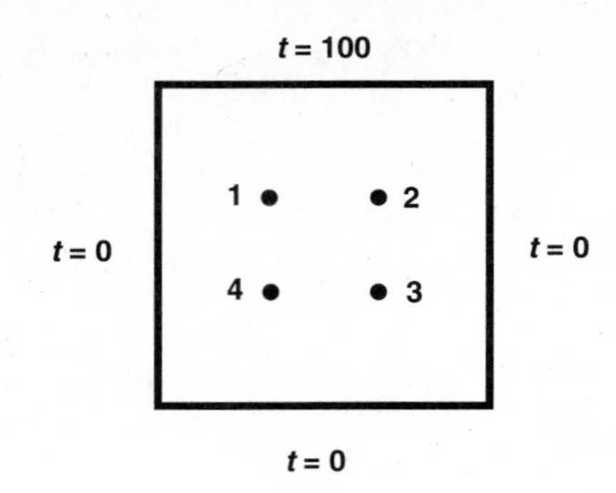


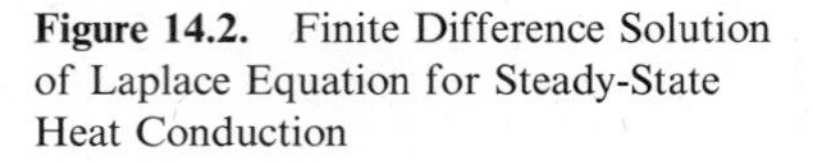
**Figure 14.2.** Finite Difference Solution of Laplace Equation for Steady-State Heat Conduction

chosen with four interior node points. At each node point we write Equation (14-10). Using the numbering system of the diagram gives the equations shown below.

$$\begin{aligned} t_1 &= (1/4)[0 + 100 + t_2 + t_4] \\ t_2 &= (1/4)[100 + 0 + t_3 + t_1] \\ t_3 &= (1/4)[0 + 0 + t_2 + t_4] \\ t_4 &= (1/4)[0 + 0 + t_1 + t_3] \end{aligned} \tag{a}$$

The solution can be verified to be $t_1 = t_2 = 37.5$ and $t_3 = t_4 = 12.5$. We note from the solution that the temperatures happen to be equal in pairs; the physical nature of the problem would have suggested this symmetry in advance. It is important to make use of any such symmetries that may exist in a problem, since this can reduce the computational labor considerably. In the present example, such a reduction hardly matters; but if there were, say, 1000 node points, it would make a very significant difference. It should be noted that Laplace-type equations (and similar steady-state models which do not contain open-ended (marching) variables) are generally not subject to the stability limitations that arise in evolutionary or time-dependent problems. For example, a simple experimental stability computation shows that errors are damped in the above finite-difference method for the Laplace equation.

### 14.4.1 Steady Conduction with Convective Boundary Conditions and Unequal Spacing

We first address the use of Equation (14-9) regarding unequal spacing for the two-dimensional Laplace equation (see Figure 14.3). Suppose we use subscripts $i$ for the $x$ (lateral) direction and $j$ for the $y$ (transverse) direction. Using this subscript notation, Equation (14-9) becomes

$$0 = \frac{t_{i+1,j} - 2t_{i,j} + t_{i-1,j}}{h_x^2} + \frac{t_{i,j+1} - 2t_{i,j} + t_{i,j-1}}{h_y^2} \tag{14-11}$$

If the boundary temperatures are known, then, by applying Equation (14-11) at each interior node point, we get a system of linear algebraic equations for the $t_{i,j}$ variables. The number of variables in this case is exactly equal to the number of interior nodes. Note that regardless of the size of the system, each equation contains no more than five unknowns. Thus, the linear algebraic system for this problem is *sparse* (the coefficient matrix contains many zeros). Because of this, it is generally

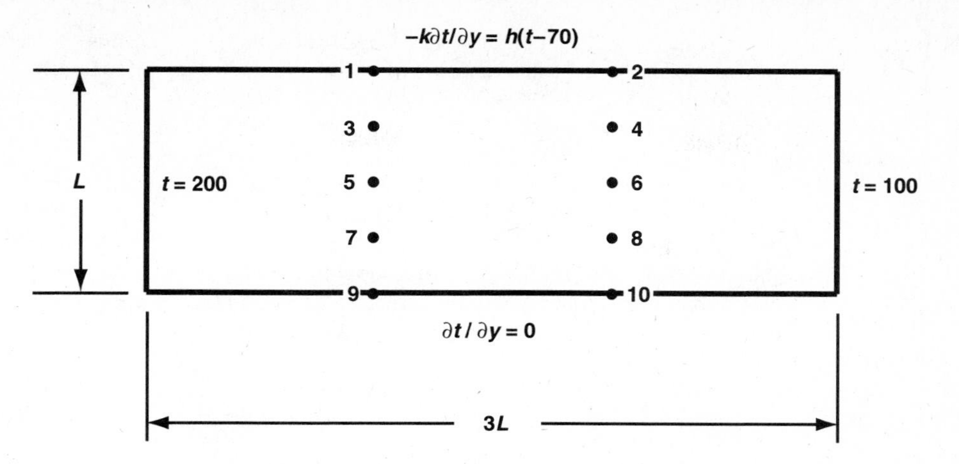


**Figure 14.3.** Finite Difference Solution of Two-Dimensional Laplace Equation for Heat Conduction

more efficient to use an iterative method to solve the equations. To do this, it is useful to solve the equation at a given node point for the node variable. Applied to Equations (14-11), this gives

$$t_{i,j} = \frac{\left[ t_{i+1,j} + t_{i-1,j} + \dfrac{h_x^2}{h_y^2}(t_{i,j+1} + t_{i,j-1}) \right]}{\left( 2 + 2\dfrac{h_x^2}{h_y^2} \right)} \tag{14-12}$$

For $h_x = h_y$, this produces the simple result (given earlier) that $t_{i,j} = (1/4)$ (sum of four neighboring values). For the three-dimensional Laplace equation $t_{xx} + t_{yy} + t_{zz} = 0$, we can proceed in the same way; in this case, the simple equal-spacing result is that $t_{i,j} = (1/6)$(sum of six neighboring values). To solve Equations (14-12) iteratively, we could use ordinary iteration in computer program DELSQ; note that we generally do *not* want to use the $\Delta^2$ option, since the equations have zero diagonal terms in the $\partial g/\partial x$ iterative matrix. If we should have derivative boundary conditions, we would generally use either three-point or fictitious-point approximations at the boundaries as we did in the unsteady-state problems. As usual, it is desirable that truncation errors be of the same order of magnitude in both the interior equations and the boundary condition equations. To monitor global errors and increase accuracy, it would be convenient to employ Richardson extrapolation by using two different spatial node distributions whose increments differ by a factor of two.

**Example 14.5 Two-Dimensional Steady-State Heat Conduction**

We now consider two-dimensional steady-state heat conduction in the rectangular slab geometry shown in Figure 14.3. The quantity $h$ is the convective heat transfer coefficient between the upper surface and the surroundings, while $k$ is the thermal conductivity of the slab. We use the dimensionless variables $x/L = X$, $y/L = Y$, and $hL/k = N$. The numerical value of $N$ is 0.1. The boundary conditions include two prescribed temperatures (the left face at 200°, the right face at 100°), one insulated face ($\partial t/\partial y = 0$

at $y = 0$), and a convective condition at $y = L$ as shown in the figure. In this problem we use six interior points, and since there are derivative boundary conditions at both the upper and lower surfaces, we have four additional node points for a total of ten. From the geometry of the node points, the spatial increments are $h_x = 1$ and $h_y = 1/4$. Using Equation (14-12), the general *interior* node equation can be written in a form ready for iteration as

$$t_{i,j} = \frac{[t_{i+1,j} + t_{i-1,j} + 16(t_{i,j+1} + t_{i,j-1})]}{34} \tag{a}$$

We use Equations (a) at the interior points together with three-point backward and forward approximations at the upper and lower surfaces, respectively, to give the following set of ten equations written in a form ready for iteration (each node equation has been solved for the variable at that node). The equations are grouped according to their similar mathematical structure. It is convenient to use a single subscript notation in reference to Figure 14.3.

$$\begin{aligned}
t_1 &= \frac{[8t_3 - 2t_5 + 70N]}{N + 6} \\
t_2 &= \frac{[8t_4 - 2t_6 + 70N]}{N + 6} \\
t_3 &= \frac{\left[t_4 + 200 + 16(t_1 + t_5)\right]}{34} \\
t_5 &= \frac{\left[t_6 + 200 + 16(t_3 + t_7)\right]}{34} \\
t_7 &= \frac{\left[t_8 + 200 + 16(t_5 + t_9)\right]}{34} \\
t_4 &= \frac{\left[100 + t_3 + 16(t_2 + t_6)\right]}{34} \\
t_6 &= \frac{\left[100 + t_5 + 16(t_4 + t_8)\right]}{34} \\
t_8 &= \frac{\left[100 + t_7 + 16(t_6 + t_{10})\right]}{34} \\
t_9 &= \frac{[4t_7 - t_5]}{3} \\
t_{10} &= \frac{[4t_8 - t_6]}{3}
\end{aligned} \tag{b}$$

For $N = 0.1$, we now use these equations in program DELSQ with the ordinary iteration option and take all first guesses equal to 150. The converged results of this calculation are shown in Table 14.6. We retain the large number of significant figures in the calculations since they will be needed in a later problem.

### 14.4.2 Effects of Other Geometries and Additional Space Dimensions

The procedures we have outlined for two-dimensional Cartesian geometry carry over directly to other geometries and additional space dimensions. Naturally, if we are

**TABLE 14.6** STEADY CONDUCTION SOLUTION FOR EXAMPLE 14.5 $N = 0.1$

| | |
|---|---|
| $t_1 = 156.220$ | $t_2 =$ 124.913 |
| $t_3 = 158.064$ | $t_4 =$ 126.102 |
| $t_5 = 159.284$ | $t_6 =$ 126.926 |
| $t_7 = 159.982$ | $t_8 =$ 127.410 |
| $t_9 = 160.215$ | $t_{10} =$ 127.571 |

dealing with cylindrical or spherical coordinates in a domain that includes $r = 0$, we must use L'hospital's rule or Taylor's theorem at $r = 0$ as necessary.

### 14.4.3 Higher-Order Approximations and Other Equations

To achieve higher accuracy in difference approximations, several approaches are possible. The brute-force approach of simply using more node points is often convenient, if it is feasible. Richardson extrapolation is also very useful for increasing accuracy and estimating errors. In addition, it is also sometimes possible to increase the order of accuracy in the difference approximation. This can be accomplished by using higher-order numerical differentiation approximations (see Appendix G) or by combining numerical differentiation approximations with the PDE itself. An interesting example of this is the so-called nine-point approximation for the two-dimensional Laplace equation Ferziger [1981]. If higher-order approximations are to be used, it should always be borne in mind that the accuracies of difference approximations in both the equation and the boundary conditions should be the same or nearly so.

It should also be mentioned that the finite difference procedures we have developed above can generally be employed in other "steady-state" equations. Two famous examples include the two-dimensional Poisson and Helmholtz equations in Cartesian coordinates, shown as follows:

$$\begin{aligned} \frac{\partial^2 c}{\partial x^2} + \frac{\partial^2 c}{\partial y^2} + f(x,y) &= 0 \qquad \text{(Poisson)} \\ \frac{\partial^2 c}{\partial x^2} + \frac{\partial^2 c}{\partial y^2} + Kc &= 0 \qquad \text{(Helmholtz)} \end{aligned} \tag{14-13}$$

In Equations (14-13), $f(x,y)$ is a known function of $x$ and $y$, while $K$ is a constant parameter. There are a number of applications of Equations (14-13) in fluid mechanics and heat or mass transfer.

## 14.5 THE METHOD OF LINES (MOL) FOR MARCHING PROBLEMS

The method of lines (MOL) is an important numerical procedure for the solution of evolutionary partial differential equations. Some useful references on the subject include Schiesser [1988] and Selibi and Schiesser [1992]. The idea is to discretize derivatives with respect to all variables except one "time-like" (or open-ended) variable, which remains continuous. This process leads to a system of ordinary differential equations (ODEs) for the discretized variables which can then, in principle, be solved by any ODE initial value method. It will be seen, for example, that the "marching" finite difference methods that we have already discussed, including the simple explicit method, the backward difference method, and the implicit trapezoidal rule (Crank-Nicholson), can all be considered special cases of the method of lines. On the other hand, it must be noted that it is *not* true that all marching finite difference methods are special cases of MOL. Thus, although the finite difference and MOL procedures for marching problems are often identical, the two procedures are, in general, distinctly different.

The discretization part of MOL can be accomplished in any way desired, such as by finite differences, a method of weighted residuals, or a finite element method. When the discretization is done with finite differences, other names, such as "Differential-Difference Method," are sometimes used. The method of lines is extremely useful for two reasons: first, it allows utilization of the benefits of sophisticated ODE initial value codes; and second, it is especially convenient for systems of coupled PDEs, ODEs, and algebraic equations. The main limitation of this method is that it may require a *stiff* ODE algorithm because of stability requirements.

The development of the MOL will be indicated for the following dimensionless form of the unsteady heat equation:

$$\frac{\partial T}{\partial \phi} + \frac{\partial^2 T}{\partial Y^2} \tag{14-14}$$

For the discussion here, the discretization of $\partial^2 T/\partial Y^2$ will be accomplished by using the customary central difference approximation. If it were necessary, we could make reference to various other numerical differentiation formulas from Appendix G that might be useful.

Using the standard three-point, second-order, central difference approximation to $\partial^2 T/\partial Y^2$ (and neglecting the error term), the ODE approximation to the PDE of Equation (14-14) becomes

$$\frac{dT_i}{d\phi} = \frac{1}{h_y^2}[T_{i+1} - 2T_i + T_{i-1}] \tag{14-15}$$

where location $Y$ corresponds to the appropriate node value $i$. Applying Equation (14-15) to all of the interior node points produces a coupled system of ODEs whose solution represents an approximation to the true unknown values $T(\phi, Y)$. As usual with difference methods, the spatial accuracy can generally be increased by using a smaller $h_y$; however, this requires an increase in the number of node equations to be solved.

At the problem boundaries, the boundary conditions must be satisfied, at least approximately. If the temperatures are specified on the boundaries, this presents no difficulty. However, if the boundary conditions involve the derivative of the temperature, then we must proceed by either using difference approximations at the boundaries or applying the interior node ODE at the boundaries and introducing appropriate "fictitious boundary" values, as we discussed in the solution of ODE boundary value problems. This latter procedure is often the most convenient and efficient, if it is done so that the order of accuracy of the boundary condition approximation is no worse than that of the interior values.

Error estimation and accuracy improvement can often be accomplished by using Richardson extrapolation applied to two or more calculations utilizing different space and/or time increments. It should be borne in mind that for any given node spacing, reduction of time integration errors will *not* reduce the spatial errors. In this connection an ODE solver such as Improved Euler/Extrapolation (IEX) or Hybrid PDE (PDEHYBD) will be particularly useful, since they can automatically monitor time errors. To assess spatial error and improve accuracy, spatial extrapolation can then be employed by running two or more solution trajectories at different node spacings (variation of node spacing $h_y$ by factors of two is generally most convenient).

### 14.5.1 Integration Methods

The method of lines employing a simple non-stiff ODE solver (such as IEX or PDEHYBD) can constitute a very efficient method for generating numerical solutions to evolutionary PDEs. However, one important limitation of this procedure must be stressed. In general, the system of ODEs generated by the method of lines becomes more and more *stiff* as the node spacing decreases. In any given situation, there will be some spatial discretization value $h_y$ below which the above procedure will not work (it will take prohibitively long). In some problems, satisfactory results can be achieved long before the limiting $h_y$ is reached; in other problems, perhaps no results can be obtained because of this stability limitation, and we must resort to a stiff ODE procedure that can deal with this problem. We shall consider below some implicit procedures that are useful for this purpose. In view of this situation, it becomes clear that the technique of spatial extrapolation may be valuable in order to achieve satisfactory results at larger spatial increments. In general, it is also true that weighted residual spatial discretization may produce an ODE system which is less stiff than that produced by finite differences, but it is not as easy to use.

Finally, it should be noted that the method of lines is valuable for problems involving some combination of PDEs, ODEs, and nonlinear algebraic equations. Also, some steady-state problems can be treated by solving an artificial transient problem that approaches the steady-state solution at large times. We will now illustrate the method of lines with a detailed example problem.

**Example 14.6 The Suddenly Heated Wall by the Method of Lines**

We now reconsider the problem of Example 14.1, which was expressed in dimensionless form in Equations (b); these are now repeated for convenient reference.

$$\frac{\partial T}{\partial \phi} = \frac{\partial^2 T}{\partial Y^2} \quad T(\phi, Y = 0) = 0; \quad \frac{\partial T}{\partial Y}(\phi, Y = 0); \quad T(\phi = 0, Y) = 1 \tag{a}$$

We now write down the method-of-lines equations for this problem. In accordance with Equation (14-15), we have $dT_i/d\phi = m^2(T_{i+1} - 2T_i + T_{i-1})$, where $h_y = 1/m$, and $m$ is the number of parts into which the interval is divided. These equations apply at the interior points ($i = 2, 3, \ldots, m$). We wish to use $m$ as a parameter, in order to monitor spatial accuracy. Since $T_1$ is constant at the value zero, its differential equation is just $dT_i/d\phi = 0$. For the upper surface where we have $\partial T/\partial Y = 0$, we proceed as follows. There are two alternatives, which seem nearly equivalent. These include the three-point backward approximation to $\partial T/\partial Y$ at $Y = 1$ or the central approximation to $\partial T/\partial Y$ at $Y = 1$, together with introduction of a fictitious point at location $Y = (1 + h_Y)$, which is node $i = (m + 2)$. Both of these approximations to the boundary condition are of accuracy $O(h_Y^2)$, which we want to maintain since the spatial discretization in the PDE is also of this accuracy.

We will consider both boundary condition procedures. For the three-point backward approximation, the proper numerical differentiation formula is $0 = \partial T/\partial Y = (T_{m-1} - 4T_m + 3T_{m+1})/h_Y^2 + O(h_Y^2)$. We then need to differentiate this equation with respect to $\phi$ and solve for $dT_{m+1}/d\phi$. This gives $dT_{m+1}/d\phi = (4dT_m/d\phi - dT_{m-1}/d\phi)/3$, which is the appropriate differential equation at node $(m + 1)$ for the three-point backward approximation. Notice that since this differential equation contains other dependent variable derivatives on the right-hand side, it must be evaluated *after* these derivatives in the derivative subroutine. If, instead, the fictitious node approach is used, we introduce a new node point $i = (m + 2)$ at $Y = (1 + h_y)$. We then apply the interior node differential equation also at $i = (m + 1)$ (which is an interior point in the new scheme) and apply the boundary condition using the central approximation as $0 = \partial T/\partial y = (T_{m+2} - T_m)/2h + O(h^2)$. In this case, it is convenient to algebraically eliminate the fictitious point variable $T_{m+2}$, using the boundary condition equation $T_{m+2} = T_m$. When this is substituted into the interior node equation for $i = (m + 1)$, we get $dT_{m+1}/d\phi = (2T_m - 2T_{m+1})/h_Y^2$. This is the required differential equation for the boundary point when we use the fictitious point approximation.

Now that we have developed appropriate MOL differential equations for our problem, we must decide on some ODE numerical method for its implementation. We could use the IEX method from the chapter on ODE initial value problems (e.g., program IEXVS). For some purposes this would be satisfactory, but this method has one important limitation in *stiff* problems, namely a severe stability requirement (for the prototype equation, this was $Kh \leq 1$). Since the PDE method-of-lines problems become more stiff as the number of nodes increases, this is a very significant problem. An alternative explicit method that is more suitable to this situation is the RK2/hybrid procedure, which was introduced in the discussion of stiff ODE methods. The hybrid method has the considerable advantage that its stability limit is $Kh \leq 7.6$, which is nearly eight times that of improved Euler. Thus, we expect that, if the calculation is stability-limited, the hybrid method will be considerably more efficient (numerical experiments have shown this to be true Ashour and Hanna [1990]). For this purpose we use program PDEHYBD (PDE hybrid), which is discussed in detail at the end of this chapter. As far as the user is concerned, program PDEHYBD is implemented in virtually an identical manner to program IEXVS, and it has the same multiple trajectory, error analysis, and error control facilities. The method requires two derivative evaluations per step and gives first-order accuracy so that Richardson extrapolation with three trajectories produces third-order accuracy in time.

It is usually desirable to use variable time-step integration in method-of-lines diffusion problems, since there is often a transition in problem behavior from very rapid changes at small times to slow changes at larger times. Having this variable time-step capability is an important advantage of MOL. Thus, our calculations here have been made utilizing error control. In this situation it is the "Tol" local relative error parameter in Sub Init that can be used to control accuracy. Also, it happens that the variable-step programs we use here often avoid instability on their own, by using step sizes small enough to remain stable, even for fairly large values of Tol. Our first calculations in this problem involved using the improved Euler, RK2, and hybrid methods to check experimental stability behavior. We used these three methods with a very loose Tol value to see the relative values of time-step that were achieved. These proved to be just as the prototype analysis predicts: RK2 could take twice the step of improved Euler, and the hybrid method could take about 7.6 times larger steps than improved Euler.

One reason for choosing this particular example problem is that besides being a typical problem, it also has a simple separation of variables analytical solution, and it should also behave close to the similarity solution for the suddenly heated semi-infinite wall at small times. These solutions are

Separation of Variables

$$T = \sum_{n=0}^{\infty} \frac{2}{\lambda_n} e^{-\lambda_n^2 \phi} \sin \lambda_n y \quad [\lambda_n = \frac{2n+1}{2}\pi;\ n = 0, 1, 2, \ldots] \tag{c}$$

and

Similarity (Small Times)

$$T = \text{Erf}(\eta) = \frac{2}{\sqrt{\pi}} \int_0^{\eta} e^{-u^2}\, du; \quad \eta = \frac{y}{2\sqrt{\phi}} \tag{d}$$

### 14.5.2 Effects of Boundary Condition Approximations

We now show the results of using the method of lines with integrator PDEHYBD. Three different spacings are considered ($m = 4, 8, 16$). The results were obtained using three trajectories, error control with Tol = 0.001, and $\Delta\phi_{\text{initial}} = 0.001$. The calculations shown are for estimated time errors of $10^{-6}$ and error ratios nearly equal to one, so that *for practical purposes there are no time integration errors*. The differences are due to different spatial discretizations. The "exact" values have been computed from the separation of variables solution, Equation (c), and they are all accurate to the number of figures shown. First, we show results for the three-point backward approximation to $\partial T/\partial Y = 0$ at $Y = 1$ in Table 14.7.

In Table 14.7 we show all six figures, since we will be doing extrapolation and making comparisons. The results of Table 14.7 show that the method of lines solution is surprisingly accurate, considering the relatively small number of nodes used. Even with $m = 4$ (three interior nodes), the errors are no more than about 1%, and the accuracy clearly increases as we increase the number of nodes.

Next, we perform the same calculations for the fictitious boundary approach and show the results in Table 14.8. Inspection of Table 14.8 and comparison of Table 14.7 shows that the fictitious-boundary calculation produces results that are good and similar to those for the three-point backward approximation.

**TABLE 14.7** SOLUTION FOR THREE-POINT BACKWARD DIFFERENCE BOUNDARY CONDITION $\phi = 0.2$

| $y$ | $T(m = 4)$ | $T(m = 8)$ | $T(m = 16)$ | $T_{\text{exact}}$ |
|---|---|---|---|---|
| 1.00 | 0.773577 | 0.771776 | 0.772032 | 0.772312 |
| 0.75 | 0.719528 | 0.716075 | 0.716043 | 0.716227 |
| 0.5 | 0.557381 | 0.553461 | 0.553145 | 0.553176 |
| 0.25 | 0.305224 | 0.302455 | 0.302127 | 0.302084 |
| 0.00 | 0.000000 | 0.000000 | 0.000000 | 0.000000 |

**TABLE 14.8** SOLUTION FOR FICTITIOUS POINT BOUNDARY CONDITION $\phi = 0.2$

| $y$ | $T(m = 4)$ | $T(m = 8)$ | $T(m = 16)$ | $T_{\text{exact}}$ |
|---|---|---|---|---|
| 1.00 | 0.764977 | 0.770530 | 0.771870 | 0.772312 |
| 0.75 | 0.710611 | 0.714856 | 0.715887 | 0.716227 |
| 0.5 | 0.551067 | 0.552638 | 0.553041 | 0.553176 |
| 0.25 | 0.302162 | 0.302072 | 0.302079 | 0.302084 |
| 0.00 | 0.000000 | 0.000000 | 0.000000 | 0.000000 |

### 14.5.3 Richardson Extrapolation in Space

By choosing to divide the interval into 4, 8, and 16 parts, we can not only see qualitatively the effect of spatial discretization, but we can also use Richardson extrapolation to attempt to increase the accuracy and estimate global errors with no further information. Since we have used $O(h_Y^2)$ accuracy for the spatial discretization both in the interior and at the boundaries, and since we have eliminated time from the comparison by keeping time errors below $10^{-6}$, we expect that the results we have obtained may follow the extrapolation formula $S = A(h_Y) + Ch_Y^2$, where $S$ is the unknown desired value, $A(h_Y)$ is the approximation for spatial increment $h_Y$, and $C$ is an unknown constant which may vary with $Y$ and $\phi$ but is *independent of* $h_Y$. If this is true, then by using results for $h_Y$ and $h_Y/2$ at given values of $Y$ and $\phi$, the improved approximation will be $S = A(h_Y/2) + [A(h_Y/2) - A(h_Y)]/3$, and the estimated error is just the bracketed term in this equation. We now show, in Table 14.9, the results of doing this single extrapolation for both the three-point backward and fictitious-point boundary condition approaches using the results for $m = 8, 16$.

A careful examination of Table 14.9 reveals some striking results. We had seen earlier that the three-point approximation seemed about the same as the fictitious-

**TABLE 14.9** RICHARDSON EXTRAPOLATION APPLIED TO THREE-POINT BACKWARD AND FICTITIOUS POINT BOUNDARY CONDITIONS ($m = 8, 16$); $\phi = 0.2$

| $y$ | $T$(three point) | Estimated Error | $T$(fictitious point) | Estimated Error | $T_{exact}$ |
|---|---|---|---|---|---|
| 1.00 | 0.772117 | $8.5e-5$ | 0.772317 | $4.5e-4$ | 0.772312 |
| 0.75 | 0.716032 | $1.1e-5$ | 0.716231 | $3.4e-4$ | 0.716227 |
| 0.50 | 0.553040 | $1.1e-4$ | 0.553175 | $1.3e-4$ | 0.553176 |
| 0.25 | 0.302018 | $1.1e-4$ | 0.302081 | $2.3e-6$ | 0.302084 |
| 0.00 | 0.000000 | 0.0000 | 0.000000 | 0.0000 | 0.000000 |

point approach. The Richardson extrapolation results show a strongly opposite conclusion. For the spatial increments used here, *the fictitious-point extrapolation results are extremely good and are far superior to those of the three-point approximation.* Presumably, this is due to the fact that the all-central difference approximation of the fictitious-point approach has error $O(h_Y^4)$, while that of the three-point approximation has error $O(h_Y^3)$. Another very important point is that the fictitious point estimated errors are generally conservative, as we desire, while those of the three-point approximation are not. Thus, the estimated errors of the three-point approximation are generally much smaller than the true errors, which is very undesirable. Although we have not shown it here, the same behavior is seen if we do extrapolation by combining the ($m = 4, 8$) results. In fact, if the two single extrapolation results for the fictitious-point calculations are compared, the estimated errors are nearly in the ratio of 4, as would be required for our extrapolation formula.

### 14.5.4 Small Time Comparison

It is also of interest to compare our numerical calculations for small times to the exact solution and to the similarity solution. We choose to make the comparison at $\phi = 0.06$, which is estimated to be approximately the time when the thermal boundary layer just reaches the upper surface of the wall. Thus, we would expect that the similarity solution for an infinitely thick wall should be a good approximation to the exact separation of variables solution for this and smaller times. The comparison is shown in Table 14.10.

The results in Table 14.10 confirm our expectations regarding the agreement of the similarity solution with the exact solution; except at $Y = 1$, there is nearly four-figure agreement. The numerical results for $m = 8$ are also quite good, with errors being less than 1%. The numerical results are for program PDEHYBD with error control and Tol = 0.001. This gives relative error estimates (in time) of $2 \times 10^{-6}$ and error ratios of 1.02. Thus, the discrepancies between the numerical and exact solution in Table 14.10 should be due only to spatial discretization.

**TABLE 14.10** COMPARISON OF NUMERICAL, SIMILARITY, AND EXACT SOLUTIONS AT SHORT TIMES; $\phi = 0.06$

| $y$ | $\eta = Y/2\phi^{1/2}$ | $T = \mathrm{Erf}(\eta)$ | $T_{num}$ Three-pt. boundary condition, $m = 8$ | $T_{exact}$ |
|---|---|---|---|---|
| 1.00 | 2.0410 | 0.9961 | 0.9919 | 0.9922 |
| 0.75 | 1.5310 | 0.9696 | 0.9663 | 0.9693 |
| 0.50 | 1.0210 | 0.8512 | 0.8486 | 0.8511 |
| 0.25 | 0.5303 | 0.5295 | 0.5306 | 0.5295 |
| 0.00 | 0.0000 | 0.0000 | 0.0000 | 0.0000 |

**TABLE 14.11** COMPARISON OF NUMERICAL AND EXACT SOLUTIONS AT LARGE TIMES; $\phi = 1.9$

| $y$ | $T_{num}$ three point boundary condition, $m = 8$ | $T_{exact}$ |
|---|---|---|
| 1.00 | $1.176e^{-2}$ | $1.172e^{-2}$ |
| 0.75 | $1.087e^{-2}$ | $1.083e^{-2}$ |
| 0.50 | $8.320e^{-3}$ | $8.287e^{-3}$ |
| 0.25 | $4.504e^{-3}$ | $4.485e^{-3}$ |
| 0.00 | 0.000000 | 0.000000 |

### 14.5.5 Large Time Comparison

We are also interested in a comparison between the exact and numerical solutions at large times, since there are many applications when the large time or nearly "steady-state" solution is the desired quantity. In this particular problem, we can verify from the boundary conditions that the steady-state (time-independent) solution is just $T = 0$. Since this will only be approached asymptotically, we are interested in observing the rates of approach of the numerical and exact solutions. In practice, in a problem such as this, the steady-state might be defined as the time when $T$ becomes within 0.01 of its final value of zero. From computational experiments, we see that this time is approximately $\phi = 1.9$, so this is where we make our long time comparisons. These are shown in Table 14.11.

The long time results of Table 14.11 show that the numerical results are quite good, with errors well below 1%. The numerical results again have been obtained so that there are no time errors; the discrepancies are all due to spatial discretization.

### 14.5.6 Effects of Other Geometries and Additional Space Dimensions

**14.5.6.1 Cylindrical and Spherical Geometry.** Solutions of PDEs are sometimes required for problems formulated in other coordinate systems, such as cylindrical, spherical, and so on. Usually, this poses no special difficulties for the method of lines, but the matter is worth a few remarks. Let us consider one-dimensional heat conduction in cylindrical coordinates (with angular symmetry) whose appropriate equation is

$$\frac{\partial t}{\partial \phi} = \frac{1}{r}\frac{\partial}{\partial r}\left(r\frac{\partial t}{\partial r}\right) = \frac{\partial^2 t}{\partial r^2} + \frac{1}{r}\frac{dt}{dr} \tag{14-16}$$

We have already incorporated the thermal diffusivity $\alpha$ in with the time variable. The question concerns how to difference the spatial (radial) coordinate, especially near the boundaries. In general, it is satisfactory to replace both radial derivatives with central approximations; and this yields, for an arbitrary interior point node, the equation

$$\frac{dt_i}{\partial \phi} = \frac{t_{i+1} - 2t_i + t_{i-1}}{h_r^2} + \left(\frac{1}{(i-1)h_r}\right)\left[\frac{t_{i+1} - t_{i-1}}{2h_r}\right] \tag{14-17}$$

The term $(i-1)h_r$ which replaces $r$ assumes that we number the nodes with $i = 1$ at $r = 0$. A sufficient condition or von Neumann stability analysis confirms the same conditional stability for the forward Euler time integration, as in the case of Cartesian geometry.

Our earlier results have suggested that if we have any boundary conditions involving derivatives, we may want to use the fictitious-point approach in order to get good behavior with Richardson extrapolation. This is done in the usual way at the $r = R$ boundary. But at $r = 0$ (node 1), where $\partial t/\partial r = 0$, we proceed as follows. For the fictitious-point formulation, we need to define an appropriate internal node ODE at $r = 0$. Near $r = 0$, the term $(1/r)\partial t/\partial r \rightarrow 0/0$; by a simple Taylor approximation about $r = 0$, this term becomes $\partial^2 t/\partial r^2$. Thus, at the point $r = 0$, the interior node equation becomes $\partial t/\partial \phi = 2\partial^2 t/\partial r^2$. The boundary condition $\partial t/\partial r = 0$ is approximated by a central difference as $(t_2 - t_0)/2h_r$, and the fictitious point $t_0$ is eliminated in the ODE at node $r = 0$. The final result of this calculation at $r = 0$ (node 1) is

$$\frac{dt_1}{d\phi} = \left(\frac{2}{h_r^2}\right)[t_2 - 2t_1 + t_0] = \left(\frac{2}{h_r^2}\right)[2t_2 - 2t_1] \tag{14-18}$$

An entirely analogous procedure is used for spherical geometry. Of course, if we use the three-point one-sided approach for the boundary conditions, no special equation for $r = 0$ is needed.

**14.5.6.2 Two and Three Spatial Dimensions.** It is straightforward to extend the method of lines to various problems involving more than two independent variables. We simply need to keep careful track of the numbering of node points, equations, and boundary conditions. It is not difficult to show—for example, in the case of the heat conduction equation—that the addition of a new independent space variable generally increases the stability restriction of the explicit calculation procedures for the method of lines. This is the principal computational difficulty

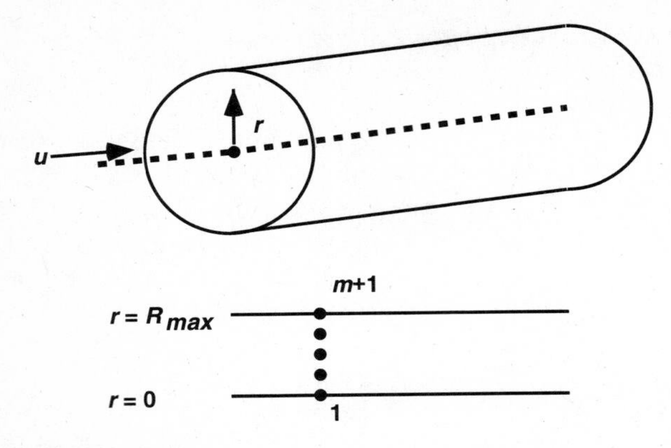


**Figure 14.4.** Finite Difference Solution for Flow and Transport in a Tube

of the method of lines. On the other hand, if we go to an implicit formulation with additional space variables, we usually encounter stable procedures that do not have a tridiagonal structure. This has led to the formulation of so-called alternating direction implicit (ADI) methods, which have found considerable success in such problems. This topic is beyond the scope of our presentation, but is discussed in Carnahan *et al.* [1969] and Ferziger [1981]. If we can tolerate fairly crude spatial accuracy, then an explicit method of lines procedure such as program PDEHYBD is useful. If we must have high spatial accuracy, the ADI methods are to be preferred.

**Example 14.7 Transport and Reaction for Flow in a Tube**

We consider steady-state laminar convective diffusion in a cylindrical tube with a first-order homogeneous chemical reaction as described by the partial differential equation

$$u\frac{\partial c}{\partial x} = D_{AB}\left[\frac{\partial^2 c}{\partial r^2} + \frac{1}{r}\frac{\partial c}{\partial r}\right] - Kc \qquad \text{(a)}$$

with boundary conditions

$$\frac{\partial c}{\partial r} = 0 \quad (\text{all } x,\ r = 0);\ c(\text{all } x,\ r = R_{\max}) = c_w;\ c(x = 0, \text{all } r) = 0 \qquad \text{(b)}$$

Here, $D_{AB}$ is the diffusion coefficient, $K$ is the reaction rate constant, $R_{\max}$ is the tube radius, and $c_w$ is the constant wall concentration. The steady parabolic laminar velocity profile is $u = u_{\max}[1-(r/R_{\max})^2]$, where $u_{\max}$ is the centerline velocity.We now introduce the following dimensionless variables: $C = c/c_w$, $R = r/R_{\max}$, $X = [xD_{AB}/(u_{\max}R_{\max}^2)]$, and $M = (KR_{\max}^2/D_{AB})$. Then the original Equations (a) and (b) become

$$(1 - R^2)\frac{\partial C}{\partial X} = \frac{\partial^2 C}{\partial R^2} + \frac{1}{R}\frac{\partial C}{\partial R} - MC \qquad \text{(c)}$$

with new boundary conditions

$$\frac{\partial C}{\partial R} = 0 \ (\text{all } X,\ R = 0);\ C(\text{all } X,\ R = 1) = 1;\ C(X = 0, \text{all } R) = 0 \qquad \text{(d)}$$

This problem is an extension of the famous Graetz problem to include first-order homogeneous chemical reaction. The quantity of interest is the gradient $\partial C/\partial R$ at $R = 1$, which is related to the rate of mass transfer at the surface. We now discretize $\partial^2 C/\partial R^2$ and $\partial C/\partial R$ using central differences and put $R_i = (i-1)h_r$ to get the following method-of-lines ODEs (see Figure 14.4). At node 1 ($R = 0$), we show two alternative equations, depending on whether we use the three-point or fictitious approximations for $\partial C/\partial R(0) = 0$.

$$\begin{aligned}
\frac{dC_1}{dX} &= 4\frac{(C_2 - C_1)}{h_r^2} - MC_1 \quad \text{(fictitious point)} \\
\frac{dC_1}{dX} &= \frac{1}{3}\left[4\frac{dC_2}{dX} - \frac{dC_3}{dX}\right] \quad \text{(three point)} \\
(1 - R_i^2)\frac{dC_i}{dX} &= \frac{(C_{i+1} - 2C_i + C_{i-1})}{h_r^2} + \frac{(C_{i+1} - C_{i-1})}{2h_r \cdot R_i} - MC_i \quad (i = 2, 3, \ldots, m) \\
\frac{dC_{m+1}}{dX} &= 0
\end{aligned} \tag{e}$$

The fictitious-point expression for node 1 was obtained from the limiting form of the original PDE for $R \to 0$, together with replacement of the fictitious point value $C_0$ by $C_2$ in accordance with the central difference expression for $\partial C/\partial R$ at $R = 0$. The appropriate initial conditions are $C_i = 0$ for $(i = 1, 2, \ldots, m)$ and $C_{m+1} = 1$.

To calculate the desired quantity $\partial C/\partial R(R = 1)$, we use the second-order accurate three-point backward approximation, which is $\partial C/\partial R(R = 1) = (3C_{m+1} - 4C_m + C_{m-1})/2h_r$. We could simply add this quantity to our list of printed variables, but it is more useful to let this quantity be a new program-dependent variable $y(m + 2)$, derive the differential equation for it, and add it to the system of equations. By differentiating the above expression we get

$$\frac{dy_{m+2}}{dX} = \left(\frac{1}{2h_r}\right)\left[3\frac{dC_{m+1}}{dX} - 4\frac{dC_m}{dX} + \frac{dC_{m-1}}{dX}\right] \tag{f}$$

Using the previous initial conditions in the definition of $y(m + 2)$ shows that its proper initial condition is $\partial C/\partial R(R = 1) = y(m + 2) = [3(1) - 4(0) + 0]/2h_r = 3/2h_r = 3m/2$. We can allow for the additional variable in program PDEHYBD very easily by simply altering the usual statement $n = m + 1$ in Sub INIT to $n = m + 2$. The advantage of converting the approximation for $\partial C/\partial R(R = 1)$ to a system differential equation is that in addition to automatic bookkeeping, we also monitor the estimated accuracy of this quantity.

We show below the results of using program PDEHYBD to calculate $\partial C/\partial R$ $(R = 1)$ both with and without chemical reaction as a function of grid spacing $m$. The results are the same (for the figures shown) for either three-point or fictitious-point boundary conditions. A variable time step is used in the integration (and is quite valuable, since the time step varies considerably from beginning to end) so as to produce estimated time errors of less than $10^{-6}$. We expect to find (i) large values of $\partial C/\partial R$ $(R = 1)$ at small $X$ (in the boundary layer part of the solution), (ii) no effect of chemical reaction at *very* small $X$, (iii) $\partial C/\partial R$ $(R = 1) \to 0$ at large $X$ if $M = 0$, and (iv) $\partial C/\partial R(R = 1) \to$ a constant (depending on $M$) at large $X$ if $M > 0$. All of these results are shown in the numerical calculations of Table 14.12.

The spatial accuracy deteriorates as $X$ becomes very small, owing to the fact that in this region there is variation only within a very thin boundary layer near $R = 1$, and our node points in this case are mostly outside this boundary layer. To deal numerically with the boundary layer (without going to similarity coordinates), we would need to operate near $R = 1$ by using a new variable $Z = (1 - R)$, transform the PDE accordingly, and then use the boundary condition $\partial C/\partial Z = 0$ at various small values of $Z$.

**TABLE 14.12** NUMERICAL RESULTS FOR PROBLEM OF DIFFUSION WITH REACTION VALUES OF $\partial C/\partial R$ $(R = 1)$

| $X$ | $m = 8, M = 0$ | $m = 16, M = 0$ | $m = 8, M = 10$ | $m = 16, M = 10$ |
|---|---|---|---|---|
| 0.001 | 8.44 | 6.72 | 8.51 | 7.07 |
| 0.010 | 2.70 | 2.53 | 3.49 | 3.48 |
| 0.020 | 1.91 | 1.85 | 2.99 | 3.05 |
| 0.050 | 1.10 | 1.16 | 2.65 | 2.73 |
| 0.100 | 0.735 | 0.733 | 2.54 | 2.62 |
| 0.200 | 0.349 | 0.347 | 2.50 | 2.57 |
| 0.300 | 0.169 | 0.167 | 2.49 | 2.57 |
| 0.400 | 0.0819 | 0.0806 | 2.49 | 2.57 |

### 14.5.7 Higher-Order Approximations and Other Equations

The application of the method of lines to other partial differential equations follows the procedure we have outlined; we discretize the partial derivatives with respect to all independent variables except one "timelike" (open-ended) variable, which is kept continuous. In this way we generate a system of ordinary differential equations for the "node" dependent variables with respect to the continuous timelike variable. We then use an ordinary differential equation algorithm to implement the solution. In unfamiliar situations it is always necessary to explore the question of stability; this can often be accomplished conveniently by using the experimental approach we outlined earlier.

There are several ways to achieve higher solution accuracy. Of course, a brute-force approach of using smaller spatial and time increments is often useful if computational requirements are not too large. As we have shown, a particular advantage of ODE algorithms which employ global Richardson extrapolation in the continuous variable is that errors with respect to this variable can be easily monitored and controlled. Richardson extrapolation for the discretized variables (e.g., spatial variables) is also useful for accuracy estimation and improvement. Another possibility for improving accuracy involves the use of higher-order numerical differentiation approximations (see Appendix G on numerical differentiation). This increases the computational complexity and involves stability questions, but sometimes the benefits are worth the trouble. If this route is followed, it must be borne in mind that the order of the boundary condition accuracy must also be increased, or the benefits may be lost.

**Example 14.8 Transient Two-Dimensional Heat Conduction**

We consider two-dimensional *transient* heat conduction in the rectangular slab geometry shown in Figure 14.3. This situation was considered earlier in Example 14.5 as a steady-state problem with $hL/k = 0.1$. Here, we take $hL/k \equiv N$. The boundary conditions include two prescribed temperatures (the left face at 200°, the right face at 100°), one insulated face, and a convective condition as shown in the figure. Initially, the slab is at steady state with the dimensionless parameter $hL/k = 0.1$. To reiterate, $h$ is the convective heat transfer coefficient, $k$ is the thermal conductivity of the slab, and $L$ is the vertical slab thickness. At some time, external conditions at the top of the slab cause a sudden, permanent change of $hL/k = N$ from 0.1 to 15. It is desired to determine both the new steady-state temperature distribution and the time required to achieve it.

The two-dimensional heat conduction PDE describing this problem is written as

$$\frac{\partial t}{\partial \theta} = \alpha\left[\frac{\partial^2 t}{\partial x^2} + \frac{\partial^2 t}{\partial y^2}\right] \tag{a}$$

We now introduce coordinate scaling transformations designed to make the independent variables dimensionless. These are, along with the new derivatives,

$$X = \frac{x}{L}, \quad Y = \frac{y}{L}, \quad \phi = \frac{\theta\alpha}{L^2}$$
$$\frac{\partial t}{\partial \theta} = \frac{\alpha}{L^2}\frac{\partial t}{\partial \phi}, \quad \frac{\partial t}{\partial x} = \frac{1}{L}\frac{\partial t}{\partial X}, \quad \frac{\partial^2 t}{\partial x^2} = \frac{1}{L^2}\frac{\partial^2 t}{\partial X^2}, \quad \frac{\partial^2 t}{\partial y^2} = \frac{1}{L^2}\frac{\partial^2 t}{\partial Y^2} \tag{b}$$

When these equations are substituted into the original Equation (a), it becomes just

$$\frac{\partial t}{\partial \phi} = \left[\frac{\partial^2 t}{\partial X^2} + \frac{\partial^2 t}{\partial Y^2}\right] \tag{c}$$

with boundary conditions

$$\begin{aligned} &X = 0, t = 200 \qquad Y = 0, \frac{\partial t}{\partial Y} = 0 \\ &X = 3, t = 100 \qquad Y = 1, -\frac{\partial t}{\partial Y} = \frac{hL}{k}(t - 70) \end{aligned} \tag{d}$$

To use the method of lines, we must derive the required ODEs by differencing in the $X$ and $Y$ directions. Since there are derivative boundary conditions at both the top and bottom surfaces, we need to introduce four additional node points there. We choose to employ three-point difference approximations for these derivative boundary conditions. For the numbering system we have chosen, it can be seen that the method-of-lines ODEs will have a similar structure for nodes (3,5,7), (4,6,8), (1,2), and (9,10). This is useful to observe for programming purposes. From the geometry of the situation, the $X$ space increment will be 1 and the $Y$ space increment (1/4). We first write down the MOL equations for nodes (3,5,7).

$$\begin{aligned} \frac{dt_3}{d\phi} &= \left[\frac{t_4 - 2t_3 + 200}{1^2}\right] + \left[\frac{t_1 - 2t_3 + t_5}{(1/4)^2}\right] \\ \frac{dt_5}{d\phi} &= \left[\frac{t_6 - 2t_5 + 200}{1^2}\right] + \left[\frac{t_3 - 2t_5 + t_7}{(1/4)^2}\right] \\ \frac{dt_7}{d\phi} &= \left[\frac{t_8 - 2t_7 + 200}{1^2}\right] + \left[\frac{t_5 - 2t_7 + t_9}{(1/4)^2}\right] \end{aligned} \tag{e}$$

The equations for nodes (4,6,8) are

$$\begin{aligned} \frac{dt_4}{d\phi} &= \left[\frac{100 - 2t_4 + t_3}{1^2}\right] + \left[\frac{t_2 - 2t_4 + t_6}{(1/4)^2}\right] \\ \frac{dt_6}{d\phi} &= \left[\frac{100 - 2t_6 + t_5}{1^2}\right] + \left[\frac{t_4 - 2t_6 + t_8}{(1/4)^2}\right] \\ \frac{dt_8}{d\phi} &= \left[\frac{100 - 2t_8 + t_7}{1^2}\right] + \left[\frac{t_6 - 2t_8 + t_{10}}{(1/4)^2}\right] \end{aligned} \tag{f}$$

At the top surface (nodes 1,2), the convective boundary condition, $-t_Y = (hL/k)(t - 70)$, is implemented with a backward three-point approximation; these equations are differentiated to give

$$-\frac{(3t_1 - 4t_3 + t_5)}{2(1/4)} = N(t_1 - 70); \quad (N + 6)\frac{dt_1}{d\phi} = 8\frac{dt_3}{d\phi} - 2\frac{dt_5}{d\phi}$$
$$-\frac{(3t_2 - 4t_4 + t_6)}{2(1/4)} = N(t_2 - 70); \quad (N + 6)\frac{dt_2}{d\phi} = 8\frac{dt_4}{d\phi} - 2\frac{dt_6}{d\phi} \tag{g}$$

At the lower surface, the insulated condition $\partial t/\partial Y = 0$ applies; this is implemented with a three-point forward approximation and differentiated to yield

$$0 = -3t_9 + 4t_7 - t_5; \quad \frac{dt_9}{d\phi} = \frac{4}{3}\frac{dt_7}{d\phi} - \frac{1}{3}\frac{dt_5}{d\phi}$$
$$0 = -3t_{10} + 4t_8 - t_6; \quad \frac{dt_{10}}{d\phi} = \frac{4}{3}\frac{dt_8}{d\phi} - \frac{1}{3}\frac{dt_6}{d\phi} \tag{h}$$

This finally completes the system of ten differential equations required for our MOL description of two-dimensional transient conduction. The results will obviously be rather crude because of the small number of node points used, but this calculation suffices to illustrate the procedure.

To calculate the transient solution, we proceed as follows. The MOL differential equations, in terms of $N$, are given in Equation (g); we must use the value $N = 15$. The values in Table 14.6 are the temperatures at the initial steady state where $N = 0.1$. These were calculated previously in Example 14.5 by a steady-state difference method. When $N$ suddenly changes at time zero from 0.1 to 15, we use these prior steady-state temperatures at nodes 3 to 10 as new initial values. But since the boundary conditions have changed at nodes 1 and 2, we need to compute initial values at those locations. These values are computed from the algebraic expressions for the boundary conditions in Equation (g), using values from Table 14.6. To show this explicitly,

$$t_1 = (8t_3 - 2t_5 + 70N)/(6 + N)$$
$$t_1 = [8(158.064) - 2(159.284) + 70(15)]/(6 + 15) = 95.0449$$

and

$$t_2 = (8t_4 - 2t_6 + 70N)/(6 + N)$$
$$t_2 = [8(126.102) - 2(126.926) + 70(15)]/(6 + 15) = 85.9507$$

Using these initial values in program IEXVS, we find that the transient solution gets to within about one degree of steady state at about $\phi = 1.2$. The final steady-state solution computed in this way is shown in Table 14.13. This solution agrees completely with the result of a purely steady-state finite difference calculation.

We should note that one useful check that can be employed here is to calculate the steady-state solution for the case of $N = 0$. In this case, both the upper and lower faces are insulated, and the steady-state solution should agree with that for the equation $t_{xx} = 0$ with boundary conditions of 200 and 100 on the left-hand and right-hand sides. The correct solution for this case is $t_1 = 166\ 2/3$ and $t_2 = 133\ 1/3$. Using $N = 0$ and all initial $t_i$ values (arbitrarily) equal to 150, we get the simple steady-state solution, as we should.

### 14.5.8 Spatial Discretization by the Method of Weighted Residuals

Our discussion of the numerical solution of partial differential equations has emphasized the method of lines. However, we do wish to mention that the method of

**TABLE 14.13**
STEADY-STATE SOLUTION
FOR EXAMPLE 14.8

| | |
|---|---|
| $t_1 = 75.40$ | $t_2 = 72.2$ |
| $t_3 = 92.50$ | $t_4 = 79.6$ |
| $t_5 = 103.7$ | $t_6 = 84.9$ |
| $t_7 = 110.1$ | $t_8 = 88.1$ |
| $t_9 = 112.2$ | $t_{10} = 89.1$ |

weighted residuals is sometimes used to solve these problems. For more detail, see Finlayson [1972, 1980] and Villadsen and Michelsen [1978].

Here we will merely illustrate the ideas for a one-dimensional transient diffusion equation in dimensionless form. Suppose we have the problem

$$\frac{\partial C}{\partial \theta} = D_{AB}\frac{\partial^2 C}{\partial x^2} \tag{14-19}$$

subject to the boundary conditions $C(\theta, 0) = 0$ and $C_x(\theta, 1) = 0$ and the initial condition $C(0, x) = 1$. This is just like the suddenly heated wall problem considered in Example 14.1, except it is now expressed in terms of concentrations instead of temperatures. We take, as our trial function for the concentration, $C_a = a_0(\theta) + a_1(\theta)x + a_2(\theta)x^2 \cdots$ and so on. This is just a polynomial of arbitrary but finite length in $x$ whose coefficients are unknown functions of time. The method of weighted residuals generates differential equations for the $a_i(\theta)$ functions, and when these are solved and put back into the trial function, we have an approximate solution.

We first fit the boundary conditions at $x = 0$ and $x = 1$. At $x = 0, C = 0$, so that $a_0(\theta) = 0$. At $x = 1$ we differentiate $C_a$ partially with respect to $x$ to get $C_x(\theta, 1) = 0 = (a_1 + a_2x + 3a_3x^2 \ldots)_{x=1} = (a_1 + 2a_2 + 3a_3 \cdots)$. By eliminating $a_1$, we get the trial function as

$$C_a(\theta, x) = a_2(x^2 - 2x) + a_3(x^3 - 3x) \ldots \tag{14-20}$$

Note that this trial function appears very similar to those we considered in the application of the method of weighted residuals to ODEs, but now the $a_i$ depend on $\theta$. We now differentiate Equation (14-20) and substitute to form the residual as follows:

$$\begin{aligned}\text{Residual} &= \frac{\partial C_a}{\partial \theta} - D_{AB}\frac{\partial^2 C_a}{\partial x^2} \\ &= \left[\frac{da_2}{d\theta}(x^2 - 2x) + \frac{da_3}{d\theta}(x^3 - 3x) \ldots\right] - D_{AB}[2a_2 + 3 \cdot 2a_3x \ldots]\end{aligned} \tag{14-21}$$

Now we apply a particular MWR criterion to get ordinary differential equations for the $a_i$. For example, use orthogonal collocation. In the case of one parameter, we retain only the $a_2$ coefficient and set the residual in Equation (14-21) equal to zero at the collocation point (Gauss quadrature point) $x = 0.5$. This gives a single ODE for the variable $a_2$. This is solved (easily in this case) for $a_2(\theta)$ subject to the initial

**TABLE 14.14** "STEADY-STATE" SOLUTION OF ARTIFICIAL FALSE TRANSIENT PROBLEM FROM EXAMPLE 14.8

| | |
|---|---|
| $T_1 = 156.22$ | $T_2 = 124.91$ |
| $T_3 = 158.06$ | $T_4 = 126.10$ |
| $T_5 = 159.28$ | $T_6 = 126.93$ |
| $T_7 = 159.98$ | $T_8 = 127.41$ |
| $T_9 = 160.22$ | $T_{10} = 127.57$ |

condition $C_a(\theta = 0, x = 0.5) = 1$. If we take more parameters (say two), we then evaluate the residual Equation (14-21) at two $x$ collocation points, and this produces two coupled linear ODEs for $a_2$ and $a_3$. In order to solve these equations numerically, the derivatives must be decoupled by the equivalent of a matrix inversion, and this is one of the disadvantages of MWR in a problem such as this. On the other hand, an advantage of MWR is that it may require fewer unknowns to get a good spatial approximation. If we integrate so that time errors are negligible, then two consecutive orthogonal collocation approximations give an indication of the spatial errors.

## 14.6 THE METHOD OF LINES FOR STEADY-STATE PROBLEMS

It is important to realize that it is often convenient to calculate steady-state solutions of PDEs by using the unsteady method of lines. This is accomplished by integrating the unsteady equations from some *consistent* arbitrary initial condition (ignoring intermediate results) to the final desired steady state (assuming the problem has a steady-state solution).

**Example 14.9 Reconsideration of Example 14.5**

For illustrative purposes, we calculate the *initial* steady-state temperature distribution corresponding to $hL/k = 0.1$ in Example 14.5 using the so-called *false transient* method. The six interior points shown in Figure 14.3 will be employed, and we use a dimensionless time that incorporates the thermal diffusivity.

Because of the convective boundary conditions, we must be very careful with the initial conditions, even though we are only interested in the steady-state solution. Consistent initial conditions are defined in the following way. We arbitrarily (for convenience) take $t_3$ through $t_8$ (interior nodes) to be initially at 100°. But the relationship between the interior and boundary point temperatures, Equations (g) in Example 14.8, must be satisfied at *all* times (including time zero). Thus we use Equations (g) to determine the appropriate initial values of $t_1$, $t_2$, $t_9$, and $t_{10}$. For the case $N = 0.1$, this yields $t_1 = t_2 = 607/6.1 = 99.5082$ and $t_9 = t_{10} = 100$. When we now solve this set of differential equations for the long-time "steady state" using program IEXVS, we arrive at the results of Table 14.14.

These values are the same as those found from a purely steady-state solution of this problem (see Example 14.8).

## 14.7 THE FINITE ELEMENT METHOD

Another important numerical procedure for solving PDE, which we will only mention, is known as the finite element method. The method involves the employment of a spatial discretization which is generally more flexible than that of finite difference approximation. This approach is especially important for problems involving irregular geometries. It is also useful for problems involving steep gradients of the dependent variable, where the finite element approach naturally allows more node points in regions of rapid function variation.

The idea of the finite element method is to break up the spatial domain into a number of simple geometrical elements such as triangles or quadrilaterals. Then a method of weighted residuals is used to approximate the PDE on each element, with care being taken to ensure continuity of the dependent variable and its first derivative. The finite element discretization leads, for marching problems in a timelike variable, to a system of ordinary differential equations for the dependent variable at the node (element intersection) points. These sets of ODEs are usually implicit, so that modifications must be made in order to use the standard ODE initial value integration algorithms. Computer programs for general finite element methods normally involve a great deal of bookkeeping. Useful references on the subject include Finlayson [1982], Scraton [1987] and Reddy [1984].

## 14.8 COMPUTER PROGRAMS

In this section we discuss the computer programs applicable to various problems in this chapter. For each program we give a brief discussion and a concise listing, which includes (i) a program description, (ii) how to run, and (iii) user-specified subroutines, including a sample problem. Detailed program listings can be obtained from the diskette.

### Method of Lines Program PDEHYBD

Instructions for running the program are given in the concise listing. This program is almost the same as the ODE program IEXVS except that it uses the RK2/Hybrid integration method (better stability, less accuracy), and it is set up to facilitate application to PDE method-of-lines problems. The program prompts for derivative check and number of trajectories (1, 2, or 3). If two trajectories are selected, there is a prompt option for either extrapolation or no extrapolation; this latter option is useful for brute-force error estimation if extrapolation seems inappropriate (as reflected by poor error ratios). The user is also prompted for the initial integration step size and whether error control is desired. In PDE method-of-lines problems, error control is often very useful. The output is the same as for program IEXVS; at each $x$ (time or open-ended independent variable), the best values of $y$ are shown, along with any global relative error estimates and error ratios that have been requested. The error ratio calculation is adjusted appropriately for the extrapolation properties of the RK2/Hybrid method, so the same interpretations apply as in the IEX method.

**Program PDEHYBD.**

```
!This is Euler/RK2 avg./extrapolation/error control for PDE MOL
!(prgm. PDEHYBD.tru) by O.T.Hanna
!n=number of differential equations
!x,y(i) are independent and dependent variables in subroutines & output
!x0,x1 are initial and final values of x
!y1(i,j) is dummy for y(i) in program
!d(i),dold(i,j),dnew(i) are derivatives of y(i)
!h=local step size;hp is print increment;
!pr is print parameter for fixed step
!j1 is number of trajectories;j1=1 for h,j1=2 for h,h/2;j1=3 for h,h/2,h/4
!*********************************************************************
!* To run,specify n,x0,x1,initial y(i),error parameters in Sub Init;    *
!* specify derivatives d(i) in terms of x,y(i) in Sub Deriv;            *
!* program prompts for (a) derivative check? (b) j1,(ext),h,ec,(pr)?    *
!*********************************************************************
!**EULER/RK2-AVG./EXTRAPOLATION/ERR-CTRL SOLUTION OF PROBLEM BELOW**
!a is "averaging parameter";a=0 is RK2,a=1 is Euler,a=.74 is Avg.
!ext is extrapol. parameter(with j1=2 only);
!ext=0,j1=2 gives direct error calc.
!error parameter tol is desired local relative error
!parameters eps,sf,are zero denominator and safety factor values
!To print every step,comment out (!DO !LOOP) in Integration section.

!This is ex. of HYBD.tru adapted to PDE soln via method of lines
SUB Init (a,n,x0,x1,y(),d(),hp,hmaxstep,tol,eps,sf)
    !***denotes user specified values
    LET a = .74          !averaging param.-Euler(a=1),RK2(a=0),(a=.74)
    LET L = 1            !***width of differenced region
    LET m = 8            !***no. of parts into which L is divided
    LET n = m + 1        !no. of ODE-can increase for extra unknowns
    LET x0 = 0           !***initial value of continuous variable
    LET x1 = 3           !***final    "          "            "
    FOR i = 1 to m
        LET y(i) = 1     !***initial values of dependent variable
    NEXT i
    LET y(m+1) = 0       !***"        "           "             "
    CALL Deriv (x0,y,d)

    !The following is used only with error control(ec=1)
    LET hp = .1               !***hprint
    LET hmaxstep = hp         !***maximum user-specified step size
    LET tol = .0001           !***desired local relative error
    LET eps = .0001           !zero denominator parameter
    LET sf = 2                !safety factors
END SUB

SUB Deriv(x,y(),d())
    !y' forward(3 pt) = [4*y(i+1)-y(i+2)-3*y(i)]/(2*h1) !For use in deriv.
```

```
    !y' backward(3 pt) = [-4*y(i-1)+y(i-2)+3*y(i)]/(2*h1) !boundary cond.
    !***denotes user specified quantity
    !This problem is dT/dt = d2T/dy2 with T=0,dT/dy=0 on bound., Tinit = 1
    LET L = 1                  !***width of differenced region
    LET m = 8                  !***no. of parts into which L is divided
    LET h1 = L/m               !spatial difference increment
    FOR i = 2 to m
        LET d(i) = (y(i+1)-2*y(i)+y(i-1))/(h1*h1)   !***internal node eqns
    NEXT i
    LET d(1) = ( -d(3) + 4*d(2) )/3             !***insulated bound., 3 pt.
    !!! LET d(1) = (2*y(2)-2*y(1))/(h1*h1)      !***insulated bound.,fic.pt
    LET d(m+1) = 0                              !***constant temp.bound.
END SUB
```

### *Implicit Tridiagonal PDE program PDEFD*

Instructions for running the program are given in the concise listing. The example shown is for a nonlinear diffusion problem with second-order chemical reaction, ($\partial C/\partial t = D\partial^2 C/\partial y^2 - kC^2$), set up for the backward Euler method. This problem illustrates how to use the $yt(i)$ variable for nonlinear problems. The program solves linear problems (tridiagonal only) with case 1; it solves nonlinear problems (ordinary iteration or delta square) with either case 2 or case 3. The quantity hprx is the space print increment; it is specified in Sub INIT, and it must be a multiple of the space increment $(x1 - x0)/m$ specified in Sub FUNCTION. The program prompts for the quantities case, $m$, $ht$ (time increment), and pr (time print parameter). For output, at time increments of $hprt = pr * ht$, we get values of $y$ at the specified $x$. The program can be used for various implicit methods if they are properly implemented.

#### Program PDEFD.

```
!This is program PDEFD.tru for simple implicit tridiagonal
!solution of PDE by O.T.Hanna
!option for nonlinear problem using ord.iteration or delta-square.
!t,x; y(i) are independent; dependent variables: difference ODE
!to get y(i) eqns; nonlinear y expressed in terms of yt(i)
!t0,t1 are initial, final values of (open-ended) variable t
!x0,x1 are initial,final values of (differenced) variable x
!Interval (x1-x0) is divided into m parts
!pr is print cycle parameter for t; hprx is print increment for x
!program prompts for case, m, ht, pr
!*****************************************************************
!*To run define parameters, init. values in Sub Init; specify    *
!*parameters, tridiagonal matrix coefficients a(i),b(i),c(i),d(i) *
!*for eqns in Sub Function; we always must have a(1)=c(n)=0.     *
!*Use yt(i) for nonlinear terms. Program prompts for case,m,ht,pr *
!*case(1 = Tridiag. only, 2 = Ord.Iteration, 3 = DeltaSquare)     *
!*To check Function use case = 1, m=2, hprx=.5 for 1 internal node *
!*****************************************************************
```

```
!This is program PDEFD.tru, Implicit Tridiagonal soln.for PDE

SUB INIT(n,mx,norm_tol,x0,x1,t0,t1,y(),hprx )
    !User specified statements are marked with ***
    LET mx = 50                    !***maximum number of iterations
    LET norm_tol = 1e-10           !***norm tolerance stop control
    LET x0 = 0                     !***initial x value
    LET x1 = 1                     !***final x value
    LET t0 = 0                     !***initial value of t
    LET t1 = 2                     !***final value of t
    LET hprx = .25                 !***print increment for x;
                                   !must be multiple of (x1-x0)/m
    LET y(1) = 0
    FOR i = 2 to n                 !***initial values of y(i)
        LET y(i) = 1
    NEXT i

END SUB

SUB FUNCTION (ht,m,x0,x1,t,a(),b(),c(),d(),y(),yt() )
    !specify coefficients a(i),b(i),c(i),d(i) in a(i)*y(i-1) +
    !b(i)*y(i) + c(i)*y(i+1) = d(i)
    !be sure to include proper boundary eqns.
    !a(1) and c(n) are not specified since they(=0) are not used
    !yt(i) is temp. sol'n in nonlin. prob.;use this to replace
    !nonlinear y(i) in eqns. below

    !Problem here is for nonlinear diffusion and reaction
    !dC/dt = d2C/dy2 - kC^2
    LET n = m + 1                          !no. of ODE-can increase for
                                           !extra unknowns
    LET h = (x1-x0)/m                      !problem parameters;they
    LET r = 1*ht/(h*h)                     !are fixed for given prob.
    LET k = 100                            !***reaction constant
    LET b(1) = 1                           !***
    LET c(1) = 0                           !***
    LET d(1) = 0                           !***
    FOR i = 2 to m
        LET a(i) = r                       !***
        LET b(i) = -2*r - 1 - k*ht*yt(i)   !***
        LET c(i) = r                       !***
        LET d(i) = -y(i)                   !***
    NEXT i
    LET a(n) = 2*r                         !***
    LET b(n) = -2*r - 1 - k*ht*yt(n)       !***
    LET d(n) = -y(n)                       ! ***

END SUB
```

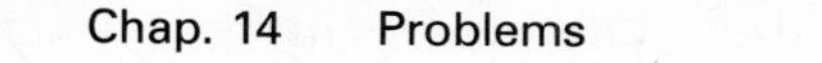



y = 1, T = 200

y

y = 0 , T = 100

Figure 14.5. Unsteady-State Heat Conduction of Problem 14.1

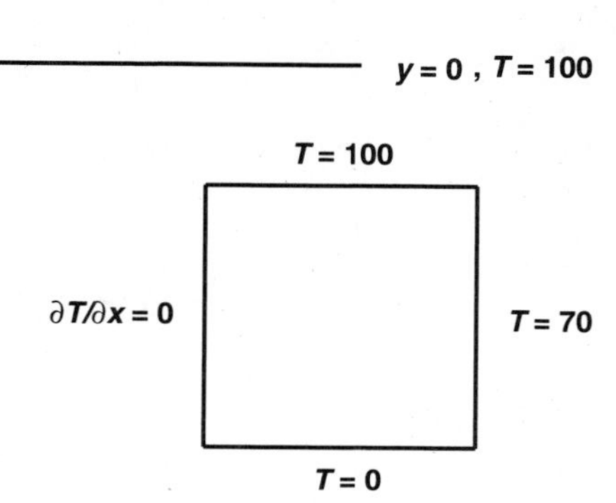



**Figure 14.6.** Steady-State Heat Conduction of Problem 14.2

## PROBLEMS

**14.1.** Consider a rectangular block of material, which is very wide in the $x$ and $z$ directions, in which transient heat conduction takes place in the $y$ direction only, as shown in Figure 14.5. For the material used and the dimensionless $y$ coordinate employed, the heat conduction equation is $\partial T/\partial\theta = \alpha\partial^2 T/\partial y^2$, where $\theta$ is in minutes and $T$ in °F. The value of $\alpha$ is 0.1 (min. $^{-1}$).

Suppose the block is initially at a uniform temperature of 100°F when the top face is suddenly changed to (and kept at) 200°F. The lower face remains at 100°F.

(i) Using three interior points, apply the simple explicit method with $r = 1/2$ to evolve the solution through 10 dimensionless time steps.

(ii) Since the simple explicit method is a special case of the method of lines where Euler integration is employed, use program PDEHYBD to check the results of part (i). Then use the program to determine the steady-state temperature distribution and how much time is required to reach it.

**14.2.** It is desired to calculate the temperature distribution for two-dimensional heat conduction in a square geometry (see Figure 14.6), where the upper surface is held at 100° and the lower surface is at 0°. The left side is insulated, while the right side is held at 70°. Use the finite difference method with nine interior points to estimate the temperature distribution in the solid. Use fictitious points to implement the derivative boundary conditions.

**14.3.** For problem 14.1, test the applicability of Richardson extrapolation in *space* at $y = 0.5$ and $\theta = 2$ min. in the following way. By keeping time integration errors below $10^{-6}$ and hence out of the problem, calculate the solution for $T$ at $y = 0.5$ using $m = 4$, 8, and 16 divisions of the spatial interval. Use these three values to estimate the value of $p$ in the Richardson extrapolation formula $T = T_a(h) + Ch^p$. Using the integer closest to your estimate of $p$, calculate the "best" value of $T$ at $y = 0.5$ and $\theta = 2$ min. together with the estimated error in this value.

**14.4.** Let us consider problem 14.1 with the following change. Now, at the time the upper face is changed to 200°F, the lower face is suddenly *insulated*, so that $\partial T/\partial y = 0$ at that location. Using program PDEHYBD with three interior points and a fictitious

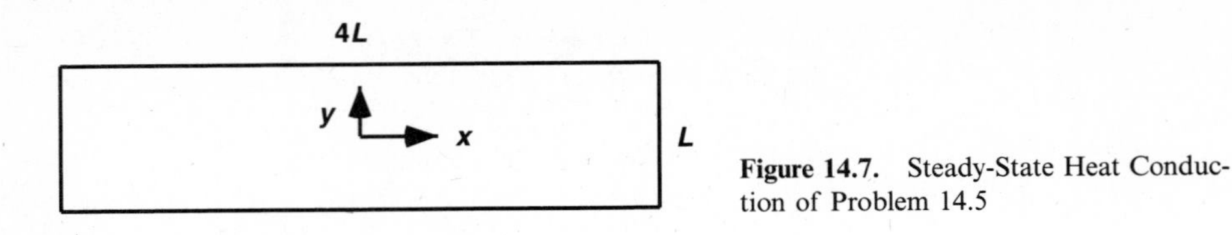


**Figure 14.7.** Steady-State Heat Conduction of Problem 14.5

boundary at the lower face, estimate how long it takes for the entire system to reach steady state.

**14.5.** We wish to consider heat conduction in a rectangular block which is very long in the third dimension and which has a cross-section as shown in Figure 14.7.

Suppose the block is initially at a uniform temperature of 500° and that the top face is suddenly changed to (and kept at) 100°. The other three faces are maintained at 500°. Using program PDEHYBD, calculate the steady-state temperature at the center of the block and how long it takes for the entire block to reach steady state. Use the dimensionless time variable $\tau = \theta\alpha/L^2$.

(i) Use three interior points in the $y$ direction, and neglect gradients in the $x$ direction.
(ii) Use three interior points in the $y$ direction only, and consider gradients in both the $y$ and $x$ directions.
(iii) Use three interior points in the $x$ direction only, and consider gradients in both the $y$ and $x$ directions.
(iv) Make comparisons as appropriate between the results of (i), (ii), and (iii).

**14.6.** The following PDE describes one-dimensional unsteady convection and diffusion.

$$\frac{\partial C}{\partial \theta} + u\frac{dC}{\partial x} = D\frac{\partial^2 C}{\partial x^2}$$

Use the central difference approximation for the spatial derivatives. For the simple explicit method (Euler) applied to time integration of this equation, determine any stability restrictions that may apply to the time or space increments. To do this, use the sufficient condition method of analysis.

**14.7.** As shown in the text, the theoretical stability limit for simple explicit (Euler) integration of the PDE $\partial C/\partial\theta = \partial^2 C/\partial y^2 - KC$ is $(\Delta\theta/h_y^2) \le 0.5/(1 + Kh_y^2/4)$. Test this limit experimentally by using program PDEHYBD ($a = 1$) with boundary conditions of $C(x = 0) = 1$, $C(x = 1) = 0$, and initial condition $C(\theta = 0, x) = 0$.

**14.8.** Referring to problem 14.7, and noting that the ODE theoretical stability limit for the hybrid procedure ($a = 0.74$) of program PDEHYBD is four times that of the simple Euler method, test experimentally whether this result applies to the PDE.

**14.9.** Consider unsteady heat conduction in a sphere (with angular symmetry) as described by the equation $\partial T/\partial\theta = \partial^2 T/\partial r^2 + (2/r)(\partial T/\partial r)$. Here, $\theta$ and $r$ are both dimensionless variables, with $r$ going from zero to unity. Initially, the sphere is at 600°F. At some time the surface of the sphere is suddenly changed to and kept at 400°F. Use the method of lines and program PDEHYBD to determine how long it takes for the center of the sphere to get within 1° of its steady-state value. Use a fictitious boundary at $r = 0$.

**14.10.** Suppose the sphere in problem 14.9 has its surface changed at time zero to the oscillatory function $400 + 50 \sin\theta$. What is the first "crossover" temperature (point where center and surface temperatures become equal), and at what time does it occur? What is the second "crossover" temperature, and at what time does it occur?

**14.11.** A particular diffusion-reaction problem is described by the PDE $\partial C/\partial\theta = \partial^2 C/\partial y^2 - kC^2$, where $\theta$ is a dimensionless time and $k$ is a dimensionless rate parameter. For this problem, we have $k = 2$ and initially $C = 0$ everywhere. At time $\theta = 0+$, $C(x = 0)$ suddenly changes from zero to unity and remains at this new value. At $y = 1$, $\partial C/\partial y = 0$ for all $\theta$.

It is desired to calculate the concentration $C$ at $y = 0.5$ when $\theta = 0.1$. Do this using three internal node points with a fictitious boundary just beyond $y = 1$ in the following two ways: (i) Use program PDEHYBD with a time step that achieves six-figure accuracy. (ii) Use program PDEFD with $\Delta\theta = 0.0005$ and $\Delta\theta = 0.0001$; use Richardson extrapolation (in time) to estimate errors and improve accuracy.

**14.12.** The PDE $\partial C/\partial\theta + \partial C/\partial x = -kf(C)$ represents a dimensionless model of an unsteady plug flow chemical reactor. We wish to solve this equation by the method of lines using program PDEHYBD. For the particular problem of interest, $k = 3$ and $f(C) = C/(1 + 2C)$. The boundary conditions are $C(\theta, x = 0) = 1$ and $C(\theta = 0, x) = 1$. From our discussion of stability for this kind of problem, we expect that a *backward* difference approximation for $\partial C/\partial x$ should be used.

Solve this problem to estimate the steady-state value of $C(x = 1)$, as well as how long it takes to achieve steady-state. Divide the spatial interval $(0 \le x \le 1)$ into 4, 8, and 16 parts. Use Richardson extrapolation (in space) to estimate improved values of $C(x = 1)$, together with error estimates. Compare with the steady-state solution obtained directly by using program IEX on the steady-state problem.

**14.13.** It is desired to calculate the velocity distribution for developed laminar flow in a duct of rectangular cross-section having one side ($2L$) equal to twice the length of the other ($L$). The PDE governing this situation is given by

$$\frac{\partial^2 u}{\partial x^2} + \frac{\partial^2 u}{\partial y^2} = \frac{1}{\mu}\frac{\partial p}{\partial z} = \text{negative constant}$$

The customary no-slip assumption requires that $u = 0$ around the periphery of the duct. In order for the results to be useful for any duct of this shape, regardless of the magnitudes of $L$ and $(1/\mu)(\partial p/\partial z)$, nondimensionalize the original equation before doing any numerical work.

(i) Employ finite difference methods to calculate the interior velocities using equal spatial increments ($h = k$). Use $h$ values of $1/2$ and $1/4$. It may be useful to make use of symmetry in the calculations.

(ii) Use the finite difference results to derive a formula relating pressure drop in the duct to the mean velocity $u_m$ given by $u_m = \int\int u\,dxdy/\int\int dxdy$. Do this double integration by successive applications of Simpson's rule (trapezoidal quadrature with one extrapolation). As a by-product of this calculation, compute numerical values of the ratio of $u_m/u_c$, where $u_c$ is the velocity at the center of the duct. From the exact solution for this problem, it is known that $u_m/u_c = 0.5$.

## REFERENCES

AMES, W. F. (1977), *Numerical Methods for Partial Differential Equations*, 2nd ed., Academic Press, New York

ANDERSON, D. A., TANNEHILL, J. C., and PLETCHER, R. H. (1984), *Computational Fluid Mechanics and Heat Transfer*, McGraw-Hill, New York

ASHOUR, S. S. (1989), *Investigation of New Methods for the Integration of Stiff Ordinary Differential Systems*, Ph D Thesis in Chemical Engineering, U. California-Santa Barbara.

ASHOUR, S. S., and HANNA, O. T. (1990), *Computers Chem. Engng.* **14**, p. 267.

CARNAHAN, B., LUTHER, H. A., and WILKES, J. O. (1969), *Applied Numerical Methods*, Wiley, New York

COLLATZ, L. (1966), *The Numerical Treatment of Differential Equations*, Springer-Verlag, New York

CRANDALL, S. (1956), *Engineering Analysis*, McGraw-Hill, New York

DAVIS, M. E. (1984), *Numerical Methods & Modeling for Chemical Engineers*, Wiley, New York

FERZIGER, J. H. (1981), *Numerical Methods for Engineering Application*, Wiley, New York

FINLAYSON, B. (1972), *The Method of Weighted Residuals and Variational Principles*, Academic Press, New York

FINLAYSON, B. (1980), *Nonlinear Analysis in Chemical Engineering*, McGraw-Hill, New York

GERALD, C. (1978), *Applied Numerical Analysis,* 2nd ed., Addison-Wesley, Reading, Mass.

HANNA, O. T. (1988), *Computers Chem. Engng.*, **12**, p. 1083.

HANNA, O. T. (1990), *Computers Chem. Engng.*, **14**, p. 273.

REDDY, J. N. (1984). *An Introduction to the Finite Element Method*, McGraw-Hill, New York

RIGGS, J. B. (1988), *An Introduction to Numerical Methods for Chemical Engineers*, Texas Tech Univ. Press, Lubbock, TX

ROACHE, P. (1976), *Computational Fluid Dynamics*, Hermosa, Albuquerque, NM

SCHEID, F. (1988), *Numerical Analysis*, 2nd ed., McGraw-Hill, New York

SCHIESSER, W. E. (1991) *The Numerical Method of Lines*, Academic Press, Inc., New York

SCRATON, R. E. (1987), *Further Numerical Methods in BASIC*, Edward Arnold, Baltimore

SILEBI, C. A., and SCHIESSER, W. E. (1992), *Dynamic Modeling of Transport Process Systems*, Academic Press, Inc., New York

SMITH, G. D. (1985), *Numerical Solution of Partial Differential Equations: Finite Difference Methods*, 3rd ed., Clarendon Press, Oxford, U.K.

STONE, H. L., and BRIAN, P. L. T. (1963), *AIChE J*, **9**, p. 681.

VILLADSEN, J., and MICHELSEN, M. L. (1978), *Solution of Differential Equation Models by Polynomial Approximation*, Prentice-Hall, Englewood Cliffs, NJ

# Appendix A

# Computer Programming

It is assumed that the reader has some familiarity with either the BASIC or FORTRAN computer languages. Programs on the enclosed diskette are given in IBM format in *both* True BASIC (TB) and FORTRAN 77 (F77), for the convenience of the reader. The only essential difference is that some TB programs contain the option for graphical output, while this is not available in the F77 programs. A reader intending to use either the True BASIC or FORTRAN 77 programs may wish to consult programming references such as those listed at the end of this Appendix. Programs in either language should run directly on IBM-compatible personal computers. The same programs (set of instructions) should also run on any other platform for which these language systems are available, such as the Apple Macintosh or UNIX workstations. As far as we know, our programs are free from errors and run correctly, but of course this can never be guaranteed. We would appreciate hearing from users regarding any significant "bugs" they may discover.

Having used both of these languages extensively, for reasons described in the preface, we personally believe that True BASIC is more useful than FORTRAN 77. At the same time, we realize that FORTRAN 77 is much better known and has been used more widely in engineering applications. Thus, we furnish our programs in both languages, so that the user can employ either one. We expect that students having a FORTRAN 77 system available may wish to use this, while others may prefer using the inexpensive student edition of True BASIC.

We have used Language Systems FORTRAN on the Apple Macintosh to develop the FORTRAN 77 programs. This was done using the ANSI F77 warning facility to be sure that the code was "generic" FORTRAN 77, without any of the many extensions that are available in various FORTRANS. In this way we are confident that our programs will run in many (we hope virtually all) FORTRAN environments which are at least of the level of ANSI FORTRAN 77. Although guarantees are never possible when using different versions of FORTRAN, tests of some of our programs with four different versions (one for a UNIX workstation, two for IBM compatibles, and the Language Systems FORTRAN for the Macintosh) proved successful.

We will give below some of the most important programming information about both True BASIC and FORTRAN 77. Subsequently, an example program will be discussed in detail for each language to illustrate the syntax.

## A.1 SOME GENERAL ADVICE AND HINTS ON PROGRAMMING

For all programs given in this book (either True BASIC or FORTRAN 77), the student should first read the instruction comments and the "how to run" box at the beginning of the program; then look at the user-defined subroutines at the end of the program. Next run the current problem specified in the subroutine to observe program behavior. Following this, edit the present program to adapt it to your problem. Using this procedure, the programs are quite easy to use.

In changing problems within a subroutine, it is often very convenient to "comment out" that part of the program which you don't want to use, yet don't want to lose. That is, to render a statement inactive, all we need to do is put either a "!" (True BASIC) or a "C" (FORTRAN 77) in column one of the statement. To "comment" (or uncomment) a number of lines at once in True BASIC, it is only necessary to mark or select the lines and press Shift-!.

It can hardly be overemphasized that it is useful to *document* a program as carefully as possible. To this end it is very useful to employ plenty of explanatory comments. True BASIC allows "tail" comments; that is, comments at the end of a program statement.

In regard to printing a hard copy of program output, it is usually a good idea to de-bug the program, if possible, prior to a lot of printing. Often the most convenient way to print hard copy is to send the output to a file called (for example) OUTPUT and then edit (if desired) and print the file. Creation of this file is accomplished in True BASIC by using the command "run >> output". Our FORTRAN 77 programs have been set up to write the output to file OUTPUT in addition to the screen.

Remember that when *saving* a program, if one simply saves using the current program name, *the previous version will generally be overlaid.* Many times this is what is desired. But if there is ever a question as to whether the old version may still be needed, save the new program under a new name so that both versions will be retained. Finally, it is useful to note that in True BASIC it is very easy to keep a collection of problems for a given program (in case it is ever desired to either re-run or re-edit a particular problem). This is accomplished by using the KEEP/INCLUDE commands. KEEP is useful for easily isolating a problem subroutine for purposes of saving it under its own name. Then, whenever it is desired to re-use this problem subroutine, we can INCLUDE its file at the end of the desired program needed to run it.

We now move ahead to individual discussions of True BASIC and FORTRAN 77.

## A.2 TRUE BASIC (TB)

The computer programs referred to in this book and given on the diskette were written in the True BASIC computer programming language which was developed by John Kemeny and Thomas Kurtz, the originators of the BASIC language. Although the authors of this text grew up on FORTRAN, True BASIC has a number of important attributes which caused us to prefer it for instructional purposes, and indeed for serious computing. In particular, True BASIC follows the new American National Standards (ANS) guidelines, it is inexpensive (there is a student edition for about $25), it contains excellent language structures, local and global variables, and both internal and external subroutines. The internal accuracy is high (16 figures for the Macintosh with math co-processor), the editing and debugging are simple but powerful, the intermediate code is very fast, and there are built-in matrix calculations and graphics. Also the same True BASIC *programs* (you need the proper language system disk) run on IBM, Apple Macintosh, Amiga and Tandy computers; a version is available for UNIX workstations. True BASIC is easily converted to other languages such as FORTRAN, C, etc., if desired. In fact, the FORTRAN 77 programs given on the diskette were converted from True BASIC.

Our experience using True Basic in various classes has been very favorable; students like it and are able to perform well using it. At UCSB we have used these programs on both IBM and Apple Macintosh systems. The programs that we use are generally organized in the following way. First, a concise description of the program is given, followed by a "how to run" box indicating the essential information needed to run the program. The user generally need only prescribe one or two subroutines at the end of the program which specify the particular problem to be solved. This is easily accomplished (through editing) by altering the problem originally contained in the program. It is recommended that copies of all programs be made before any alterations are made.

An introductory discussion and example of how True BASIC works will be discussed next. For further information, consult the True BASIC Reference Manual or the True BASIC User Guide for the particular computer system being used. First, we give some brief tables of information for using True Basic on IBM or Macintosh computers.

True BASIC control structures include the FOR-NEXT loop for loops to be executed a specific number of times, and the DO-LOOP (WHILE, UNTIL) structure for either infinite or conditional loops. The IF-THEN-ELSE structure is probably the most-used control structure, while the SELECT CASE statement is also valuable.

In contrast to FORTRAN 77, True BASIC does *not* require line numbers or special double precision calculations (TB generally uses a minimum of 10 figure

**TABLE A.1** MOST COMMON COMMANDS FOR TRUE BASIC (TB)

| To | On MAC use | On IBM use |
|---|---|---|
| get into TB | click TB icon | type HELLO from DOS |
| leave TB | Quit/File menu | BYE |
| load stored program | Open/File menu | OLD filename |
| run current program | Run/Run menu | RUN |
| stop current program | Stop/Run menu | Stop key |
| restart program | Continue/Run menu | CONTINUE |
| print listing of current program | Print/File menu | LIST |
| create new program | type in | type in Edit mode |
| save a program | Save As/File menu | SAVE filename |
| send output to a file for edit/printing | Run >> filename | Run >> filename |
| turn printer on/off for program output | on-command ECHO off- "ECHO OFF | on-command ECHO off- "ECHO OFF |

**TABLE A.2** PROGRAM SYNTAX

| Operation | Example |
|---|---|
| Constants—scientific notation | $2e - 12$ is $2x10^{-12}$ |
| Assignment | Let $y = x$ assigns value of $x$ to $y$ |
| Arithmetic operators | (+, −, *, /, ^) |
| Built-in functions | Exp($x$), Sin($x$), many more |
| `READ, DATA,` | To input data |
| `INPUT, INPUT PROMPT"":` | To input data |
| Functions | Internal, external |
| Subroutines | Internal, external |
| Variable names | Up to 31 characters |

accuracy without a math co-processor and 16 figures with a co-processor). Also, it does not use variable types with mixed-mode arithmetic (real vs. integer) or rigid positioning of statements in particular columns.

Some further comments about editing include:

(i) Typing at cursor squeezes in new text
(ii) To make room for a new line at the cursor, press Enter
(iii) For comments we can use REM or !—everything from ! to the end of the line is a comment; thus "tail" comments after a statement are allowed.
(iv) On IBM, command SPLIT $n$ divides edit and command windows at line $n$ ($0 < n < 24$) and shows current program name and function key designation.

**TABLE A.3** SIMPLE EDITING

| Editing Key | Action |
|---|---|
| Backspace key | Erase to left of cursor, squeeze up |
| Delete key | Erase at cursor, squeeze up; deletes line on line-tag |
| Arrow key | Moves cursor (hold down the keys to repeat) |
| Insert key | Overtype (switches on and off) |
| Home key | Goes to start of file |
| End key | Goes to end of file |
| Page Up key | Goes up one screen-full |
| Page Down key | Goes down one screen-full |
| Fl key (IBM only) | Moves cursor to edit window |
| F2 key (IBM only) | Moves cursor to command window |

(iv) To get color on IBM (each window separate) use Ctrl-f (foreground), Ctrl-b (background), Ctrl-e (edge)

**Example A.1 A Sample Computer Program in True BASIC**

We use the following program to illustrate most of the simple features of True BASIC. The program is for the solution of a system of nonlinear algebraic equations using the Newton method, where the user must furnish the required partial derivatives as well as the equations to be solved. This general problem is discussed in Chapter 7 on nonlinear algebraic equations. We first show the program and then discuss it.

Note that we have put in line numbers for convenient reference in this discussion. *Line numbers are not used in the actual program.*

## Program NEWTNLE.

```
1000 !This is program NEWTNLE.TRU for a system of nonlinear eqns. by
1010 !O.T.Hanna. The program solves a system of nonlinear algebraic
1020 !equations by the Newton method.
1030 !**********************************************************
1040 !*To run we must specify the n equations f(i) as a function   *
1050 !*of x(j) in Sub Func, as well as the partial derivatives      *
1060 !*df(i)/dx(j) = f1(i,j), max. number of iterations(mx) and    *
1070 !*convergence tolerance parameter norm_tol. Specify initial   *
1080 !*values of x(j) in Sub Init. Program prompts for option to   *
1090 !*over-ride initial values or verify f(i),df(i)/dx(j).        *
1100 !*Program is stopped by either convergence or max. iterations.*
1110 !**********************************************************
1120 DO
1130    DIM x(1),f(1),f1(1,1),a(1,1),xc(1),a_inv(1,1)
1140    CALL Init(n,x())
1150    INPUT prompt"Wish to over-ride Sub initial values(0-No,1-Yes)?":q
1160    IF q = 1 THEN
1170       MAT INPUT prompt"Initial values of x(1).x(2)..? ":x(?)
```

```
1180    END IF
1190    MAT redim x(n),f(n),f1(n,n),a(n,n),xc(n)
1200    LET m = 0                          !initialize iteration counter
1210    PRINT
1220
1230    !Option to verify correctness of functions f(i),df(i)/dx(j)
1240    INPUT prompt"Verify f(i),df(i)/dx(j); No(0),Yes(1) ? ":ver
1250    IF ver <> 0 then
1260       CALL Func(x,f,f1,mx,norm_tol)
1270       FOR i = 1 to n
1280           PRINT"f(";Str"$(i);") = ";f(i)
1290       NEXT i
1300       PRINT
1310       FOR i = 1 to n
1320           FOR j = 1 to n
1330               PRINT"f1(";Str$(i);",":Str$(j);") =";f1(i,j)
1340           NEXT j
1350       NEXT i
1360       STOP
1370    END IF
1380    CLEAR
1390
1400    DO
1410       !Newton calculation
1420       CALL Func(x(),f(),a(,),mx,norm_tol)
1430       MAT a_inv = inv (a)
1440       MAT xc    = a_inv * f     !x = x - {[df(i)/dx(j)]^(-1)}*f(j)
1450       MAT  x    = x - xc
1460
1470       FOR j = 1 to n
1480           PRINT"x(";Str$(j);") =";x(j);Tab(25);
1490           PRINT"xc(";Str$(j);") =";xc(j)
1500       NEXT j
1510       LET m = m + 1                   !increment iteration counter
1520       LET norm = Sqr(Dot(xc,xc))  !(sum{[xc(j)^2)]^(1/2)
1530       PRINT"number of iterations =";m,"norm =";norm
1540       PRINT
1550    LOOP until m >= mx or norm < norm_tol
1560    IF norm > norm_tol then PRINT"NO CONVERGENCE"
1570 LOOP
1580 !End of main program
1590 END
1600
1610 !This is NEWTNLE.tru of problem below
1620 SUB Init(n,x())
1630      !***denotes user-specified statement
1640      LET n = 2                        !***number of equations
1650      MAT redim x(n)
1660      LET x(1) = 1                     !***initial values of variables
1670      LET x(2) = 5
```

```
1680 END SUB
1690
1700 SUB Func(x(),f(),f1(,),mx,norm_tol)
1710      !***denotes user-specified quantity
1720      LET mx = 100                    !***max allowed iterations
1730      LET norm_tol = 1e-7             !***convergence tolerance
1740
1750      !***Specify Eqns
1760      LET f(1) = x(1)^2 + x(2) - 7
1770      LET f(2) = x(2)^2 + x(1) - 11
1780
1790      !***Specify n x n partial derivatives as df(i)/dx(j) = f1(i,j)
1800      LET f1(1,1) = 2*x(1)
1810      LET f1(1,2) = 1
1820      LET f1(2,1) = 1
1830      LET f1(2,2) = 2*x(2)
1840 END SUB
```

We now give a brief explanation of the program. First, let us look at the function subroutine SUB Func beginning at line 1700, where the user defines the equations to be solved, the partial derivatives of the equations with respect to the variables, and the mx and norm_tol parameters. As in all of our user subroutines, the asterisks (***) in the tail comments after a statement generally signify that this statement is one which is to be specified by the user. In many instances, the user may simply wish to use some nominal value that is already set. For example, the values mx = 100 and norm_tol = 1e-7 ($10^{-7}$) are useful in many problems and could probably be used as they are. The functions f(1) and f(2) in lines 1760 and 1770 represent the two equations whose common roots are to be found. The partial derivatives df(i)/dx(j) (lines 1800-1830) must be specified by the user as part of the Newton algorithm. The arguments of the subroutine (line 1700) work as follows. The parentheses denote the number of dimensions of an array variable; thus f() is a one-dimensional array variable (vector), and f1(,) is a two-dimensional array variable (matrix). The subroutine arguments (or parameters) transfer information both ways; when we CALL Func from the main program (line 1420), the call transmits the current x( ) values to the subroutine for use in evaluating the functions and partial derivatives. On the other hand, information regarding the current computed values of f( ), f(,), mx, and norm_tol is transmitted back to the main program via the return.

Looking now at the beginning of the main program (line 1000), we see several lines of informational comments followed by the "How to Run" box (line 1040), which is enclosed by asterisks (*). The first DO LOOP (lines 1120 to 1570) is an *infinite* loop, which can only be exited by stopping the program with the appropriate break command. This is a convenient way to set up the program to keep repeating (using new initial values) as long as desired. The dimension statement (line 1130) has entries of unity since we will subsequently redimension (in statement MAT redim, line 1190) to provide proper dimensions for the array variables as soon as they are known from input by the user. The proper initial values of $n$ and x(1), x(2) are all specified by the user in SUB Init (line 1620), where it is noted that it is necessary to re-dimension the x(i) after *n* has been specified. To transmit the initial information from SUB Init to the main program, we use the CALL Init statement in line 1140. Following this, we use an INPUT-prompt statement to allow the user to over-ride the initial values specified in SUB Init, if desired. If that option is not selected, the input values remain as specified in SUB Init. Next, the iteration counter m is initialized (line 1200), after which we provide the option for verifying the functions f(i) and the partial derivatives df(i)/dx(j) (lines 1230 to 1370), which will be computed with the specified initial values. This is important for discovering programming errors. We are now ready to implement the Newton algorithm.

The CLEAR statement (line 1380) clears the screen for the subsequent output. The DO LOOP-until construct (lines 1400 to 1550) allows us to have the Newton algorithm continue until either the required convergence criterion is met (norm < norm_tol) or the specified maximum number of iterations (mx) is exceeded. This is very convenient, since it can't be known in advance how many iterations may be necessary. The Newton calculation itself involves first a call to subroutine Func (lines 1700 to 1840) in order to obtain mx, norm_tol, and the function and partial derivative values based on the initial user-specified x( ). Next, the matrix of partial derivatives a(,) undergoes a matrix inversion in the statement MAT a_inv = inv (a) (line 1430, built-in True BASIC matrix inversion), followed by a matrix multiplication and addition (lines 1440 and 1450) to form the Newton iteration as shown in the tail comment.

The rest is mostly bookkeeping. Thus, we print the calculated new x and correction (xc) values (lines 1480, 1490), increment the iteration counter m (line 1510), and calculate the norm of the corrections (line 1520) using the built-in function Sqr (square root) and Dot (dot product of the vector xc), defined as the sum of the squares of the components of xc, as we attempt to depict in the tail comment. At the LOOP-until (line 1550), we stop because either we have exceeded the maximum number of iterations (unsuccessful), or because the computed norm is less than the specified norm_tol value (successful); otherwise, we continue. If we were unsuccessful, we print "NO CONVERGENCE" (line 1560). The output of the calculation appears on the screen. If printed hard-copy results are desired, the most convenient procedure is to use the "run >> filename" command discussed earlier.

## A.3 FORTRAN 77 (F77)

We now give some review information regarding the rules and syntax of FORTRAN 77, as well as certain programming structures that we have generally utilized. As necessary, the user should consult a FORTRAN 77 reference book, such as those cited at the end of this Appendix. It was convenient for us ( and we hope also for the reader) to develop the FORTRAN 77 programs in a way that closely parallels the True BASIC programs. Since generic FORTRAN 77 does not contain an editor, any required editing must be carried out in accordance with whatever edit facilities are available to the user. This might be a DOS, or a FORTRAN which contains an editor, or a word processing application, etc. We will concentrate our discussion on the rather rigid (and unforgiving) rules and syntax required by a successful FORTRAN 77 program.

### A.3.1 Fundamentals

Names for variables, programs, arrays, and subroutines must contain from 1 to 6 alphanumeric characters (numbers or letters), with the first character a letter. One is allowed to use only columns 1 through 72 for F77 statements. Aside from comments, statement numbers, and line continuation markers, F77 statements cannot begin before column 7. We have used a comment line of 72 columns to be sure that statements do not exceed this length. For statements longer than 72 columns, an F77 line can be continued by use of a + in column 6. A "C" in column 1 denotes a comment line, which is not processed. Statement numbers, when necessary, go in columns 1-5.

The assignment statement X = Y is not an algebraic statement , but rather means "assign to X the value of Y." Mixed-type arithmetic (integer and real) must be used with care. Thus, we use the DOUBLE PRECISION AND INTEGER type statements as necessary.

It has been our experience that in extensive calculations, very often roundoff errors affect the results so much that double precision arithmetic (14 places) is necessary. Since we normally wish to obtain four, five, or six significant-figure accuracy, we have adopted the policy of doing all of our FORTRAN 77 calculations with double precision. To accomplish this in an efficient way, we generally use the IMPLICIT DOUBLE PRECISION (A-H, O-Z) statement near the beginning of each main program. In this way, the usual convention of variables starting with I,J,K,L,M,N being of INTEGER type is followed. For exceptions to this rule and to provide dimensioning for array variables, we use additional DOUBLE PRECISION and INTEGER type statements as necessary. To maintain the desired double precision arithmetic within all calculations, we must be sure that all required variables are taken as double precision in the main program and also in all subroutines. In addition, any constants should be specified as double precision. For example, a double precision constant of unity should be programmed as 1.D0, and so on. A reminder of this is given in the instructions to all of the F77 programs. Double precision functions begin with a prefix of D, as in DSQRT for the square-root function.

We use the PARAMETER statement in general to define an output unit number to write to the file OUTPUT (see Output Formatting section), and also to specify any problem-dependent number of equations. These are values that the user may need to specify, and a reminder of this is given in the instructions to all of the F77 programs.

Do *not* raise a real expression to an integer power. For example, do not use A**(1/3), but instead use either A**(1./3.) for single precision real arithmetic or A**(1.D0/3.D0) for double precision real arithmetic. Also, a negative expression must *not* be raised to a real power. Use arguments of correct type in functions; for example, DSQRT(2) and DSQRT(I) are both wrong, since 2 and I are not real.

Array declarations must be made; they can include lower and upper bounds as in the one dimensional array A(−3:9) where the lower bound is −3 and the upper bound 9, or the two dimensional array C(0:6,0:6) where each dimension has a lower bound 0 and upper bound 6. We cannot use dummy (variable) dimensions for an array unless it is an argument in a subroutine (for example, see F77 program DELSQ.F).

The procedures for compiling, running, and stopping an F77 program are *system dependent*, and therefore cannot be specified once and for all. Some possible stop commands (which may be useful for breaking out of an "infinite" loop) include Control-Break, Control-S, and Control-Z. In some systems it is necessary to delete any "old" OUTPUT file before running an F77 program which opens a "new" OUTPUT file.

### A.3.2 Control Structures and Loops

The most common control and loop structures in F77 are the IF-THEN-ELSE structure and the DO loop. The IF-THEN-ELSE structure is virtually the same in F77 as in True BASIC, except that the "conditions" to be tested for truth are expressed in a slightly different way. The ordinary DO loop in F77 corresponds to the FOR-NEXT loop in TB.

We have found the use of an "infinite" loop in TB (DO-LOOP) to be very convenient in many applications. In F77, a practical form of the infinite loop can be obtained from the following code:

```
DO 100 J=1,1E6
..............
100 CONTINUE
```

This is actually a finite loop which may be traversed up to one-million times; in all the practical programs where we employ this construct, the loop will be stopped or exited prior to the one-million cycles.

A slight extension of the above "infinite" loop corresponds to the very useful DO-LOOP UNTIL construct in TB. We have used this often and a practical F77 version of this can be constructed by the subsequent code:

```
DO 1000 J=1,1E6
...............
IF (Condition) GO TO 1001
1000 CONTINUE
1001 CONTINUE
```

In this case the loop continues to cycle (up to as many as one-million times) until the "Condition" is found to be true.

The SELECT CASE structure is sometimes useful in TB; in F77 we use instead the COMPUTED GO TO statement as in GO TO (10,17,41,70) i. When i has the value 1, control is transferred to statement number 10, when i = 2 control is transferred to statement number 17, and so on. We use this construct particularly in applications where we are calculating numerical partial derivatives of functions in a subroutine.

### A.3.3 Output Formatting

Formatting output to either a monitor screen or to a hard copy device is one rather notorious aggravation of F77. There are various ways that this can be accomplished;

we have chosen a method that in spirit is very similar to the convenient output formatting of TB. Instead of using PRINT statements, we generally use appropriate forms of WRITE statements to output both to the screen and to a file called OUTPUT. One output difficulty in F77 is that the "standard" unit number in the WRITE statement (which is, for example, often unit number 6 as in WRITE (6,23), where 23 is a typical FORMAT statement number) sometimes varies from one facility to another. To minimize difficulties of this kind, we have employed a dummy unit number (NU) for writing to the file OUTPUT in our programs. This number (which we have generally taken to be 8) must be specified in the PARAMETER statement near the beginning of the main program. If it should happen that the reader cannot use NU = 8 on his/her system, then the NU value in the parameter should be changed accordingly. We show examples of the output formats that we use:

For list directed format (for writing a heading ) we use, for example

```
WRITE (*,*) 'Error Control' (prints to screen)
WRITE (NU,*) 'Error Control' (prints to file OUTPUT)
```

This would simply print the heading "Error Control"
For formatted output we use, for example

```
WRITE (*,23) 'X =',X1,'Y =',Y2 (prints to screen)
WRITE (NU,23) 'X =',X1,'Y =',Y2 (prints to file OUTPUT)
23 FORMAT (' ',A,D14.6,T35,A,D14.6)
```

This would print (on the next line) in either column 1 or 2, "X = (value of X1)", then skip to column 35 and print "Y = (value of Y2)." The reason for the first entry of 'space' in the FORMAT specification is that some systems will print the first character in the FORMAT specification and others will not. This first character is sometimes used for carriage control (then it is not printed). Thus by taking the first entry as 'space' we can be sure that if the first character is printed, it will be a blank and not interfere with our printed output. If our first "blank" entry is used for carriage control, printing begins on the next line. The FORMAT specification in statement 23 above is interpreted as follows. We have already discussed the advantage of taking the first entry to be a character blank. The first "A" specification refers to the character variable 'X =' in the WRITE statements; the "A" allows just enough space for however many characters are contained between the single quote (') marks. Next, the D14.6 specification indicates 14 spaces for the variable X1, with 6 of those spaces after the decimal point. The "D" refers to double precision (we could also use E or F instead) and is a reminder that we generally wish to have all calculations in double precision. The specification T35 (Tab 35) places the next entry (Y =) in column 35. Finally, the last two specifications A and D14.6 are merely repetitions of those used earlier.

**Example A.2 A Sample Computer Program in FORTRAN 77**

We now use the same algorithm employed in Example A.1 above (for True BASIC) to illustrate most of the simple features of FORTRAN 77. The program is for the solution of a system of nonlinear algebraic equations using the Newton method, where the user must furnish the required partial derivatives as well as the equations to be solved. This general problem is discussed in Chapter 7 on nonlinear algebraic equations. We first show the program and then discuss it. The extension ".F" in our program NEWTNLE.F stands for FORTRAN; this extension depends on the system being used. The most common FORTRAN extensions are ".F" and ".FOR".

Note that we have put in *extra* line numbers for convenient reference in this discussion. *The extra line numbers are not used in the actual program.*

## Program NEWTNLE.

```
1000       PROGRAM NWTNLE
1010 C...THIS IS PROGRAM NEWTNLE.F, A FORTRAN 77 PROGRAM FOR A SYSTEM OF
1020 C...NONLINEAR ALGEBRAIC EQUATIONS BY O.T.HANNA.THE PROGRAM SOLVES
1030 C...A SYSTEM OF NONLINEAR ALGEBRAIC EQUATIONS BY THE NEWTON METHOD.
1040 C...USER MUST SPECIFY NO. OF EQNS. N IN PARAMETER STATMENT BELOW.
1050 C...FOR DESIRED DOUBLE PRECISION REMEMBER TO (I) GIVE CONSTANTS AS
1060 C...(FOR EXAMPLE) 1.D0/7.D0, NOT 1./7., AND (II) SPECIFY ANY NEW
1070 C...VARIABLES AS DOUBLE PRECISION TYPE.
1080 C...IF REQUIRED FOR YOUR SYSTEM, CHANGE THE WRITE UNIT NO. (NU)
1090 C...LOCATED IN THE *PARAMETER* STATEMENT BELOW.
1100 C...TO EXIT FROM INFINITE LOOPS, USE CONTROL-BREAK ON IBM,
1110 C...COMMAND . ON APPLE MACINTOSH.
1120 C*************************************************************
1130 C*USER MUST SPECIFY NO. OF EQNS. N IN PARAMETER STATEMENT BELOW   *
1140 C*TO RUN WE MUST SPECIFY THE N EQUATIONS F(I) AS A FUNCTION      *
1150 C*OF X(J) IN SUB FUNC, AS WELL AS THE PARTIAL DERIVATIVES        *
1160 C*DF(I)/DX(J) = F1(I,J), MAX. NUMBER OF ITERATIONS(MX) AND       *
1170 C*CONVERGENCE TOLERANCE PARAMETER NRMTOL. SPECIFY INITIAL VALUES  *
1180 C*OF X(J) IN SUB INIT. PROGRAM PROMPTS FOR OPTION TO OVER-RIDE    *
1190 C*INITIAL VALUES OR TO VERIFY F(I),DF(I)/DX(J).                  *
1200 C*PROGRAM IS STOPPED BY EITHER CONVERGENCE OR MAX. ITERATIONS.   *
1210 C*************************************************************
1220
1230      IMPLICIT DOUBLE PRECISION (A-H,O-Z)
1240
1250 C...***USER MUST SPECIFY NO. OF EQNS. N IN PARAMETER STATEMENT BELOW
1260 C...**********************************************************
1270      PARAMETER (NU = 8,N = 2)
1280 C...**********************************************************
1290 C...***USER MAY NEED TO ALTER UNIT NO. NU IN PARAMETER STATEMENT ABOVE
1300
1310      DOUBLE PRECISION X(N),F(N),F1(N,N),DIFF(N),NORM,NRMTOL
1320      INTEGER Q,VER
1330      OPEN (UNIT=NU,FILE='OUTPUT',STATUS='NEW')
1340
1350 C...THE LINE BELOW HAS 72 CHARACTERS;THIS ALLOWS CHECK OF <= 72
        COLUMNS
1360 C...AS REQUIRED FOR FORTRAN 77
1370
C12345678901234567890123456789012345678901234567890123456789012345678901
```

```
1380
1390 C...START OF "INFINITE" LOOP
1400      DO 1000 I1 = 1,1E6
1410         CALL INIT(N,X)
1420         WRITE (*,*) 'WISH TO OVER-RIDE SUB INITIAL VALUES(O-NO,1-YES)?'
1430         WRITE (NU,*)'WISH TO OVER-RIDE SUB INITIAL VALUES(O-NO,1-YES)?'
1440         READ*,Q
1450         IF (Q .EQ. 1) THEN
1460            DO 1 I=1,N
1470               WRITE (*,2)  'INPUT VALUE OF X(',I,')'
1480               WRITE (NU,2) 'INPUT VALUE OF X(',I,')'
1490 2             FORMAT (' ',A.I2,A)
1500               READ*,X(I)
1510 1          CONTINUE
1520         END IF
1530         WRITE (*,*)
1540         WRITE (NU,*)
1550 C...INITIALIZE ITERATION COUNTER
1560         M = 0
1570 C...OPTION TO VERIFY CORRECTNESS OF FUNCTIONS F(I),DF(I)/DX(J)
1580         WRITE (*,*)  'VERIFY F(I),DF(I)/DX(J); NO(O),YES(1) ? '
1590         WRITE (NU,*) 'VERIFY F(I),DF(I)/DX(J); NO(0),YES(1) ? '
1600         READ*,VER
1610         IF (VER.NE.0) THEN
1620            CALL FUNC(N,X,F,F1,MX,NRMTOL)
1630            DO 10 I = 1,N
1640               WRITE (*,15)  'F(',I,')=',F(I)
1650               WRITE (NU,15) 'F(',I,')=',F(I)
1660 15            FORMAT (' ',A,I2,A,D14.6)
1670 10         CONTINUE
1680            WRITE (*,*)
1690            WRITE (NU,*)
1700            DO 20 I = 1,N
1710               DO 30 J = 1,N
1720                  WRITE (*,31)  'F1(',I,',',J,') =',F1(I,J)
1730                  WRITE (NU,31) 'F1(',I,',',J,') =',F1(I,J)
1740 31               FORMAT (' ',A,I2,A,I2,A,E14.6)
1750 30            CONTINUE
1760 20         CONTINUE
1770         END IF
1780         WRITE (*,*)
1790         WRITE (NU,*)
1800 C...START OF 2ND "INFINITE" LOOP
1810         DO 500 I2=1,1E6
1820 C...NEWTON CALCULATION
1830            CALL FUNC(N,X,F,F1,MX,NRMTOL)
1840 C...SOL'N OF 0 = FA + (DF/DX)A*(X-XA) = F + F1*DIFF; X = XA - DIFF
1850            CALL LNEQ(N,F1,F,DIFF,DET,RELERR)
1860 C...-DIFF IS XCORRECTION,XCORR SO THAT X = XA + XCORR
1870            DO 40 I=1,N
```

```
1880                X(I) = X(I) - DIFF(I)
1890 40          CONTINUE
1900 C...INCREMENT ITERATION COUNTER
1910             M = M + 1
1920 C...SUM OF [(DIFF(J)**2.DO)]**(1.DO/2.D0)
1930             CALL DOT(N,DIFF,DIFF,PROD)
1940             NORM = DSQRT(PROD)
1950             WRITE (*,55)  'NUMBER OF ITERATIONS =',M,'NORM =',NORM
1960             WRITE (NU,55) 'NUMBER OF ITERATIONS =',M,'NORM =',NORM
1970 55          FORMAT (' ',A,I4,T35,A,D14.6)
1980             DO 50 J=1,N
1990                WRITE (*,51) 'X(',J,') =',X(J),'XCORR(',J,') =',-DIFF(J)
2000                WRITE (NU,51)'X(',J,') =',X(J),'XCORR(',J,') =',-DIFF(J)
2010 51             FORMAT (' ',A,I2,A,D14.6,T25,A,i2,A,D14.6)
2020 50          CONTINUE
2030             IF ((M.GE.MX) .OR. (NORM.LT.NRMTOL)) GO TO 501
2040 500      CONTINUE
2050 501      CONTINUE
2060          IF (NORM.GT.NRMTOL) THEN
2070             WRITE (*,*)  '**NO CONVERGENCE**!!'
2080             WRITE (NU,*) '**NO CONVERGENCE**!!'
2090          END IF
2100          WRITE (*,*)
2110          WRITE (NU,*)
2120 1000 CONTINUE
2130 C...END OF MAIN PROGRAM
2140      END
2150
2160       SUBROUTINE LNEQ(N,A,B,X,DET,RELERR)
2170 C...THIS IS SUB FOR SOLVING A SET OF LINEAR EQUATIONS
2180      IMPLICIT DOUBLE PRECISION (A-H,O-Z)
2190 C...SUB IS DIMENSIONED FOR UP TO 50 EQUATIONS
2200      DOUBLE PRECISION A(N,N),A1(50,50),B(N),B1(50),X(N),X1(50,2),
2210     + RELERR(N),R(50),DET
2220 C...THE REST OF THIS SUBROUTINE IS NOT SHOWN FOR REASONS OF SPACE
2230       .
2240       .
2250       .
2260      RETURN
2270      END
2280
2290      SUBROUTINE DOT(N,Y,Z,PROD)
2300 C...SUBROUTINE FOR DOT PRODUCT OF VECTORS Y(I) AND Z(I)
2310      DOUBLE PRECISION Y(N),Z(N),PROD
2320      PROD = 0.D0
2330      DO 10 I = 1,N
2340         PROD = PROD + Y(I)*Z(I)
2350 10   CONTINUE
2360      RETURN
2370      END
```

```
2380
2390       SUBROUTINE INIT(N,X)
2400 C...THIS IS NEWTNLE.FOR SOLUTION OF PROBLEM BELOW
2410       DOUBLE PRECISION X(N)
2420 C...***USER SPECIFIES INITIAL VALUES OF VARIABLES
2430       X(1) = 1.DO
2440       X(2) = 5.DO
2450       RETURN
2460       END
2470
2480       SUBROUTINE FUNC(N,X,F,F1,MX,NRMTOL)
2490       DOUBLE PRECISION X(N),F(N),F1(N,N),NRMTOL
2500 C***DENOTES USER-SPECIFIED QUANTITY
2510 C***MX IS MAX ALLOWED ITERATIONS AND NRMTOL IS CONVERGENCE TOLERANCE
2520       MX = 100
2530       NRMTOL = 1.D-7
2540 C***SPECIFY EQNS
2550       F(1) = X(1)**2.D0 + X(2) - 7.D0
2560       F(2) = X(2)**2.D0 + X(1) - 11.D0
2570
2580 C***SPECIFY N X N PARTIAL DERIVATIVES AS DF(I)/DX(J) = F(I,J)
2590       F1(1,1) = 2.D0*X(1)
2600       F1(1,2) = 1.D0
2610       F1(2,1) = 1.D0
2620       F1(2,2) = 2.DO*X(2)
2630       RETURN
2640       END
```

We now give a brief explanation of the program. First let us look at the function subroutine FUNC beginning at line 2480, where the user defines the equations to be solved, the partial derivatives of the equations with respect to the variables, and the MX and NRMTOL parameters. As in all of our user subroutines, the asterisks *** in comment statements generally signify user-specified quantities. In many instances, the user may simply wish to use some nominal value that is already set. For example, the values MX = 100 and NRMTOL = 1.D-7 ($10^{-7}$ in double precision) are useful in many problems and could probably be used as they are. The functions F(1) and F(2) in lines 2550 and 2560 represent the two equations whose common roots are to be found. The partial derivatives df(i)/dx(j) (lines 2590-2620) must be specified by the user as part of the Newton algorithm.

The arguments of the subroutine (line 2480) work as follows. The dimensions of the array variables are specified in the DOUBLE PRECISION statement 2490 in terms of the integer variable N which is passed on from the main program through the argument list. Thus F is a one-dimensional array variable (vector) and F1 is a two-dimensional array variable (matrix). The subroutine arguments (or parameters) transfer information both ways; for example, when we CALL Func from the main program (line 1830), the call transmits the current X values to the subroutine for use in evaluating the functions and partial derivatives. On the other hand, information regarding the current computed values of F, F1, MX, and NRMTOL is transmitted back to the main program via the return.

The FORTRAN program title NWTNLE at line 1000 is abbreviated from the filename NEWTNLE.F, since a maximum of six characters are allowed for the title. Looking now down from the beginning of the main program, we see several lines of informational comments followed by the "How to Run" box (line 1130), which is enclosed by asterisks (*). The first DO LOOP (lines 1400 to 2120) is an "*infinite*" loop, which can only be exited by stopping the program with the appropriate "break" command. This is a convenient way to set up the program to keep repeating (using new initial values) as long as desired. The IMPLICIT DOUBLE PRECISION (A-H,O-Z) statement (line 1230) specifies that all variables beginning with any letter *except* I,J,K,L,M, or N will be real double precision variables. By default, unless otherwise specified in a type statement, any variables beginning with I,J,K,L,M, or N will be integer variables. We see examples of type statements over-riding the default conditions in the DOUBLE PRECISION and INTEGER statements in lines 1310 and 1320. The dimensioning of real double precision arrays is accomplished in the DOUBLE PRECISION statement. Note that arrays that occur in subroutines must be dimensioned there; an array occurring in both the main program and in a subroutine must be dimensioned in both places (see line 2490).

In the PARAMETER statement (line 1270), the user must specify the number of equations N, and *possibly* an alternative unit number NU for the file OUTPUT, which is opened in the OPEN statement of line 1330. As explained in the section on Output Formatting, we use NU = 8; if this does not work on the user's system, try a different integer value.

The proper initial values of X(1), X(2) are all specified by the user in SUB INIT at line 2390. To transmit the initial information form SUB INIT to the main program we use the CALL INIT statement in line 1410. Following this, we use a prompt structure (lines 1420 tp 1520) to allow the user to over-ride the initial values specified in SUB INIT, if desired. If that option is not selected, the input values remain as specified in Sub INIT. Next the iteration counter M is initialized (line 1560), so we then provide the option for verifying the functions F(I) and the partial derivatives df(i)/dx(j) = F1(I,J) (lines 1570 to 1770), which will be computed with the specified initial values. This is important for discovering programming errors. We are now ready to implement the Newton algorithm.

The DO LOOP-until construct (lines 1810 to 2050) allows us to have the Newton algorithm continue until the required convergence criterion is met (NORM < NRMTOL) or the specified maximum number of iterations (MX) is exceeded. This is very convenient, since it can't be known in advance how many iterations may be necessary. The Newton calculation itself involves first a call to subroutine FUNC (lines 2480 to 2640) in order to obtain MX, NRMTOL, and the function and partial derivative values based on the initial user-specified X(I). Next, we solve the Newton linear algebraic equations shown in comment line 1840 by calling the subroutine LNEQ (linear equations), which is indicated in abbreviated form in lines 2160 to 2270. We finally calculate the new Newton values at line 1880.

The rest is mostly bookkeeping. Thus, we increment the iteration counter M (line 1910), and calculate the norm of the corrections (line 1940) using the built-in function DSQRT (square root) and subroutine DOT (dot product of the vector xc, lines 2290 to 2370), defined as sum of the squares of the components of DIFF, as we attempt to depict in the comment of line 1920. We then print the number of iterations and the calculated new x and correction (xc) values (lines 1980 to 2020). At the LOOP-until condition (line 2030), we stop because either we have exceeded the maximum number of iterations (unsuccessful), or the computed norm is less than the specified NRMTOL value (successful); otherwise, we continue. If we were unsuccessful, we print "NO CONVERGENCE" (line 2070). The output of the calculation appears on the screen. If printed hard-copy results are desired, the most convenient procedure is to print the file OUTPUT, which was produced by the program.

## REFERENCES

KEMENY, J. G., and KURTZ, T. E. (1988), *True BASIC Reference Manual*, True BASIC, Inc., West Lebanon, NH

KOFFMAN, E. B., and FRIEDMAN, F. L. (1993), *FORTRAN with Engineering Applications*, Addison-Wesley, Reading, MA

Language Systems FORTRAN, 100 Carpenter Drive, Sterling VA 20166

MEISSNER, L. P., and ORGANICK, E. I. (1984), *FORTRAN 77*, Addison-Wesley, Reading, MA

# Appendix B

# Difference Equations

Difference equations, which are sometimes also called recurrence relations, can be considered to be discrete forms of differential equations. That is, the argument or independent variable in a difference or recurrence relation is an integer. In modeling and computation, difference equations are used principally to model discrete stage-wise processes (such as separation processes), to conduct error-propagation analyses for various numerical algorithms, and in various approximation problems. It should not be surprising that there is a theory of difference equations that closely parallels that of the better-known (and more widely used) subject of differential equations. For our purposes it will be sufficient to briefly discuss the solution technique for linear, constant-coefficient, first- and second-order difference equations.

## B.1 FIRST-ORDER DIFFERENCE EQUATIONS

The standard first-order, constant-coefficient difference equation can be written as

$$y_{n+1} + ay_n = c \tag{B-1}$$

Here the coefficient $a$ is constant and $c$, the non-homogeneity, is also assumed to be constant. The simplest procedure to solve Equation (B-1) is analogous to that for solving the comparable differential equation $y' + ay = c$. The idea is that the general solution of Equation (B-1) can be expressed as the sum of the general solution of the homogeneous equation $y_{n+1} + ay_n = 0$, plus any particular solution of the full equation. Just as for the differential equation, we assume, for the case where $c$ is a constant, a particular solution of $y_p$ = constant = $D$. When we substitute this into Equation (B-1) to find $D$, we get $(D + aD) = c$, or $D = c/(1 + a)$. Thus, the particular solution is $y_p = c/(1 + a)$ (for $a \neq -1$). To get the solution of the homogeneous equation we try $y_n = K\lambda^n$, where $K$ and $\lambda$ are to be determined. This gives $\lambda + a = 0$, or $\lambda = -a$ (the constant $K$ cancels). Adding the two solutions together gives

$$y_n = K(-a)^n + \frac{c}{1+a} \tag{B-2}$$

Equation (B-2) represents the general solution of Equation (B-1); the constant $K$ must be determined from a boundary condition of the form $y = y_1$ for $n = 1$.

## B.2 SECOND-ORDER DIFFERENCE EQUATIONS

The standard second-order, constant-coefficient difference equation can be written as

$$y_{n+2} + ay_{n+1} + by_n = c \tag{B-3}$$

Here again, the values $a$, $b$, and $c$ are taken to be known constants. The general solution procedure is the same as for the first-order equation; namely, form the sum of the solution of the homogeneous equation $y_{n+2} + ay_{n+1} + by_n = 0$ and any particular solution of the full equation. For the particular solution we take $y_n =$ constant $= D$, and substitution into Equation (B-3) gives $(D + aD + bD) = c$. Thus, $D = c/(1 + a + b)$. For the homogeneous solution we try $y_n = K\lambda^n$, and this yields the characteristic equation for $\lambda$ of the form $\lambda^2 + a\lambda + b = 0$. The solution of this quadratic equation for $\lambda$ yields the following solutions of the complementary equation:

$$\begin{aligned}
&\lambda_1, \lambda_2 \text{ real, distinct:} && y_n = K_1\lambda_1^n + K_2\lambda_2^n \\
&\lambda_1, \lambda_2 \text{ real, equal:} && y_n = K_1\lambda_1^n + K_2 n\lambda_2^n \\
&\lambda_1, \lambda_2 \text{ complex conjugate:} && y_n = (K_1 \cos n\theta + K_2 \sin n\theta)r^n
\end{aligned} \tag{B-4}$$

$$\lambda = \alpha \pm \beta i \quad r = (\alpha^2 + \beta^2)^{1/2} \quad \theta = \tan^{-1}(\beta, \alpha)$$

To get the general solution of the above non-homogeneous, linear, constant-coefficient difference equation (B-3) (or any other one for that matter), we add any particular solution of the full non-homogeneous equation to the general solution of the homogeneous (complementary) equation. In the case of Equation (B-3), we add the particular solution $D = c/(1 + a + b)$ to the appropriate complementary solution from Equations (B-4). The arbitrary constants $K_1$ and $K_2$ are evaluated from two known boundary conditions.

**Example B.1** $\mathbf{y_{n+2} - y_{n+1} - 6y_n = 3}$**, with** $\mathbf{y_1 = 1, y_2 = 0}$

First, we determine a particular solution *assuming* $y_p$ = constant = $D$. Substitution into the full equation gives the requirement $D = -1/2$. Note that our simple particular solution of $y_p$ = constant works, in general, only when the non-homogeneity (right-hand side) is a constant independent of $n$. In other cases, different trial functions of $n$ can be used. To get the complementary solution we substitute $y_n = K\lambda^n$ into the homogeneous form of the equation, $y_{n+2} - y_{n+1} - 6y_n = 0$. This yields the quadratic equation for $\lambda$ of the form $\lambda^2 - \lambda - 6 = 0$, which has the real roots $\lambda_1 = -2$, $\lambda_2 = 3$. Using the complementary solution for real, distinct roots in Equations (B-4) and adding in the particular solution gives the general solution

$$y_n = K_1(-2)^n + K_2(3)^n - \frac{1}{2} \tag{B-5}$$

When we apply the given boundary conditions, $y_1 = 1$, $y_2 = 0$, to determine $K_1$ and $K_2$, we obtain the final form of the solution,

$$y_n = -\frac{4}{10}(-2)^n + \frac{7}{30}(3)^n - \frac{1}{2} \tag{B-6}$$

To help establish the correctness of the solution, Equation (B-6), it is useful to *directly* verify that it produces the known boundary conditions upon which it is based; Equation (B-6) does satisfy this check. It is also of interest to compare one or more particular values of $y_n$ based on the original difference equation and the solution, Equation (B-6). For instance, the reader might verify that $y_3 = 9$ from either representation.

A final word should be said regarding the reason for preferring the explicit solution (B-6) versus using the original difference equation directly. The essential point is that the original difference equation can be used to calculate any finite number of $y_n$ values (subject to roundoff limitations), but only the analytical solution (B-6) can show limiting behavior as $n \rightarrow \infty$.

## REFERENCES

BENDER, C., and ORSZAG, S. (1978), *Advanced Mathematical Methods for Scientists and Engineers*, McGraw-Hill, New York

JENSON, V. G., and JEFFREYS, G. V. (1977), *Mathematical Methods in Chemical Engineering*, 2nd ed., Academic Press, New York

# Appendix C

## Review of Infinite Series

Here we wish to review some of the most important aspects of infinite series, since this subject is so important in applications where a Taylor expansion has such an interpretation. Admittedly this subject is rather dry, so we will try to make it as painless as possible. The infinite series $\sum_{i=1}^{\infty} a_1$ is said to converge to $A$ if $|\sum_{i=1}^{\infty} a_1 - A|$ is less than an arbitrarily small positive number if $n$ is sufficiently large. In words, we may say that a series converges if the sequence of partial sums of its terms approaches a limit as the number of elements in the sequence becomes arbitrarily large. A *necessary* condition for convergence of a series is that $a_n \to 0$ as $n \to \infty$, since this is necessarily true if the sequence of partial sums approaches a limit. A series $\sum a_i$ converges provided the series of absolute values, $\sum |a_i|$, converges. This is referred to as absolute convergence. One can add or take away any finite number of terms at the beginning of a series without affecting the convergence.

It is usually impossible to directly apply the definition of convergence, so several simple tests are often useful for determining whether a given series converges. These tests include the ratio test, the integral test, the comparison test, and the alternating series test. We review some of the ideas involved in these tests. To do this we first need to discuss one of the most important simple series, the so-called *geometric* series. The geometric series is the basis of the well-known ratio test.

The geometric series can be written as $S = 1 + a + a^2 + a^3 + \cdots$, where each new term is just a factor $a$ times the preceding term. The geometric series is so simple that we can actually sum it in finite form and explicitly take the limit as the number of terms $\to \infty$. To do this, write $S_n = 1 + a + a^2 + \cdots + a^n$. If we then multiply this expression by $a$ and subtract this expression for $aS_n$ from $S_n$, we get, after solving for $S_n$,

$$S_n = \frac{1-a^{n+1}}{1-a} \tag{C-1}$$

The expression for $S_n$ in Equation (C-1) is well-defined for any value of $a$ except $a = 1$. If $|a| < 1$, then $a^{n+1} \to 0$ for $n \to \infty$ and $S_n \to 1/(1-a)$. Similarly, if $|a| > 1$, $a^{n+1}$ becomes arbitrarily large in magnitude and $S_n$ does not approach a limit. Thus, the geometric series converges if $|a| < 1$ and diverges if $|a| > 1$. If $|a| = 1$, no limit is approached.

## C.1 THE RATIO TEST

The ratio test is based on a comparison with a geometric series. For the series $S = a_0 + a_1 + a_2 + \cdots$, the idea is to consider the limit of $|a_{n+1}/a_n|$ as $n \to \infty$. If this limit is less than unity, the series converges; if it is greater than unity, it diverges; and if it is *equal* to unity, we don't know about convergence. To show the above results, one considers the "tail" of the series, $|a_{n+1}| + |a_{n+2}| + \cdots$, that is, the sum of the absolute values of the terms from $(n + 1)$ onward. If the ratio $|a_{n+1}/a_n|$ is less than unity for all $n$ sufficiently large, then it is easy to show that a dominating geometric tail series converges, and therefore, the original series converges.

**Example C.1** $e^x = 1 + x + x^2/2! + \cdots + x^n/n! + \cdots$

If we consider $r = a_{n+1}/a_n$ for this series, we find that $r = x/(n + 1)$. This ratio value approaches zero for any finite $x$ as $n \to \infty$. Thus, the above series for $e^x$ converges for any $x$, although it may not be *useful* for larger values.

## C.2 THE INTEGRAL TEST

The integral test for series is more powerful than the ratio test, but its implementation takes more effort. This test is for *positive* terms and has the following requirements. If $f(x)$ is positive, continuous, and *monotonically decreasing* (that is, $f'(x)$ is negative) for $x \geq N$, and is such that $f(n) = a_n$ for $n = N, N+1, N+2, \cdots$, then the infinite series $a_N + a_{N+1} + \cdots$ converges or diverges according to whether $\int f(x)\,dx$ from $N$ to $\infty$ converges or diverges. To see this we can write, for the above-specified conditions, that $a_{n+1} = f(n + 1) \leq f(x) \leq f(n) = a_n$ for $(n \leq x \leq n + 1)$. By integrating this inequality from $x = n$ to $x = n + 1$ and adding the results for $n = N, N+1, N+2, \ldots$, we finally get

$$a_{N+1} + a_{N+2} + a_{N+3} \cdots \leq \int_N^\infty f(x)\,dx \leq a_N + a_{N+1} + a_{N+2} \cdots \tag{C-2}$$

This gives the desired results. The above result also has a simple geometrical interpretation in terms of the integral $\int f(x)\,dx$ and right- and left-hand rectangular approximations to it. This is illustrated in Figure C.1

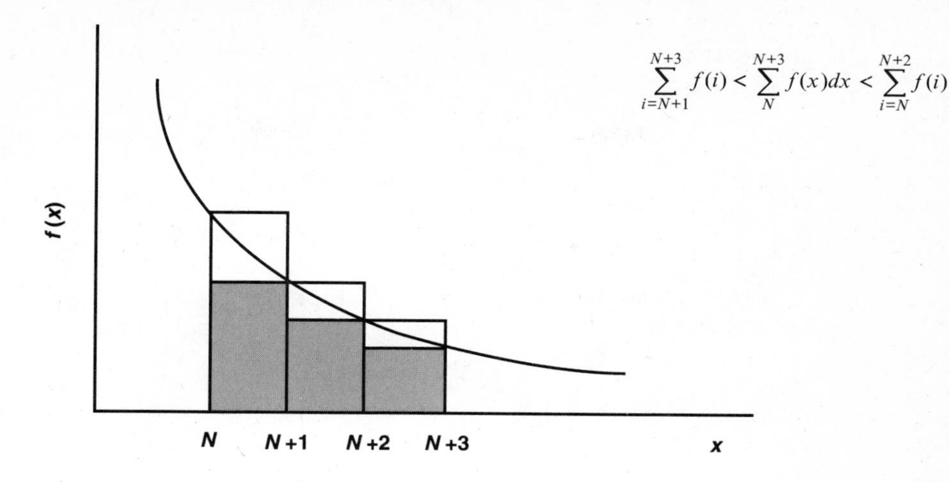


**Figure C.1.** Ilustration of the Integral Test

**Example C.2** $S = 1 + 1/2^2 + 1/3^2 + 1/4^2 + \cdots$

Here, $a_n = 1/n^2$, and the ratio test yields the limit as $n \to \infty$ of $n^2/(n+1)^2$, which is unity; this tells us nothing. However, using the integral test, we have that the integral of $1/x^2$ from any $N$ to $\infty$ is finite (it is just equal to $1/N$), so the above series does indeed converge.

## C.3 THE COMPARISON TEST

The comparison test is for positive terms and states that if the terms of the desired series, $a_n$, are *smaller* than those of some comparison series, say $b_n$, then, if the $b_n$ series converges, so does the $a_n$ series. This is because if the $b_n$ series converges, its tail becomes arbitrarily small, and therefore the tail of the $a_n$ series, which is even smaller, must do likewise.

**Example C.3** $S = (\cos x)/1^2 + (\cos 2x)/2^2 + (\cos 3x)/3^2 + \cdots$

For this series $a_n = (\cos nx)/n^2$, and the $a_n$ values are *not* of one sign. However, since $|\cos nx| \leq 1$ for all $x$, the $a_n \leq 1/n^2 = b_n$ for all $n$ in the sum. Thus since $\sum b_n$ converges by the integral test (Example C.2), the sum $\sum a_n$ necessarily converges.

## C.4 THE ALTERNATING SERIES TEST

This test applies to a series of *monotonically decreasing* terms $a_n$ that alternate in sign and for which $a_n \to 0$ when $n \to \infty$. A simple example of this type of series is $\ln(1+x) = x - x^2/2 + x^3/3 - x^4/4 + \cdots$ *for positive* $x \leq 1$. Note that in this series (i) if $x$ is *negative*, all terms are negative, and (ii) if $x > 1$, the $a_n$ term eventually grows without bound rather than going to zero, as required. If the above conditions of the alternating series test are met, then the series converges and the |error| in computing the sum is bounded by the magnitude of the first term neglected. The sign of the remainder is the same as that of the first neglected term.

To obtain this result, consider the general alternating series $S = a_0 - a_1 + a_2 - a_3 + a_4 - \cdots$, where all $a_n$ are positive and where $a_n$ decreases monotonically with $n$. By grouping the terms as $S = (a_0 - a_1) + (a_2 - a_3) + \cdots$, we see that $S$ is composed of a sum of *positive* terms. On the other hand, if we group the terms of $S$ according to $S = a_0 - (a_1 - a_2) - (a_3 - a_4) - \cdots$, we see that $S < a_0$ since the remaining terms are all negative. Thus, we see that the sum $S$ is a monotonic increasing sequence which is bounded above and therefore converges. Very similar arguments are used to show the results stated above concerning the error in computing the sum.

**Example C.4** $\mathbf{S = \ln 2 = 1 - 1/2 + 1/3 - 1/4 + \cdots}$

This series corresponds to that of the foregoing section evaluated for $x = 1$. For $x < 1$, the ratio test applied to this series shows convergence; however, for $x = 1$ the ratio test is inconclusive. To study the case $x = 1$ considered in this example, we note that $a_n = 1/n$, so that $a_n$ is monotonically decreasing and $a_n \to 0$ for $n \to \infty$. Thus, the alternating series test applies, and the series above for ln 2 converges. Also, in computing the value of this series, the sum of the first three terms ($= 5/6 = 0.833$) overestimates the true value ($\ln 2 = 0.693$) by less than the next term $|1/4| = 0.25$, as required by the alternating series test.

## C.5 POWER SERIES

Power series of the form $a_0 + a_1x + a_2x^2 + a_3x^3 + \cdots$ are used so much in applications that a few results should be stated here. We normally determine the radius of convergence of a power series by means of the ratio test. Power series can be differentiated or integrated within their radius of convergence to produce new series which also have the same radius of convergence as the original.

## REFERENCES

SCHEID, F. (1988), *Numerical Analysis*, 2nd ed., McGraw-Hill, New York

SPIEGEL, M. (1963), *Advanced Calculus*, Schaum, New York

# Appendix D

# Advanced Series Manipulations

In many situations it is desirable to use various power series techniques which would require prohibitive hand calculations. Applications include problems in approximation, evaluation of integrals, solutions of differential equations, and so on. The advent of the ubiquitous personal computer now allows many practical calculations of this type that were heretofore nearly impossible. We consider here the algebraic basis for the following manipulations of power series:

(i) Multiplication of series
(ii) Division of series
(iii) Natural logarithm of a series
(iv) Exponential of a series
(v) Raising a series to a power
(vi) Reversion of a series

The above manipulations can be carried out within computer program SERIES, which is discussed at the end of this appendix, using any desired number of terms. Here, we will merely indicate the mathematical development of the first few terms of the desired operations in order to show the procedure involved. Additional series manipulations which are designed to accelerate convergence are discussed in the appendix on acceleration of convergence and in the chapter on acceleration and summation of series.

## D.1 MULTIPLICATION OF SERIES

Given the two truncated power series $S1 = (a_0 + a_1x + a_2x^2 \cdots)$ and $S2 = (b_0 + b_1x + b_2x^2 \cdots)$, we wish to form the product series $(S1)(S2) = (d_0 + d_1x + d_2x^2 \cdots)$ whose coefficients depend on $a_i$ and $b_i$. Thus, we wish to express the $d_i$ in terms of the $a_i$ and $b_i$. The multiplication operation is fundamental to several other series operations, so we discuss it here in some detail. The series multiplication algorithm yields the following result:

$$(a_0 + a_1x + a_2x^2 \cdots)(b_0 + b_1x + b_2x^2 \cdots) \\ = a_0b_0 + x(a_0b_1 + a_1b_0) + x^2(a_0b_2 + a_1b_1 + a_2b_0) + \cdots \tag{D-1}$$

This result is obtained systematically by first collecting all of the constant terms in the product, then all of the terms linear in $x$, then all of the terms quadratic in $x$, and so on. To accomplish this, we proceed first by multiplying $a_0$ by all terms in the $b_i$ series to find the constant terms (there is only one, $a_0b_0$). Then we multiply $a_0$ by the $b_i$ series to find a linear term in $x$ (namely $x \cdot a_0b_1$), followed by multiplying $a_1$ by the $b_i$ series to get another linear $x$ term, $x \cdot a_1b_0$. The quadratic terms in $x$ correspond to multiplying $a_0$ by $b_2x^2$, $a_1x$ by $b_1x$, and $a_2x^2$ by $b_0$. As is easily seen by putting pencil to paper, the coefficient $d_k$ of the $x^k$ term in the product series will contain the sum of $(k + 1)$ separate product terms. The first three coefficients $d_i$, which are the coefficients of $x^i$ in the above equation, are shown in Table D.1.

It is important to note that in the multiplication algorithm given above, as well as in the others to follow, the calculations always assume that both the original series and the new desired series are specified to contain exactly the same powers of $x$. That is, to get terms of the new series of degree $x^k$, we must have terms up to $x^k$ in the original series.

## D.2 DIVISION OF SERIES

We now consider the problem of finding the unknown series $(d_0 + d_1x + d_2x^2 \cdots)$ that is equal to the quotient $(a_0 + a_1x + a_2x^2 \cdots)/(b_0 + b_1x + b_2x^2 \cdots)$ of the two known series. This problem is solved by the series multiplication algorithm as follows. We simply *cross-multiply* the known $b_i$ series with the unknown $d_i$ series and equate the result to the known $a_i$ series. The multiplication is as follows:

$$(d_0 + d_1x + d_2x^2 \cdots)(b_0 + b_1x + b_2x^2 \cdots) \\ = d_0b_0 + x(d_0b_1 + d_1b_0) + x^2(d_0b_2 + d_1b_1 + d_2b_0) + \cdots \tag{D-2}$$

The above equation is merely a repetition of the earlier equation for series multiplication, but with $a_i$ replaced by $d_i$. The difference is that now the $d_i$ are unknown, whereas before the $a_i$ were known. When we equate coefficients of powers of $x$, we get recursive linear equations for the $d_i$ in terms of $a_i$, $b_i$, and previous $d_j$. The first three of these relations are given in Table D.1.

## D.3 NATURAL LOGARITHM OF A SERIES

Suppose $y = \ln(a_0 + a_1x + a_2x^2 + \cdots)$ and we wish to find the power series expansion of this as $y = (d_0 + d_1x + d_2x^2 + \cdots)$. First, we note that for the series to exist, we must have $a_0 > 0$; for otherwise, if $x = 0$, the logarithm is not defined. Then putting $x = 0$ in both series above yields $d_0 = \ln a_0$. To find the remaining $d_i$ in terms of the $a_i$, it is convenient to *differentiate* the two equal series to get

$$y' = \frac{a_1 + 2a_2x + 3a_3x^2 \cdots}{a_0 + a_1x + a_2x^2 \cdots} = d_1 + 2d_2x + 3d_3x^2 \cdots \tag{D-3}$$

Now we simply cross-multiply the $d_i$ series above with that of the denominator series and equate powers of $x$ to get recursive linear equations for the $d_i$. The first three terms of the series are shown in Table D.1.

## D.4 EXPONENTIAL OF A SERIES

The problem here is to find $y = (d_0 + d_1x + d_2x^2 \cdots) = \exp[(a_0 + a_1x + a_2x^2 \cdots)]$ where $\exp(z)$ is the exponential function, $e^z$. Again we wish to express the $d_i$ in terms of the $a_i$. First, we see that for $x = 0$, $d_0 = \exp(a_0)$. Then we take natural logarithms of both sides and differentiate to get recursive linear equations for the $d_i$. The calculations are so similar to those for the logarithm of a series that we won't repeat them here. The first three coefficients of the series are given in Table D.1.

## D.5 RAISING A SERIES TO A POWER

In this case, we wish to find the series $(d_0 + d_1x + d_2x^2 \cdots)$ which is equal to $(a_0 + a_1x + a_2x^2 \cdots)^p$, where $p$ is a real number. Here we require for convenience that $a_0 \neq 0$; if $a_0$ happens to be zero, we can factor out a power of $x$ so that the *new* $a_0$ is not zero. We could take logarithms, differentiate, and cross-multiply, but since we have already developed algorithms for the logarithm and exponential of a series, it is convenient to make use of these results. Note that we can write

$$d_0 + d_1x + d_2x^2 \cdots = (a_0 + a_1x + a_2x^2 \cdots)^p = e^{p \ln(a_0 + a_1x + a_2x^2 \cdots)} \tag{D-4}$$

In this form, we see that we can get the series for the right-hand side (and hence the $d_i$) by taking the logarithm of the $a_i$ series, multiplying each coefficient by $p$, and then exponentiating that series. As a simple *necessary* (but not sufficient) check on the new coefficients $d_i$, we can see that for $p = 1$ we must recover the original coefficients $a_i$. The results for the first three $d_i$ values are given in Table D.1.

## D.6 REVERSION OF A SERIES

Certain procedures for solving nonlinear algebraic equations and for determining asymptotic expansions of integrals require the reversion of power series. The problem is, given that $y = (a_1x + a_2x^2 + a_3x^3 \cdots)$ where the $a_i$ are known, find the *reverted* form $x = (d_1y + d_2y^2 + d_3y^3 \cdots)$. That is, we wish to express the unknown $d_i$ values in terms of the known $a_i$'s. We should explain that this computation only makes good sense if $a_0 = 0$, so we have explicitly indicated this in the above notation (the curious

**TABLE D.1** GENERAL RESULTS FOR THREE TERMS OF VARIOUS SERIES OPERATIONS;

Result Is $d_0 + d_1 x + d_2 x^2$

| Operation | Result |
|---|---|
| Multiplication<br>$(a_0 + a_1 x + a_2 x^2 \cdots)(b_0 + b_1 x + b_2 x^2 \cdots)$ | $d_0 = a_0 b_0$<br>$d_1 = a_0 b_1 + a_1 b_0$<br>$d_2 = a_0 b_2 + a_1 b_1 + a_2 b_0$ |
| Division<br>$(a_0 + a_1 x + a_2 x^2 \cdots)/(b_0 + b_1 x + b_2 x^2 \cdots)$ | $d_0 = a_0 / b_0$<br>$d_1 = (a_1 - d_0 b_1)/b_0$<br>$d_2 = (a_2 - d_0 b_2 - d_1 b_1)/b_0$ |
| Logarithm<br>$\ln(a_0 + a_1 x + a_2 x^2 \cdots)$ | $d_0 = \ln(a_0)$<br>$d_1 = a_1 / a_0$<br>$d_2 = (2a_2 - d_1 a_1)/2a_0$ |
| Exponentiation<br>$\mathrm{Exp}(a_0 + a_1 x + a_2 x^2 \cdots)$ | $d_0 = \mathrm{Exp}(a_0)$<br>$d_1 = d_0 a_1$<br>$d_2 = (2d_0 a_2 + d_1 a_1)/2$ |
| Raise to a Power $p$<br>$(a_0 + a_1 x + a_2 x^2 \cdots)^p$ | $d_0 = a_0^p$<br>$d_1 = p d_0 a_1 / a_0$<br>$d_2 = [2p d_0 a_2 + (p-1) d_1 a_1]/2a_0$ |
| Reversion<br>$x = (a_1 y + a_2 y^2 + a_3 y^3 \cdots)$<br>$y = (d_1 x + d_2 x^2 + d_3 x^3 \cdots)$ | $d_1 = 1/a_1$<br>$d_2 = -d_1 a_2 / a_1^2$<br>$d_3 = -(d_1 a_3 - 2 d_2 a_1 a_2)/a_1^3$ |

reader should try leaving in the $a_0$ term). If the original series has a nonzero $a_0$, then we form the difference $(y - a_0)$ and call this the new $y$ variable. To accomplish the reversion, we substitute the known series for $y$ into the series for $x$ expressed in terms of $y$. This gives

$$x = d_1(a_1 x + a_2 x^2 \cdots) + d_2(a_1 x + a_2 x^2 \cdots)^2 + d_3(a_1 x + a_2 x^2 \cdots)^3 + \cdots \quad \text{(D-5)}$$

The next step is to use the raising-to-a-power algorithm for each of the terms of the form $(a_1 x + a_2 x^2 \cdots)^k$, where $k = 2, 3, \cdots$ as far as is needed. Note that if the original series is known through terms of $x^p$, then the reverted series goes up to terms of $y^p$, and each series raised to a power only uses terms through $x^p$. When the raising to a power has been accomplished, we equate powers of $x$ on both sides of the equation (on the left-hand side, only the power $x$ is present). This produces recursive equations for the $d_i$ values. The first three terms of the reversion process are indicated in Table D.1.

## D.7 SERIES MANIPULATION PROGRAM SERIES

The various series manipulations described above, as well as some others, can all be carried out using program SERIES. The algorithms used are the ones given above, extended so that any finite number of terms can be obtained. The results of program SERIES express the various series coefficients in decimal (rather than fractional) form, so that normally six significant digits are obtained. Since the results are calculated internally in double precision if a math co-processor is used, greater accuracy is available by employing a PRINT USING command.

The instructions for running the program are given in the concise listing. The program prompts for (i) $n$ = number of terms in both the given and computed series, (ii) case (1–10 depending on the application), (iii) the $x$ value to be used in computing the new series, as well as any other needed parameters described below. The output gives an option for printing the original input series coefficients (for checking purposes), followed by the computed coefficients of the new series, together with an option to calculate the new series for any desired $x$ values. In computing the sum of the new series for any $x$, the last term is also given as an indication of the error.

Program SERIES is a powerful program, which is actually a collection of ten different programs put together for convenience. The user subroutine Coefficients applies to all ten programs (cases). For every case except $j = 10$ (direct error bounds), the user subroutine Function can be ignored. The coefficients of the input series may be expressed individually in terms of assignment statements, or, if the general term of the series is known, a loop can be used. Since there are slight variations in the requirements for different cases, we discuss these now individually. The series shown specified in subroutine Coefficients of the concise listing below is that for the function $\ln(1 + x)$ (which is Log $(1 + x)$ in True BASIC). Since we know the general term of this series, it is specified within a loop so that we can easily use any desired number of terms by responding appropriately to the prompt for $n$.

Considering first cases 1 and 2, addition and subtraction, note that we need *two* series for these operations. Thus, the second series for the $b(i)$ must also be specified in subroutine Coefficients. One convenient and common application for case 1 is to use the second series as $b(i) = 0$ (which for convenience is presently programmed) so that the result is simply the original series, which can then be summed directly for any $x$.

Cases 3 and 4, for multiplication and division, respectively, also require two series. In the case of division we cannot have $b_0 = 0$, since this would produce a zero denominator. Case 5 for the natural logarithm of a series requires that $a_0 \neq 0$, since this would make $d_0 = \infty$. Case 7 for raising a series to a power requires the user to specify the desired power $p$ in response to a prompt, and it also does *not* allow $a_0 = 0$. Case 8 for reversion of series requires that both $a_0 = 0$ and $a_1 \neq 0$. Case 9 for the Euler transformation of a series requires specification of the positive parameter $p$ in response to a prompt; the Euler transformation of a series is discussed in Chapter 4, "Acceleration and Summation of Series."

Case 10, for direct error bounds of a series, requires specification of user subroutine Function in addition to subroutine Coefficients. The general procedure of the calculation, as well as numerical examples, are given in the chapter on Taylor expansions. As indicated in the concise listing below (where we employ the function Log $(1 + x)$ for which the corresponding $a(i)$ are given), the required function can be specified either analytically (if it is known) or numerically. The interval in $x$ to be considered must be specified by $x$min and $x$max. The number of parts into which the specified interval is to be divided in the calculation must be indicated as no_intervals. The calculation produces an approximate error bound for the prescribed Taylor expansion in the desired interval.

### Program SERIES.

```
!This is program SERIES.tru by O.T.Hanna
!For the Taylor series a(0)+a(1)x+a(2)x^2..+a(n)x^n
!(and b(0)+..b(n)x^n,if needed)
!the program can do addition,subtraction,multiplication,
!division,Logarithm,
!Exponential,raising to a power,reversion,Eulers transformation
!and direct error bounds for a Taylor series.
!n = highest Taylor power in all series involved in calculation.
!Program determines coefficients of new series,values of series
!for any x, and last term of series for error indication.
!**************************************************************
!* To run,input a(),[b()if necessary] in Sub Coefficients;     *
!* To use Direct Error Bounds option(j=10),function must be    *
!* specified in Sub Function either analytically or via data.  *
!* program prompts for n,case(j),x, as well as other needed    *
!* parameters; check individual Subs for more information      *
!**************************************************************

Sub Function(n,x,f,f1(),no_intervals,xmin,xmax) !***user defined
   !This Sub is used only for direct error bounds option and can be
   !ignored if this option is not being used
   !To use specify f(x) either analytically or numerically
   !Maximum error is determined in range xmin <= x <= xmax using a
   !number of intervals equal to no_intervals.

   Let no_intervals = 10 !***
   Let xmin = 0          !***
   Let xmax = 1          !***

   Mat redim f1(no_intervals + 1)

   Let f = Log(1+x)      !***analytically defined function

   !The following is for function defined by data !***
   !define f1(1)=f(xmax),f1(2)=f(xmax-d),f1(3)=f(xmax-2d), ...,
   !f1(no_intervals+1)=f(xmin)

   !The example below is for function defined by data with
   !xmin=0,xmax=1,no_intervals=5 as illustrated using
   f(x)=Log(1+x)
```

```
    ! Let f1(1)=Log(2)
    ! Let f1(2)=Log(1.8) !d=(xmax-xmin)/no_intervals = (1-0)/5 = 0.2
    ! Let f1(3)=Log(1.6)
    ! Let f1(4)=Log(1.4)
    ! Let f1(5)=Log(1.2)
    ! Let f1(6)=Log(1)

End Sub

Sub Coefficients(n,a(),b()) !***User specified subroutine
   !User must set Taylor coefficients
   !a(0) to a(n) by assignment or recurrence.
   !If b(i) are needed, they must also be specified.
   !To get original Taylor sum, use Sum(j=1) with b(i) = 0
   !***denotes user specified quantities

   !The following operations are available for case = j
   !j= 1 Sum of two series
   !j= 2 Difference of two series
   !j= 3 Product of two series having same n value
   !j= 4 Quotient of two series having same n value
   !j= 5 Logarithm of a series
   !j= 6 Exponential of a series
   !j= 7 Series raised to a power
   !j= 8 Reversion of a series
   !j= 9 Euler transformation of a series
   !j=10 Direct error bounds for a series

   Mat redim a(0 to n),b(0 to n)

   ! !Coeff. for tan (x) = x + x^3/3 + (2/15)*z^5 ...
   ! Let a(0) = 0        !***a(0) to a(n) by assignment
   ! Let a(1) = 1
   ! Let a(2) = 0
   ! Let a(3) = 1/3
   ! Let a(4) = 0
   ! Let a(5) = 2/15

   !Coeff. for ln (1 + x)
   Let a(0) = 0          !***a(0) to a(n) in loop
   For i = 1 to n
       Let a(i) = (-1)^(i+1)/i
   Next i
   For i = 0 to n        !this is useful for simple sum of Taylor series
       Let b(i) = 0
   Next i

   !Let b(0) = 1         !***set b(0) to b(n);if necessary,this will
End Sub                  !override b(i) = 0
```

## REFERENCES

BENDER, C., and ORSZAG, S. (1978), *Advanced Mathematical Methods for Scientists and Engineers*, McGraw-Hill, New York

# Appendix E

# Limits and Convergence Acceleration

Many numerical computational problems involve approximating to some unknown limiting quantity. Examples include finding the sum of an infinite series, solving a nonlinear algebraic equation by iterative methods, numerical evaluation of a definite integral, finding the numerical solution of a differential equation, and so on. In all of these problems, we generally obtain some kind of sequence of approximations which we hope converges to the desired solution. Sometimes the rate of convergence is quite satisfactory, but in many situations the convergence of the original numerical sequence is so slow that convergence-acceleration techniques are needed. There exist a number of such techniques, but here we focus attention on two of the simplest, most general, and most widely used techniques, which are known as Shanks transformation (also known as Aitken acceleration, the delta-square process, etc.) and Richardson extrapolation. These methods are used in a variety of contexts, but here we concentrate on their general development and the similarities between them. But before proceeding, we need to introduce "asymptotic" notation.

## E.1 LIMITS AND ASYMPTOTIC NOTATION

Here we introduce the so-called "big oh" asymptotic notation, which is very convenient in the discussion of limits, particularly for the use of Richardson extrapolation. Use of the notation $f(h) = O(h^p)$ as $h \to 0$ means that if $h$ is sufficiently small, then there exists a constant $K$ for which the inequality $|f(h)| \leq Kh^p$ is satisfied. This gives us a convenient notation for "sizing" the dependence of a function with respect to its $h$ dependence. As an example, if $f(h) = O(1)$ as $h \to 0$, this refers to the case $p = 0$, so that $f(h)$ is less than some constant for all sufficiently small $h$. The most common situation that allows a sizing calculation of the above type to be made is Taylor's theorem. Thus, we could say that $\sin x = O(x)$ for $x \rightarrow 0$, since we know from Taylor's theorem that near $x = 0$, $\sin x \approx x - x^3/3!$. Another important result, which is sometimes confusing, is that $e^{-x} = O(1/x^k)$ as $x \rightarrow \infty$ for *any* integer $k$ if $x$ is positive. This can be shown by noting that $e^{-x} = 1/(1 + x + x^2/2! + \cdots + x^k/k! + \cdots)$, and therefore, by deleting all the positive terms in the denominator except $x^k/k!$, we see that $e^{-x} < k!/x^k$. An additional useful result is that if $f(x) = O(x^p) + O(x^q)$ for $x \rightarrow 0$ with $p < q$, then $f(x) = O(x^p)$. This is true since $O(x^q)$ is relatively small when $x$ is sufficiently small.

## E.2 SHANKS TRANSFORMATION

The calculation procedure known as "Shanks transformation" is often useful in speeding up the convergence of a sequence. This process also goes by a variety of other names, including Aitken acceleration and the delta-square process. The Shanks process is generally useful in a variety of problems; *it can sometimes even cause an apparently divergent sequence to converge!* It is noteworthy that in the technical literature, application of the technique to various problems may sometimes *appear* to be different. We therefore give a generic discussion of the procedure and try to point out its similarity to Richardson extrapolation as well as to other acceleration models.

Shanks transformation can be presented in a general context in terms of an attempt to "model" the behavior of the error in a sequence. Thus, if the sequence $S_0$, $S_1$, $S_2$, and so on is presumed to converge to the unknown limit $S$, the Shanks model for the error is expressed as $S = S_n + K\lambda^n$. Here the error, $K\lambda^n$, depends on the unknown modeling constants $K$ and $\lambda$, which are assumed to be independent of $n$. Since the model equation involves three unknowns ($S$, $K$, $\lambda$), it would be expected that three known values of $S_n$ would be required to establish a determinate system of equations, and this is indeed the case. The above so-called Shanks model for the error assumes a "geometric" convergence of the error, such as occurs in the geometric infinite series. For this reason the Shanks transformation is ideally suited to sum "nearly" geometric infinite series, such as many power series. It is very important to note that an entirely equivalent Shanks error model can be expressed by a first-order, constant-coefficient difference equation for the error $e_n = (S - S_n)$, which is written as $e_{n+1} = \lambda e_n$. This error equation follows directly from the earlier model equation by eliminating $K$. It should be mentioned that other error models are sometimes useful; in particular, the technique known as Richardson extrapolation is associated with the model $S = S_n + K/n^p$.

The fundamental equations needed to implement Shanks transformation are now derived. It is convenient to start with the following error difference equation form of the model:

$$(S - S_{n+1}) = \lambda(S - S_n) \tag{E-1}$$

By writing this equation with $n$ replaced by $(n + 1)$ and rearranging, it is easy to show that

$$S = S_{n+2} + \left(\frac{\lambda}{1-\lambda}\right)(S_{n+2} - S_{n+1}) \tag{E-2}$$

To determine the quantity $\lambda$, we write Equation (E-1) twice, once as it stands, and once with $n$ replaced by $(n + 1)$. When these two equations are subtracted, we find that $\lambda$ can be expressed as

$$\lambda = \frac{(S_{n+2} - S_{n+1})}{(S_{n+1} - S_n)} \tag{E-3}$$

Equations (E-2) and (E-3) together constitute implementation of the simplest form of Shanks transformation. It is interesting to observe that the transformation requires only three elements of the sequence $(S_n, S_{n+1}, S_{n+2})$; it does *not* depend explicitly on the value of $n$. If the model is reasonably valid, the second term of Equation (E-2) should represent a conservative estimate of the error in the Shanks approximation to $S$. Naturally, how well Shanks transformation does in any particular problem depends on the degree to which the Shanks model is accurate.

Experience shows that the Shanks model does very well in many important applications. These include (i) sums of series, (ii) iterative procedures in problems such as nonlinear algebraic equations (where the method is often called the delta-square process), algebraic eigenvalues, and linear algebraic equations, and (iii) Gauss quadrature.

Extensions of the Shanks transformation can be made in several directions. First, it is often useful to *repeat* the transformation. An example of this will be shown below. Higher-order Shanks transformations, for instance, of the form $S = S_n + K_1\lambda_1^n + K_2\lambda_2^n$, are also possible but more cumbersome in practice. For further discussion of Shanks transformation, see Bender and Orszag [1978].

**Example E.1 Sum of the Series $\ln 2 = 1 - 1/2 + 1/3 - 1/4$**

Here we illustrate the Shanks transformation (and one *repetition*) for the first five terms ($a_0 = 0$) of the Maclaurin series for $\ln(1 + x)$ with $x = 1$. The true value of $\ln 2 = 0.693147$ to six figures. To implement Shanks transformation, we must first calculate the partial sums of the series. Then we use the formulas given above to calculate the Shanks values. For example, the first three partial sums of the above series are $S_1 = 0$, $S_2 = 1$, $S_3 = 0.5$ (see Table E.1). Then $\lambda = (S_3 - S_2)/(S_2 - S_1) = (0.5 - 1)/(1 - 0) = -0.5$. Finally, the first Shanks value is $S = S_3 + \lambda/(1 - \lambda)(S_3 - S_2) = 0.5 + (-0.5/1.5)(0.5 - 1) = 2/3 = 0.666667$. This is shown as $S_1'$ in Table E.1 under the column entitled "One Shanks Transformation." Note that this value is *considerably* closer to the true value than the individual partial sums from which it was constructed. The remaining results of the calculation are shown in the table. These were determined using program SHANKS, which is discussed in the chapter on acceleration techniques.

We observe that each Shanks transformation requires *three* sequence elements. Thus, $S_1'$ of column two (one Shanks transformation) is constructed from $(S_1, S_2, S_3)$ of column one, $S_2'$ of column two from $(S_2, S_3, S_4)$ of column one, and $S_3'$ of column two from $(S_3, S_4, S_5)$ of column one. Five terms of the original series give the five partial sums of column one, and these in turn yield three Shanks values. To *repeat* Shanks transformation we use the three sequence elements in column two to produce one new Shanks value, $S_1'' = 0.692593$, in column three. We observe that as we go down in column two, the higher Shanks values produce better accuracy, as expected, since they are based on more information. In the last row of the table, we see that the best once-transformed sequence is much better than the non-transformed partial sum, and

**TABLE E.1** SHANKS TRANSFORMATION FOR SUM OF $\ln 2 \approx 0 + 1 - 1/2 + 1/3 - 1/4$ ($\ln 2 = 0.693147$)

| Partial Sums of Original Series | One Shanks Transformation | Two Shanks Transformations |
|---|---|---|
| $S_1 = 0.0$ | | |
| $S_2 = 1$ | | |
| $S_3 = 0.5$ | $S'_1 = 0.666667$ | |
| $S_4 = 0.833333$ | $S'_2 = 0.7$ | |
| $S_5 = 0.583333$ | $S'_3 = 0.690476$ | $S''_1 = 0.692593$ |

the twice-transformed sequence is again considerably better than the once-transformed value. In particular, the Shanks transformations in this problem have produced an *amazing* improvement in the original partial sums.

We should note that it is possible, and often desirable, to perform further repetitions of the Shanks transformation. These are accomplished just as indicated above. For instance, if we used nine terms of the series in the above example, we could perform seven single Shanks, five double Shanks, three triple Shanks, and one quadruple Shanks. Such repetitions are accomplished automatically by program SHANKS.

## E.3 RICHARDSON EXTRAPOLATION

In addition to Shanks transformation, another very useful acceleration technique is known as Richardson extrapolation. The Richardson method models the error in the approach to a limit $S$ by the equation $S = S_n + K/n^p$. In this case the decay of the error is algebraic, as opposed to the geometric decay of the error in the Shanks method. Since the model contains three unknowns $(S, K, p)$, three elements of the sequence $S_n$ are generally required in order to solve for these values. Thus, the Shanks and Richardson techniques indeed appear very similar. But since the model errors vary differently with $n$ for these two approaches, it is natural that some processes follow one model and some the other (and many follow *neither* model!). In particular, problems where the Richardson method is applicable include numerical differentiation, numerical quadrature, and the numerical solution of ordinary and partial differential equations. The Richardson method is used as an integral part of several programs given in the chapter on numerical solution of differential equations.

The simplest and most common form of Richardson extrapolation occurs when it is possible to develop an approximation formula for the desired quantity, $S$, of the form

$$S = A(h) + Kh^p + O(h^q), \quad h \to 0 \tag{E-4}$$

where $A(h)$ is the approximation (which depends on the value of $h$), $K$ is an unknown constant, and both $p$ and $q$ are *known*, with $p < q$. The model of Equation (E-4) is often expected, based on a Taylor expansion of the approximation $A(h)$ about $h = 0$. In Equation (E-4), $Kh^p$ is the dominant contribution to the error, and the $O(h^q)$ term is of higher order in $h$. From Equation (E-4), we see that $S \to A(h)$ for $h \to 0$. If an approximation $A(h)$ is calculated for a particular $h$, then the error of the approximation is $O(h^p) + O(h^q) = O(h^p)$ since $p < q$. However, if a second approximation $A(h/2)$ is determined, for $h$ replaced by $h/2$, the second approximation satisfies

$$S = A(h/2) + K(h/2)^p + O(h^q), \quad h \to 0 \tag{E-5}$$

If we now solve Equations (E-4) and (E-5) for $K(h/2)^p$, our estimate of the leading error contribution, we get

$$K(h/2)^p = \frac{A(h/2) - A(h)}{2^p - I} + O(h^q), \quad h \to 0 \tag{E-6}$$

Substitution of (E-6) into (E-5) then yields

$$S = A(h/2) + \frac{A(h/2) - A(h)}{2^p - 1} + O(h^q) \tag{E-7}$$

We thus see that the error of Equation (E-7), which is $O(h^q)$ for $h \to 0$, will be smaller than that of Equation (E-5) for $h$ sufficiently small. This suggests that rather than decreasing $h$ to make the error small, it might be better to use the extrapolation procedure outlined. In practice, this usually turns out to be the case, since smaller $h$ values are normally associated with a higher cost, or roundoff error, or perhaps both. Note also from Equation (E-7) the important result that the quantity $[A(h/2) - A(h)]/(2^p - 1)$ is an *estimate* (not a bound) for the error incurred by using $A(h/2)$ as the best approximation to $S$. This quantity may be used as a conservative estimate of the error $O(h^q)$ in Equation (E-7), since this will be true when $h$ is sufficiently small. Thus, not only does extrapolation usually provide an improved approximation, but more important it also gives some information about the error.

### E.3.1 Experimental Determination of p in Richardson Extrapolation

We should mention that if the quantity $K$ in Equation (E-4) happens to be zero, then extrapolation does not improve the approximation and, indeed, may make it worse. Observe, however, that in situations where the exponent $p$ of the leading error term is in doubt, computational experiments may help establish it. Thus, we could determine three approximations (e.g., $A(h/4)$, $A(h/2)$, and $A(h)$) and use Equation (E-6) to derive an expression for $p$. This is accomplished by evaluating Equation (E-6) twice (neglecting the term $O(h^q)$), once as it stands and once with $h$ replaced by $h/2$. This yields the following equation for $p$, which is solved explicitly for $p$ by taking logarithms on both sides:

$$2^p = \frac{A(h/2) - A(h)}{A(h/4) - A(h/2)} \tag{E-8}$$

In the above development we have used $h$ values that differ from each other by a factor of two. This is usually the most convenient arrangement, but it is straightforward to use other combinations.

It is very useful in some applications to employ *repeated* Richardson extrapolation. A way this can be accomplished conveniently is as follows. We start by extending the error model to read

$$S = A(h) + K_1 h^p + K_2 h^q + O(h^r), \quad h \to 0 \tag{E-9}$$

In this case, assuming both $p$ and $q$ to be known with $p < q < r$, we have the three unknowns $(S, K_1, K_2)$ in Equation (E-9), and we therefore need three different $A(h)$ values to solve for them. Again, it is usually convenient to have $h$ values that differ by a factor of two, so we consider the values $h$, $h/2$, $h/4$. If we now proceed with model (E-9) just as we did earlier with Equations (E-6) and (E-7), we find that

$$S = \overline{A}(h/2, h) + D_1 h^q + O(h^r), \quad h \to 0 \tag{E-10}$$

where $\overline{A}(h/2, h) = A(h/2) + [A(h/2) - A(h)]/(2^p - 1)$, and $D_1$ is independent of $h$. This shows that Equation (E-10) is of the same form as Equation (E-4), with the appropriate replacements. Thus, the once-extrapolated quantity $\overline{A}(h/2, h)$ can also be extrapolated in the same way as before, but with the exponent $p$ replaced by $q$. The case of this double extrapolation gives the expression

$$S = \overline{A}(h/4, h/2) + \frac{[\overline{A}(h/4, h/2) - \overline{A}(h/2, h)]}{2^q - 1} + O(h^r), \quad h \to 0 \tag{E-11}$$

As in the earlier single extrapolation formula, we can take the correction term in Equation (E-11) to be a conservative estimate of the error $O(h^r)$ when $h$ is sufficiently small.

**Example E.2 Numerical Solution of $y' = y$ with $y(0) = 1$**

We wish to calculate $y(1)$. The known solution of the differential equation is $y = e^x$, so that the true value of $y(1)$ is 2.71828. For the Improved Euler numerical method (see the chapter on ODE initial value problems) illustrated here, it is known that $y = y(h) + Kh^2$, where $y$ is the true value and $y(h)$ is the approximate value for step size $h$. Two integrations produce for $y(1)$ the approximate values $y(h = 0.1) = 2.70838$ and $y(h = 0.05) = 2.71563$. Then the Richardson extrapolation calculation gives $y \approx 2.71563 + [2.71563 - 2.70838]/(2^2 - 1) = 2.71805$. We see that the Richardson value is much closer to the true value than the original numerical approximations. It is also interesting to note that the Richardson correction, 0.00241667, is larger than the *actual* error $(2.71828 - 2.71805 = 0.00023)$ of the Richardson value. Thus, *in this example*, $h$ is small enough for the Richardson correction $[A(h/2) - A(h)]/3$ to be a conservative estimate of the actual error. This error characteristic of Richardson extrapolation is very important in practice, since (of course) the true solution is always unknown. The matter of Richardson error estimation is discussed in detail in the chapter on ODE initial value problems.

## REFERENCES

BENDER, C., and ORSZAG, S. (1978), *Advanced Mathematical Methods for Scientists and Engineers*, McGraw-Hill, New York

# Appendix F

# The Euler-Maclaurin Summation Formula

The Euler-Maclaurin summation formula is a fundamental and powerful relationship between the sum and integral of a function. The classical Euler-Maclaurin formula may be viewed as an extension of trapezoidal integration to include further approximations. This is also equivalent to approximating the difference between a sum and an integral. The Euler-Maclaurin formula can be derived by using Taylor's theorem or through continued integration by parts, and also in various other elegant ways. The Taylor approach is simplest and is easily implemented; this is the approach we shall follow.

The uses of the Euler-Maclaurin formula include summation of series and acceleration of convergence, numerical evaluation of integrals, and development of approximate relations and inequalities. The Euler-Maclaurin formula is the basis of the well-known Romberg algorithm for the evaluation of integrals Dahlquist *et al.* [1974].

## F.1 DERIVATION OF THE EULER-MACLAURIN FORMULA

We consider first the integral $I$.

$$I = \int_a^{a+h} f(x)\, dx = h \int_0^1 f(a + ht)\, dt = h \int_0^1 \phi(t)\, dt \tag{F-1}$$

In Equation (F-1) we have introduced, for convenience, the change of variable $x = a + ht$. This permits us to work with an integral that has limits of zero and unity. Since $\phi(t) = f(a + ht)$, we have $\phi'(t) = hf'(a + ht)$, $\phi''(t) = h^2 f''(a + ht)$, and so on. Now we extract the trapezoidal approximation to the $\phi$ integral, which is just $[\phi(1) + \phi(0)]/2$. We do this by integrating Equation (F-1) by parts, to get

$$\int_0^1 \phi(t)\,dt = (t + C)\phi(t)\Big|_0^1 - \int_0^1 (t + C)\phi'(t)\,dt \tag{F-2}$$

The constant of integration $C$ is taken to be $C = -1/2$ in order to obtain the trapezoidal approximation and leave behind an integral expression for the error as follows:

$$\int_0^1 \phi(t)\,dt = \frac{\phi(1) + \phi(0)}{2} - \int_0^1 (t - 1/2)\phi'(t)\,dt \tag{F-3}$$

The change of variable $u = (t - 1/2)$ is now used in the integral remainder of Equation (F-3) to give

$$J = \int_0^1 (t - 1/2)\phi'(t)\,dt = \int_{-1/2}^{1/2} u\phi'(1/2 + u)\,du \tag{F-4}$$

In Equation (F-4) we expand $\phi'(1/2 + u)$ in a Taylor expansion about $u = 0$ and integrate term by term to obtain

$$\begin{aligned} J &= \int_{-1/2}^{1/2} u[\phi'(1/2) + \phi''(1/2)u \cdots]\,du \\ &= \frac{\phi''(1/2)}{2^2 \cdot 3 \cdot 1!} + \frac{\phi^{iv}(1/2)}{2^4 \cdot 5 \cdot 3!} + \frac{\phi^{vi}(1/2)}{2^6 \cdot 7 \cdot 5!} \cdots \end{aligned} \tag{F-5}$$

Note that in the integral of Equation (F-5), all of the terms involving $u^k$ where $k$ is odd vanish since these are integrals of odd functions over a *symmetric* interval. Similarly, all integrals of $u^k$ where $k$ is even equal twice the integral from 0 to 1/2. Equation (F-5) shows that the corrections to the trapezoidal approximation can be expressed as a sum of even-order derivative terms. But what is really subtle is that this sum can be converted to a new sum of the *differences* of the odd-order derivatives, and this leads to a marvelous simplification of the final result. To show how this is accomplished, we indicate in detail the procedure for the $\phi''(1/2)$ term. We write the following Taylor expansions:

$$\begin{aligned} \phi'(1) &= \phi'(1/2) + \phi''(1/2)(1/2) + \frac{\phi'''(1/2)(1/2)^2}{2!} + \frac{\phi^{iv}(1/2)(1/2)^3}{3!} \cdots \\ \phi'(0) &= \phi'(1/2) - \phi''(1/2)(1/2) + \frac{\phi'''(1/2)(1/2)^2}{2!} - \frac{\phi^{iv}(1/2)(1/2)^3}{3!} \cdots \end{aligned} \tag{F-6}$$

When we subtract the expressions of (F-6), we find that

$$\Delta\phi'(0) \equiv \phi'(1) - \phi'(0) = 2\left[\phi''(1/2)(1/2) + \frac{\phi^{iv}(1/2)(1/2)^3}{3!} \cdots\right] \tag{F-7}$$

In the same way, each $\phi^{(k)}(1/2)$, with $k$ even, can be expressed as $[\Delta\phi^{(k-1)}(0)]/2$ plus a linear combination of higher even-order derivatives, for which this could be done again, and so on. When we now back-substitute these various relationships, we finally recover the equation

$$\int_0^1 \phi(t)\,dt = \frac{\phi(0) + \phi(1)}{2} - \frac{\Delta\phi'(0)}{12} + \frac{\Delta\phi'''(0)}{720} - \frac{\Delta\phi^{v}(0)}{30240} + \text{remainder} \tag{F-8}$$

where $\Delta\phi^{(k)}(0) \equiv \phi^{(k)}(1) - \phi^{(k)}(0)$. We now consider the original integral $\int f(x)\,dx$ between the limits of $a$ and $b$. The interval of integration $(a, b)$ is then broken up into subintervals of width $h$ according to

$$\int_a^b f(x)\,dx = \int_a^{a+h} f(x)\,dx + \int_{a+h}^{a+2h} f(x)\,dx + \int_{a+2h}^{a+3h} f(x)\,dx + \cdots \int_{b-h}^{b} f(x)\,dx \tag{F-9}$$

When each of these subintegrals is replaced by the fundamental results from Equations (F-1) and (F-8), and account is taken of the cancellation in the sum of the differences of the derivatives, we finally obtain the general trapezoidal form of the Euler-Maclaurin summation formula.

$$\int_a^b f(x)\,dx = h\left[\frac{f(a)}{2} + f(a+h) + f(a+2h) + \cdots + f(b-h) + \frac{f(b)}{2}\right] - \frac{h^2[f'(b)-f'(a)]}{12} + \frac{h^4[f'''(b)-f'''(a)]}{720} - \frac{h^6[f^v(b)-f^v(a)]}{30240} + R \tag{F-10}$$

The remainder $R$ in Equation (F-8), regardless of how many terms we take, can be shown by brute force to be of the form of some constant times the bound on the interval $(a, b)$ of the second derivative higher than the last term retained. Thus, for example, we could write in Equation (F-10) that $R = O(h^8)$. This is useful for purposes of numerical integration, but not so useful for summation of series. An alternative error result, which is especially useful for summation, can be shown to be true. This result states that if an even-order derivative has a uniform sign on the interval $(a, b)$, then the error $R$ in Equation (F-10) is less in magnitude than the term of the expansion just before this derivative. Thus, for example, in Equation (F-10), if $f(x)$ has a fourth derivative which has a uniform sign on the interval $(a, b)$, then the magnitude of $R$ will be less than that of the term $h^4[f'''(b) - f'''(a)]/720$. Also, if $f(x)$ has a second derivative that has a uniform sign on the interval $(a, b)$, the magnitude of $R$ will be less than the term $h^2[f'(b) - f'(a)]/12$.

The Euler-Maclaurin formula is used to sum series in the chapter on acceleration and summation of series, and it is used as a basis for Richardson extrapolation in the chapter on numerical integration.

## REFERENCES

DAHLQUIST, G., BJORCK, A., and ANDERSON, N. (1974), *Numerical Methods*, Prentice-Hall, Englewood Cliffs, NJ

# Appendix G

# Numerical Differentiation

In the applications of calculus to practical problems, it is often required to compute the derivatives of various functions at particular points. It frequently happens that the standard analytical approach to this problem (i.e., the use of differentiation formulas plus the chain rule) is either very inefficient or else impossible. For example, if we wish to differentiate a very complicated function at only certain selected points, it may be far easier to use numerical, rather than analytical, methods. On the other hand, if the desired function is defined only at discrete points, as in the case of a table or computer output, then analytical methods cannot be used at all. Thus, we need to develop methods from which a derivative can be computed approximately, using only known values of the function. For this purpose, depending on the particular situation, we use Taylor's formula, interpolation procedures, or least squares.

In addition to the need expressed above for calculation of numerical derivatives, another important application of numerical differentiation formulas is to the area of finite difference approximations in the solution of ordinary and partial differential equations. The methods discussed here can be used to derive such formulas, and for convenient reference we indicate a number of these formulas in Table G.2.

## G.1 SIMPLE APPROXIMATIONS FOR $y'(x)$ AND $y''(x)$

We begin by temporarily assuming that we know the function values $y(x)$ at any points we wish. From now until the section on unequal spacing, if we are dealing with three or more points, it is assumed that they are *equally spaced* in $x$ with increment $h$. We then write Taylor's formula twice, once to relate $y(x+h)$ to $y$ and its derivatives at $x$, and once to relate $y(x-h)$ to these same quantities. This gives

$$y(x+h) = y(x) + y'(x)h + \frac{y''(x)h^2}{2!} + \frac{y'''(x)h^3}{3!} + \frac{y^{iv}(x)h^4}{4!} + \cdots \qquad \text{(G-1)}$$

and

$$y(x-h) = y(x) - y'(x)h + \frac{y''(x)h^2}{2!} - \frac{y'''(x)h^3}{3!} + \frac{y^{iv}(x)h^4}{4!} + \cdots \qquad \text{(G-2)}$$

These two equations, taken either individually or in combination, give much information concerning numerical differentiation. It should be kept in mind that the above expressions can be terminated at any point, with the corresponding remainder being equal to the next term, except for the derivative being evaluated at an intermediate point. In more complicated cases, to be considered later, we once again start with a Taylor expansion, but there may be more than two equations, and the increments may differ from each other.

We first consider the development of simple expressions for $y'(x)$ and $y''(x)$. If we consider only Equation (G-1), terminating with the $y''(x)$ term, then by a simple rearrangement we have

$$y'(x) \cong \frac{y(x+h) - y(x)}{h} - \frac{y''(x)h}{2} = y'_F(x) - \frac{y''(x)h}{2} \qquad \text{(G-3)}$$

Here, we use the symbol $y'_F(x)$ to stand for the forward difference approximation to the true unknown value $y'(x)$. This equation shows that the leading error term in this approximation is $-y''(x)h/2$ when $h$ is small, and this shows that the error of the forward approximation goes to zero proportional to $h$. Thus, we say that the error of this approximation is of order $h$ as $h \to 0$, and this is written as $O(h)$. In a similar way, if only Equation (G-2) is used, we see that we have, for small $h$, the approximation

$$y'(x) \cong \frac{y(x) - y(x-h)}{h} + \frac{y''(x)h}{2} = y'_B(x) + \frac{y''(x)h}{2} \qquad \text{(G-4)}$$

In this case, the quantity $[y(x) - y(x-h)]/h = y'_B(x)$ is referred to as the backward difference approximation to $y'(x)$. The preceding equation shows that the error in the backward approximation to $y'(x)$ is $O(h)$ as $h \to 0$. It is useful to note that the leading error terms of the forward and backward approximations have equal magnitudes but opposite signs. Thus, for sufficiently small $h$, the forward and backward approximations bracket the true value, and this provides a useful error estimate. By taking the approximation to be the arithmetic average of these two, we generate the so-called central difference approximation to $y'(x)$, which we would expect to be more accurate. Another way to develop the central approximation, which better illustrates its error properties, is to return to Equations (G-l) and (G-2) and subtract one from the other, this time retaining terms up to $y'''$ in both equations. This gives

$$y'(x) \cong \frac{y(x+h) - y(x-h)}{2h} - \frac{y'''(x)h^2}{6} = y'_C(x) - \frac{y'''(x)h^2}{6} \qquad \text{(G-5)}$$

**TABLE G.1** $y'_F$ VALUES FOR $y = e^x$ AT $x = 2 \quad y'(2) = 7.389$

| $h$ | $y'_F$ | $\lvert y'_{exact} - y'_{approx} \rvert$ |
|---|---|---|
| $10^{-1}$ | 7.771 | 0.382 |
| $10^{-2}$ | 7.426 | 0.037 |
| $10^{-3}$ | 7.393 | 0.004 |
| $10^{-4}$ | 7.389 | 0.0 |
| $10^{-5}$ | 7.389 | 0.0 |
| $10^{-8}$ | 7.400 | 0.011 |
| $10^{-9}$ | 7.000 | 0.389 |
| $10^{-10}$ | 0.0 | 7.389 |

This shows that the error of the central difference approximation to $y'(x)$ is $O(h^2)$ for $h \rightarrow 0$. Thus, we expect that for small $h$ values, the central approximation for $y'(x)$ will be much more accurate that either the forward or backward values. We continue by deriving the central difference approximation for the second derivative, $y''(x)$. This is obtained by adding Equations (G-l) and (G-2), in each case retaining terms up to $y^{iv}(x)$. This procedure yields

$$y''(x) \cong \frac{y(x+h) - 2y(x) + y(x-h)}{h^2} - \frac{y^{iv}(x)h^2}{12} = y''_C(x) + \frac{y^{iv}(x)h^2}{12} \tag{G-6}$$

The error of the central approximation for $y''(x)$ is therefore seen to be $O(h^2)$ as $h \rightarrow 0$.

Roundoff error is a hazard in numerical differentiation. This is because the process generally involves subtraction of nearly equal numbers. Even for functions defined analytically, the possibility of roundoff error must be carefully monitored. For example, in Table G.1 we indicate the results of the computation of $y'_F$ for $y = e^x$ at $x = 2$, determined by a nine-figure calculator using various values of $h$. Results are shown to four figures.

The results of Table G.1 are qualitatively typical of what happens in the numerical differentiation of an analytically defined function, where we can use any increment $h$ that we please. According to the *theory* of calculus, we wish to use a very small value of $h$. However, with finite arithmetic, we would always eventually get a numerical derivative of zero when $h$ is small enough, since roundoff will ultimately cause cancellation in the numerator, while the denominator remains finite. In this example we see that, fortunately, the correct limiting value of 7.389 is produced before $h$ gets so small that significant roundoff error is incurred. It is quite possible in some situations for roundoff error to contaminate a calculation before the truncation error has been reduced enough for the correct limit to be reached. In such cases, either more accuracy must be given for the tabular values or an acceleration method must be used to arrive at the limiting value for $h$ values large enough to avoid roundoff error. This example makes it clear that there are two basic problems in numerical differentiation: choice of a proper value of $h$, and estimation of the error of approximation. These two problems are *not* independent; they will be addressed below, following a discussion of higher approximations.

## G.3 A GENERAL PROCEDURE FOR HIGHER APPROXIMATIONS

It is sometimes necessary to develop higher-order accuracy approximations for first and second derivatives, as well as to calculate higher derivatives, such as $y'''$ or $y^{iv}$. In either case, it is possible to proceed using Taylor's theorem as above; the results of such calculations are simplest in the case of equal $x$ spacing $h$.

To compute higher derivatives numerically, we may extend the procedure we have used up to now for calculating $y'$ and $y''$. If we reconsider Equation (G-1), for example, we saw that by terminating before the $y''$ term, we could solve for $y'$ in terms of $y(x)$ and $y(x+h)$ to get the forward approximation to $y'$. It is apparently impossible to derive an approximation for $y''(x)$ using Equation (G-l) *alone*, since this would correspond algebraically to having one equation with the two unknowns $y'(x)$ and $y''(x)$. Thus, to determine $y''(x)$, we wrote Equation (G-2), and this in turn provided the second equation in the two unknowns $y'(x)$ and $y''(x)$. This came about when Equations (G-l) and (G-2) were truncated before the $y'''(x)$ terms. If we now wish to compute $y'''(x)$, it is clear that by retaining the $y'''$ terms in Equations (G-l) and (G-2), we generate two equations in the three unknowns $y'(x)$, $y''(x)$, and $y'''(x)$. However, we need one more equation in the same three unknowns, and this is provided by writing the following Taylor expansion:

$$y(x+ah) = y(x) + y'(x)ah + \frac{y''(x)(ah)^2}{2!} + \frac{y'''(x)(ah)^3}{3!} + \cdots \qquad \text{(G-7)}$$

In Equation (G-7) we cannot select the value of $a$ equal to either $+1$ or $-1$, since this would merely duplicate Equations (G-l) and (G-2). Rather, we would normally select $a$ so that $(x+ah)$ coincides with the next nearest point where $y$ is known. Thus, if the function is known at equal $x$ increments, we would choose $a = +2$ or $-2$. This would give three linear algebraic equations for $y'(x)$, $y''(x)$, and $y'''(x)$ expressed as a linear combination of known $y$ values. The error terms for $y'(x)$, $y''(x)$, and $y'''(x)$ in the approximations resulting from this procedure would be $O(h^3)$, $O(h^2)$ and $O(h)$, respectively. Thus, this process produces not only an approximation for $y'''$, but also improved approximations for $y'$ and $y''$.

The foregoing procedure for computing a numerical approximation to $y'''(x)$ can be extended in a straightforward way to the computation of higher derivatives. This can be done by writing enough Taylor expansions to provide a determinate algebraic system for the desired derivative plus all of the lower-order derivatives. Since the algebra is somewhat lengthy, it is useful to verify that the final formula reduces to correct values for $y =$ constant, $y = x$, and so on. This can be accomplished most easily by choosing simple values for $x$ and $h$ (such as $x = 0$ and $h = 1$, for example). It should also be noted that the procedure described here is applicable for the determination of higher-order formulas for various derivatives, for either forward, backward, or central approximation. Some common results of this type are shown in Table G.2, together with their order of accuracy, for the case of equal spacing $h$ in $x$.

**TABLE G.2** COMMON NUMERICAL DIFFERENTIATION FORMULAS (EQUAL SPACING)

| Forward Approximation |
|---|
| $y'(x) = [y(x+h) - y(x)]/h + O(h)$ |
| $y'(x) = [-y(x+2h) + 4y(x+h) - 3y(x)]/2h + O(h^2)$ |
| $y'(x) = [2y(x+3h) - 9y(x+2h) + 18y(x+h) - 11y(x)]/6h + O(h^3)$ |
| $y''(x) = [y(x+2h) - 2y(x+h) + y(x)]/h^2 + O(h)$ |
| $y''(x) = [-y(x+3h) + 4y(x+2h) - 5y(x+h) + 2y(x)]/h^2 + O(h^2)$ |
| **Central Approximation** |
| $y'(x) = [y(x+h) - y(x-h)]/2h + O(h^2)$ |
| $y'(x) = [-y(x+2h) + 8y(x+h) - 8y(x-h) + y(x-2h)]/12h + O(h^4)$ |
| $y''(x) = [y(x+h) - 2y(x) + y(x-h)]/h^2 + O(h^2)$ |
| $y''(x) = [-y(x+2h) + 16y(x+h) - 30y(x) + 16y(x-h) - y(x-2h)]/12h^2 + O(h^4)$ |
| $y'''(x) = [y(x+2h) - 2y(x+h) + 2y(x-h) - y(x-2h)]/2h^3 + O(h^2)$ |
| $y^{iv}(x) = [y(x+2h) - 4y(x+h) + 6y(x) - 4y(x-h) + y(x-2h)]/h^4 + O(h^2)$ |
| **Backward Approximation** |
| $y'(x) = [y(x) - y(x-h)]/h + O(h)$ |
| $y'(x) = [3y(x) - 4y(x-h) + y(x-2h)]/2h + O(h^2)$ |
| $y'(x) = [11y(x) - 18y(x-h) + 9y(x-2h) - 2y(x-3h)]/6h + O(h^3)$ |
| $y''(x) = [y(x) - 2y(x-h) + y(x-2h)]/h^2 + O(h)$ |
| $y''(x) = [2y(x) - 5y(x-h) + 4y(x-2h) - y(x-3h)]/h^2 + O(h^2)$ |

## G.4 RICHARDSON EXTRAPOLATION IN NUMERICAL DIFFERENTIATION

We now apply Richardson extrapolation to the numerical computation of derivatives for the purpose of error estimation and accuracy improvement. For example, if we refer back to our forward approximation to $y'(x)$ in Equation (G-3), we see that $y' = y'_F(h) - y''(x)h/2 + O(h^2)$. According to the discussion in Appendix E (which should be reviewed if necessary at this time), this is exactly the kind of relationship ($S = A(h) + Kh^p$) for which Richardson extrapolation is applicable. Here $K$ happens to equal the unknown value $-y''(x)/2$, which is just an unknown constant for a particular $x$. Thus, with $p = 1$, and using two forward approximations with $h$ and $2h$, we get

$$y' = y'_F(h) + \frac{y'_F(h) - y'_F(2h)}{2-1} + O(h^2) \tag{G-8}$$

Thus, we may compute a forward approximation to $y'(x)$ of higher accuracy by simply combining the two simple forward approximations for $h$ and $2h$ according to the above equation. In addition, the correction term on the right-hand side of Equation (G-8), $[y'_F(h) - y'_F(2h)]$, represents a conservative estimate of the error in the approximation (G-8), provided that $h$ is sufficiently small. A virtually identical

**TABLE G.3**
VALUES OF $e^x$
(SIX FIGURES)

| $x$ | $e^x$ |
|---|---|
| 1.8 | 6.04965 |
| 1.9 | 6.68589 |
| 2.0 | 7.38906 |
| 2.1 | 8.16617 |
| 2.2 | 9.02501 |

result holds for a second-order backward approximation to $y'(x)$. In the case of central approximations for $y'(x)$ and $y''(x)$, the result changes, but only slightly. Since in each of these two cases we have $S = A(h) + Kh^2 + O(h^4)$, we now use $p = 2$ and $q = 4$ in the formula $y' = y'_C(h) + Kh^p + O(h^q)$ with both $h$ and $2h$ to get

$$y' = y'_C(h) + \frac{y'_C(h) - y'_C(2h)}{2^2 - 1} + O(h^4) \tag{G-9}$$

An identical formula holds for $y''(x)$ if we replace $y'_C$ by $y''_C$.

The extrapolation technique is very powerful, since it is so simple to apply, and it allows increased accuracy for larger $h$ values (and hence helps avoid the roundoff problem). Also, it allows *estimation of accuracy* (which is very important) without any additional calculation. To illustrate how the single extrapolation technique works in practice, we now consider a tabulation of the function $y = e^x$, which is shown in Table G.3. We compute both ordinary and extrapolated values of $y'_F$, $y'_C$ and $y''_C$ at $x = 2$ in Table G.4.

The extrapolation results using both $h = 0.1$ and $0.2$ are compared to the ordinary unextrapolated results for $h = 0.2$ and $h = 0.1$ in Table G.4. We can see several patterns from the results. As expected, for no extrapolation the results with smaller $h$ values are more accurate. Also, the extrapolated results are seen to be much more accurate than the unextrapolated ones. As far as the errors are concerned, we can note the following. The magnitudes of the errors estimated by extrapolation are conservative; that is, they are larger than the magnitudes of the actual errors. Also, the actual errors vary with step size in accordance with the appropriate leading error expression, $Kh^p$. Thus, for the first-order forward approximation ($p = 1$), the ratio of the errors is about two, as expected, while the ratio of the second-order central approximation errors ($p = 2$) is about four, as theory predicts.

## G.5 NUMERICAL DIFFERENTIATION WITH UNEQUAL SPACING

All of the preceding discussion has revolved about the situation where numerical differentiation is carried out at a point where $[x, y(x)]$ are known, and where other $(x, y)$ pairs are known or can be calculated at equal increments $h$ with respect to $x$. The preceding results are particularly convenient in such a case. It often happens, however, especially when dealing with tabulated data, that either the spacing of the

**TABLE G.4** NUMERICAL DIFFERENTIATION ESTIMATES FOR $y'(x = 2)$ AND $y''(x = 2)$

| Forward Approximation | | | |
|---|---|---|---|
| $h$ | Approximation | Actual Error | \|Error\| Estimated by Extrapolation |
| 0.2 | $y' = 8.180$ | −0.79 | |
| 0.1 | $y' = 7.771$ | −0.38 | |
| 0.2, 0.1 | $y'_{\text{ext}} = 7.363$ | +0.026 | 0.41 |
| **Central Approximation** | | | |
| $h$ | Approximation | Actual Error | \|Error\| Estimated by Extrapolation |
| 0.2 | $y' = 7.438$ | −0.049 | |
| 0.1 | $y' = 7.401$ | −0.012 | |
| 0.2, 0.1 | $y'_{\text{ext}} = 7.389$ | 0.0 | 0.012 |
| 0.2 | $y'' = 7.414$ | −0.025 | |
| 0.1 | $y'' = 7.394$ | −0.005 | |
| 0.2, 0.1 | $y''_{\text{ext}} = 7.388$ | +0.001 | 0.0065 |

$y = e^x, y'(x = 2) = y''(x = 2) = e^2 = 7.38906$

data (in $x$) is *unequal*, or it is desired to know a derivative at a location where $y$ is unknown. It is possible to use the Taylor series technique used above to derive appropriate formulas, but except for the cases of central approximations to $y'$ and $y''$, this approach is rather tedious. Since they are sometimes used in practice, we now quote these results for $y'$ and $y''$.

$$y'(x) = \frac{k^2 y(x + h) - (k^2 - h^2) y(x) - h^2 y(x - k)}{hk(k + h)} + O(h^2) \tag{G-10}$$

and

$$y''(x) = \left[\frac{ky(x + h) + hy(x - k)}{h + k} - y(x)\right]\left(\frac{2}{hk}\right) + O(h) \tag{G-11}$$

In Equations (G-10) and (G-11), it is assumed that $y$ is known or can be evaluated at $(x - k)$, $x$, and $(x + h)$.

A more common, and usually easier, approach to perform numerical differentiation with unequally spaced data is to use interpolation of some type and then differentiate the interpolation equation. To accomplish this we might fit the data to an ordinary polynomial, a rational approximation, a spline approximation, or perhaps to some particular model equation the data is known to follow. These methods are all discussed in Chapter 5 on interpolation, along with the appropriate computer programs to implement them.

## G.6 NUMERICAL DIFFERENTIATION OF ANALYTICALLY DEFINED FUNCTIONS

In principle, we can compute the derivative of virtually any differentiable, analytically defined function through the use of basic differentiation formulas plus the chain rule. However, it is often desirable to have a simple means of performing the differentiation of an analytical function automatically, by numerical means. A facility for doing this is contained in computer programs CUBSPL (discussed in Chapter 5 on interpolation), and LSTSQGR (discussed in Chapter 6 on least squares). The numerical procedure employed makes use of Richardson extrapolation to adjust the spatial increment $h$ so as to conform to a user-specified error tolerance.

To compute the *partial derivatives* for one or more functions of several variables, computer program NEWTONVS can be used. This program (which is discussed in Chapter 7 on nonlinear algebraic equations) contains an automatic variable-increment subroutine for calculating the partial derivatives needed to implement the Newton nonlinear equation algorithm. The program has an option for determining numerical derivatives of any specified functions. This option can be exercised *independent* of a nonlinear equation solution. If one is dealing with a problem involving the same number of variables as equations (square system), the system of equations is programmed and the partials can be obtained. If the desired set of equations is not "square" in the variables, it is only necessary to specify enough artificial functions or variables to make the system square. For example, if one wants the partials for one function of four variables, then we program the desired equation as $f(1)$ and introduce the three other "dummy" functions $f(2) = f(3) = f(4) = 0$. On the other hand, if there are more equations than variables, we simply program the equations and set the system size ($n$) equal to the number of equations. In either case, the values of the dummy partials will all be computed to be zero.

## G.7 USE OF LEAST SQUARES FOR NOISY DATA

Numerical differentiation of data that may be contaminated with experimental error is especially hazardous. In general, when we are confronted with such a situation, it is desirable to use some kind of least-squares approximation to fit the data with a smooth function. The smooth function is then differentiated either analytically or numerically to obtain any desired derivatives. Methods for accomplishing this by means of a polynomial, a rational function, or any specific model equation are discussed in the chapter on least-squares approximation.

## REFERENCES

FERZIGER, J. H. (1981), *Numerical Methods for Engineering Application*, Wiley, New York

# Index

## LICENSE AGREEMENT AND LIMITED WARRANTY

READ THE FOLLOWING TERMS AND CONDITIONS CAREFULLY BEFORE OPENING THIS DISK PACKAGE. THIS LEGAL DOCUMENT IS AN AGREEMENT BETWEEN YOU AND PRENTICE-HALL, INC. (THE "COMPANY"). BY OPENING THIS SEALED DISK PACKAGE, YOU ARE AGREEING TO BE BOUND BY THESE TERMS AND CONDITIONS. IF YOU DO NOT AGREE WITH THESE TERMS AND CONDITIONS, DO NOT OPEN THE DISK PACKAGE. PROMPTLY RETURN THE UNOPENED DISK PACKAGE AND ALL ACCOMPANYING ITEMS TO THE PLACE YOU OBTAINED THEM FOR A FULL REFUND OF ANY SUMS YOU HAVE PAID.

1. **GRANT OF LICENSE:** In consideration of your payment of the license fee, which is part of the price you paid for this product, and your agreement to abide by the terms and conditions of this Agreement, the Company grants to you a nonexclusive right to use and display the copy of the enclosed software program (hereinafter the "SOFTWARE") on a single computer (i.e., with a single CPU) at a single location so long as you comply with the terms of this Agreement. The Company reserves all rights not expressly granted to you under this Agreement.

2. **OWNERSHIP OF SOFTWARE:** You own only the magnetic or physical media (the enclosed disks) on which the SOFTWARE is recorded or fixed, but the Company retains all the rights, title, and ownership to the SOFTWARE recorded on the original disk copy(ies) and all subsequent copies of the SOFTWARE, regardless of the form or media on which the original or other copies may exist. This license is not a sale of the original SOFTWARE or any copy to you.

3. **COPY RESTRICTIONS:** This SOFTWARE and the accompanying printed materials and user manual (the "Documentation") are the subject of copyright. You may not copy the Documentation or the SOFTWARE, except that you may make a single copy of the SOFTWARE for backup or archival purposes only. You may be held legally responsible for any copying or copyright infringement which is caused or encouraged by your failure to abide by the terms of this restriction.

4. **USE RESTRICTIONS:** You may not network the SOFTWARE or otherwise use it on more than one computer or computer terminal at the same time. You may physically transfer the SOFTWARE from one computer to another provided that the SOFTWARE is used on only one computer at a time. You may not distribute copies of the SOFTWARE or Documentation to others. You may not reverse engineer, disassemble, decompile, modify, adapt, translate, or create derivative works based on the SOFTWARE or the Documentation without the prior written consent of the Company.

5. **TRANSFER RESTRICTIONS:** The enclosed SOFTWARE is licensed only to you and may not be transferred to any one else without the prior written consent of the Company. Any unauthorized transfer of the SOFTWARE shall result in the immediate termination of this Agreement.

6. **TERMINATION:** This license is effective until terminated. This license will terminate automatically without notice from the Company and become null and void if you fail to comply with any provisions or limitations of this license. Upon termination, you shall destroy the Documentation and all copies of the SOFTWARE. All provisions of this Agreement as to warranties, limitation of liability, remedies or damages, and our ownership rights shall survive termination.

7. **MISCELLANEOUS:** This Agreement shall be construed in accordance with the laws of the United States of America and the State of New York and shall benefit the Company, its affiliates, and assignees.

8. **LIMITED WARRANTY AND DISCLAIMER OF WARRANTY:** The Company warrants that the SOFTWARE, when properly used in accordance with the Documentation, will operate in substantial conformity with the description of the SOFTWARE set forth in the Documentation. The Company does not warrant that the SOFTWARE will meet your requirements or that the operation of the SOFTWARE will be uninterrupted or error-free. The Company warrants that the media on which the SOFTWARE is delivered shall be free from defects in materials and workmanship under normal use for a period of thirty (30) days from the date of your purchase. Your only remedy and

the Company's only obligation under these limited warranties is, at the Company's option, return of the warranted item for a refund of any amounts paid by you or replacement of the item. Any replacement of SOFTWARE or media under the warranties shall not extend the original warranty period. The limited warranty set forth above shall not apply to any SOFTWARE which the Company determines in good faith has been subject to misuse, neglect, improper installation, repair, alteration, or damage by you. EXCEPT FOR THE EXPRESSED WARRANTIES SET FORTH ABOVE, THE COMPANY DISCLAIMS ALL WARRANTIES, EXPRESS OR IMPLIED, INCLUDING WITHOUT LIMITATION, THE IMPLIED WARRANTIES OF MERCHANTABILITY AND FITNESS FOR A PARTICULAR PURPOSE. EXCEPT FOR THE EXPRESS WARRANTY SET FORTH ABOVE, THE COMPANY DOES NOT WARRANT, GUARANTEE, OR MAKE ANY REPRESENTATION REGARDING THE USE OR THE RESULTS OF THE USE OF THE SOFTWARE IN TERMS OF ITS CORRECTNESS, ACCURACY, RELIABILITY, CURRENTNESS, OR OTHERWISE.

IN NO EVENT, SHALL THE COMPANY OR ITS EMPLOYEES, AGENTS, SUPPLIERS, OR CONTRACTORS BE LIABLE FOR ANY INCIDENTAL, INDIRECT, SPECIAL, OR CONSEQUENTIAL DAMAGES ARISING OUT OF OR IN CONNECTION WITH THE LICENSE GRANTED UNDER THIS AGREEMENT, OR FOR LOSS OF USE, LOSS OF DATA, LOSS OF INCOME OR PROFIT, OR OTHER LOSSES, SUSTAINED AS A RESULT OF INJURY TO ANY PERSON, OR LOSS OF OR DAMAGE TO PROPERTY, OR CLAIMS OF THIRD PARTIES, EVEN IF THE COMPANY OR AN AUTHORIZED REPRESENTATIVE OF THE COMPANY HAS BEEN ADVISED OF THE POSSIBILITY OF SUCH DAMAGES. IN NO EVENT SHALL LIABILITY OF THE COMPANY FOR DAMAGES WITH RESPECT TO THE SOFTWARE EXCEED THE AMOUNTS ACTUALLY PAID BY YOU, IF ANY, FOR THE SOFTWARE.

SOME JURISDICTIONS DO NOT ALLOW THE LIMITATION OF IMPLIED WARRANTIES OR LIABILITY FOR INCIDENTAL, INDIRECT, SPECIAL, OR CONSEQUENTIAL DAMAGES, SO THE ABOVE LIMITATIONS MAY NOT ALWAYS APPLY. THE WARRANTIES IN THIS AGREEMENT GIVE YOU SPECIFIC LEGAL RIGHTS AND YOU MAY ALSO HAVE OTHER RIGHTS WHICH VARY IN ACCORDANCE WITH LOCAL LAW.

**ACKNOWLEDGMENT**

YOU ACKNOWLEDGE THAT YOU HAVE READ THIS AGREEMENT, UNDERSTAND IT, AND AGREE TO BE BOUND BY ITS TERMS AND CONDITIONS. YOU ALSO AGREE THAT THIS AGREEMENT IS THE COMPLETE AND EXCLUSIVE STATEMENT OF THE AGREEMENT BETWEEN YOU AND THE COMPANY AND SUPERSEDES ALL PROPOSALS OR PRIOR AGREEMENTS, ORAL, OR WRITTEN, AND ANY OTHER COMMUNICATIONS BETWEEN YOU AND THE COMPANY OR ANY REPRESENTATIVE OF THE COMPANY RELATING TO THE SUBJECT MATTER OF THIS AGREEMENT.

Should you have any questions concerning this Agreement or if you wish to contact the Company for any reason, please contact in writing at the address below.

Robin Short
Prentice Hall PTR
One Lake Street
Upper Saddle River, New Jersey 07458